Unravelling the Mystery of the Atomic Nucleus

T0215644

Bernard Fernandez

Unravelling the Mystery of the Atomic Nucleus

A Sixty Year Journey
1896 — 1956

English version by Georges Ripka

 Springer

Bernard Fernandez
rue Gabrielle d'Estrées 17
Vanves
France

Georges Ripka
La Croix du Sauveur
Queyssac les Vignes
France

Title of the original French edition, De l'atome au noyau. Une approche historique et de la physique nucléaire—published by Ellipses—Copyright 2006, Édition Marketing S.A.

ISBN 978-1-4899-8562-0   ISBN 978-1-4614-4181-6 (eBook)
DOI 10.1007/978-1-4614-4181-6
Springer New York Heidelberg Dordrecht London

Foreword to the French Edition

Throughout my life as a nuclear physicist, spent in the laboratory probing the properties of the atomic nuclei, I was repeatedly confronted with the question: how did this idea, this concept, this understanding arise, and by what path was it reached? The question obviously concerns our understanding and formulation of physical theory but also, and this is all too often forgotten, by the development of instrumentation. The revolutionary changes in our understanding of physical phenomena, which took place in the span of a few decades of the first half of the twentieth century, concern both equally. In fact, momentous upheavals of physical theory, such as the formulation of quantum mechanics, were forced upon physicists, often against their will, by a variety of experimental data which obstinately refused to be accounted for by prevailing theories.

Curiously, I never found a book which really answered this question. The book of Abraham Pais, *Inward Bound*, is a wonderful work and an inexhaustible source of references, written more for specialists. But it is a history of the physics of elementary particles and not of nuclear physics which preceded it. It highlights the evolution of the theory, casting somewhat aside the history of instrumentation. The two-volume work of Milorad Mladjenović is well documented, but it addresses mainly physicists without really answering the question. Upon scrutinizing paper after paper, upon following the tracks of progress, dead-ends, questioning and controversy, which form the matter upon which science breads, I observed that every step forward, be it modest or fundamental, was the fruit of a necessity. It never entered ready-made into the mind of a physicist, even if he was a genius, and we shall encounter several. It was almost always the answer to a concrete problem.

This book describes how atomic nuclei were discovered, progressively probed and understood. It begins with the discovery of radioactivity by Becquerel in 1896. It is written in a nontechnical language, without mathematical formulas. However, it is not intended to be a popularization of a scientific work, which might attempt to convey the essentials by means of analogies. I wish each sentence to be legible by both full-fledged physicists and non-specialists. The latter may occasionally consult the glossary at the end of the book for words marked by the sign $\diamond$. Footnotes offer punctual explanations and comments. References are listed at the end of each chapter. A detailed bibliography of all the cited books can be found at the end of this volume.

As far as possible, the narrative uses terms and concepts, such as rays, atoms, elements,... *in the sense they were used and conceived at the time,* and it follows their progressive and occasionally abrupt changes in meaning. Terms which were used at a given time were the most suitable and plausible working tools. It would be both silly and unbecoming to comment or criticize them from the point of view of one "who knows the end of the story." The reader, who knows more and better today, may find it occasionally surprising to be faced with a hypothesis considered to be a verified truth, only to find it discarded later.

I should add what this book *is not.* It describes only briefly the technical applications of atomic and nuclear physics. For example, it does not describe the history of nuclear power plants. However, a chronology of the development of the atomic bomb is given because its development caused a qualitative change in the research facilities after 1945.

It all started with the discovery of radioactivity by Becquerel in 1896. Radioactivity confirmed the reality of atoms and produced a profound change in the very concept of atoms. It later provided insight in to their structure and the existence of an inner nucleus. What at first appeared to be a simple black blur on a photographic plate prompted physicists to discover more in order to "lift a corner of the veil," according to the expression of Einstein. Progressively and due to relentless work and fertile imagination, new concepts were forged. Our knowledge of the atom greatly expanded during the 1930–1940 decade. The theoretical schemes upon which our present understanding is based were developed shortly before and shortly after the Second World War. That is where the history covered by this book ends, although it is a pursuing adventure.

*

*   *

This work has benefited from the encouragement and active help of my close collaborators, particularly of my friends at the *Service de Physique Nucléaire* of the French Atomic Energy Commission, as well as of the *Direction des Sciences de la Matière.* I spent endless hours and days in numerous libraries searching for documentation and original publications. It is a pleasure to acknowledge the warm and friendly welcome of the librarians, whose competence and devotion were a great help.

Some faithful friends not only encouraged me but also accepted the task of making a critical reading of this work, namely the nuclear physicists Jean Gastebois and Georges Ripka as well as the nonphysicist Maurice Mourier and the nonspecialist scientist Philippe Lazar. The translation of Russian texts is due to Anne-Emmanuelle Lazar. Finally Bernard Gicquel took the trouble to read and correct the translations of the German texts. A hearty thanks to them all!

Vanves, France   Bernard Fernandez
February 2006

Foreword to the English Edition

The present English version of the original book is the result of 3 years of fruitful collaboration between us. All the sections have been revised and often rewritten. Many references as well as the glossary have been reviewed and rewritten with English readers in mind. Indeed it should be considered as a second edition.

We would like to express our gratitude to Aron Bernstein and Philippe Lazar for their critical reading of the manuscript.

Vanves, France   Bernard Fernandez
Queyssac les Vignes, France   Georges Ripka

Contents

Radioactivity: The First Puzzles

Leurs métamorphoses sont soumises à des lois stables, que vous ne sauriez comprendre.

A. France, *La Révolte des anges*.

Their transformations are subject to stable laws which you could not comprehend.

The "Uranic Rays" of Henri Becquerel

Henri Becquerel, while searching for X-rays, discovers a radiation emitted by uranium. The scientific community shows no interest in such a weak and incomprehensible phenomenon with no practical applications.

On this Sunday morning, March 1, 1896, Henri Becquerel is working in his laboratory at the *Muséum d'Histoire Naturelle* in Paris. He is waiting in vain for the sun to come out [1–3] because he needs the intensity of sunlight in order to confirm some interesting observations made a week earlier and communicated to the *Académie des Sciences* on February 24. But in this never ending winter, the sky remains obstinately covered, day after day.

Becquerel is a distinguished physicist, born in a family with several generations of scientists [4, 5]. His grandfather, Antoine César, born in 1788, was admitted to the *École Polytechnique* in 1806. He distinguished himself as an officer in the Napoleonic armies. After the final fall of Napoleon in 1815, he left the army and began a successful scientific career, working on electricity, optics, phosphorescence, and electrochemistry. In 1829, he constructed the first constant current electric cell.

B. Fernandez and G. Ripka, *Unravelling the Mystery of the Atomic Nucleus: A Sixty Year Journey 1896 — 1956*, DOI 10.1007/978-1-4614-4181-6_1,
© Springer Science+Business Media New York 2013

He was awarded the prestigious Copley Medal of the *Royal Society* in London in 1837, and in 1838, he became member of the *Académie des Sciences*. In 1838, he held the first physics chair in the *Muséum d'Histoire Naturelle* in Paris. When he died in 1878, Henri Becquerel, his grandson, was 26 years old.

Becquerel's father was the second son of Antoine César, Alexandre Edmond Becquerel, born in 1820. Although he passed successfully the admittance examinations to both the *École Polytechnique* and the *École Normale Supérieure*, he chose to work as an assistant to his father in the *Muséum d'Histoire Naturelle*. In 1852, he became Professor at the *Conservatoire National des Arts et Métiers* and he was elected member of the *Académie des Sciences* in 1863. Upon the death of his father, he succeeded him as professor in the *Muséum d'Histoire Naturelle,* where he specialized in electricity, magnetism, and optics. His works on phosphorescence and luminescence [6] were published in 1959 and assembled in two books [7, 8], published in 1859 and in 1867. They remained a standard reference for half a century. He invented a device, called the phosphoroscope, with which he proved that fluorescence, which had been discovered by G. G. Stokes in 1852, was nothing but phosphorescence lasting for a very short time. Alexandre Edmond Becquerel died in 1891.

Henri (Antoine Henri Becquerel, according to his birth certificate) was born on December 15, 1852, in the *Muséum*, the home of his parents. In 1872, he was admitted to the *École Polytechnique*, where he met Henri Poincaré, who was to become one of the most famous scientists of the time. They develop a long-lasting friendship. In 1876, he graduated from the *Écoles des Ponts et Chaussées*. First, he became an instructor at the *École Polytechnique* and later an assistant naturalist in the *Muséum*. In 1889, at the age of 37, he was elected member of the *Académie des Sciences*, and in 1895, he became physics professor at the *École Polytechnique*.

Henri Becquerel, polite and friendly, is a clever and rigorous experimentalist. Akin to many French physicists at that time, he is more inclined to observation than to theoretical speculation. His research, so far, is devoted to optics, a family tradition. In 1876, Lucie Jamin, the daughter of the Academician J. C. Jamin, becomes his wife and gives birth to a son, Jean, in 1878. She dies a few weeks later at the age of 20. On August 1890, Louise Désirée Lorieux becomes the second wife of Henri and Jean is brought up as her son. True to the family tradition, Jean will later also be admitted to the *École Polytechnique* and elected member of the *Académie des Sciences*.

The Discovery

The experiments, which Becquerel is performing in 1896, are motivated by the discovery of "X-rays," which Wilhelm Conrad Röntgen [9–11] had made a few months earlier. Röntgen had studied the "cathode rays" produced by electrical discharges in gases. When a voltage exceeding a 1,000 V is created between two conductors placed in a container of gas maintained at low pressure, an electrical discharge occurs. The discharge consists of *cathode rays* emanating from the negatively charged conductor, called the cathode (We know today that cathode rays are electrons). Röntgen discovered that, when the cathode rays hit the glass wall of the container, they emit an unknown radiation which has a greater penetration power than light. He called them "X-rays." This discovery caused quite a stir and physicists, among whom Henri Becquerel, were quite excited. In the session of January 20, 1896 of the *Académie des Sciences,* two medical doctors, Paul Oudin and Toussaint Barthélémy, displayed X-ray photographs. Poincaré received a reprint of the paper of Röntgen. He and Becquerel were particularly impressed by the fact that the X-rays were emitted from the luminescent spot which was produced on the glass container by the impinging cathode rays. In a paper devoted to X-rays and published on January 30, 1896 in the *Revue Générale des Sciences,* Poincaré wrote:

> It is the glass which emits the Röntgen rays and it emits them by becoming phosphorescent. Are we not then entitled to ask whether all bodies, whose phosphorescence is sufficiently intense, emit X-rays of Röntgen, in addition to light rays, whatever the cause of the fluorescence is? [12].

This is precisely what Becquerel is investigating in his laboratory of the *Muséum d'Histoire Naturelle.* He is quite familiar with luminescence which he had studied at length with his father. Luminescent bodies are not spontaneously luminous but, when they are exposed to light, they radiate their own light, almost immediately[1] in the case of fluorescence, or within a variable laps of time, in which case the phenomenon is called phosphorescence.[2] Becquerel possesses thin strips of double uranium and potassium sulfate, and he is quite familiar with their phosphorescence which is intense but lasts only about a hundredth of a second. He then performs the following experiment, which he later described in a communication to the *Académie des Sciences,* dated February 24:

> We wrap a Lumière photographic plate, composed of a bromide gel, between two sheets of very thick black paper, such that the photographic plate does not become veiled when exposed to sunlight during a whole day. On top of the paper sheet, we place a strip of a phosphorescent substance, and the lot is exposed to the sun during several hours. When the photographic plate is subsequently developed, the silhouette of the phosphorescent substance appears in black on the photograph [...] We are led to conclude from these experiments, that the phosphorescent substance emits a radiation capable of passing through the paper which is opaque to light [13].

Becquerel exposes this assembled package to sunlight, the most intense source of light at his disposal. The following Wednesday, February 26, he attempts to make an X-ray photograph. He repeats the experiment, but this time, he slips a thin strip

---

[1]That is to say, within a time delay of the order of one hundred millionth of a second.

[2]The laps of time can vary from a thousandth of a second to several thousand seconds.

of copper, in the shape of a Maltese cross, between the phosphorescent uranium sulfate sheet and the photographic plate, the latter being again wrapped in thick black paper. He knows that the copper strip is opaque to X-rays, and he expects that, after a similar exposure to sunlight, a Maltese cross will appear in white on the developed photographic plate. He proceeds to expose this newly assembled package to sunlight in order to produce the phosphorescence. The sky is clear until 10 a.m. but obstinately remains clouded thereafter. The following day, the sun shines only between 3 p.m. and 7 p.m. when new clouds appear. Becquerel then puts the package into a drawer, pending better weather. The following 2 days remain grey. No sign of improvement on the following Sunday, March 1, when it even begins to rain [14].

Rather than wait, possibly several days more, Becquerel decides to develop the photographic plate in his drawer. He expects to obtain a weak picture because the plate was exposed to sunlight for a short time only, and the induced phosphorescence was expected to be weak. However, contrary to his expectations, the developed photographic plate shows that it had been intensely exposed. It also displays a somewhat blurred shape of the Maltese cross! Becquerel is surprised and, true to the clear-sighted and rigorous physicist he was, he repeats the experiment maintaining this time the assembled package in complete darkness. The photographic plate is again strongly exposed! On Monday, March 2, 1896, he presents the following note to the *Académie des Sciences*:

> *I insist on the following feature, which I consider very important and not in accord with the phenomena we might have expected to observe: the same crystalline strips, placed upon the photographic plates, under the same conditions and with the same screens, but protected from incident radiation and maintained in darkness, produce the same exposure on the photographic plate [...] I immediately thought that this action had necessarily continued in darkness [15].*

Henri Becquerel has just discovered what we call today *radioactivity*.

Is It Really Phosphorescence?

At first, Becquerel believes that the physical process which he is observing is phosphorescence produced by exposure to light and that it should therefore die out in time. In order to make sure, doubt being the physicist's best advisor, from March 3 onwards, Becquerel maintains his strips in darkness, and, from time to time, he checks their radiative power. Month after month, it persists, showing no sign of weakening. In November 1896, Becquerel notes:

> *... protected from any known radiation, [...] the substances continued to emit active radiation which penetrated glass and black paper, and this has been going on for 6 months for some samples and 8 months for others [16].*

He makes another strange observation: similar experiments performed with other luminescent substances fail to produce the effect [17]. However:

All the uranium salts which I have studied, whether they are, or not, phosphorescent, exposed to light, crystallized, melted or dissolved, gave similar effects; I was therefore led to conclude that the effect was due to the presence of the element uranium in the salts[1], and that the metal would produce a stronger effect than its compounds. The experiment was performed [...] and it confirmed this prediction; the photographic effect is notably more intense than that produced by a uranium salt [18].

Becquerel insists that it does not matter whether the uranium salts are crystallized, melted, or dissolved because only the crystallized form is phosphorescent. The relation between the phenomenon he discovered and phosphorescence becomes increasingly doubtful. In other words, the "radiant" activity appears to bear no relation to the exposure of the substance to sunlight.

Although he continues to use the word "phosphorescence," Becquerel gradually gives up the original idea which led him to the discovery. To be faced with such a phenomenon, which occurs in a similar fashion independently of the chemical compound of uranium, was quite an extraordinary experience for a physicist or a chemist at the end of the nineteenth century. One thing, which chemistry had shown since Lavoisier, was precisely the fact that properties of chemical substances did not reflect the properties of the elements from which the substances are formed. Kitchen salt, for example, is sodium chloride and its properties are quite different from those of either sodium or chlorine. The radiant activity of uranium was both strange and unique.

What Is the Nature of the Radiation?

The terms "ray" or "radiation" are used to describe something which emanates from a source and propagates in a straight line, as sun rays do. In the paper announcing his discovery of X-rays, Röntgen wrote:

The reason why I allowed myself to call "rays" the agent which emanated from the wall of the discharge vessel, is partly due to the systematic formation of shadows which were observed when more or less transparent materials were placed between the apparatus and the fluorescent body (or the sensitive plate) [9].

According to the theory of Maxwell, brilliantly confirmed experimentally in 1888 by Hertz, any sudden electric or magnetic disturbance becomes the source of an electromagnetic field$^\diamond$ which propagates in a straight line at the speed of light.

---

[1]Emphasized by the author.

This electromagnetic field is in fact light, visible light being nothing but a particular instance. Röntgen showed that X-rays propagate in a straight line and, in spite of the fact that they could neither be reflected nor refracted, he believed that they were electromagnetic waves, that is, a kind of light which is invisible to our eyes but which can be detected on a photographic plate (or on a luminescent screen).

In his second communication on the discovery of X-rays, Röntgen noted that they had the power of discharging electrified bodies [10], that is, that they allowed an electric current to pass through air, a feature which was confirmed by numerous other works [19–22]. Becquerel subjects his "uranium rays" to similar tests. For this purpose, he uses a gold leaf electroscope$^\diamond$. When they are electrically charged, the gold leaves repel each other. But when Becquerel places a piece of uranium in their vicinity, they gradually coalesce: the electroscope discharges itself, indicating that some electricity has escaped through the air:

> I have recently observed that the invisible radiation emitted under these conditions has the property of being able to discharge electrified bodies which are subject to their radiation [23].

This property will play a major role, as we shall see. Since it manifests itself by a measurable electric process, the radiation becomes detectable. This became the first detector other than the photographic plate.

A Limited Impact on Scientists and the Public

Whereas the discovery of X-rays aroused considerable interest among both physicists and the public, the "radiant activity of uranium" made a very limited impact on physicists and none on the general public. In the year 1896, more than 1,000 publications were devoted to X-rays, but barely a dozen to the radiation of uranium [24]. Indeed, X-rays provided the possibility to see the interior of the human body, the dream of medical doctors, who would not even have imagined such a possibility a year earlier. Furthermore, X-rays are easy to produce. They required a Crookes tube and a Rühmkorff coil which could be found in practically any lab. The 1897 issue of the *Almanach Hachette*, subtitled *Petite Encyclopédie populaire de la vie pratique*[1] noted:

> It is truly the invisible which is displayed by the mysterious X-rays, which we all have heard about. To show the bone hidden under the flesh, the weapon or projectile buried in a wound; to read all the inside of the human body—perhaps even thoughts!—to count the coins through a carefully closed purse; to seek the most intimate confessions hidden in a sealed envelope; it all becomes child's play for any amateur. And what is required to perform such miracles? Precious little: an induction coil, a glass bulb and a simple photographic plate [25].

---

[1]Little encyclopædia of practical life.

The radiation of uranium was far less interesting. For one thing, it was very weak: exposures lasting hours were required whereas, in 1897, 10 min were sufficient to produce an X-ray photograph (the first X-ray photograph, which showed the hand of Bertha, the wife of Röntgen, was obtained in 1 h). But most of all, nobody could see what the uranium rays could be used for. The case of the English physicist Sylvanus P. Thomson is quite instructive in this respect. He was also interested in X-rays, and, like Becquerel, he thought that they were linked to phosphorescence. He even observed, at about the same time as Becquerel, that phosphorescent uranium salts emitted a radiation, which he proposed to call "hyperphosphorescence." But Becquerel was the first to publish his observations. Thomson published his a few months later [26], in June 1896, and then he abandoned their study in order to devote his research to the study of X-rays. After November 1896, even Becquerel abandoned the study of uranium radiation for several years. With the experimental means available to him at the time, he could not see how to progress further.

Why 1896?

Becquerel used to say that radioactivity was bound to be discovered at the *Muséum*. He considered that his discoveries were "daughters of his father and grandfather; they would have been impossible without them." [27] However, in a lecture delivered at the University of Yale in March 1905, Ernest Rutherford claimed that the discovery could well have been made a century earlier:

> In this connection it is of interest to note that the discovery of the radioactive property of uranium might accidentally have been made a century ago, for all that was required was the exposure of a uranium compound on the charged plate of a gold-leaf electroscope. Indications of the existence of the element uranium were given by Klaproth in 1789, and the discharging property of this substance could not fail to have been noted if it had been placed near a charged electroscope. It would not have been difficult to deduce that the uranium gave out a type of radiation capable of passing through metals opaque to ordinary light. The advance would probably have ended there, for the knowledge at that time of the connection between electricity and matter was far too meagre for an isolated property of this kind to have attracted much attention [28].

Was Radioactivity Discovered by Chance?

When he developed his photographic plate on March 1, 1896, Becquerel certainly did not expect to see what he saw. Can we say that he discovered radioactivity by chance? Becquerel had designed an experiment with a well defined goal, namely, to observe a radiation, if it exists, similar to X-rays and emitted by phosphorescent substances. The lack of sunlight as well as his decision to develop the photographic plate admittedly played an important role. But his experiments would have led him, sooner or later, to the same discovery. The nature of a true physicist consists in being surprised by the right thing. In this respect, Becquerel left nothing to chance [29]. Better still, by mounting successive and rigorous experiments, he gradually showed that his initial idea was wrong, that the radiation was not linked to phosphorescence, but that instead, it was a truly new phenomenon linked to the presence of uranium. It is in this respect that he truly discovered radioactivity. Sylvanus Thomson had made the same observation in a similar fashion, but without persevering. Similarly, Abel Niepce de Saint-Victor, a French officer and amateur chemist, had observed that uranium salts could leave a trace on a photographic plate long after it had been exposed to sunlight, and he observed the same effect with tartaric acid. He published a number of papers between 1857 and 1867 on what he called "A new action of light." [30] But he always linked the observed effects to exposure to light: he did not discover radioactivity.

The discovery made by Becquerel was truly unexpected. But is that not the nature of every true discovery?

Polonium and Radium

A young Polish student and her French husband, working outside the French university establishment, discover two new elements, polonium and radium, which are considerably more radioactive than uranium. Their discovery rekindles research on radioactivity. Pierre and Marie Curie ask the crucial question: where do radioactive elements find the energy required for them to radiate?

Two years after the discovery of radioactivity by Henri Becquerel, the study of the "radiating activity" of uranium had ceased. But on the April 12, 1898, a young Polish woman, married to a French Physicist, delivers a communication to the *Académie des Sciences* which ignites a fire of interest which, this time, is likely to last.

Marya Skłodowska

Marya Skłodowska [31–34] was born in Warsaw in 1868 into a family with already three daughters, Sofia, Bronisława and Helena, and a son, Joseph. Her father, Władysław Skłodowski, teaches physics at the *Gymnasium* in Nowolipki street. Marya was born at a particularly dark time of Polish history. The defeat of the January 1864 uprising against Russian rule is followed by a ferocious repression. The Tsar decides to Russianize the country. Russian becomes the official language and the use of Polish is forbidden, even in schools. Władysław loses his job. After considerable difficulties, he succeeds in becoming a monitor in a boarding school with a small teaching duty. The family lives in poverty. Sofia dies from typhus in 1876 and Mrs. Skłodowska catches tuberculosis. She dies May 9, 1878, when Marya is barely 11 years old.

On June 12, 1883, at the age of 15, Marya graduates brilliantly from secondary school, earning a gold medal. But universities are closed to women. Her elder sister Bronia would also like to attend university and so the two sisters decide to make a deal: Marya will help Bronia financially to go to Paris by becoming a primary school teacher. Once Bronia gets the required diploma, she will in turn help Marya to join her in Paris. Seven years pass before Bronia, who has almost finished her medical studies and is married, can welcome her sister.

In the fall of 1891, in Paris, Marya attends the lectures of Gabriel Lippmann, Edmond Bouty, and Paul Appell at the *Sorbonne*. In July 1893, after living in considerable poverty for 2 years, she obtains a bachelor's degree in physics; she is the best student in her class. She goes back home to Poland for a vacation, fearing that she might not find the money to return to Paris. But, thanks to a heaven-sent subsidy (an *Aleksandrovič* grant of 600 rubbles), she returns to Paris and, in July 1894, she obtains a bachelor's degree in mathematics, graduating as second best in her class.

While preparing her bachelor's degree in mathematics, Marya begins to work in the laboratory of Gabriel Lippmann where she receives an assignment which pleases her: the *Société d'Encouragement de l'Industrie Nationale*[1] asks her to study magnetic properties of various steels. However, she lacks both the necessary funds and know-how. Then 1 day she mentions this to a Polish friend, Jósef Kowalski, physics professor in Freiburg, who was passing through Paris. He proposes to present her to Pierre Curie, a physicist who had done important work on magnetization.

Pierre Curie

Born on May 15, 1859, Pierre Curie is then 35 years old [35–38]. His brother Jacques is 4 years older. His father, Eugne Curie, was a medical doctor. Pierre never went to school: he was educated by his parents, some friends, and private tutors. He was described as a dreamy person who loved to walk in the country, where, thanks to his father, he could name every plant and animal he would come across. At the age of 14, his father entrusted him to a mathematics teacher, Albert Bazille. He passed the *baccalauréat*[2] at the early age of 16. The following year, he became an assistant to Paul Desains, a specialist of infrared radiation, after which he began to work in the laboratory of Charles Friedel, where he joined his brother Jacques. The two brothers discovered that some crystals, when compressed or elongated, emit electricity. Ten years, later the phenomenon was called piezoelectricity [39]. Pierre used this property to construct an extremely sensitive and precise electrometer.

In 1882, Pierre becomes an assistant at the newly founded *École Municipale de Physique et de Chimie Industrielle*.[3] Strictly, he does not have a lab at his disposal because the school's lab is reserved for the students. Fortunately, however, the director, Léon Schützenberger, a chemist who is also professor at the *Collge de France*, is an intelligent and liberal minded man who permits Pierre to pursue his personal research there. Pierre continues to work on crystallography. He believes that the symmetries displayed in the beautiful geometrical figures of crystals reflect deeper symmetries of the constituent atoms [40]. The importance which Pierre Curie attached to symmetry makes him appear today as a precursor [40, 41].

In 1891, he begins to study magnetization. He discovers and formulates what we call today the "Curie law"$^\diamond$ which exhibits a critical temperature (the Curie

---

[1]The society for the encouragement of national industry.

[2]Equivalent to the GCE both O and A levels.

[3]The municipal school of industrial physics and chemistry.

temperature$^\diamond$) above which ferromagnetic substances lose their magnetization [42]. In spite of the fact that he holds no university position and has no official laboratory to work in, he becomes a well known scientist, especially abroad. It is therefore quite logical for Jósef Kowalski to suggest that Marya Skłodowska should consult him for her work on magnetization. They meet 1 day in the spring of 1894. The meeting becomes a mutual discovery and they are married a year later, on July 25, 1895, after some hesitation of Marya, to whom marriage means that she must give up the idea of returning to her father in her home country. She has the feeling of somehow betraying her country by getting married to a Frenchman and settling in France. But Pierre insists on the fact that she can continue her scientific work in France. And, after all, they are in love...

Polonium and Radium: Pierre and Marie Curie Invent Radiochemistry

Following the advice of Pierre, Marya, who now bears the name of Marie, completes her work on magnetization [43, 44] and searches for a subject for her PhD. This by itself is exceptional: so far, no woman in France had defended a PhD thesis in physics. Pierre suggests studying the "Becquerel rays" a subject that had been neglected for about 2 years. He even offers her a quartz piezoelectric electrometer with which she can measure the extremely weak electric current produced by the radiation of uranium. Although quadrant electrometers were available, his electrometers made it possible to measure the absolute value of the current in units of amperes (in fact tiny fractions of amperes). As Marie later stated:

> We obtain thus not only an indication but a number which accounts for the amount of active substance [45].

Where should she begin? Together with Pierre, Marie decides to find out whether substances other than uranium emit similar radiations. She soon discovers that thorium also radiates [46]. By coincidence, the German physicist Gerhard Schmidt published only a week earlier his observation that thorium was "active," that is, it emitted radiations [47]. However, the attention of Marie is attracted to a small detail. In practically all the cases she had studied, the activity of the uranium compound was precisely that which she could calculate, knowing the amount of uranium in the sample. She finds, however, one exception: two uranium minerals, namely, pitchblende (uranium oxide) and chalcolite (a copper and uranyl phosphate), are more active than what their uranium content would grant. She sees in this remarkable feature a hint that these minerals contain an element which is far more active than uranium. This is where the electrometer of Pierre turns out to be useful because it makes it possible to measure precisely weak currents of the order of 10^{-11} amps,[1] in order to detect such anomalies. Marie Curie is surprised by the right thing. The mineral certainly contains another active substance, but the amount is far too small to be measured by a chemical analysis. Marie Curie has a brilliant idea. Since this substance can only be detected by its radiation, why not use its radiation to follow its trace? With the help of Pierre who discontinues (for a while only, he believes) his work on crystals, she begins with a chemical separation, or at least a concentration of a special kind. She proceeds with several successive chemical reactions with the aim of progressively separating the elements while retaining the most radioactive ones. The radioactivity increases each time she makes a chemical reaction which concentrates bismuth, as if she was extracting bismuth from the mineral:

> *We obtain more and more active products. We finally obtained a substance which is about 400 times more active than uranium. We therefore believe that the substance which we have extracted from the pitchblende contains a metal which has not yet been reported, similar to bismuth in its analytic properties. If the existence of this new metal is confirmed, we propose to call it* polonium, *from the name of the country one of us originates from [45].*

Pierre and Marie Curie have just invented what we call today *radiochemistry.* It is in the title of their publication, "About a new radioactive substance contained in pitchblende," that the term *radioactive* appears for the first time. The terms *radioactive* and *radioactivity* will soon be adopted worldwide.

They soon make a new discovery, which is communicated to the *Académie des Sciences* on December 26, 1898:

> *We have discovered a second substance which is strongly radioactive and which differs from the first [polonium] by its chemical proper-ties [48].*

As in the case of polonium, they perform chemical separations guided by measuring the radioactivity. This time, they find that:

> *The new chemical substance which we found has all the chemical appearances of barium.*

In fact, they cannot separate the new substance from barium, but:

> *Barium and its compounds are usually non radioactive; however, one of us has shown that radioactivity appears to be an atomic property which persists in all the chemical and physical states of the matter. If we adopt this view, the radioactivity of our substance is not due to barium and has to be attributed to a new element.*

---

[1]That is, a hundred thousandth of a micro-ampere, a hundredth of a nano-ampere, or $10\,pA$.

This is *the first time that radioactivity is considered to be an atomic phenomenon.* Oddly reference is made to the first publication of Marie Curie on the subject [46], where the word "atomic" is only used in the term "atomic weight."

To make sure, they ask the expert Eugène Demarçais to make a spectroscopic analysis of their substance. His results confirm their hypothesis: Demarçais observes an unknown optical spectral line which becomes more intense when the sample is more radioactive [49]. They conclude:

The various reasons mentioned above make us believe that the new radioactive substance contains a new element which we propose to call radium.

This is the discovery of radium which was soon to become famous. They also note:

The new radioactive substance certainly contains a strong fraction of barium; in spite of this, its radioactivity is considerable.

There is more: a platinum-cyanide screen, known to become luminous when exposed to X-rays, becomes also luminous when it is placed in the vicinity of the substance. But they believe that this raises a problem:

We obtain this way a source of light, an admittedly very weak source, but which works without a source of energy. This at least appears to contradict the law of Carnot.

They refer to the second law of thermodynamics. They would not have questioned the first law which states that energy is conserved and that energy cannot be created from nothing. But the second law states that the energy of the luminous source cannot be extracted from the surroundings by cooling it, for example. So where does the energy come from? This is the first time that the question is clearly raised.

For months, even years, Pierre and Marie Curie extract and purify radium. Finally, in 1900, after painstaking labor, they succeed in extracting a few decigrams of pure radium from 2 T of mineral [50]! The task was made more difficult by the fact that the room, which the good Schützenberger allowed them to use, was no longer suitable. Charles-Marie Gariel, the new director of the *École*, allowed them to use an abandoned shed in the courtyard. The shed was hot in the summer and dead cold in the winter, but most of the chemical treatments had to be performed outside in the open. At each step of the purification process, Marie Curie made a chemical measurement of the atomic weight$^\diamond$ of the radium. In 1902, she obtained a value of 225 with an uncertainty of one unit [51], a value confirmed by a later measurement [52], in 1907, which gave the value of 226.18 (the value measured today is 226.097 in units used at that time, namely, 1/16 of the mass of oxygen). Radium is indeed a new element, several million times more radioactive than uranium.

Enigmas

What is the nature of the Becquerel rays and where does their energy come from? The problem is reviewed in a paper published in 1899 by Marie Curie in the *Revue Générale des Sciences* [53]:

> *Becquerel radiation is spontaneous; it is not sustained by any known agent. [...]What is more remarkable is the constancy of the radioactivity of uranium in its various physical and chemical states. [...] The uranic radiation appears therefore to be a molecular property inherent in the uranium substance itself.*

In a new paper published in 1900 in the *Revue Scientifique*, known as the *Revue Rose*, she is more specific:

> *Radioactivity is therefore a property which is tagged to uranium and thorium in all their states—it is an atomic property of these elements [54].*

But the deep enigma of radioactivity is the origin of its energy:

> *The emission of uranic rays is spontaneous, meaning that it is not produced by any known cause. For a long time, Mr Becquerel believed that it was caused by light, that uranium somehow absorbed light and that the energy thus absorbed was re-emitted in the form of uranic rays. [...] But experiment does not confirm this interpretation [...]. The emission of uranic rays is remarkably constant and does not change either with time, nor with its illumination nor with its temperature. This is its most troubling feature. When we observe Röntgen rays, we furnish electrical energy to the tube; this energy is provided by batteries, which have to be renewed, or by machines which are set into motion by work which we supply. But the matter is not modified when it emits, admittedly weakly, uranic rays continuously.*

Marie Curie finally raises the question of the nature of the radioactivity. Is it "materialistic" like cathode rays, which J. J. Thomson had shown to be material particles, with measurable mass, charge, and velocity? If such is the case, she claims, we must face the consequences of upsetting several laws of chemistry:

> *The materialistic theory of radioactivity is very tempting. It explains many features of radioactivity. However, if we adopt this theory, we are forced to admit that a radioactive material is not in a usual chemical state; its atoms are not in a stable state, since particles, smaller than the atoms are radiated. The atom, which is an indivisible unit in chemistry, is divisible in this case and sub-atoms arc in motion. The radioactive substance therefore undergoes a chemical transformation which is the source of the radiated energy; but it is not an ordinary chemical transformation, because usual chemical transformations leave the atom unchanged. In a radioactive*

material, if anything changes, it is necessarily the atom, because the radioactivity is attached to the atom. The materialistic theory of radioactivity leads us therefore quite far.

Marie Curie concludes thus that the materialistic theory leads inevitably to the transformation of atoms, therefore to *transmutations*, which she is not ready yet to admit. She continues:

Even if we refuse to admit its consequences, we cannot avoid being embarrassed. If the radioactive material is not modified, where does the energy of radioactivity come from? If we are unable to find the source of energy, we contradict the law of Carnot, which is a fundamental law of thermodynamics, according to which a substance, at a given temperature cannot furnish energy if it does not receive some from the outside. We are then forced to conclude that the law of Carnot is not a general law, that is does not apply to certain molecular phenomena, and that radioactive substances possess means of producing work from the heat of its surroundings.

And Marie Curie concludes. . . that it is difficult to conclude:

Such a hypothesis bears a blow which is as serious to the ideas admitted in physics as to the hypothesis of the transformation of elements in chemistry, and we see that the problem is not easy to solve.

Emanation from Thorium

A young physicist from New Zealand, Ernest Rutherford, begins to study radioactivity. He shows that radioactivity consists of two distinct radiations which he calls α and β rays and the latter are identified to electrons. He discovers that a radioactive gas, belonging to the family of argon, is continuously produced by radioactive thorium. This leads him to the discovery of exponential decay, the fundamental law of radioactivity. But the energy of radioactivity remains an enigma.

In those years, the Cavendish Laboratory in Cambridge is the stage of intense activity: J. J. Thomson [55] and his team are studying cathode rays and the flow of electricity through gases. In November 1895, X-rays are discovered and it is observed that air conducts electricity when it is exposed to X-rays. J. J. Thomson immediately suggests an explanation: the X-rays ionize the molecules of air, splitting them into two "ions," one positively charged the other negatively. The explanation needs to be confirmed. A couple of months earlier, a young physicist arrives from New Zealand.

Ernest Rutherford

Ernest Rutherford [56–58] was born on September 30, 1871, in a family of New Zealand farmers. His father arrived there at the age of 3. The country was occupied by the British. The home in which the young Ernest was raised was governed by his mother, a woman with a strong character who remained active until she died in 1935 at the age of 92. A former teacher, she loved reading and playing the piano. During all his life, Ernest Rutherford was an eager reader, particularly of detective novels. He had just reached the age of 6 when New Zealand made education compulsory for children between the ages of 6 and 13. He was a brilliant pupil. At the age of 15, he was granted a scholarship which enabled him to study at the Nelson College, which today, *noblesse oblige*, is called the Rutherford College. He excelled in English and French literature, history, Latin, mathematics, and rugby. Two years later, he obtained a scholarship from the University of New Zealand, which enabled him to enter the Canterbury College, in Christchurch. He graduated M.A. in 1893 and received the B.Sc. degree in 1894.

That same year, he began to study magnetism and the detection of the recently discovered Hertzian waves. In 1894, he published his first papers in the local scientific journal, the *Transactions of the New Zealand Institute* [59, 60]. He was endowed with a strong personality and was always ready to help his colleagues. Thanks to his charm and power of persuasion, he was always able to obtain help for his projects. In 1894, he ranked second in a competition, which had been initiated during the 1851 "Great Exhibition" in London, and which granted the winner a scholarship allowing him to study in England for 2 or 3 years. It so happened that personal reasons prevented the winner from going, so that the scholarship was given to Rutherford. He chose the Cavendish Laboratory where he arrived in October 1895. He wrote letters to his mother, roughly every 2 weeks, until her death. The letters are a precious testimony in spite of several being lost.

At the Cavendish, he first pursues his research on the detection of electromagnetic waves. However, in February 1896, J. J. Thomson suggests that he should join him in the study of the mechanism which makes air an electric conductor when it is exposed to X-rays. Rutherford rapidly confirms the ideas put forth by J. J. Thomson: the X-rays decompose the molecules of the gas into pairs of "ions" with opposite electric charges, in the same way as dissolved salts do during electrolysis$^\diamond$. An electrically charged neighboring body attracts electric charges of opposite sign, thereby producing an electric current in the gas [61]. Rutherford studies this "ionization" of the gas by X-rays with the meticulous care which characterizes his work throughout his life: he measures the rate of production and recombination of the ions, as well as their velocity in the gas [62, 63].

Rutherford Studies Radioactivity: α-and β-Rays

In 1898, Rutherford turns his attention to the "rays of Becquerel." He wants to find out if they produce the same "electrification of air" as X-rays do and he quickly confirms that this is indeed the case. This leads him to make a detailed study of the penetration of the rays in different substances. He discovers that they are in fact composed of two very different kinds of rays: some ionize strongly the gas which they pass through (meaning that they produce a large number of ions) and they can be stopped by a piece of cardboard; the others have a much stronger penetrating power while ionizing less:

> *These experiments show that the uranium radiation is complex, and that there are present at least two distinct types of radiation—one that is very readily absorbed, which will be termed for convenience the α radiation, and the other of a more penetrative character, which will be termed the β radiation [. . .] [64].*

α- and β-rays are born. The former are often referred to as "weakly penetrating radiation" and the latter as "strongly penetrating radiation."

β-Rays Are Electrons

The discovery of radium marks a new stage in the study of radioactivity: indeed, a tiny sample of radium is an intense and almost point like radioactive source, thereby making it possible to perform a much finer study of the radiation than with the weakly radiating uranium.

As early as 1899, Friedrich Giesel [65], in Braunschweig, Germany, as well as Stefan Meyer and Egon von Schweidler [66] showed that some of the rays emitted by radium could be deflected by a magnetic field, while others could not. Independently, Becquerel observed the same thing and he noted that the rays which could be deflected had properties similar to cathode rays, that is, to electrons. Pierre Curie undertook a more quantitative study. He noticed that the rays which can be deflected have a greater penetrating power than those which cannot [67]. Working with Marie Curie, he showed that the transported electric charge was negative [68]. The rays which can be deflected appear to be the same as those which Rutherford had coined as β-rays. By measuring the deviation produced by magnetic and electric fields, Becquerel was able to measure the ratio of the mass and of the electric charge of the deflectable rays: it turned out to be the same as that of cathode rays [69, 70]. Finally, in 1902, the German physicist Walter Kaufmann made a careful measurement of this ratio for the rays emitted by radium, and he confirmed that it was identical to that of cathode rays, that is, of electrons [71]. The conclusion was that β-rays were very fast electrons with a velocity certainly higher than that of cathode rays. But the value of the velocity of the β-rays was badly known.

Rutherford in Montreal: The Radiation of Thorium, the Exponential Decrease

In 1899, the scholarship of Rutherford expires. A research professor position becomes available at the McGill University in Montreal. This position as well as the attached laboratory are funded by a tobacco millionaire, named MacDonald. The salary is modest but the laboratory has the best equipment in the world. J. J. Thomson is consulted and he strongly recommends Rutherford, who thus becomes MacDonald Professor of Physics in the University of McGill at the age of 28.

Rutherford immediately resumes his research on radioactivity. As he later explains in a letter to his mother:

> I have to keep going, as there are always people on my track. I have to publish my present work as rapidly as possible in order to keep in the race. The best sprinters in this road of investigation are Becquerel and the Curies in Paris, who have done a great deal of very important work in the subject of radioactive bodies during the last few years [72].

He begins a collaboration with R. B. Owens, another professor at the university who studied the ionization of air by thorium. Then, 1 day, they discover a phenomenon, surprising at first:

The sensitiveness of thorium oxide to slight currents of air is very remarkable. The movement of the air caused by the opening or closing of a door at the end of the room opposite to where the apparatus is placed, is often sufficient to considerably diminish the rate of discharge [73].

Rutherford can think of only one explanation:

Thorium compounds continuously emit radio-active particles of some kind, which retain their radioactive power for several minutes. This "emanation," as it will be termed for shortness, has the power of ionizing the gas in its neighborhood. . . .

Always very careful, Rutherford does not claim the agent to be a gas, but he proceeds with careful experiments to show that the emanation is neither due to a fine dust of thorium particles nor to thorium vapor. Furthermore, he notes an essential feature: the activity of the emanation decreases *geometrically*, that is, *exponentially* as we would say today. This means that if the activity of the sample diminishes by a factor of 2 after a certain time, which we call today the *radioactive half-life*$^\diamond$, the activity again diminishes by a factor of 2 in the following same interval of time and continues to do so. The measurements of Rutherford showed that the radioactive half-life of the "emanation from thorium" was 60 s. This means that it is reduced to 1/2 in 1 min, to 1/4 in 2 min, to 1/8 in 3 min, and so on. It becomes a 1,000 times weaker after 10 min and a million times weaker after 20 min.

This law of radioactive decay has a great importance: it will soon be observed that the radioactive half-life is a property of each radioactive substance and that it can be used to characterize a radioactive substance, to detect its presence, even in very small quantities.

"Induced" and "Excited" Radioactivity

Pierre and Marie Curie make an observation which they communicate to the *Académie des Sciences* on November 6, 1899:

While studying the strongly radioactive samples which we prepared (polonium and radium), we noticed that the rays emitted by these substances were able to transfer the radioactivity to otherwise inactive substances and that this radioactivity lasts for quite a long time [74].

Like Rutherford, they observe that this "induced radioactivity" decreases with time:

If one isolates the activated sample from the influence of the radioactive substance, it remains radioactive for several days. However, its radioactivity decreases, fast at first and then progressively more slowly. It appears to disappear asymptotically.

This is a qualitative description of the exponential law of Rutherford. As good experimentalists, Pierre and Marie Curie make sure that this is not due to a trivial cause, or to an illusion:

The aim of this work was mainly to find out whether this induced radioactivity was not due to traces of radioactive material which could have been transported in the form of vapor or of dust onto the exposed strip. [. . .] we believe that we can claim that it is not so and that there exists an induced radioactivity.

This radioactivity has a surprising feature:

We examined the effect of the Becquerel rays on various substances: zinc, aluminum, brass, lead, platinum, bismuth, nickel, paper, barium carbonate, sulphuric bismuth. We were very surprised not to discover order of magnitude differences in the radioactivity induced in these various substances, which all appeared to behave in a similar fashion.

How could one explain this phenomenon? Can it be compared to the emission of electrons when the X-rays of Röntgen impinge on a substance?

The induced radioactivity is a kind of secondary radiation, caused by the Becquerel rays. However it differs from that which is known to occur with the rays of Röntgen. Indeed the secondary rays of Röntgen, which have been studied so far, are created at the instant when the rays of Röntgen impinge on the substance, and they cease as soon as the they are suppressed. In view of the facts reported above, we may ask whether radioactivity, which is apparently spontaneous, is not an induced effect for certain substances.

On November 22, 1899, Rutherford, who had not yet read the communication of Pierre and Marie Curie on induced radioactivity, sends a second paper to the *Philosophical Magazine*, describing his observations of induced radioactivity:

Thorium compounds under certain conditions possess the property of producing temporary radioactivity in all solid substances in their neighborhood. The substance made radioactive behaves, with regard to its photographic and electric actions, as if it were covered with a layer of radio-active substance like uranium or thorium [75].

Rutherford shows that this "excited" radioactivity is always associated to an "emanation." He measures its rate of decay and finds that the half-life is about 11 h. It is the same for all the substances which are exposed to the radioactivity, and the half-life is considerably longer than that of the emanation itself (1 min). After a thorough discussion, a prime example of rigor and imagination, he proposes the only plausible explanation he can think of: the "excited" radioactivity[1] must be caused by radioactive particles originating in the thorium and most likely transported by the "emanation":

> The power of producing radioactivity is closely connected with the presence of the "emanation" from thorium compounds, and is in some way dependent upon it.

Rutherford observes one more thing: in the absence of an electric field, the induced radioactivity is uniformly distributed on the surface of the surrounding material. But if a body carries a negative electric charge, the radioactivity becomes concentrated on this body, suggesting that the induced radioactivity is associated to positive electric charges:

> All thorium compounds examined produce radioactivity in substances in their neighborhood, if the bodies are uncharged. With charged conductors the radioactivity is produced on the [negatively] charged body.

Elster and Geitel: The Radioactivity of the Air and of the Earth

Two German physicists, Julius Elster and Hans Geitel, provide further data. Elster was born on December 24, 1854, in Bad Blankenburg, Germany. In school, he becomes a friend of Geitel, who is a few months younger (born on July 26, 1855, in Brunswick). After studying in Heidelberg from 1875 to 1877, and in Berlin in 1878, Elster returns to Heidelberg where he obtains his Ph.D. He passes successfully an examination allowing him to become a high school professor and he obtains a position in Wolfenbüttel, where Geitel, who passed the same examination in Berlin in 1879, had been teaching for a year. When Elster got married, he had a house constructed and Geitel came to live there. They installed a laboratory and soon embarked on their research. At first, they were interested in the conduction of electricity by gases, bearing particular attention to electrical phenomena in the atmosphere, well before the theory of ionization could explain it. In 1889, they studied the photoelectric effect and made important contributions to the field. As soon as the discovery of radioactivity becomes known in 1896, they begin to work on it.

In an address to the British Association on September 7, 1898, Crookes makes a daring suggestion:

---

[1]Rutherford preferred to use the term "excited" radioactivity rather than the term "induced" radioactivity. He will later explain [76] that the latter could suggest that the phenomenon is due to the action of passing through the air, whereas he shows that the "excited" radioactivity is transported by positively charged particles. It is, however, the same phenomenon which Pierre and Marie Curie call "induced" radioactivity.

It has long been to me a haunting problem how to reconcile this apparently boundless outpour of energy with accepted canons[...] It is possible to conceive a target capable of mechanically sifting from the molecules of the surrounding air the quick from the slow movers[...] Let uranium or polonium, bodies of densest atoms, have a structure that enables them to throw off the slow moving molecules of the atmosphere, while the quick moving molecules, smashing on to the surface, have their energy reduced and that of the target correspondingly increased [77].

This conjecture appears to contradict the second principle of thermodynamics because it implies that heat could flow from one body to another at the same temperature.

Three weeks later, Elster and Geitel disprove the conjecture of Crookes by showing that the radioactivity of uranium is the same in air at the atmospheric pressure, in vacuum, or in a vessel under pressure [78]. The air is therefore not responsible for the occurrence of radioactivity.

They then investigate whether the radioactivity of a sample varies under different conditions: when it is subject to cathode rays, when it is heated to different temperatures, and when it is taken to a high altitude or even to the bottom of a mine (852 m deep). Since the radioactivity seems to remain obstinately insensitive to these conditions, they conclude that:

Since the property of emitting Becquerel rays belongs, as it seems, to all chemical compounds of an active element, it is difficult to interpret it as the sign of a chemical process in the true sense; indeed one should rather seek the source of energy in the atom of the element concerned. One is not far from the idea that a radioactive element, like the molecule of an unstable compound, turns into a stable state. In fact this idea leads to assume a gradual transition of the active substance toward an inactive substance, and therefore, logically, an alteration of its elementary properties [79].

In 1901, they discover that when an electric conductor, in the air, is connected to the negative pole of a battery thereby becoming negatively charged, it becomes radioactive. It has attracted positively charged radioactive particles present in the air, which is therefore weakly radioactive, and so is the earth, as they soon discover [80, 81].

Within a few years, Elster and Geitel acquired a great scientific reputation. Known as the "Castor and Pollux" of physics, they did all their research together. In 1899, the University of Breslau offered a professorship to both of them. However, fearing for their independence, they preferred to remain professors at the *Gymnasium* of Wolfenbüttel. The respect and esteem which they enjoyed in the scientific community was expressed in 1915 by the edition, for their 60th birthday, of a voluminous commemorative edition with contributions from the greatest German physicists of the time, namely, Max Born, Max von Laue, Philip Lenard, Max Planck, and Arnold Sommerfeld. Elster died on April 6, 1920, in Bad Harzburg, and Geitel died on April 15, 1923, in Wolfenbüttel [82, 83].

A Third Type of Ray: γ-Rays

On April 9, 1900, at a session of the *Académie des Sciences*, Paul Villard presented a communication on "The reflection and refraction of cathode rays and of deflectable uranium rays." Under this somewhat trivial title, he included a "remark on the radiation of uranium":

> *I almost always observed that, in addition to the refracted beam, a beam propagating in a straight line was superposed. [. . .] These observations lead us to admit that the emission of radium contains a very penetrating radiation, which can pass through metallic strips and which the photographic method is able to detect [84].*

A few months later, pursuing his study, Villard showed that these "non deflectable" rays have a penetrating power about 160 times larger than that of β-rays [85]. He believed that they are similar to X-rays and he called them "radium X-rays." They will soon be called γ-rays, the third Greek letter after α and β.

The Emanation of Thorium Is a Gas Belonging to the Argon Family

When Rutherford returns to New Zealand for a vacation, an important event takes place. Some 6 years earlier, still a student in Christchurch, he rented a room in the house of Mrs. Arthur de Renzy Newton, a widow with four children. The young Ernest fell in love with the eldest daughter, Mary. They became engaged, but there was no question of marriage as long as he was unable to provide for the needs of the family. Now, in 1900, the time is ripe: they are married and they return to Montreal.

Rutherford continues his detailed observations of the "emanation" of thorium. He proves not only to be an outstanding physicist but also a leader. He builds a small research team which includes Frederick Soddy, a young chemist from the University of Oxford. This is the beginning of a fruitful collaboration. In a paper, published in 1901, they study the chemical properties of the "emanation of thorium [. . .], [which] behaves in every way as a temporarily radioactive gas." This leads them to an important conclusion:

> *It will be noticed that the only known gases capable of passing in an unchanged amount through all the reagents employed are the recently discovered gases of the argon family [86].*

Note the caution and the art with which Rutherford expresses himself: he incites the reader to note and to conclude for himself that *the emanation is a gas belonging to the argon family*.

A Proliferation of "X" Radiations

Rutherford and Soddy have hardly finished writing their paper on the radioactivity and the emanation of thorium compounds, when they discover a new phenomenon which, yet again, leaves them puzzled. They do not rewrite their paper. Instead, they write:

... developments have been made in the subject which completely alter the aspect of the whole question of emanation power and radioactivity.

They observe that the amount of emanation of a given quantity of thorium varies from one chemical compound to another! Rutherford and Soddy make a fractional chemical analysis, similar to the one used by Marie Curie, and they reach the only possible conclusion (because they cannot question the atomic nature of radioactivity):

There seems little doubt of the actual existence of a constituent ThX to which the properties of radioactivity and emanating power of thorium must be ascribed [86].

Further down, they add:

The manner in which it makes its appearance, associated with each precipitate formed in its concentrated solution, resembles the behavior of Crookes' UrX.

They refer to a paper written by the English chemist William Crookes who, at the age of 68, embarked on a study of the radioactivity of uranium and showed that it was possibly not due to uranium proper but to another constituent:

... the radioactive property ascribed to uranium and its compounds is not an inherent property of the element, but resides on some outside body which can be separated from it [87].

Crookes calls this substance "uranium X," UrX in short, in order to specify that it is associated to uranium (although it is not uranium). Eighteen months earlier, Becquerel had performed similar experiments: he succeeded in separating a uranium salt from a substantially more radioactive substance mixed with barite sulfate [88, 89]. However, he kept the inactive uranium preparation and he discovered that the uranium had recovered its original radioactivity:

I studied again the progressively weakened products which I had prepared 18 months ago and, as I expected, I found that all the products were identical [...] Thus the lost radioactivity was spontaneously recovered. On the other hand, the precipitated barite sulfate, which before was more radioactive than uranium, is completely inactive today. The loss of radioactivity, which is a property of activated or induced substances, shows that barium did not decrease the essentially active and permanent part of uranium [90].

The mystery grows.

"An Enigma, a Deeply Astonishing Subject"

1900 is the year of the universal exposition in Paris, and on this occasion, the *Société Française de Physique* (the French Physical Society) decides for the first time to organize an international physics meeting. The first circular is mailed to physicists throughout the world in June 1899. There is a large response and over 800 physicists attend the opening session in the *Grand Palais* on Monday, August 6, 1900. It is presided by Alfred Cornu and Lord Kelvin is named honorary president, by acclamation. The meeting lasts 6 days. It is divided into seven sessions, each one devoted to an important problem at the time. One session is devoted to cathode and uranic rays. The physicists are invited to the *Élysée Palace* by the president of the Republic Émile Loubet and the prince Roland Bonaparte offers a reception in his private mansion in *avenue d'Iéna*. They are also guided to the top of the new Eiffel Tower. Visits to laboratories are organized as well as some general talks, among which two, devoted to radioactivity, are delivered by Becquerel and Pierre Curie. Becquerel speaks about the radioactivity of uranium [91], and the talk of Pierre Curie, signed also by Marie Curie, is devoted to "new radioactive substances" and to general problems of radioactivity [92]. Their conclusion bears yet another question mark:

> *But the spontaneous nature of the radiation is an enigma, a deeply astonishing subject. What is the source of the energy of the Becquerel rays? Should one search for it within the radioactive substance itself or outside? [. . .] In the first case, the energy could be drawn from the heat of the surrounding matter, but such a hypothesis would contradict the Carnot principle. In the second case, [. . .] radium would continuously emit very small particles carrying negative electric charge. The available energy, stored as potential energy would progressively dissipate, and this view would necessarily lead us to abandon the idea that atoms are invariable.*

The Puzzle Is Disentangled

Rutherford shows that radioactivity is the transformation of an atom into another: the atom explodes while violently ejecting microscopic particles. Radioactivity draws its energy from within the atom, in an enormous quantity.

In this new year of 1902, Rutherford and Soddy pursue their patient and obstinate study in their laboratory at McGill University. The situation is really confusing. The simple and constant radioactivity of uranium or thorium (or still radium) is presently replaced by a multitude of further phenomena: the induced radioactivity, as Pierre Curie calls it, the emanation, loss and recovery of the activity of thorium and of uranium. Furthermore Rutherford and Soddy recently discovered "thorium X," coined ThX. It is not thorium since it can be separated by chemical methods. Nonetheless, Rutherford calls it "thorium X" as a reminder that it is associated with thorium. In a similar fashion "uranium X" is not the same element as uranium. They make a hypothesis:

It therefore follows that [. . .] the experimental curve will be explained if two processes are supposed to be taking place:

1. That the active constituent ThX is being produced at a constant rate.
2. That the activity of ThX decays geometrically with time [93].

Several observations corroborate this idea. Paying attention to the question recurrently raised by Pierre and Marie Curie, they carefully discuss the origin of the energy involved:

Energy considerations require that the intensity of radiation from any source should die down with time unless there is a constant supply of energy to replace that dissipated. This has been found to hold true in the case of all known types of radioactivity with the exception of the "naturally" radioactive elements [. . .] In the case of the three naturally occurring radioactive elements, however, it is obvious that there must be a continuous replacement of the dissipated energy, and no satisfactory explanation has yet been put forward to account for this.

Rutherford and Soddy then propose the following explanation of all the observations: by its radioactivity, thorium constantly produces "thorium X," which in turn progressively disappears by radioactivity. But since it remains mixed to the thorium, an equilibrium is reached between the production and disappearance of "thorium X."

The material constituent responsible for the radioactivity, when separated from the thorium which produces it, behaves in the same *way as the other typically radioactive substances. Its activity decays geometrically with time, [. . .]. The normal radioactivity is, however, maintained at a constant value by a chemical change which produces fresh radioactive material. . .*

They consider that this explanation applies equally well to uranium and radium, which allows them to conclude:

All known types of radioactivity can thus be brought into the same category.

This is not the end yet, but they hold the thread which will lead them out of the labyrinth. *Radioactivity is not a simple phenomenon but a cascade of superimposed events.* Radioactive substances are subject to continuous transformations, while emitting radiation and producing new substances which are themselves transformed. And to complicate things further, each transformation occurs at a different rate. What is observed is a mixture of the radiations of these substances. A well-entangled process! Rutherford insists that it is an atomic phenomenon:

*All the most prominent workers in this subject are agreed in considering radioactivity an atomic phenomenon. M. and Mme Curie, the pioneers in the chemistry of the subject, have stated (*Comptes Rendus *1902, 134, 85) that this idea underlies their whole work from the beginning and created their method of research.*

Furthermore, the radiation consists of *material* particles and not of waves similar to electromagnetic waves or X-rays:

M. Becquerel, the original discoverer of the property for uranium, [. . .] points out the significance of the fact that uranium is giving out cathode rays. These, according to the hypothesis of Sir William Crookes and Professor J. J. Thomson, are material particles of mass one-thousandth that of the hydrogen atom.

Induced radioactivity behaves as a deposit of a certain kind of radioactive material:

The present researches had their starting point in the fact that had come to light with regard to the emanation produced by thorium compounds and the property it possesses of exciting radioactivity on surrounding objects. In each case, the radioactivity appeared as the manifestation of a special kind of matter in minute amount. The emanation behaved in all respects like a gas, and the excited radioactivity it produces as an invisible deposit of intensely active material independent of the nature of the substance on which it was deposited, and capable of being removed by rubbing or by the action of acids.

After having carefully assessed these observations, Rutherford and Soddy inevitably conclude:

The position is thus reached that radioactivity is at once an atomic phenomenon and the accompaniment of a chemical change in which new kinds of matter are produced. The two considerations force us to the conclusion that radioactivity is a manifestation of a subatomic chemical change.

This is quite an extraordinary conclusion! New forms of matter and possibly new elements can be created. At this point, Rutherford avoids being explicit and he avoids the word "transmutation" although that is what it is all about: he does not want to be coined an alchemist. He restricts himself to a minimal formulation which is, however, strongly stated as "the two considerations force us to the conclusion. ..." Such a personal formulation is extremely rarely used by him. Finally, Rutherford and Soddy observe that "disactivated" thorium (meaning distinct from ThX) is not really inactive: it remains endowed with a residual activity of its own. Soddy makes a similar observation for uranium [94].

α-Rays Revisited

At the same time, Rutherford studies the weakly penetrating α-rays. They had not been the subject of as much attention as the penetrating β-rays. Becquerel had shown that the radiation of radioactive substances was similar to cathode rays, that is, to electrons but with a velocity which is much higher than that of cathode rays [69, 95–98]. The radioactivity discovered in 1896 by Becquerel was in fact composed of β-rays because the α-rays emitted by uranium were stopped by the cardboard which he placed between the uranium and the photographic plate in order to protect it from sunlight. It is not easy for Rutherford to study α-rays for this same reason: they have a weak penetrating power. A thin layer of matter is enough to stop them so that an extremely thin sample of radioactive material needs to be used if the α-rays are to be emitted into the air before becoming absorbed. By using a magnet, more powerful than the one of Stefan Meyer and Egon R. von Schweidler,[1] he notices that α-*rays are indeed also deflected by a magnetic field*. They therefore carry an electric charge. Rutherford succeeds in deflecting them by an electric field and this enables him to estimate their mass [99, 100] which he discovers to be much larger than the mass of the electron. However, they have a considerably smaller velocity of about 25,000 km/s. The α-rays are thus more akin to ionized atoms. They could be hydrogen or helium atoms. Their higher mass allows them to transport a higher energy in spite of their slower velocity.

---

[1] See page 19.

Radioactivity Is an Atomic Decay

In the spring of 1903, the University College of the University of London offers a position to Soddy in the laboratory of Sir William Ramsay, the famous chemist, who had isolated, purified and identified several new elements, namely, argon in 1894 with Rayleigh as well as helium, neon, krypton, and xenon [101]. These elements formed a new column which was added to the Mendeleev table$^\diamond$. They are "rare" or "noble" gases which neither combine with themselves nor with any other elements. Ramsay is particularly interested in the "emanations" which bear a striking resemblance to "his" rare gases.

Before Soddy embarks to England, he and Rutherford publish several papers in which they summarize the work done at McGill University. They first rewrite two papers, originally published in the *Journal of the Chemical Society,* and send them to the *Philosophical Magazine* [102, 103], which enjoys a much wider audience, among physicists in particular. In their second paper, they add a very important remark:

> So far it has been assumed, as the simplest explanation, that the radioactivity is preceded by chemical change, the products of the latter possessing a certain amount of available energy dissipated in the course of time. A slightly different view is at least open to consideration, and in some ways preferable. Radioactivity may be an accompaniment of the change, the amount of the former at any instant being proportional to the amount of the latter. On this view the non-separable radioactivity of thorium and uranium would be caused by the primary change in which ThX and UrX are produced.

Thus, the radiation would be emitted during the transformation of the element thorium (or uranium) into the mysterious ThX (UrX). The radioactive transformation would be a kind of an explosion during which α-particles would be expelled: a real *decay*!

The Puzzle Is Unravelled: Radioactive Families

In the fall of 1903, Rutherford and Soddy publish the last paper belonging to their collaboration in the laboratory of McGill. They first make a systematic review of their results. Their view of the nature of radioactivity has matured and they take another step forward:

> There is every reason to suppose, not merely that the expulsion of a charged particle accompanies the change, but that this expulsion actually is the change [104].

In other words, radioactivity is the way in which the transformation of an atom is visible to us. They give a simple formulation of the law of radioactive decay:

Uranium	Thorium	Radium
⇓	⇓	⇓
Uranium X	Thorium X	Radium emanation
⇓	⇓	⇓
?	Thorium emanation	Radium-excited activity I
	⇓	⇓
	Thorium-excited activity I	Ditto II
	⇓	⇓
	Ditto II	Ditto III
	⇓	⇓
	?	?

Fig. 1 The three radioactive families proposed by Rutherford and Soddy in 1903. Each known element (uranium, thorium, and radium) is the starting point of a succession of radioactive transformations: uranium transforms into uranium X, what follows is not known; thorium transforms into thorium X, which transforms into the "thorium emanation," etc. [104]

The proportional amount of radioactive matter that changes in unit time is constant.

This is an alternative way to express the law of exponential decay. They add:

The complexity of the phenomena of radioactivity is due to the existence as a general rule of several different types of matter changing at the same time into one another, each type possessing a different radioactive constant.

The nature of radioactivity can only be a decay:

Since radioactivity is a specific property of the element, the changing system must be the chemical atom, and since only one system is involved in the production of a new system and, in addition, of heavy charged particles, in radioactive change the chemical atom must suffer disintegration.

The situation is clear: radioactivity is the decay of an atom which expels a particle. The residual atom is different, a chemical transformation has taken place.

However, Rutherford and Soddy are not content to simply state that the observed radioactivity is the mixture of several radioactive decays. They also classify radioactive substances into three groups, corresponding to the decay of uranium, thorium, and radium (see Fig. 1). This is the first example of what will soon be called "radioactive families." True, there still remain substances and emanations of unknown nature, but the direction is defined. Physicists must now complete this table, identify the various elements, measure their decay rate, etc. The general framework will not change. Rutherford and Soddy ask what name should be given to the intermediate atomic fragments:

. . . which remain in existence only a limited time, continually under-going further change. Their instability is their chief characteristic. On

one hand, it prevents the quantity from accumulating, and in conse-
quence it is hardly likely that they can ever be investigated by ordinary
methods. On the other hand, the instability and consequent ray-
expulsion furnishes the means by which they can be investigated. We
would therefore suggest the term metabolon *for this purpose [104].*

The term *metabolon* did not survive. Clearly, the prudent Rutherford did not wish to expand the list of elements too hastily, even with question marks. After all, there existed no experimental evidence, neither chemical nor spectroscopic, which would point to new elements. He said that they were fragments of atoms with specific features; their short lifetime implies that they appear in too small quantities to be identified by usual chemical methods or even by spectroscopic methods. Each one, has however, a unique feature: its radioactive half-life. But to conclude from this that they should be new chemical elements is a step which Rutherford carefully avoids.

Where Does the Energy of Radioactivity Come from?
The Conjecture of Rutherford

Rutherford paid much attention to the problem of the energy involved in radioactivity. It appeared to stem from nowhere. This question was also a major concern for Pierre and Marie Curie. Rutherford conjectured that atoms possess and internal "latent" energy which is released during the radioactive decay. In a similar fashion, heat is released during the combustion of hydrogen and oxygen, the process during which two atoms of hydrogen stick to an atom of oxygen so as to form a water molecule. However, this process is not a decay but rather the coalescence of three atoms which form a molecule. In the fall of 1902, Rutherford and Soddy attempt to measure the energy released during the radioactive transformation of radium. They know roughly the mass and the velocity of the emitted "rays" which they assume to be material particles. They estimate their number which in turn allows them to estimate the amount of energy which is released. They find an enormous number: a hundred million (10^8) calories are released by 1 g of radium (assuming it decays completely) and that is a minimal estimate:

10^8 *gramme-calories per gramme may [. . .] be accepted as the least*
possible estimate of the energy of radioactive change in radium. The
union of hydrogen and oxygen liberates approximately 4×10^3 *gramme-*
calories per gramme of water produced, and this reaction sets free
more energy for a given weight than any other chemical change known.
The energy of radioactive change must therefore be at least 20 000
times, and may be a million times, as great as the energy of any
molecular change [104].

They calculate that, in the span of 1 year, a single gram of radium releases at least 15 000 g-cal. This leads them to explain the apparent constancy of the radioactivity of radium and thorium:

Since the α radiation of all the radio-elements is extremely similar in character, it appears reasonable to assume that the feebler radiations of thorium and uranium are due to these elements disintegrating less rapidly than radium. [. . .] We obtain the number of 6 × 10⁻¹⁰ as a maximum estimate for the proportionate amount of uranium or thorium undergoing change per year. Hence in 1 "gramme" of these elements less than a "milligramme" would change in a million years. In the case of radium, however, the same amount must be changing per year. The "life" of the radium cannot be in consequence more than a few thousand years.

In other terms, uranium and thorium do not last forever, but their decay rate is very weak: only one part of a sample in a thousand will disintegrate in the span of one million years. About a billion years (10^9 years) will pass before half of the sample disappears. This is why it appears to be permanent. But radium decays a million times faster, so that its half-life is of the order of a 1000 years. The order of magnitude of these estimates is correct. The present-day measured half-lives are 4.47 billion years (4.47×10^9) for uranium, 14 billion years (14×10^9) for thorium, and 1600 years for radium.

There is more. These results suggest an explanation of a completely different phenomenon, namely, the source of energy of the sun:

The energy latent in the atom must be enormous compared with that rendered free in ordinary chemical change[. . .] It must be taken into account in cosmic physics. The maintenance of solar energy, for example, no longer presents any fundamental difficulty if the internal energy of the component elements is considered to be available, i.e. if processes of sub-atomic change are going on.

This is a bold conjecture. Rutherford had little inclination towards abstract or adventurous speculation. He simply acknowledged the fact that the internal energy of atoms was probably of the right order of magnitude to provide for the energy of the sun, a controversial subject at the time.

At the same time, Pierre Curie, together with his young assistant Albert Laborde, measured the increase in temperature of a water bath in which they immersed a vessel containing a small amount of radium (in the form of a chloride mixed with some barium chloride). They obtained a huge number:

1^g of radium releases a quantity of heat which is of the order of 100 small calories per hour.

1 gramme-atom of radium (225^g) would release, during each hour, $22,500^{cal}$, a number comparable to the heat released by the combustion of 1 gramme-atom of hydrogen in oxygen.

The release of such an energy cannot be explained by an ordinary chemical transformation. If we seek the origin of this production in an internal transformation, this transformation must be of a deeper nature and it must be a modification of the radium atom itself [. . .] Thus if the preceding hypothesis were exact, the energy involved in the transformation of atoms would be extraordinarily large.

> *The hypothesis of a continual transformation of the atom is not*
> *the only one which is compatible with the heat released by radium.*
> *The heat release can also be explained by assuming that radium uses*
> *an external energy, the nature of which is unknown [105].*

Let us ignore units such as the gramme-atom, the small calorie$^\diamond$, and gramme-calories$^\diamond$, which are no longer used today, in order to retain this essential fact: 225 g of radium release as much energy *per hour, and at a seemingly constant rate*, as the total combustion of 22 l of hydrogen. Just like Rutherford, Pierre Curie notes that this energy is much larger than energies involved in chemical reactions and that radioactivity must correspond to a deep transformation of the radium atom. But he does not yet abandon the hypothesis of a cosmic energy flux which radium (as well as other radioactive substances) would be able to pick up. Pierre Curie maintains this hypothesis because radium releases energy permanently and that one cannot detect a corresponding decrease of its weight. He is not yet aware of Rutherford's explanation which is the key to this enigma: it is precisely because the energy released by the decay of each atom is so large that this energy can manifest itself at our scale while only a tiny fraction of atoms actually decay. Indeed barely 0.04 % of the radium atoms and only one out of ten billion (10^{10}) uranium or thorium atoms decay in 1 year. *The constancy of these elements is an illusion.*

How do the measured values of Pierre Curie compare with those of Rutherford? It takes 1,600 years for half a gram of a radium sample to decay so that, according to the estimate of Rutherford, it would release at least a dozen gramme-calories per hour, possibly 10 or a 100 times more. This estimate is quite compatible with the measurement of Pierre Curie, who quotes 100 cal/h.

The theory of Rutherford and Soddy was not immediately approved by everyone. J. J. Thomson was readily convinced, but the idea of a transmutation, even if the word wasn't used, was difficult to swallow, especially by chemists and also by Lord Kelvin, who found it hard to believe that so much energy could be stored in an atom. He preferred to think, like Pierre Curie, that some atoms could absorb some radiation propagating in space. Pierre Curie was somewhat reluctant to believe in the existence of such substances whose existence is only revealed by a radioactivity of a given half-life.

But the theory of Rutherford was effective and it explained the great complexity of radioactive phenomena. By 1904 it was accepted. In that same year, Rutherford published his first book, *Radioactivity* [106], which he dedicated to J. J. Thomson "to acknowledge his respect and admiration." In the preface, he states his philosophy:

> *The phenomena exhibited by the radioactive bodies are extremely*
> *complicated, and some form of theory is essential to connect in*
> *an intelligible manner the mass of experimental facts that have*
> *now been accumulated. I have found the theory that the atoms*
> *of the radioactive bodies are undergoing spontaneous disintegration*
> *extremely serviceable, not only in correlating the known phenomena,*
> *but also in suggesting new lines of research.*

It must be admitted that his theory does explain the observations and succeeds in unravelling what appeared to be an inextricable puzzle.

Experimental Evidence of Transmutation

Rutherford and Soddy noticed that uranium ore always contained a certain amount of helium. The element helium was first observed in 1868 by Lockyer [107], the founder of *Nature,* in the form of a bright yellow line in the spectrum$^\diamond$ of solar light in a prominence which he observed during the total eclipse of August 18, 1868. He attributed this spectral line to a new element which he called *helium,* in order to stress its solar origin. Twenty seven years later, William Ramsay detected its presence in cleveite, a rare variety of pitchblende which is found in Sweden. He showed that it was an inert gas with atomic weight$^\diamond$ 4, four times heavier than hydrogen. Crookes showed that it gave rise to the spectrum observed by Lockyer.

In 1903, Frederick Soddy joins William Ramsay in the laboratory of University College in London. Soon, they begin to study the emanation of radium. They enclose a 20 mg sample of radium in a closed vessel and, after a few months, *they discover that the vessel contains helium* [108]. This is a proof that the radium atom splits while emitting a helium atom. What remains of the radium atom is then necessarily *another element.* This observation adds considerable credit to Rutherford's general theory of radioactive transformations which becomes generally accepted.

Radioactivity is Understood. Radioactive Families

The physical phenomenon is now understood. It is still necessary to identify each one of the successive transformations of naturally radioactive elements. On May 19, 1904, Rutherford is invited to deliver the prestigious annual Bakerian Lecture at the Royal Society in London [109]. In 1775, Henry Baker donated £100 with the instructions that a Fellow of the Royal Society should deliver a talk on *"some part of natural history or of experimental philosophy at a date and in a fashion found suitable to the President and the Council of the Royal Society."* Previous speakers included William Thomson (Lord Kelvin), James Dewar, Norman Lockyer, William Crookes, James Clerk Maxwell, Lord Rayleigh and Michael Faraday. Rutherford reports on progress in the theory of radioactive transformations. The list of radioactive families has grown and become more precise (see Fig. 2).

Considerably more work was required to complete the radioactive families. The "metabolons," as Rutherford calls them, need to be identified. Some are radium A,B,C, and others thorium A,B,C, denominations which simply indicating their origin. But the path is clear.

Radium	Thorium	Uranium	Actinium
⇓	⇓	⇓	⇓
Radium	ThX	UrX	Actinium X?
emanation	⇓	⇓	⇓
⇓	Thorium	Final product	Actinium
Radium A	emanation		emanation
⇓	⇓		⇓
Thorium A	Thorium A		Actinium A
⇓	⇓		⇓
Radium C	Thorium B		Actinium B
⇓	⇓		⇓
&c.	Thorium C		Actinium C
	(final product)		(final product)

Fig. 2   There are four known radioactive families in 1904, together with the actinium family. The number of "metabolons" has nearly tripled [109]

Rutherford stresses what he considers an essential point: each radioactive substance is characterized by its half-life (as it is called today) by means of which it can be unambiguously identified.

Rutherford remained in Montreal for three more months during which he confirmed and consolidated his results. His reputation grew to the extent that he had to refuse most invitations. The laboratory in McGill attracted many researchers among whom a young German chemist, Otto Hahn who will soon appear on the scene.

In September 1906, Rutherford received a letter from Arthur Schuster, physics professor at the University of Manchester. He was of German origin (born in Frankfurt in 1851 into a well-to-do Jewish family) and, since 1887, he was Langworthy Professor of Experimental Physics at Owens College in Manchester, where he acquired a reputation in the fields of spectroscopy and the conduction of electricity by gases. At the age of 55, he wanted to retire, enjoy his fortune and write a book. He wished to be succeeded by whom he considered to be the greatest physics experimentalist, namely Ernest Rutherford. After exchanging letters for 6 months, Rutherford finally accepted the offer much to the dismay of his colleagues in McGill. It was an occasion for him to return to England, and to be closer to the best physics laboratories. When he retired, Arthur Schuster created a scholarship allowing to invite a young mathematical physicist. The scholarship will be used later by a young Danish physicist, named Niels Bohr. On May 17, 1907, Ernest Rutherford embarks on a boat headed to the old continent.

Consecrations and Mourning: The End of an Era

First Henri Becquerel, Pierre and Marie Curie, then Ernest Rutherford are awarded the Nobel Prize. The deaths of Pierre Curie and of Henri Becquerel mark the end of an era.

On November 27, 1895, the Swedish industrialist Alfred Bernhard Nobel, aged 63, signed his last will in Paris. He endowed most of his fortune to a fund

... the interest on which shall be annually distributed in the form of prizes to those who, during the preceding year, shall have conferred the greatest benefit on mankind. The said interest shall be divided into five equal parts [...]: one part to the person who shall have made the most important discovery or invention within the field of physics; one part to the person who shall have made the most important chemical discovery or improvement; one part to the person who shall have made the most important discovery within the domain of physiology or medicine; one part to the person who shall have produced in the field of literature the most outstanding work in an ideal direction; and one part to the person who shall have done the most or best work for fraternity between nations, for the abolition or reduction of standing armies and for holding the promotion of peace congresses [...] It is my express wish that in awarding the prizes, no consideration whatever shall be given to the nationality of the candidates, but that the most worthy shall receive the prize, whether he be a Scandinavian or not [110].

Alfred Nobel died on December 10, 1896. The first Nobel prize in physics was attributed in 1901 to Conrad Röntgen for his discovery of X-rays. The 1902 Nobel Prize was attributed to Hendrik Lorentz and Pieter Zeeman. The Nobel Prize quickly acquired a fame unequalled by any other.

1903: Henri Becquerel Shares the Nobel Prize with Pierre and Marie Curie

On November 1903, telegrams sent from Stockholm announce that the Nobel Prize in Physics is shared: one half is attributed to Henri Becquerel "in recognition of the extraordinary services he has rendered by his discovery of spontaneous radioactivity"; the other half to Pierre and Marie Curie "in recognition of the extraordinary services they have rendered by their joint researches on the radiation phenomena discovered by Professor Henri Becquerel" [111].

Just a few months earlier, on June 25, 1903, Marie Curie defended her Ph.D. thesis in front of a jury composed of Gabriel Lippmann, Edmond Bouty and Henri Moisan. The thesis bore a simple title: "Research on radioactive substances." *It was the first Ph.D. thesis defended by a woman in France.* That same evening, Paul Langevin invited the Curie couple to dinner, together with Jean Perrin, his wife as well as Ernest Rutherford and his wife, who were passing through Paris. Paul Langevin had met Rutherford at the Cavendish where he spent the academic year 1897–1898, after graduating from the *École Normale Supérieure*. That was to be the only encounter between Rutherford and Pierre Curie.

The Nobel Prize was officially awarded on December 11. According to the statutes of the Nobel Fund, the laureate was invited to deliver a lecture to the Swedish Academy, within 6 months after receiving the prize. Becquerel went to Stockholm with his wife and, in his lecture, he gave a very precise account of his observations. However, Pierre and Marie Curie were unable to go. For some time already they felt very tired and they feared the 2-day train trip to Stockholm. A few months before, Marie gave birth to a premature child who died within a few hours. Pierre suffered from pain in his joints, which he thought were due to rheumatism. They thought they had worked too hard, which was true, but they did not suspect that a continuous exposure to radioactive substances could be the cause.

In fact, biological effects of radioactive substances had been observed as early as 1900 by Giesel [112], who noted a red spot which developed into a wound on his arm which had been exposed to a radioactive source. One day in 1901, Pierre Curie lent Henri Becquerel some barium chloride enclosed in a test tube, itself wrapped in a cardboard box. After keeping it for about 6 h in his waistcoat, Becquerel noticed that his skin had reddened and that a wound formed which took a month to heal. Then Pierre Curie voluntarily attached a sample to his arm. The resulting wound took about 52 days to heal. In a communication to the *Académie des Sciences*, Becquerel and Pierre Curie wrote:

The rays of radium act strongly on the skin. The effect is similar to the one produced by the rays of Röntgen [113].

They stress the effect produced on the hands of experimentalists who handle radioactive substances:

In addition to these strong effects, various other effects are noticeable on our hands while we manipulated strongly radioactive substances. The hands have a tendency towards desquamation; the tips of fingers, which held test tubes or capsules containing very radioactive substances, harden and become occasionally very painful; the inflammation of the finger tips of one of us lasted two weeks after which the skin pealed off; the pain persisted for 2 months.

Pierre and Marie Curie must have believed that they were only local injuries. Pierre wrote to the Swedish Academy of Sciences that his teaching duties made it difficult for him to travel to Sweden. It was the French ambassador in Stockholm who received the Nobel Prize from the King of Sweden. After postponing the trip to

Sweden several times, they finally went in June 1905 and the trip was very pleasant. In his talk, Pierre exhibited several experiments using radium and other apparatus brought from Paris. He also recalled what opposed him to Rutherford. In order to account for the huge energy released by radioactivity, Pierre and Marie Curie made the hypothesis that radioactive atoms captured an as yet undiscovered radiation which uniformly permeated space. Pierre Curie then recognized that Rutherford made a better hypothesis which he considered bold and even somewhat daring at first. This is an example of two outstanding physicists, with different temperaments and different cultures, who have a different vision. Pierre Curie preferred his hypothesis because, in the absence of convincing experimental evidence, it seemed more plausible, natural and the least daring. The opposite was true for Rutherford. In all fairness, Pierre Curie admitted that the second hypothesis explained more results, a touchstone of any theory. Pierre Curie ended his talk with premonitory considerations marked by his idealism and faith in science:

It is conceivable that, placed in the wrong hands, radium could become very dangerous and we can wonder whether humanity is ripe enough to benefit from the knowledge of nature's secrets, or if this knowledge will be harmful. The discoveries made by Nobel are characteristic in this respect. His powerful explosives enable wonderful constructions to be made. They are also means of terrible destruction in the hands of criminals who lead people to war. I belong to those who believe, as Nobel did, that humanity will derive more good than evil from new discoveries [111].

The Death of Pierre Curie

There remained less than a year for Pierre Curie to live. On April 19, 1906, while crossing the *rue Dauphine*,

. . . he was struck by a truck coming from the Pont Neuf *and fell under its wheels. A concussion of the brain brought instantaneous death.*
 So perished the hope founded on the wonderful being who thus ceased to be. In the study room to which he was never to return, the water buttercups he had brought from the country were still fresh [114].

He was buried in intimacy on Saturday, April 21. On April 23, Henri Poincaré presided the session of the *Académie des Sciences* and opened the session with an eulogy:

You all know what a kind and reliable person he was; you are all familiar with the delicate charm which emanated, so to speak, from his gentle modesty. One would not have believed that, behind this gentleness, hid an uncompromising mind. He made no compromise with the generous principles under which he was brought up, with the moral ideal which he conceived, this ideal of absolute sincerity, probably too elevated for the world we live in [115].

Upon which Poincaré closed the session, an exceptional step, in sign of mourning.

1908: Rutherford is Awarded the Nobel Prize

A telegram sent in the end of September 1908 informed Rutherford and the world that the Swedish Royal Academy had attributed to him the Nobel Prize for Chemistry "for his investigation into the disintegration of the elements and the chemistry of radioactive substances."

He was surprised to be awarded the Nobel Prize for chemistry, instead of physics, since his whole work was that of a physicist and he joked about that for some time. But it is true that he modified the way chemists conceived atoms. He travelled to Stockholm with his wife Mary and he delivered the traditional Nobel lecture on "The chemical nature of α-particles emitted by radioactive substances." Like Pierre Curie, he illustrated his lecture with experimental demonstrations.

The Death of Henri Becquerel

A few months before, on the 29th of June, 1908, Henri Becquerel was elected Perpetual Secretary of the *Académie des Sciences*, and he set off, as he did each year, to his summer residence at *Le Croisic*. He died there on the 29th of August, after a short illness.

The first page of the history of radioactivity was turned.

References

1. Becquerel, J., "La découverte de la radioactivité", in *Conférences prononcées à l'occasion du 50ᵉ anniversaire de la Découverte de la Radioactivité*, pp. 2–14, Muséum d'Histoire Naturelle et École Polytechnique, Paris, 1946.

2. Ranc, A., *Henri Becquerel et la découverte de la radioactivité*, Édition de la liberté, Paris, 1946.

3. Badash, L., "The Discovery of Radioactivity", *Physics Today* pp. 21–26, February 1996.

4. Barbo, L., *Les Becquerel, une dynastie scientifique*, Belin, Paris, 2003.

5. de Raspide, S., *Les Becquerel ou le devoir de transmettre*, L'Harmattan, Paris, 2003.

6. Becquerel, A. E., "Recherches sur divers effets lumineux qui résultent de l'action de la lumière sur les corps", *Annales de Chimie et de Physique* **55**, 5–119; **57**, 40–124, 1959.

7. Becquerel, A. E., *Recherches sur divers effets lumineux qui résultent de l'action de la lumière sur les corps. Premier, deuxième et troisième mémoires*, Mallet-Bachelier, Paris, 1859.

8. Becquerel, A. E., *La lumière, ses causes et ses effets*, Firmin-Didot, Paris, 1867 (vol. 1) and 1868 (vol. 2).

9. Röntgen, W. C., "Über eine neue Art von Strahlen. Vorläufige Mittheilung", *Sitzungsberichte der physikalisch-medicinischen Gesellschaft zu Würzburg* pp. 132–141, session of December 30, 1895.

10. Röntgen, W. C., "Über eine neue Art von Strahlen. 2. Mittheilung", *Sitzungsberichte der physikalisch-medicinischen Gesellschaft zu Würzburg* pp. 17–19, 1896.

11. Röntgen, W. C., "On a new kind of rays", *Nature* **53**, 274–276, January 23, 1896.

12. Poincaré, H., "Les rayons cathodiques et les rayons Röntgen", *Revue Générale des Sciences* **7**, 52–59, January 30, 1896.

13. Becquerel, A. H., "Sur les radiations émises par phosphorescence", *Comptes Rendus de l'Académie des Sciences* **122**, 420–421, session of February 24, 1896.

14. *Annales du Bureau Central Météorologique de France*, E. Mascart, Paris, 1898. This was pointed out by L. Badash, *op. cit.*

15. Becquerel, A. H., "Sur les radiations invisibles émises par les corps phosphorescents", *Comptes Rendus de l'Académie des Sciences* **122**, 501–503, session of March 2, 1896.

16. Becquerel, A. H., "Sur diverses propriétés des rayons uraniques", *Comptes Rendus de l'Académie des Sciences* **123**, 855–858, session of November 23, 1896.

17. Becquerel, A. H., "Sur les radiations invisibles émises par les sels d'uranium", *Comptes Rendus de l'Académie des Sciences* **122**, 689–694, session of March 23, 1896.

18. Becquerel, A. H., "Émission de radiations nouvelles par l'uranium métallique", *Comptes Rendus de l'Académie des Sciences* **122**, 1086–1088, session of May 18, 1896.

19. Benoit, L. and Hurmuzescu, D., "Nouvelles propriétés des rayons X", *Comptes Rendus de l'Académie des Sciences* **122**, 235–236, session of February 3, 1896.

20. Thomson, J. J., "Röntgen rays", *Electrician* **36**, 491, 1896.

21. Righi, A., "Sulla produzione di fenomeni elettrici par mezzo dei raggi di Röntgen", *Rendiconto delle Sessioni, Reale Academia delle Scienze, Bologna* p. 45, February 14, 1896.

22. Righi, A., "Phénomènes électriques produits par les rayons de Röntgen", *Comptes Rendus de l'Académie des Sciences* **122**, 376–378, session of February 17, 1896.

23. Becquerel, A. H., "Sur quelques propriétés nouvelles des radiations invisibles émises par divers corps phosphorescents", *Comptes Rendus de l'Académie des Sciences* **122**, 559–564, session of March 9, 1896.

24. Badash, L., "Radioactivity before the Curies", *American Journal of Physics* **33**, 128–135, 1965.

25. *Almanach Hachette*, Libraire Hachette, Paris, 1897.

26. Thomson, S. P., "On Hyperphosphorescence", *Philosophical Magazine* **42**, 103–107, July 1896.

27. Perrier, E., "Discours prononcé aux obsèques d'Henri Becquerel", *Comptes Rendus de l'Académie des Sciences* **147**, 445–448, session of August 31, 1908.

28. Rutherford, E., *Radioactive transformations*, Charles Scribner's Sons, New York, 1906.
29. Badash, L., " 'Chance favors the prepared mind': Henri Becquerel and the discovery of radioactivity", *Archives internationales d'histoire des sciences* **18**, 55–66, January–June 1965.
30. Niépce de Saint-Victor, C. F. A., "Mémoire sur une nouvelle action de la lumière", *Comptes Rendus de l'Académie des Sciences* **45**, 811–815, session of November 15, 1857; **46**, 448–452, session of March 1, 1858; **46**, 489–491, session of March 8, 1858; **47**, 866–869, session of November 29, 1858; **47**, 1002–1006, session of November 20, 1858; **53**, 33–35, session of July 1, 1861; **65**, 505–507, session of September 16, 1867.
31. Curie, È., *Madame Curie*, Gallimard, Paris, 1938.
32. Reid, R., *Marie Curie*, Collins, London, 1974.
33. Pflaum, R., *Grand Obsession: Madame Curie and her world*, Doubleday, New York, 1989.
34. Quinn, S., *Marie Curie, a life*, Simon and Schuster, New York, 1995.
35. Langevin, P., "Pierre Curie", *La Revue du Mois* **2**, 5–36, July–December 1906.
36. Curie, M., *Pierre Curie*, Macmillan, 1923. With an introduction by Mrs. William Brown Meloney and autobiographical notes by Marie Curie.
37. Hurwic, A., *Pierre Curie*, Flammarion, Paris, 1995.
38. Barbo, L., *Pierre Curie 1859–1906. Le rêve scientifique*, Belin, Paris, 1999.
39. Curie, J. and Curie, P., "Développement, par pression, de l'électricité polaire dans les cristaux hémièdres à faces inclinées", *Comptes Rendus de l'Académie des Sciences* **91**, 294–295, session of August 2, 1880.
40. Curie, P., "Sur la symétrie dans les phénomènes physiques, symétrie d'un champ électrique et d'un champ magnétique", *Journal de Physique* **3**, 393–415, 1894.
41. de Gennes, P.-G., "Pierre Curie et le rôle de la symétrie dans les lois de la physique", in *Symmetries and broken symmetries in condensed matter physics. Proceedings of the Colloque Pierre Curie held at the École supérieure de physique et de chimie industrielles de la ville de Paris, Paris, September 1980*, edited by Boccara, N., pp. 1–9, ISDET, Paris, 1981.
42. Curie, P., "Propriétés magnétiques des corps à différentes températures", *Annales de Chimie et de Physique, Paris* **5**, 289–405, 1895.
43. Curie-Skłodowska, M., "Propriétés magnétiques des aciers trempés", *Comptes Rendus de l'Académie des Sciences* **125**, 1165–1169, Session of December 27, 1897.
44. Skłodowska-Curie, M., "Propriétés magnétiques des aciers trempés", *Bulletin de la Société d'Encouragement à l'Industrie Nationale* **97**, 36–76, 1898.
45. Curie, M. and Curie, P., "Sur une substance nouvelle radioactive contenue dans la pechblende", *Comptes Rendus de l'Académie des Sciences* **127**, 175–178, session of July 18, 1898.
46. Curie, M., "Rayons émis par les composés de l'uranium et du thorium", *Comptes Rendus de l'Académie des Sciences* **126**, 1101–1103, session of April 12, 1898.
47. Schmidt, G. C., "Über die von den Thorverbindungen und einigen anderen Substanzen ausgehende Strahlung", *Annalen der Physik, Leipzig* **65**, 141–151, 1898.
48. Curie, P., Curie-Skłodowska, M. and Bémont, G., "Sur une nouvelle substance fortement radioactive contenue dans la pechblende", *Comptes Rendus de l'Académie des Sciences* **127**, 1215–1217, session of December 26, 1898.
49. Demarçay, E., "Sur le spectre d'une substance radioactive", *Comptes Rendus de l'Académie des Sciences* **127**, 1218, session of December 26, 1898.
50. Curie, P. and Curie, M., "Les nouvelles substances radioactives et les rayons qu'elles émettent", in *Travaux du Congrès international de physique réuni à Paris en 1900*, edited by Guillaume, C.-É. and Poincaré, L., Vol. 3, pp. 79–114, Gauthier-Villars, Paris, 1901.
51. Curie, M., "Sur le poids atomique du radium", *Comptes Rendus de l'Académie des Sciences* **135**, 161–163, session of July 21, 1902.
52. Curie, M., "Sur le poids atomique du radium", *Comptes Rendus de l'Académie des Sciences* **145**, 422–425, session of August 19, 1907.
53. Curie, M., "Les rayons de Becquerel et le polonium", *Revue Générale des Sciences* **10**, 41–50, January 30, 1899.

54. Curie, M., "Les nouvelles substances radioactives", *Revue Scientifique* [4] 14, 65–71, July 21, 1900.
55. Rayleigh, John William Strutt, Baron, *The life of Sir J. J. Thomson*, Cambridge University Press, Cambridge, 1942.
56. Feather, N., *Lord Rutherford*, Priory Press, London, 1973.
57. Wilson, D., *Rutherford simple genius*, Hodder & Stoughton, London, 1983.
58. Campbell, J., *Rutherford: scientist supreme*, AAS, Christchurch, N.Z., 1999.
59. Rutherford, E., "Magnetization of Iron by high frequency discharges", *Transactions of the New Zealand Institute* 27, 481–513, 1894.
60. Rutherford, E., "Magnetic Viscosity", *Transactions of the New Zealand Institute* 28, 182–204, 1895.
61. Thomson, J. J. and Rutherford, E., "On the Passage of Electricity through gases exposed to Röntgen rays", *Philosophical Magazine* [5] 42, 392–407, November 1896.
62. Rutherford, E., "On the Electrification of Gases exposed to Röntgen Rays, and the Absorption of Röntgen Radiation by Gases and Vapours", *Philosophical Magazine* [5] 43, 241–255, April 1897.
63. Rutherford, E., "The velocity and Rate of Recombination of the Ions of Gases exposed to Röntgen radiation", *Philosophical Magazine* [5] 44, 422–440, November 1897.
64. Rutherford, E., "Uranium Radiation and the Electrical Conduction Produced by It", *Philosophical Magazine* 47, 109–163, January 1899.
65. Giesel, F., "Über die Ablenkbarkeit der Becquerelstrahlen im magnetischen Felde", *Annalen der Physik und Chemie, Leipzig* 69, 834–836, 1899.
66. Meyer, S. and von Schweidler, E. R., "Über das Verhalten von Radium und Polonium in magnetischen Felde", *Physikalische Zeitschrift* 1, 90–91, November 25, 1899.
67. Curie, P., "Action du champ magnétique sur les rayons de Becquerel. Rayons déviés et non déviés.", *Comptes Rendus de l'Académie des Sciences* 130, 73–76, session of January 8, 1900.
68. Curie, P. and Curie, M., "Sur la charge électrique des rayons déviables du radium", *Comptes Rendus de l'Académie des Sciences* 130, 647–650, session of March 5, 1900.
69. Becquerel, A. H., "Déviation du rayonnement du radium dans un champ électrique", *Comptes Rendus de l'Académie des Sciences* 130, 809–815, session of March 26, 1900.
70. Becquerel, A. H., "Note sur la transmission du rayonnement du radium au travers des corps", *Comptes Rendus de l'Académie des Sciences* 130, 979–984, session of April 9, 1900.
71. Kaufmann, W., "Die elektromagnetische Masse des Elektrons", *Physikalische Zeitschrift* 4, 54–57, October 10, 1902.
72. Rutherford, E., letter to his mother, dated January 5, 1902, in A. S. Eve, *Rutherford*, p. 80.
73. Rutherford, E., "A radio-active substance emitted from thorium compounds", *Philosophical Magazine* 49, 1–14, January 1900.
74. Curie, P. and Curie, M., "Sur la radioactivité provoquée par les rayons de Becquerel", *Comptes Rendus de l'Académie des Sciences* 129, 714–716, session of November 6, 1899.
75. Rutherford, E., "Radioactivity produced in substances by the action of thorium compounds", *Philosophical Magazine* 49, 161–192, February 1900.
76. Rutherford, E., "Excited Radioactivity and the Method of its Transmission", *Philosophical Magazine* 5, 95–117, January 1903.
77. Crookes, W., "The British association", *Nature* 58, 436–460, September 8, 1898.
78. Elster, J. and Geitel, H., "Versuche an Becquerelstrahlen", *Annalen der Physik und Chemie, Leipzig* 66, 735–740, 1898. Received on September 26, 1898.
79. Elster, J. and Geitel, H., "Ueber den Einfluss eines magnetischen Feldes auf die durch die Becquerelstrahlen bewirkte Leitfähigkeit der Luft", *Verhandlungen der deutschen physikalischen Gesellschaft* 1, 136–138, session of May 5, 1899.
80. Elster, J. and Geitel, H., "Über eine fernere Analogie im elektrischen Verhalten der natürlichen Luft und der durch Becquerelstrahlen abnorm leitend gemachten Luft", *Physikalische Zeitschrift* 2, 590–593, July 6, 1901.
81. Elster, J. and Geitel, H., "Über die Radioaktivität der im Erdboden enthaltenen Luft", *Physikalische Zeitschrift* 4, 574–577, September 15, 1902.

82. Gerlach, W., "Elster, Johann Philipp Ludwig Julius", in *Dictionary of Scientific biography*, edited by Gillipsie, C. C., pp. 354–357, Scribner, 1970.
83. Gerlach, W., "Geitel, F. K. Hans", in *Dictionary of Scientific biography*, edited by Gillipsie, C. C., p. 341, Scribner, 1970.
84. Villard, P., "Sur la réflexion et la réfraction des rayons cathodiques et des rayons déviables du radium", *Comptes Rendus de l'Académie des Sciences* **130**, 1010–1012, session of April 9, 1900.
85. Villard, P., "Sur le rayonnement du radium", *Comptes Rendus de l'Académie des Sciences* **130**, 1178–1182, session of April 30, 1900.
86. Rutherford, E. and Soddy, F., "The Radioactivity of Thorium Compounds. I. An Investigation of the Radioactive Emanation", *Transactions of the Chemical Society* **81**, 321–350, 1902.
87. Crookes, W., "Radio-Activity of Uranium", *Proceedings of the Royal Society of London* **66**, 409–422, 1899–1900.
88. Becquerel, A. H., "Note sur le rayonnement de l'uranium", *Comptes Rendus de l'Académie des Sciences* **130**, 1583–1585, session of June 11, 1900.
89. Becquerel, A. H., "Sur le rayonnement de l'uranium", *Comptes Rendus de l'Académie des Sciences* **131**, 137–138, session of July 16, 1900.
90. Becquerel, A. H., "Sur la radioactivité de l'uranium", *Comptes Rendus de l'Académie des Sciences* **133**, 977–980, session of December 9, 1901.
91. Becquerel, A. H., "Sur le rayonnement de l'uranium et sur les diverses propriétés physiques du rayonnement des corps radioactifs", in *Travaux du Congrès international de physique réuni à Paris en 1900*, Vol. 3, pp. 47–78.
92. Curie, P. and Curie, M., "Les nouvelles substances radioactives et les rayons qu'elles émettent", in *Travaux du Congrès international de physique réuni à Paris en 1900*, Vol. 3, pp. 79–114.
93. Rutherford, E. and Soddy, F., "The Radioactivity of Thorium Compounds. II. The Cause and Nature of Radioactivity.", *Transactions of the Chemical Society* **81**, 837–860, 1902.
94. Soddy, F., "The Radioactivity of Uranium", *Journal of the Chemical Society* **81**, 860–865, 1902.
95. Becquerel, H., "Note sur quelques propriétés du rayonnement de l'uranium et des corps radioactifs", *Comptes Rendus de l'Académie des Sciences* **128**, 771–777, session of March 27, 1899.
96. Becquerel, H., "Influence d'un champ magnétique sur le rayonnement des corps radioactifs", *Comptes Rendus de l'Académie des Sciences* **129**, 996–1001, session of December 11, 1899.
97. Becquerel, H., "Contribution à l'étude du rayonnement du radium", *Comptes Rendus de l'Académie des Sciences* **130**, 206 211, session of January 29, 1900.
98. Becquerel, H., "Sur la dispersion du rayonnement du radium dans un champ magnétique", *Comptes Rendus de l'Académie des Sciences* **130**, 372–376, session of February 12, 1900.
99. Rutherford, E., "The Magnetic and Electric Deviation of the easily absorbed Rays from Radium", *Philosophical Magazine S.6.* **5**, 177–187, February 1903.
100. Rutherford, E., "Die magnetische und elektrische Ablenkung der leicht absorbierbaren Radiumstrahlen", *Physikalische Zeitschrift* **4**, 235–240, March 5, 1903.
101. Romer, A., *The discovery of radioactivity and transmutations*, Dover, New York, 1964.
102. Rutherford, E. and Soddy, F., "The Cause and Nature of Radioactivity. Part I.", *Philosophical Magazine* **4**, 370–396, September 1902.
103. Rutherford, E. and Soddy, F., "The Cause and Nature of Radioactivity. Part II.", *Philosophical Magazine* **4**, 569–585, November 1902.
104. Rutherford, E. and Soddy, F., "Radioactive Change", *Philosophical Magazine* **5**, 576–591, May 1903.
105. Curie, P. and Laborde, A., "Sur la chaleur dégagée spontanément par les sels de radium", *Comptes Rendus de l'Académie des Sciences* **136**, 673–675, session of March 16, 1903.
106. Rutherford, E., *Radio-activity*, University Press, Cambridge, 1904.
107. Lockyer, J. N., "Notice of an Observation of the Spectrum of a Solar Prominence", *Proceedings of the Royal Society, London* **17**, 91–92, 1868.

108. Ramsay, W. and Soddy, F., "Gas Occluded by Radium Bromide", *Nature* **68**, 246, July 16, 1903.
109. Rutherford, E., "The succession of changes in radioactive bodies", *Philosophical Transactions of the Royal Society of London* **A204**, 169–219, 1904.
110. Alfred Nobel's Will, see : http://nobelprize.org/alfred_nobel/will/will-full.html.
111. *Les Prix Nobel en 1903*, Imprimerie Royale, P. A. Norsted & Söner, Stokholm, 1906.
112. Giesel, F., "Ueber radioactive Stoffe", *Berichte der deutschen chemischen Gesellschaft* **33**, 3569–3571, 1900.
113. Becquerel, A. H. and Curie, P., "Action physiologique des rayons du radium", *Comptes Rendus de l'Académie des Sciences* **132**, 1289–1291, session of June 3, 1901.
114. Curie, M., *Pierre Curie*, Macmillan, 1923, p. 137–138.
115. Poincaré, H., "Discours à la mémoire de Pierre Curie", *Comptes Rendus de l'Académie des Sciences* **142**, 939–941, session of April 23, 1906.

A Nucleus at the Heart of the Atom

Les atomes, petits dieux. Le monde n'est pas une façade, une apparence. Il est: ils sont. Ils sont, les innombrables petits dieux, ils rayonnent. Mouvement infini, infiniment prolongé.

Henri Michaux, *Difficultés*, *Plume*.

Atoms, little gods. The world is not a façade, an appearance. It *is*: they *are*. They are innumerable little gods, they radiate. An infinite motion, infinitely extended.

Prehistory of the Atom

The concept of atoms has existed for centuries, ever since the speculations of Epicurus, until the hesitations of the abbot Nollet. The first experimental manifestations and the conjectures of Dalton, Avogadro and Prout. The atom, a modest point-like object at first, becomes adorned with spectral lines and claims the right to possess an internal structure.

We owe the idea that matter should be composed of small indivisible parts, atoms, to the Greek philosophers of the fourth and fifth centuries before Christ, namely Leucippus, Democritus, and Epicurus. Their writings have been lost for the most part. But their ideas are expressed by the Latin poet Lucretius in his *De Rerum Natura* [1]. He lays down several general principles:

B. Fernandez and G. Ripka, *Unravelling the Mystery of the Atomic Nucleus:*
A Sixty Year Journey 1896 — 1956, DOI 10.1007/978-1-4614-4181-6_2,
© Springer Science+Business Media New York 2013

We start then from her first great principle
That nothing ever by divine power comes from nothing
[. . .]
The next great principle is this: that nature
Resolves all things back into their elements
And never reduces anything to nothing.

These principles lead Lucretius to consider that matter is composed of *primordia rerum*, primeval bodies, in terms of which all others are constructed. When a body is destroyed, the original bodies are dispersed and used to form other bodies. The original bodies must therefore be indestructible and eternal. Many centuries later, the *chemical elements* of Lavoisier express the same idea. The question of divisibility is raised. Can it be repeated forever? Lucretius claims that not:

To proceed with the argument: in every body
There is a point so small that eyes cannot see it.
This point is without parts, and is the smallest
Thing that can possibly exist.
[. . .]
Besides, unless there is some smallest thing,
The tiniest body will consist of infinite parts
Since these can be halved, and their halves halved again,
Forever, with no end to the division.
So then what difference will there be between
The sum of all things and the least of things?
There will be none at all [. . .] [1].

The existence of a lower limit of size, of indivisible original elements, atoms, appears to be a question of common sense: to imagine that matter could be indefinitely divisible leads to what Lucretius considers to be an absurdity.

Eighteenth Century: The Abbot Nollet

Debates concerning the existence of atoms lasted for centuries. In the middle of the eighteenth century, the abbot Nollet is a known physicist, a member of the Royal Academy of Science, of the Royal Society of London and a professor of experimental physics in the college of Navarre. In 1743, he uses his last lectures to write a *Traité de physique expérimentale*[1] for which he becomes famous [2]. He asks whether matter is indefinitely divisible, and he concludes:

Although I am more inclined to admit the existence of atoms
or indivisible small bodies, than to suppose that matter can be

---

[1] A treatise of experimental physics.

indefinitely divided, I cannot hide the fact that the argument which I presented, specious as it might seem, is not strong enough to decide upon the matter and that we do not have a valid answer.

The conclusion of the abbot Nollet is that of a modern physicist: lacking experimental evidence, he does not conclude.

Beginning of the Nineteenth Century: John Dalton, William Prout, Gay-Lussac, Avogadro, and Ampère

Until the eighteenth century, the concept of the atom remained the same. Some believed in atoms, others did not, and it was impossible to decide who was right or wrong. A major development occurred at the end of the century. Antoine Laurent de Lavoisier formulated what was to become the basis of modern chemistry, based on the conservation of elements and of mass during chemical reactions. He proved this by a systematic use of scales to measure weights. In 1792, the German chemist Jeremias Benjamin Richter showed that, when a salt was formed, the acid and the base always combined in a definite and constant proportion [3]. In 1794, the French chemist Joseph Louis Proust showed that substances always combine in constant and definite proportions [4, 5]. It is this constancy of proportions which led a British meteorologist and chemical theorist, John Dalton, to formulate a theory, exposed in his book *New System of Chemical Philosophy* [6] published in 1810 and 1812 and which he completed in 1827. According to Dalton, the law of constant proportions can naturally be explained if simple elements are composed of invariable atoms which combine to make composite bodies. All matter is composed of indestructible particles, which are identical in a given element. Assuming what today we call a chemical formula, which specifies the respective proportions of the simple substances which combine to make composite substances, he was able to calculate the relative masses of atoms. For example, he assumed that an atom of water (today we would call it a *molecule*) is composed of an atom of oxygen and an atom of hydrogen. This allowed him to conclude that the oxygen atom is eight times heavier than the hydrogen atom.[1] In 1808, Joseph Louis Gay-Lussac made another discovery: in a chemical reaction, the *volumes* of the gases involved are in a simple ratio. For example, 1 liter of oxygen combines with exactly 2 liters of hydrogen in order to produce water and, if the latter is in form of vapor, exactly 2 liters of water vapor are formed.

It seemed difficult to reconcile this strange law with the law of Proust, which states that the *masses* of elements involved occur in simple ratios. The puzzle was resolved by Amedeo Avogadro [7] who expressed in 1811 his famous law:

---

[1] Dalton was making the simplest assumption. We know today that the water molecule is composed of one oxygen atom and *two* hydrogen atoms, which we write as H_2O. In this case, the mass of an oxygen atom is 16 times higher than the mass of a hydrogen atom.

equal volumes of gas (at the same temperature and pressure) always contain the same number of molecules, whatever the gas is [8]. Sometime later, Ampère independently came to the same conclusion [9]. This hypothesis, which was needed in order to understand the experimental results in terms of the atomic hypothesis, was truly astounding. It was not easily accepted, and since it was impossible to count the number of molecules in a given volume, one might as well forget the atomic theory!

William Prout, a medical doctor who was also a chemist, took a further step. Basing himself on what he called the "volume doctrine" of Gay-Lussac, he noted [10] that the masses of known elements, which either he or other chemists had measured, were proportional to the masses of a given volume of these elements (which today we call the *density*) when they are in the gaseous phase. He naturally concluded that the mass of each atom of a given element was proportional to his *density* which he called the *specific weight*. He thus arrived at the same conclusion as Avogadro, expressed in another form. When he calculated the masses of atoms of 14 elements, ranging from carbon to iodine, he noted that they are multiples of the mass of hydrogen: carbon is 12 times heavier than hydrogen, nitrogen 14 times heavier, oxygen 16 times, etc. This led him to propose a new idea: hydrogen could be the basic substance from which all atoms are formed:

> If our way of understanding, which we take the risk of proposing, is correct, we can almost assume that the [basic substance] of the ancients is accounted for by hydrogen [11].

For the first time, the hypothesis is put forth that even atoms have an internal structure.

Do Atoms Really Exist?

The reality of atoms was slowly accepted, but chemists took longer than physicists to do so [12]. In 1853, Adolphe Ganot wrote in his *Traité élémentaire de Physique*, published in Paris [13] and translated into English in 1862:

> The properties of substances indicate that they do not consist of a continuous matter, but of infinitely small elements, so to speak, which cannot be physically split and which are assembled without touching [. . .] These elements are called atoms [14].

This does not imply that there existed a general consensus at the time. In a book published in 1895, the Scottish physicist Peter Guthrie Tait (author, with William Thomson [lord Kelvin] of a treatise of physics which remained a reference until the beginning of the twentieth century) writes, on the atomic theory of matter:

> This theory must be regarded as a mere mathematical fiction, very similar to that which (in the hands of Poisson and Gauss) contributed so much to the theory of statical Electricity [. . .]

> *A much more plausible theory is that matter is continuous (i.e. not made up of particles situated at a distance from one another) and compressible, but intensely heterogeneous; [. . .] The finite heterogeneousness of the most homogeneous bodies, such as water, mercury, or lead, is proved by many quite independent trains of argument based on experimental facts. If such a constitution of matter be assumed, it has been shown (W. Thomson, Proc. R.S.E., 1862.) that gravitation alone would suffice to explain at least the greater part of the phenomena which (for want of knowledge) we at present ascribe to the so-called Molecular Forces [15].*

It is true that, before the kinetic theory of gases was formulated, the atomic hypothesis had no incidence on physics. It was only an hypothesis, an interesting philosophical speculation, not necessarily true. But in chemistry, chemical formulas differed according to whether the atomic notation was used or not and so did the reference masses ("atomic" or "equivalent"). Since no agreement was reached, considerable confusion ensued. It is precisely in order to put an end to this confusion that the German chemist August Kekulé took the initiative to organize the first international chemistry meeting [16]. It was held in Karlsruhe on September 3, 4, and 5, 1860. Although a general agreement was not reached, the Italian chemist Stanislao Cannizzaro succeeded in convincing a number of his colleagues that the atomic notation had a pedagogical advantage and, from then on, it began to dominate. In spite of this, the conjecture of Prout was progressively given up because increasingly precise mass measurements of Berzelius showed that *the masses are not multiples of the hydrogen mass*. For example, the Belgian chemist Jean Stas believed in 1860 that the conjecture of Prout was only an illusion [17].

1865: Loschmidt Estimates the Size of Air Molecules

The atomic theory had two handicaps: both the size and the number of atoms in a given quantity of matter were unknown. The Austrian physicist Johann Joseph Loschmidt was the first to propose a method to measure the size. In a communication presented to the Academy of Sciences in Vienna, Loschmidt used the value of the mean free path calculated by Maxwell [18] who deduced it from the viscosity of air.[1] The mean free path is the average distance which a molecule of air travels before colliding with another. Loschmidt showed how the mean free path allowed an estimate of the diameter of a molecule: he found one millionth of a millimeter. He knew that this was only an estimate, but he claimed that the order of magnitude

---

[1]The viscosity of a fluid determines the velocity with which it escapes out of a small orifice: hydrogen escapes much faster from a balloon than a heavy gas such as krypton. Gases have a much lower viscosity than liquids: air has a viscosity a 100 times smaller than water, and water a 1,000 times smaller than glycerine.

was correct [19]. Those who opposed the atomic hypothesis found a flaw in his argument: it was based on the kinetic theory of gases, a theory which was not accepted by all and which assumed the existence of atoms from the outset.

Spectral Lines: A First Indication of an Internal Structure of Atoms

From 1859 onwards, the German physicists Gustav Kirchhoff and Robert Bunsen showed that the spectrum of light emitted by a given element, when viewed in a spectroscope, consists of lines with definite wavelengths: each element possesses its own set of spectral lines, by means of which it can be identified. The existence of such spectra is necessarily caused by the structure of the atoms which radiate them. This was the first experimental indication of an internal structure of the atom.

Jean Perrin Advocates the Reality of Atoms

On February 16, 1901, students and members of the *Amis de l'Université de Paris* are gathered to hear a lecture of Jean Perrin on the "molecular hypothesis." At the age of 30, Jean Perrin [20,21] was the author of a brilliant thesis in which he showed, in 1895, that cathode rays carried a negative electric charge [22]. He had been a lecturer at the Sorbonne for 2 years. Perrin explains what he still calls the "molecular hypothesis":

> The hypothesis, which is generally assumed to be the simplest, consists in assuming that any pure substance, water for example, is in fact composed of a very large number of distinct material particles, all completely identical, which mark the point beyond which water can no longer be divided. Those are the molecules of the body [23].

Jean Perrin goes on to explain how the measurement of the viscosity of a gas gives a first estimate of the size of molecules and, as a consequence, of their number:

> The number N of molecules contained in a liter of gas, at normal temperature and pressure, is, as we might have expected, extraordinarily large. One finds that it is equal to 55 billion trillions (10^{22}), a number which is so large that it surpasses our imagination.

During the following years, Jean Perrin devoted his efforts to the determination of the number of atoms in matter, to be more exact, to the number N of atoms in one mole$^\diamond$ of an element, called one "gram-atom" at the time: it was the number of atoms in 1 g of hydrogen, the lightest of all elements.

One gram of hydrogen gas contains only half as many molecules because each molecule is composed of two atoms of hydrogen. Thus 2 g of hydrogen contain

N molecules, the same number contained in 32 g of oxygen, 28 g of nitrogen, because their atomic mass, chemically determined, are in this ratio. Perrin proposed to call this number N the "Avogadro constant" [24]. Today, we call it the *Avogadro number* or *Avogadro constant*. To measure this number, Perrin makes use of Einstein's interpretation of Brownian motion. In a book, *Les Atomes*,[1] published in 1913, he compares a dozen methods, which are all independent, the intensity of the blue sky, radioactivity, Brownian motion, the Planck constant, and more. He finds that they all give consistent results,[2] and he concludes:

We can only admire the miracle which yields such a precise agreement, starting from so different phenomena. First, we obtain the same order of magnitude, with each of the methods, [...] second, the numbers which are thus unambiguously defined by so many methods, coincide; this gives to the reality of molecules a likelihood close to a certainty [25].

A precise determination of the Avogadro number would reveal the real mass of atoms. The chemical methods used in the nineteenth century could determine the ratios of the masses, so that the masses were only known one relative to another. If one single mass could be measured, the others could be deduced. And this would reveal the size of molecules, assuming that, in a liquid, the molecules would be in contact with each other. For example, if N molecules were known to be contained in 18 g of water, the volume of which is 18 cm^3, one could calculate the average volume occupied by each molecule of water and therefore estimate its size. The Avogadro number was a key to many unknowns!

In 1926, Jean Perrin received the Nobel Prize of physics for his determination of the Avogadro number.

---

[1] Atoms.

[2] These values ranged from about 6.2×10^{23} to about 6.8×10^{23}. The presently adopted value is $6.022\,141\,29 \times 10^{23}$. See the glossary.

1897: The Electrons Are in the Atom

> *The discovery of the electron, a universal constituent of matter,*
> *leads to a burst of speculations in Cambridge, Paris, Heidelberg,*
> *Tokyo... J. J. Thomson describes the atom as consisting of*
> *thousands of electrons and proposes a "plum-pudding" model of*
> *the atom. Barkla succeeds in counting the number of electrons*
> *in an atom. Nobody understands what prevents the atom from*
> *radiating.*

Electric Discharges in Gases, Cathode Rays and the Electron

The study of cathode rays culminated in 1897 with the measurements of Emil Wiechert, Walter Kaufmann, and J. J. Thomson, to whom history attributes the discovery of the electron. Furthermore, the passage of electricity, caused by either X-rays or particles emitted by radioactive substances passing through a gas, was interpreted as an ionization of the atoms in the gas. The transformation of atoms into ions was a phenomenon familiar in electrolysis: the atom splits into two ions, one positively and the other negatively charged. The positively charged ions had a mass comparable to the mass of the original atom, and the so-called "negative ions" had a much smaller mass: they displayed an uncanny resemblance to cathode rays. J. J. Thomson measured their mass and electric charge: they were electrons. Thus atoms, the existence of which was not yet universally acknowledged, were not indivisible and eternal objects, because they were able to emit these tiny electrons! The presence of electrons in the atom was further confirmed when Lorentz explained the Zeeman$^\diamond$ effect. The existence of an inner structure of the atom thus became convincing, and the structure involved electrons. In addition, it was known that the β-rays emitted by radioactive substances were high-velocity electrons. By 1903, it became obvious that the atom contained electrons.

"Dynamids": The Atoms of Philipp Lenard

In 1903, Philipp Lenard proposes another model of the atom. Born in Hungary in 1862, Philipp Lenard obtained the position of *Privatdozent* in Bonn and became the assistant of Hertz. Hertz suggested that he should send a beam of cathode rays through a thin film of aluminum which was just strong enough to maintain the vacuum. The famous "Lenard window" was constructed, and it led to a series of original experiments. Lenard made several crucial observations concerning the photoelectric effect, which had been discovered by Hertz, namely, that the *number* of electrons emitted by the metal depends only on the intensity of the incident light, whereas their *energy* depends only on its *frequency*. This will be explained by Einstein 2 years later, in 1905. Lenard's model relies on the fact that cathode rays, namely electrons, pass through matter very easily. This led Lenard to postulate that the atom is an assembly of what he called "dynamids," which are very small particles (more than 10,000 times smaller than the size of the atom) which are separated by empty space. According to Lenard, "the space occupied by $1\,m^3$ of platinum is empty, just like the astronomical empty space." Indeed, it is only filled with "dynamids," the total volume of which does not exceed $1\,ml^3$. This is why cathode rays pass through matter so easily. He believed that a dynamid was an electrically neutral particle, consisting of an electron associated to a positive charge. Lenard was awarded the Nobel Prize in 1905 for his work on cathode rays. After that, he became truly paranoic and he accused everyone of stealing his ideas and his discoveries. He could never forgive Einstein for having explained the photoelectric effect which he considered to be "his" effect. An ardent Nazi, he occupied an important position in the Nazi party, and in 1933, he became one of the "theoreticians" of "Aryan science" as opposed to "Jewish science" which included relativity and quantum mechanics.

Numeric Attempts to Describe Spectral Rays: Balmer and Rydberg

There was one perfectly clear and yet completely enigmatic indication of the internal structure of atoms: namely, the spectral rays which became both a strange and precise signature of every element. The permanent structure of rays of different colors defied all understanding. Failing a physical explanation, several physicists attempted to discern a logical pattern of the rays. In the report he delivered in the International Physics Congress of 1900, the Swedish physicist Johannes Robert Rydberg, specialized in spectroscopy, quoted about 40 such numerical patterns [26]. Only one entered history, the one of the Swiss mathematician Johann Jacob Balmer. In 1885, he showed that the wavelengths λ of the four known hydrogen lines, called H_α, H_β, H_γ, and H_δ, obeyed the following simple mathematical formula [27]:

$$\lambda = h\frac{m^2}{m^2 - n^2},$$

where h is a factor which is determined empirically and where m and n are small integers, equal to 2, 3, 4, and 5. Rydberg generalized this pattern by stating that the frequencies of the hydrogen spectral rays are equal to the differences of the inverse squares of small integers, multiplied by a constant which today is still called the *Rydberg constant*. This was an entirely empirical law, and nobody understood the meaning of the integers nor the value of the Rydberg constant.

J. J. Thomson's First Model: An Atom Consisting Entirely of Electrons

After having discovered that cathode rays consist of small particles which everybody will soon call "electrons," J. J. Thomson attempts to measure their mass m and their electric charge e. He first measures the ratio e/m which is determined by their deflection in a magnetic field. He then makes a direct measurement of their electric charge by measuring the velocity of small electrically charged water droplets, falling in the air and by measuring the change in their velocity when they are exposed to an electric field [28]. This is a very difficult experiment which the American physicist Robert Millikan repeated later [29] using oil droplets which had the advantage of not evaporating in the air, as water droplets do, thereby modifying somewhat the results of the measurements. J. J. Thomson thought that atoms were composed of electrons. He knew that electrons were 2000 times lighter than the lightest atoms, which should therefore contain thousands of electrons. The snag was that electrons carry a negative electrical charge, whereas the atom is electrically neutral. In spite of this, he believed that it was the most plausible hypothesis:

> *I regard the atom as containing a large number of smaller bodies which I will call corpuscles; these corpuscles are equal to each other; the mass of a corpuscle is the mass of the negative ion[1] of the gas at low pressure, i.e. about 3×10^{-26} of a gramme. In the normal atom, this assemblage of corpuscles forms a system which is electrically neutral. Though the negative corpuscles behave as negative ions, yet when they are assembled in a neutral atom the negative effect is balanced by something which causes the space through which the corpuscles are spread to act as if it had a charge of positive electricity equal in amount to the sum of the negative charges of the corpuscles [30].*

J. J. Thomson is visibly embarrassed by the negative charge of the electrons in the atom. He formulates a vague hypothesis according to which "something" would modify space which would behave "as if" it was positively charged. In fact, in 1904, he wrote a letter to Oliver Lodge [31] expressing the hope that he could do without this positive electric charge.

A Speculation of Jean Perrin: The Atom Is Like a Small Scale Solar System

In the talk delivered to the *Amis de l'Université de Paris*,[2] Jean Perrin stresses an important point:

---

[1] What Thomson refers to here as a negative ion is an electron.

[2] "Friends of the University," see p. 52.

What is essential to note is that the negative particles appear always to be absolutely identical, *whatever the atoms they are ejected from*[1].

And he imagines a possible structure of the atom:

For the first time a path leads to the intimate structure of the atom. We will, for example, make the following hypothesis, which corroborates the preceding facts. Each atom would consist, on one hand of one or several positively charged masses—a kind of positive sun the electric charge of which would greatly exceed that of a particle—and, on the other hand, by numerous particles acting as tiny negatively charged planets orbiting under the action of the electric forces, their negative total charge balancing exactly the total positive charge, thereby making the atom electrically neutral.

He underlines the fact that this is but one of the models which one can conceive. However, he calculates the velocity with which the electrons rotate around their "sun," and he notes that the rotations have frequencies similar to the frequencies of the observed spectroscopic rays. This planetary model of the atom has however a serious drawback which Perrin fails to mention: according to the laws of classical electrodynamics, the rotating electrons should radiate electromagnetic waves. The energy which they would spend radiating would have to originate from their motion, and they should therefore eventually fall onto what Perrin calls their "positive sun." One could calculate that it would take hydrogen atom only one ten-millionth of a second to disappear! Jean Perrin will not make a deeper study of his speculation nor will he publish anything on the subject. His talk is published in the *Revue Scientifique (Revue Rose)*, which maintains a good standard but which is not a specialized scientific journal [23].

The "Saturn" Model of Hantaro Nagaoka

A few years later, the Japanese physicist Hantaro Nagaoka, professor in the Imperial University of Tokyo, proposes a somewhat similar model. He explains that he is not the only one to seek an explanation of spectral rays:

Since the discovery of the regularity of spectral lines, the kinetics of a material system giving rise to spectral vibrations has been a favorite subject of discussion among physicists. The method of enquiry has been generally to find a system which will give rise to vibrations conformable to the formulæ given by Balmer, by Kayser and Runge, and by Rydberg [32].

He then explains his model in which the electrons form rings similar to the rings of Saturn:

---

[1] The emphasis on "absolutely identical" is by Jean Perrin.

*I propose to discuss a system whose small oscillations agree qualita-
tively with the regularities observed [. . .] The system consists of large
number of particles of equal mass arranged in a circle at equal angular
intervals and repelling each other with forces inversely proportional to
the square of the distance; at the centre of the circle, place a particle
of large mass attracting the other particles according to the same law
of force.*

He is perfectly conscious of the objection which one could have already raised
to the model of Jean Perrin:

*The objection to such a system of electrons is that the system must
ultimately come to rest, in consequence of the exhaustion of energy
by radiation, if the loss be not properly compensated.*

Without replying to this objection, Nagaoka pursues and writes down the
equations which determine the stability of the system. His calculations remain
however formal, and he makes no numerical evaluation. He simply states that the
central charge must be "very large" compared to that of the electron. He studies the
form of the vibrational modes. Unfortunately, a simple calculation suffices to show
that the rings becomes unstable as soon as an electron moves out of the plane of the
rings. This instability should be added to the one caused by the radiation.

The "Plum-Pudding" Atom of J. J. Thomson

Two months earlier, Joseph John Thomson, following an idea which had been put
forth by Lord Kelvin [33], published a paper in the *Philosophical Magazine*, in
which he proposed another model, consisting of electrons embedded in a positively
charged sphere [34]. The electrons are subject to two opposing forces: the repulsive
forces which attempt to separate the negatively charged electrons and the attraction
which is exerted on them by the positively charged sphere, as soon as they attempt
to escape from it. The positively charged sphere has the size of the atom. The
main problem addressed by Thomson is the stability of such a system. Does the
system have an equilibrium shape such that, if it is perturbed, it returns to its
equilibrium shape, or does the atom fall apart? To answer this question, Thomson
is inspired by the experiments of Alfred Mayer, a physicist who became known for
designing some simple, pedagogical, and yet spectacular experimental setups. In his
"floating magnet" experiment [35], Mayer stuck a few magnetic needles vertically
into a floating material and let them float freely on water, their north pole pointing
vertically upwards. Left as such, the magnetic needles would repel each other with
a force which decreases as the square of the distance separating them. However,
Mayer then placed a strong negative magnetic pole above the water. The floating
needles are all attracted to this pole by a force which decreases as the inverse of
the distance to the negative pole. The magnetic needles, which are forced to move
on the water surface, are subject to two opposing forces, similar to the electrons on

the "plum-pudding" model of J. J. Thomson. Mayer noticed a curious phenomenon: the magnetic needles had a tendency to form concentric rings on the water surface. This observation led Thomson to formulate a model in which the electrons rotate in concentric rings. He shows that this is a stable configuration provided that the electrons rotate fast enough. The atom is akin to a merry-go-round!

However, as in the other models, the electrons should radiate and the atom therefore cannot remain stable. In a previous publication, Thomson showed that the energy radiated by a large number of electrons, rotating and equally spaced on a circle, decreases as the number of electrons increases. For example, two electrons, rotating at a speed equal to 1/100 of the speed of light, radiate $1000 \left(10^3\right)$ times less energy than a single electron; six electrons would radiate 10^{17} times less energy. And because Thomson believes that the atom contains thousands of electrons, his argument carries some weight, even if his atom still suffers from some instabilities. Thomson is quite aware of the fact that his model assumes that the electrons all move in a plane, as planets almost do in the solar system, but what forces the electrons to do so in a real atom? He admits to not having a mathematical solution to this problem. However, he attempts to imagine a solution by extrapolating what he considered to be natural:

When the corpuscles are not constrained to one plane, but can move about in all directions, they will arrange themselves in a series of concentric shells; for we can easily see that, as in the case of the ring, a number of corpuscles distributed over the surface of a shell will not be in stable equilibrium if the number of corpuscles is large, unless there are other corpuscles inside the shell, while the equilibrium can be made stable by introducing within the shell an appropriate number of other corpuscles.

This is his *plum-pudding* model of the atom, named after the very British Christmas pudding. It becomes a new version of the electronic theory of matter. It will remain a reference until the work of Rutherford,[1] in 1911.

Charles Barkla Measures the Number of Electrons in an Atom

J. J. Thomson thought that the mass of the atom was the sum of the masses of its constituent electrons. This is why he assumed that the atom contained thousands of electrons. A crucial step consisted therefore in measuring the number of electrons.

As soon as they were discovered in 1895, X-rays fascinated physicists. As early as 1897, Jean Perrin noticed that, in addition to ionizing the gas they passed through, X-rays caused a particular phenomenon when they impinged on a metal [22, 36]. The French physicist Georges Sagnac then showed that the gas,

---

[1]See p. 74.

through which X-rays had passed, became the source of a secondary emission, which he called a proliferation and which he thought was due either to a scattering process or to luminescence [37]. He was particularly interested on the effect of X-rays impinging on metals [38]. In a series of experiments which he performed with Pierre Curie [39,40], he showed that the secondary emission carried negative electric charge: it consisted of electrons. The other emitted rays were X-rays probably scattered by the atoms, just as light is scattered by particles of smoke.

In 1902, a 25-year-old physicist begins to study the secondary X-rays produced by gases and he will continue to do so all his life. Charles Barkla studied in the University of Liverpool, after which he worked at the Cavendish under the supervision of J. J. Thomson. It is at King's College that he tackles the problem. He observes the secondary X-rays emitted by various gases (air, hydrogen, sulfuric hydrogen, carbon dioxide, sulfur dioxide) exposed to a primary X-ray beam. He has an electrometer with which he measures the intensity of the secondary rays and its progressive attenuation as it passes through matter. The rate at which the secondary radiation is absorbed allows him to estimate, roughly at least, its energy. With this electrometer, together with a few absorptive sheets, Barkla obtains an important result: *all gases subject to X-rays, emit secondary X-rays which are identical to the primary X-rays. The secondary radiation is more intense when the gas is composed of heavy atoms* [41]. Barkla makes a careful measurement of the rate at which the X-rays are absorbed by air. He finds that their intensity is reduced by 0.024 % per cm of air they pass through [42]. This attenuation had in fact been calculated by J. J. Thomson with his electronic theory of matter [43]. Thomson assumed that the X-rays were scattered (or disseminated, as Sagnac would say) by the electrons contained in the atoms of the gas. He could relate the attenuation of X-rays to the number of electrons contained in a given volume of gas. Barkla uses the formula given by Thomson, and he concludes that the number of electrons contained in $1 \, cm^3$ of gas is equal to 0.6×10^{22}. This corresponds to about a 100 electrons per atom of nitrogen. He concludes that the X-rays are indeed reemitted by the electrons and not by the molecules of the gas, which would need to be a hundred billion times more numerous to produce the observed effect. J. J. Thomson then uses the results of Barkla to measure the *number of electrons per atom*. He invents two further methods to measure this number, which are completely independent of the method used by Barkla: the index of refraction of the gas (in other words its capacity to deflect light) and the attenuation of β-rays passing through the gas. The three methods give compatible results:

> The evidence at present available seems sufficient [. . .] to establish the conclusion that the number of corpuscles is not greatly different from the atomic weight [44].

The atom does not therefore contain thousands of electrons as Thomson himself had suggested. Thomson does not mention this nor does he discuss the problem of the stability of the atom, in spite of the fact that with so few electrons, his atom cannot be stable. Nonetheless, as noted by Abraham Pais [45], the paper of J. J. Thomson is a decisive step forward because it gives a tangible information about the structure of the atom, namely, its number of electrons. For several years, Barkla continues to study X-rays. In 1911, he discusses his 1904 measurements in the light of more precise values available for the mass and the electric charge of electrons. He estimates that:

> The theory of scattering as given by Sir J. J. Thomson leads to the conclusion that the number of scattering electrons par atom is about half the atomic weight in the case of light atoms [46].

He adds the footnote:

> This applies to atomic weights not greater than 32, with the possible exception of hydrogen.

The picture becomes clearer: carbon, for example, with atomic mass 12, would have six electrons; oxygen, with atomic mass 16, would have eight electrons, and hydrogen, a single electron.

The Scattering of α Particles Makes It Possible to "See" a Nucleus in the Atom

The genius in Manchester, namely Ernest Rutherford, observes deviations of the trajectories of α-rays, which were believed never to deviate. He actually sees them rebound from a collision with an incredibly small nucleus. The stability of the atom remains a mystery.

An Observation of Marie Curie

In 1900, Marie Curie studied the penetration power of α-particles which she still considered to be rays which could not be deviated [47]. She used a radioactive source consisting of polonium, which was known to radiate only α-particles. She was surprised to discover that the α-rays, which could not be deviated, would lose their ionization power after traveling a distance of only 4 cm in the air. Furthermore:

The rate of absorption of the non-deviable rays increases as they pass through increasing thickness of matter. This singular absorption law is opposite to the one known for other radiation; it suggests rather the behavior of a projectile which looses part of its kinetic energy by passing through obstacles [47].

In their report to the 1900 International Congress of Physics, Pierre and Marie Curie emphasize the curious absorption law of the rays which could not be deviated [48].

William Henry Bragg: The Slowing Down of α-Particles in Matter

A few years later, in 1904, a physics professor in the University of Adelaide, South Australia, happens to read the paper of Marie Curie. William Bragg was born, July 2, 1862, in Westward, Cumberland, a county in North-East England. After brilliant studies in mathematics in Cambridge, he studied physics at the Cavendish Laboratory, under the supervision of J. J. Thomson. But that same year, he accepted a professor's chair at the University of Adelaide, in Australia. His first 20 years there were devoted to teaching. But the paper he reads changes everything:

It was known that when the radium atom broke up into two parts, one large and one small, the latter, which was really an atom of helium, was driven into the surrounding air, and these particles constituted what was called the 'alpha' radiation. Mme Curie described experiments which implied that all the alpha particles thus expelled travelled about the same distance.

This interested me greatly. All ordinary radiations fade away gradually with distance; the alpha particles seemed to behave like bullets fired into a block of wood. But, if this were so, the particle must travel in a straight line through the air, as the bullet does through the block. Now some hundreds of thousands of air atoms would necessarily be met with on its journey. How did it get past? [. . .]

There was only one answer to the problem. The particle must go through the air atoms that it met [49].

Together with his assistant Richard Kleeman, he studies α-particles emitted by various radioactive substances. Bragg speculates: *the α-particle must pass through atoms without being deviated from its straight-line trajectory.* He assumes that the only difference between α- and β-rays is that β-particles are deflected by collisions, whereas α-particles are not. We are in 1904, and Bragg believes in the model of J. J. Thomson, according to which the α-particle consists of thousands of electrons, which explains its behavior:

The α ray is a very effective ionizer, and rapidly spends its energy on the process. It is of course far more likely than the β ray to ionize an atom through which it is passing, because it contains some thousands of electrons and the ionizing collision is so much more probable. But a collision between an electron of the flying atom of the α ray and an electron of the atom traversed, can have very little effect on the motion of the α atom as a whole [50].

In the experiments which he performs with Kleeman, he shows that the ionization is roughly constant throughout its trajectory and suddenly stops at the end. He uses a very thin radium source because he does not have access to a polonium source, as Marie Curie did. He observes a complex curve which suggests the existence of three or four groups of α-particles with different velocities. He proposes the following interpretation:

The atom passes through several changes, and it is supposed that at four of these an α atom is expelled. Probably the α particles due to one change are all projected with the same speed. We ought therefore to expect four different streams of α particles, differing from each other only in initial energy.

The fact, that all the α-particles emitted by a given radioactive substance have the same energy, will play an important role later. The paper ends by a detailed study performed with Kleeman and dated September 8, 1904. As Marie Curie had observed 4 years earlier, they note that:

The alpha particle is a more efficient ionizer towards the extreme end of its course [51].

They attempt to explain this phenomenon:

The disturbing influence of the α particle in its transit through an atom must become greater as the speed diminishes. The diminution is not likely to be great except at the end. [. . .] Theoretical considerations based on a somewhat insufficient hypothesis show that the effect should be inversely proportional to the energy of the moving particle [. . .]. It is possible that it is only at the end, when the change of velocity is very great in proportion of what remains, that the influence of this cause is perceptible. It is also conceivable that the particle, as its speed approaches the critical value below which it looses the power of penetration, may leave its rectilinear path and be buffeted about, causing a considerable amount of ionization without getting much further away from its source.

They present a last paper on the subject to the Royal Society of South Australia. It is published in September 1905 in the *Philosophical Magazine*. Bragg and Kleeman repeat their previous conclusions, and they add more precise data on the trajectories of the α-particles emitted by radium and of the emanation of radium A and C, which they determine to, within half a millimeter [52]. The slowing down of α-particles passing through matter is the subject of a series of subsequent experiments in other laboratories [53–55]. They confirm the results of Bragg, namely, that:

- α-particles emitted by a given radioactive substance all have the same velocity.
- The slowing down of α-particles is roughly constant (slightly increasing in fact) along its observed trajectory, and it suddenly increases considerably towards the end of its trajectory, and drops to zero soon after.

The "Scattering" of α-Particles

By that time, Rutherford, then at the University of McGill, is determined to ascertain the nature of the α-particle. He believes that it is an ionized helium atom, but he has no decisive proof of this. He makes precise measurements of the charge to mass ratio e/m of α-particles and makes sure that it does not depend on their velocity. In a talk delivered to the Royal Society of Canada, he praises the results of Bragg and Kleeman and he calls *range* the distance to which the α-particles penetrate into matter [56]. The word is used by the military to designate the "range" of firearms. We return to the concept of a projectile, suggested by Marie Curie. According to Rutherford, the α-particle emitted by a radioactive substance is a projectile of "exceptional violence." Rutherford begins to measure the velocity of α-particles which pass through successive thin aluminum strips, by deviating them with a magnetic field. He finds that the range measured by Bragg is also the distance beyond which the α-particles fail to leave a trace on a photographic plate or to cause phosphorescence on a zinc sulfide screen. Beyond this range, the α-particles can no longer be detected; they disappear. In a paper dated November 15, 1905, Rutherford reports for the first time that the spot formed by α-particles, after passing through a given thickness of matter, becomes broader and loses some of its definition:

The greater width and lack of definition of the α-lines have been noticed in all other experiments and show evidence of an undoubted scattering of the rays, in their passage through air [57].

It follows that α-particles may well be deviated when they pass through matter. On February 27, 1906, Rutherford sends a letter to the *Philosophical Magazine* in which he reports on his most recent observations of the slowing down of α-particles. The slowest observed α-particles had 43 % of their initial velocity, and he confirmed that they retained their mass and charge, no matter what their velocity was [58]. In a complete paper, dated June 14, 1906, he describes this "scattering":

There is [...] an undoubted slight scattering or deflection of the path of the α particle in passing through matter [...]
   From measurements of the width of the band due to the scattered α rays, it is easy to show that some of the α rays in passing through the mica have been deflected from their course through an angle of about $2°$. It is possible that some were deflected through a considerably greater angle; but, if so, their photographic action was too weak to detect on the plate [...]
   It can easily be calculated that the change of direction of $2°$ in the direction of the motion of some of the α particles in passing through the thickness of mica (0,003 cm) would require over that distance an average transverse electric field of about 100 million volts per cm. Such a result brings out clearly the fact that the atoms of matter must be the seat of very intense electrical forces—a deduction in harmony with the electronic theory of matter [59].

Rutherford considered this to be an important result which was thereafter studied in various laboratories [60–62]. In 1907, when he became professor of physics at the University of Manchester, Rutherford marked on his notebook a list of "possible researches" [63]. It consisted of ideas for experiments, among which were the "scattering of α-rays" and the "number of α-rays from radium," experiments which eventually completely modified the physics of the atom.

The Nature of the α-Particle: An Unresolved Question

Rutherford still does not consider that the nature of the α-particle is firmly established; its electric charge, for example, is not measured. On January 31, 1908, he delivers a talk at the Royal Institution:

We may regard the α-particle as a projectile traveling so swiftly that it plunges through every molecule in its path, producing positively and negatively charged ions in the process. On an average, an α particle before its career of violence is stopped breaks up about 100,000 molecules. So great is the kinetic energy of the α-particle that its collisions with matter do not sensibly deflect it, and in this respect it differs markedly from the β-particle, which is apparently easily deflected by its passage through matter. At the same time, there is undoubted evidence that the direction of motion of some of the α-particles is slightly changed by their passage through matter [64].

Although its charge to mass ration e/m is known, the mass of the α-particle is unknown. The observed deviations produced by a magnetic field suggest that its mass is similar to that of a light atom. It could have two units of charge and four units of mass (i.e., a mass equal to four times the mass of hydrogen), in which case it could be a helium atom which has lost two electrons. But it could equally well have a single unit of charge and two units of mass. To decide, it was necessary to devise an instrument able to count the α-particles *one by one*. With such an instrument, Rutherford would be able to count the number of particles emitted by a given quantity of radium, measure their total electric charge, and deduce the electric charge of a single α-particle.

The First Geiger Counter

On his first visit to the laboratory in Manchester, Rutherford was guided by a young physicist, Hans Geiger, who showed him the available equipment and described the ongoing experiments. Geiger was finishing a one-year postdoc position and was about to return to Erlangen, in Germany.

Hans Geiger was born on September 30, 1882, in Neustadt, in the Palatinate, which was then part of the kingdom of Bavaria [65, 66]. He studied in Erlangen where his father was professor of ancient languages. In 1906, he defended his Ph.D. thesis on the discharge of gases. His thesis advisor was Gustav Wiedemann [67]. He then spent one year in Manchester. He must have pleased Rutherford, who suggested that he should stay longer and work with him. Geiger must have been impressed by the reputation and the personality of Rutherford, and he gave up the idea of returning to Erlangen.

Geiger was an exceptionally gifted and rigorous experimentalist with an insatiable desire to work [68]. He acquired both prestige and popularity in the lab. In a speech delivered in 1950, the chemist Alexander Russel spoke of him in the following terms:

Geiger was too much of an Olympian for me to know him well. Gentle, without being docile, and aloof, he seemed in laboratory hours to live entirely for the work. He was a beautiful experimenter of the Sir James

Dewar type, splendid with his hands. Like many Germans, he loved good music and good dinners [69].

Rutherford still intends to count the α-particles, one by one, but the amount of electricity induced by one particle is far too small to induce a signal in an electrometer. He calculates that the ionization caused by an α-particle produces about 20,000 pairs of ions, which could produce a potential of $6\,\mu V$ with the available electric fields and capacities. That was too small to be observed.

John Townsend, who was professor in Oxford and whom Rutherford had met before in the Cavendish, discovered and studied a potentially interesting phenomenon [70–73]. It was well understood (and Rutherford had contributed to this) that the electric current, which passes through a gas exposed, for example, to X-rays, is due to the ionization of the molecules of the gas which liberate electrons and positive ions. The observed current is due to the motion of the electrons. When the applied electric field$^\diamond$ is progressively increased, the current increases at first and then it reaches a constant value. It is said to "saturate." Indeed, when the applied electric field is weak, the velocity of the electrons is small and some of them can recombine with positive ions and form neutral atoms. This has the unfortunate consequence that the induced electric current is not a measure of the initial number of liberated electrons and therefore of the intensity of the X-rays. When the applied electric field is increased, the electrons move faster and eventually none of them recombine. This is why the induced electric current attains a constant "saturation" value. However, Townsend, who was studying gases maintained at a low pressure, showed that, when the applied electric current is further increased above a certain value, the induced current increases again, because the electrons then acquire a velocity large to enough to ionize other atoms which in turn emit electrons. In other words, the number of electrons proliferates, just as a falling snowball can grow into an avalanche. More charge is then collected than had originally been liberated by the X-rays. There is an amplification of the signal.

This phenomenon became known as the "Townsend avalanche," and it attracted the attention of Rutherford. It might provide a way to detect a single α-particle. Rutherford probably knew about the experiment performed by P. J. Kirkby, a student of Townsend, who constructed an apparatus which consisted of an aluminum cylinder, which played the role of a cathode, and of a wire running along its axis which was the anode [74]. Rutherford suggests that Geiger should attempt to use this phenomenon in order to detect individual α-particles. A year later, in 1908, Rutherford and Geiger succeed in making the first *counter* work. After a short communication sent to the Manchester Literary and Philosophical Society on February 11, they send a detailed report [75, 76] to the Royal Society on June 18. It is a decisive step towards the detection of radiation.

This first counter is the ancestor of "Geiger counters." It is a metallic tube, 25 cm long and 17 mm in diameter. A metal wire runs through the center. A potential of 1,200–1,300 V is established between the wire which is connected to the positive pole of an electric battery (or rather to a rack of hundreds of batteries connected in series), while the external cylinder is connected to the negative pole. The cylinder contains a gas (air or carbon dioxide) maintained at a low pressure, a few percent of normal atmospheric pressure. When an α-particle passes through the tube, electron-ion pairs are created along its trajectory. The electrons are attracted to the positive central wire, while the ions travel towards the external cylinder. The electrons, being much lighter, acquire a considerable velocity. Furthermore, the cylindrical shape causes the electric field$^\diamond$, which drives the electrons towards the central wire, to increase as they approach the wire. They acquire sufficient velocity to ionize the molecules they collide with, and new pairs of electrons are thus created. This is the avalanche phenomenon described above. In the first counter constructed by Rutherford and Geiger, one electron could produce a 1,000 electrons, thereby amplifying considerably the electric signal on the wire. The goal was attained: a single α-particle, passing through the counter produced an observable effect on the electrometer. Rutherford finally succeeded in counting individual α-particles. This ancestor of Geiger counters, which are still used today, was not so easy to use. Between 10 and 20 s were required for the "avalanche" to die out before another particle could be detected. However, Rutherford set out immediately to count the number of α-particles emitted by a sample of radium-C.

The Nature of the α-Particle

That same day, Rutherford and Geiger send another communication to the Royal Society, concerning "the electric charge and the nature of the α-particle." From the outset, they state:

In a previous paper, we have determined the number of α-particles expelled per second per gramme of radium by a direct counting method. Knowing this number, the charge carried by each particle can be determined by measuring the total charge carried by the α-particles expelled per second from a known quantity of radium [77].

After a detailed description of their experiment and of their results, they quote the result they worked so hard to achieve:

Considering the data as a whole, we may conclude with some certainty that the α-particle carries a charge 2e, and that the value of e is not very different from $4 \cdot 65 \times 10^{-10}$ E.S. unit.

They add a footnote:

It is of interest to note that Planck deduced a value of $e = 4 \cdot 69 \times 10^{-10}$ E.S. unit from a general optical theory of the natural temperature-radiation.

The agreement cannot be fortuitous. Rutherford finally reaches his goal:

We may conclude that an α-particle is a helium atom, *or, to be more precise*, the α-particle, after it has lost its positive charge, is a helium atom.

But that is not all. Once the electric charge of the α-particle is known, it becomes possible to calculate other fundamental quantities. Thus, the radioactive half-life of radium is estimated to be 1760 years (today, the value of 1600 years is retained). They also determine the famous Avogadro number, by a completely independent method. The value they find agrees with the other determinations. They stand on sure ground.

Another Way to Count α-Particles: Scintillations

There was in fact another way to detect α-particles one by one. Fluorescent substances, such as zinc sulfide, were known to emit a weak light signal when they were placed in the proximity of a radioactive source. In 1903, William Crookes observed the luminous surface of zinc sulfide through a magnifier. He noticed that the luminosity was not constant but that it was caused by a large number of punctual and very short flares. He even constructed a simple device to observe them, which he named the "spinthariscope," from the Greek word *spintharis*, spark, scintillation [78]. Elster and Geitel made the same observation [79]. It was quite tempting to attribute these flares to a passing particle, but how could one be sure that it was one and only one particle which caused the substance to scintillate? In his lectures on "radioactive transformations" delivered in 1905 in Yale University, Rutherford reviews the ways in which radioactive radiation could be detected and measured:

There are three general properties of the rays from radioactive substances which have been utilized for the purpose of measurements, depending on (1) the action of the rays on a photographic plate, (2) the phosphorescence excited in certain crystalline substances, (3) the ionization produced by the rays in a gas [57].

He is however cautious concerning the counting of flares in a scintillating substance:

The property of the α rays of producing scintillations on a screen covered with zinc sulphide is especially interesting, and it has been found possible by this method to detect the α rays emitted by feebly active substances like uranium, thorium and pitchblende. Screens of zinc sulphide have been used as an optical method for demonstrating the presence of the emanations from radium and actinium. Speaking generally, the phosphorescent method, while interesting as an optical means for examining the rays, is very limited in its application and is only roughly quantitative.

But as soon as he could use his counter to evaluate the accuracy of the scintillation method, Rutherford and Geiger compare results obtained by the two methods. Good news:

The number of scintillations observed on a properly-prepared screen of zinc sulphide is, within the limit of experimental error, equal to the number of α-particles falling upon it, as counted by the electrical method. It follows from this that each α-particle produces a scintillation [76].

The two independent methods give the same results. During the following 20 years, Rutherford will exclusively use the scintillation method, which was much easier to use. The young German physicist Erich Regener, who was trying to measure the elementary electric charge$^\diamond$, made a careful study of the method. He perfected an operating procedure to count particles using a scintillating material [80]. He recommended to work in a weakly illuminated room, and to use a microscope with a medium magnifying power and a large aperture, in order to increase the luminosity of the flares. The screen was dimly illuminated. Each time a scintillation occurred, the physicist would press an electric contact which caused a deviation of the line on the paper strip of a chronograph. The counting required great concentration in order not to miss a flare. Eyes would quickly tire, and this made it necessary to make short observation periods and take frequent breaks. A well-trained observer could register 95 % of the flares, provided that no more than about 20 flares would occur per minute. The method had many advantages. First, it allowed to detect only α-particles because neither β- nor γ-rays would produce visible scintillations. Furthermore, the device was very simple to set up and reliable. For 20 years, it remained the supreme method of counting α-rays.

Back to the Scattering of α-Particles

While they were perfecting their counter, Rutherford and Geiger noticed a weak scattering of the α-particles which had to pass through a 4 m tube before being detected. The scattering was caused by the gas remaining in the tube. Rutherford then proposed that Geiger should make a systematic study of the scattering of α-rays using the scintillation method which had proved to be so accurate. On July 17, 1908, Rutherford reports to the Royal Society on the first observations made by Geiger, who placed a source of α-rays at the extremity of a tube 114 cm long. After passing through the tube, the α-particles pass through a narrow slit before being detected on a screen situated 54 cm from the extremity of the tube. The apparatus is maintained in vacuum. No scattering of the α-particles is observed. The particles move in a straight line. But if a gold or aluminum sheet is placed in front of the slit, the image becomes blurred:

The observations just described give direct evidence that there is a very marked scattering of the α-rays in passing through matter, whether gaseous or solid. It will be noticed that some of the α-particles after passing through the very thin leaves—the stopping power of one leaf corresponded to about 1 mm of air—were deflected through quite an appreciable angle [81].

Hans Geiger makes a systematic study of these deviations. He studies the deviations of α-particles by various metals of different thickness. He uses gold sheets because gold is the metal which produces the largest deviations. After observing the deviation caused by up to 35 gold sheets, he concludes that the most likely angle by which an α-particle is deviated is proportional to the atomic weight of the atoms of the sheet placed in front of the slit. In the case of gold, the angle is, on the average, about 1/200 of a degree after passing through a single atom [82]. The deviation can be considerably larger after passing through a large number of atoms:

The probable angle through which an α-particle is turned in passing through an atom is proportional to its atomic weight. The actual value of this angle in the case of gold is about 1/200 of a degree.

Geiger observes deviations as large as 15°. About 3,000 deviations, *all in the same direction*, would be required to attain such a large deviation. This is most unlikely since successive deviations have random values. These large deviations puzzle Rutherford.

The Experiments of Geiger and Marsden

In a talk given in Cambridge in 1936,[1] Rutherford recalls:

One day Geiger came to me and said, "Don't you think that young Marsden, whom I am training in radioactive methods, ought to begin a small research?" Now I had thought that too, so I said, "Why not let him see if any α-particles can be scattered through a large angle?" I may tell you in confidence that I did not believe that they would be, since we knew that the α-particle was a very fast massive particle, with a great deal of energy, and you could show that if the scattering was due to the accumulated effect of a number of small scatterings the chance of an α-particle being scattered backwards was very small [84].

---

[1] The University of Cambridge organized in 1936 a series of lectures on the history of science. Rutherford delivered two lectures entitled "Forty Years of Physics," the first one on "The History of Radioactivity" and the second one on "The Development of the Theory of Atomic Structure." These lectures were published in a book, *Background to Modern Science [83]*, but Rutherford died before being able to write them in a form suitable for publication. The text was therefore prepared by John Ratcliffe, a radiophysicist from Cambridge, on the basis of the verbatim taken by a stenographer.

The young Ernst Marsden was a 20-year-old student. Rutherford continues:

Then I remember two of three days later Geiger coming to me in great excitement and saying, "We have been able to get some of the α-particles coming backwards[. . .]". It was quite the most incredible event that ever happened to me in my life. It was almost as incredible as if you fired a 15-inch shell at a piece of tissue paper and it came back and hit you [85].

On June 17, 1909, Rutherford presents to the Royal Society the work of Geiger and Marsden:

When β-particles fall on a plate, a strong radiation emerges from the same side of the plate as that on which the β-particles fall [. . .] For α-particles a similar effect has not been previously observed, and is perhaps not to be expected on account of the relatively small scattering that α-particles suffer in penetrating matter. In the following experiments, however, conclusive evidence was found of the existence of a diffuse reflection of the α-particles. A small fraction of the α-particles falling upon a metal plate have their directions changed to such an extent that they emerge again at the side of incidence [86].

Geiger and Marsden used sheets consisting of various metals: aluminum, iron, tin, gold and lead. They noticed that the effect is larger with heavier metals. They stack a number of plates of various thicknesses and they observe that, to some extent, the effect increases with the thickness of the target, which proves that the effect is not a reflection on its surface but a phenomenon which occurs inside the sheets.

It seems very surprising that some of the α-particles, as the experiment shows, can be turned within a layer of 6×10^{-5} cm of gold through an angle of $90°$, and even more [. . .] Three different determinations showed that, of the incident α-particles, about 1 in 8 000 was reflected, under the described conditions.

Are the Large Deviations Caused by Multiple Small Deviations?

How could one understand that a particle as massive and rapid as an α-particle could rebound on a surface which was expected to be soft? According to the plum-pudding model of Joseph John Thomson, the α-particle would pass through a kind of positively charged jelly (the pudding) where it would collide only with negatively charged particles which had a 7 000 times smaller mass and which could therefore not scatter them backwards. A first possibility would be multiple scattering during which the α-particle would undergo a large number of small deviations. However, in a different context, Lord Rayleigh (William Strutt) had calculated the mean deviation caused by a succession of random deviations [87]. According to his calculation, the fraction of α-particles which would suffer deviations larger than 90°would be much smaller than the fraction observed by Geiger and Marsden.

Rutherford Invents the Nucleus

Rutherford ponders for a long time. A year later, on December 14, 1910 he writes to his friend Bertram Boltwood:

I have been doing a good deal of calculation on scattering. I think I can devise an atom much superior to J. J.'s, for the explanation of and stoppage of α and β particles, and at the same time I think it will fit extraordinarily well with the experimental numbers. It will account for the reflected α particles observed by Geiger, and generally, I think, will make a fine working hypothesis. Altogether I am confident that we are going to get more information from scattering about the nature of the atom than from any other method of attack [88].

He is of course referring to the large angle scattering of α-particles and "J. J." is no other than J. J. Thomson. And indeed, a short time later in 1911, as Geiger recalls:

He entered my office visibly in a good mood and he told me that he now knew what the atom looked like and how one could understand the large deviations. That same day, I began research to check the relation which Rutherford had established between the deviation angle and the number of particles. The strong variation of this function with angle made my work relatively easy and it was possible to check rather quickly at least the approximate validity of his model [89].

Rutherford reached what he believed to be an inescapable conclusion. If the α-particle bounces back, it must have undergone a strong thrust from a sufficiently massive object. And this must have occurred *in a single collision*, because too few particles would scatter backwards in the case of multiple collisions. However, with the plum-pudding model of J. J. Thomson, backward scattering would be impossible because, although the sphere carrying the positive charge may have sufficient mass, it dilutes the positive charge in a too large volume. There remains one other possibility, the only one Rutherford can think of: *the sphere must be much smaller*. At first he even imagines that the positive charge is concentrated at one point. An α-particle which gets close enough to the tiny positively charged sphere can scatter backwards and Rutherford calculates the fraction of particles which scatter at a given angle. The first measurements of Geiger confirm his calculation. What might at first have appeared to be a crazy idea, acquires some substance and in a two-page note read to the Manchester Literary and Philosophical Society, on March 7, 1911. Rutherford presents a first description of his model of the atom:

The scattering of the electrified particles is considered for a type of atom which consists of a central electric charge concentrated at a point and surrounded by a uniform spherical distribution of opposite electricity equal in amount [90].

This atom is quite different from the model of J. J. Thomson! In April, Rutherford writes a paper which is published in the May issue of the *Philosophical Magazine*. From the outset, he states:

Since the α and β particles traverse the atom, it should be possible from a close study of the nature of the deflection to form some idea of the constitution of the atom to produce the effects observed. In fact, the scattering of high speed charged particles by the atoms of matter is one of the most promising methods of attack of this problem. The development of the scintillation method of counting single α particles affords unusual advantages of investigation, and the researches of H. Geiger by this method have already added much to our knowledge of the scattering of alpha rays by matter [91].

He then presents a detailed account of his model. He describes the trajectories of the scattered particles and he estimates the value of the positive electric charge concentrated at a point in the center of the atom. He also discusses the possible size of the central positive charge:

It is of interest to examine how far the experimental evidence throws light on the question of the extent of the distribution of the central charge.

He then shows that one cannot explain the observed results if the α-particles pass *through* the volume occupied by the positive charge (otherwise they would suffer smaller deviations). The central region is therefore very small. Rutherford estimates that for a substance consisting of atoms with a central charge equal to 100 units of elementary charge, which he believes is the case of gold,[1] some of the α-particles come as close as 3×10^{-12} cm to the central charge, which is 30,000 times smaller than the atom itself! Furthermore not only is the total positive charge concentrated there, but *so is practically all the mass of the atom*! This is all which Rutherford claims at the time. Such a structure is reminiscent of the model proposed by Perrin or Nagaoka, the latter being simply quoted in the paper. He does not dwell further on the question of the stability of atom:

The question of the stability of the atom proposed need not be considered at this stage, for this will obviously depend upon the minute structure of the atom, and on the motion of the constituent charged parts.

Indeed, an atom with such a structure is *a priori* unstable. According to a theorem proved by Samuel Earnshaw, there is no static equilibrium state for particles which interact with forces which are proportional to the inverse square of the distance separating them [92]. But if electrons wander around the central positive charge as planets do around the sun, then they should radiate energy and the system would not be stable.

---

[1] It was measured shortly afterwards to be equal to 79 units of elementary charge.

So far, Rutherford only mentions a "small region" of charge concentrated at one point. In a paper dated August 16, 1912 and published in October, he uses, apparently for the first time, the Latin (and therefore learned!) term *nucleus* [93]. The word was used in biology to designate the nucleus of a cell. The strange and incredibly small object which concentrates all its positive charge and practically all its mass now bears a name: *the nucleus of the atom*.

As mentioned above, Rutherford did not dwell on the structure of the atom and he was well aware of the fact that his model was unsatisfactory. He simply stated what his experiments suggested. For him, the role of the experimentalist consisted in *reading nature*. And he read beyond doubt that the atom consisted of a tiny nucleus surrounded by electrons. Why did it not radiate? A young Danish physicist would soon resolve the problem.

A Last Ingredient: Moseley Measures the Charge of the Nucleus in the Atom

Max von Laue succeeds in diffracting X-rays. Bragg invents the X-ray spectrometer. The young Moseley measures the characteristic radiation of elements. The atomic number of Mendeleev is no longer simply a catalogue number, but is identified, instead, to a fundamental physical constant, namely the number of positive electric charges in the nucleus of the atom.

Barkla Creates X-ray Spectroscopy

We described above the fundamental work of Barkla on X-rays, which allowed the determination of the number of electrons in each kind of atom.[1] Now Barkla makes another important observation: part of the secondary X-ray radiation, caused by atoms irradiated by primary X-rays, does not have the same energy as the primary X-rays. This secondary radiation is not a simple scattering because its energy is independent of the energy of the primary radiation. It is characteristic of the irradiated element [46, 94–101]. Barkla lists, element by element, not the energy (which he could not measure) but the absorption coefficient (a number which measures how quickly the radiation is absorbed as it passes through matter) which depends sensitively on the energy. He notes also that this characteristic X-ray radiation behaves exactly as fluorescence$^\diamond$, and he notices that these "fluorescent Röntgen radiations" can be divided into two families, which he first calls *A* and *B* but which he later prefers to call *K* and *L* because, he claims, there must certainly be radiation which is more penetrating that the one he labels *K* and less penetrating than the one he labels *L*. One can only wonder and admire how Barkla managed to obtain such a wealth of data, using simply an X-ray source, an electrometer and a few aluminum sheets. The road is now clear for a precise measurement of the wavelength of the X-rays.

---

[1] See p. 60.

The Diffraction of X-rays: Max von Laue, William Henry and William Lawrence Bragg

As soon as X-rays were discovered, physicists attempted to obtain diffraction$^\diamond$ or interference$^\diamond$ which would have immediately proved the wave-character of the X-rays. However these early attempts failed. The French physicist Gouy [102] X-rayed a fine grating of wires, with no result and this allowed him to fix an upper limit of 50 Å on the wavelength of X-rays.[1] In 1903, two Dutch physicists form Gronigen, Hermanus Haga and C. H. Wind used a 15 μm (15/1000 of a millimeter) slit as a source and a triangular slit for diffraction, the width of which was 27 μm at one end and zero at the other. However, their results were not conclusive [103]. In 1909, two physicists from Hamburg, B. Walter and R. Pohl, made a similar attempt. They could only give an upper limit to the wavelength of the X-rays, namely somewhere between one and one tenth of an angstrom [104].

In 1912, the young Paul Ewald was preparing his Ph.D. thesis in Munich, under the leadership of Arnold Sommerfeld. It is for this work that he sought the advice of Max von Laue, who was then a *Privatdozent* (assistant) in Munich, on the mathematical analysis he was making of light passing through a crystal. Max von Laue asked him what would happen in the wavelength of the light was smaller than the distance separating the atoms in the crystal, and Ewald replied that his analysis would still remain valid. Laue then wondered if that would not be a way to diffract X-rays and he suggested to Walter Friedrich, an assistant of Sommerfeld, to do the experiment. In spite of the skepticism of Sommerfeld, Friedrich performed the experiment with the help of the student Paul Knipping. The roughly 1-mm wide X-ray beam was focused by a series of slits. It was incident on a crystal. Photographic plates were placed behind and on the side of the crystal. At first, they used a copper sulphide crystal and the image was blurred. Then, with a blende (a zinc sulphide cubic, waxlike and bluish crystal) they obtained a much finer picture consisting of geometrically placed spots which clearly showed that the X-rays were diffracted by the crystal [105]. Von Laue made a mathematical analysis of the diffraction pattern which showed that the incident beam had a definite wavelength [106]. His analysis enabled him to determine the ratio between the wavelength of the incident X-rays, and the distance between the atoms in the crystal. He found thus that the wavelength of the X-rays was very small, between 3 % and 14 % of the distance separating the atoms, which is about 1 Å.

The next step was taken by William Henry Bragg, whom we mentioned in connection with his work on the slowing down of α-particles,[2] and his son William Lawrence Bragg. The two succeeded in obtaining the diffraction of X-rays which were *reflected by the crystal* [107]. Their X-ray spectrometer was similar to an optical spectrometer, except for the fact that the prism was replaced by the crystal,

---

[1]One angstrom is equal to one ten-millionth of a millimeter.

[2]See p. 63.

on which the X-rays were reflected before impinging on a photographic plate or an ionization chamber [108], which enabled them to measure also the intensity of the reflected beam. Their first results were spectacular: by displacing the detector, they found that the intensity of the beam varied with the reflection angle and that it displayed well defined peaks, which indicated monochromatic rays on top of a white background, consisting of a mixture of several wavelengths. They had discovered a way to measure the wavelength, which was much simpler than the method used by von Laue. This new spectrometer gave birth to two new domains of physics: the study of X-rays emitted by various elements and the study of the structure of crystals. William Henry, the father, worked mostly on the first, whereas William Lawrence, the son, mostly on the second.

Henry Moseley Measures the Charge of Nuclei

As soon as the results of von Laue, Friedrich and Knipping were published, a young physicist from Manchester begins an active study of X-rays. Henry Gwyn Jeffreys Moseley [109] was born in 1887. He obtained his Ph.D. in Oxford after which he joins Rutherford in Manchester. He first uses the experimental setup of Friedrich et Knipping, and then the one of Bragg. He publishes a first paper in 1913, with Charles Galton Darwin, a theoretician in Manchester, the grandson of the founder of the theory of the evolution of species. Moseley is mainly interested in the "selective reflection" which had been observed by the two Braggs. He describes the experimental difficulties [110]:

The ratio of the selective to the general reflection is greatly increased by limiting the breadth of the slits and so increasing the parallelism of the primary beam. Unfortunately, a very small rotation of the crystal will then remove all traces of the selective effect. It therefore proved necessary to take readings with the crystal set at every 5′ of arc between 10° and 14°.

This preliminary work allows him to measure the scope of the problem. Moseley then begins to measure, for each element, the wavelengths of the X-ray radiation, which Barkla called the "characteristic radiation." He measures the wavelength of the radiation of eleven elements: calcium, scandium, titanium, vanadium, chromium, manganese, iron, cobalt, nickel, copper and zinc [111]. In each case he finds two high-frequency rays, which he calls α and β, and he notices that the wavelengths decrease regularly as the atomic masses increase.

Moseley then makes a big discovery. He draws a table in which he notes, for each element, the famous characteristic angles, the wavelengths of the α and β radiation, the atomic mass (which was called the "atomic weight" at the time), and a quantity Q, which he calculates using the wavelength.[1]

Until that time, the atomic number was simply an index which denoted the order of the elements in the Mendeleev table and by which the elements were ordered by increasing mass (with a few exceptions). It is upon ordering the elements that Mendeleev noticed some regularities in the chemical properties of various elements$^\diamond$. In 1913, the atomic number had no physical or chemical meaning. But Moseley notices something curious:

> It is at once evident that Q increases by a constant amount as we pass from one element to the next, using the chemical order of the elements in the periodic system. Except in the case of nickel and cobalt, this is also the order of the atomic weights. While, however, Q increases uniformly, the atomic weights vary in an apparently arbitrary manner, so that an exception in their order does not come as a surprise. We have there a proof that there is in the atom a fundamental quantity, which increases by regular steps as we pass from one element to the next. This quantity can only be the charge on the central positive nucleus, of the existence of which we already have definite proof [. . .]
>
> We are therefore led by experiment to the view that N is the same as the number of the place occupied by the element in the periodic system [111].

Moseley made a major discovery: *the atomic number of an element is equal to the number of elementary charges in the nucleus of the atom*! He can even measure it. He works very hard, assembling and disassembling apparatus to improve it. He constructs a second spectrometer and soon he publishes a second paper [112]. He has measured the K-rays of 44 elements, ranging from aluminum to gold and he determined their atomic number. From then on, the atomic number will be understood as the number of charges in the nucleus.

A year after the onset of the First World War, Moseley is drafted into the navy and is killed in the battle of the Dardanelles. In a letter, Rutherford announces this sad news to his friend Boltwood:

> You will be very sorry to hear that Moseley was killed in the Dardanelles on Aug. 10th. You will see my obituary notice of him in Nature. He was the best of the young people I ever had and his death is a severe loss to science [113].

Clearly, Rutherford considered Moseley to be one of the best physicists of his generation.

---

[1]In fact he used the frequency which he called ν, related to Q by the expression $Q = \sqrt{\frac{\nu}{\frac{3}{4}\nu_0}}$, where ν_0 is a reference frequency related to the Rydberg constant $N_0 = \frac{\nu_0}{c} = 109.72$.

A Paradox

Through the work Rutherford, Barkla and of Moseley, experiment had spoken: the atom appears to consist of a central nucleus, with a known positive charge and mass (almost equal to the mass of the atom). The nucleus is surrounded by electrons, the number of which is equal to the charge of the nucleus, so that the atom is electrically neutral. But there is a flaw in this model. As discussed above,[1] Newton's and Maxwell's laws forbid the existence of such a stable and static structure. The electrons would always end up collapsing into the nucleus. If electrons describe orbits around the nucleus, as planets do around the sun, then Maxwell's equations also state that this is an unstable state, because a rotating electron would radiate, lose its energy and finally collapse into the nucleus.

There is no solution to this problem, within the framework of the dynamics of Newton and of Maxwell. Who will have the imagination, the clear-sightedness and the audacity to solve this puzzle?

---

[1] See p. 75.

References

1. Lucretius, *On the Nature of the Universe (De rerum natura)*, Oxford University Press, Oxford, 1997. A verse translation by Ronald Melville.
2. Nollet, A., *Leçons de Physique Expérimentale*, chez Durand, Paris, 1743.
3. Richter, J. B., *Anfangsgründe der Stöchyometrie oder Meßkunst chymischer Elemente*, J. F. Korn, Breslau & Hirschberg, 1792–94.
4. Proust, J. L., "Extrait d'un Mémoire intitulé : Recherches sur le Bleu de Prusse", *Journal de Physique, de Chimie, d'Histoire naturelle et des Arts* **2**, 334–341, novembre 1794.
5. Proust, J. L., "Recherches sur le bleu de Prusse", *Journal de Physique, de Chimie et d'Histoire naturelle* **6**, 241–251, 1799.
6. Dalton, J., *New System of Chemical Philosophy. Vol. 1*, Bickerstaff, Manchester, Part I, 1808; Part II, 1810.
7. Morselli, M., *Amedeo Avogadro, a scientific biography*, D. Reidel, Dordrecht, 1984.
8. Avogadro, A., "Essai d'une manière de déterminer les masses relatives des molécules élémentaires des corps et les proportions selon lesquelles elles entrent dans ces combinaisons", *Journal de Physique, de Chimie, d'Histoire naturelle et des Arts* **73**, 58–76, 1811.
9. Ampère, A. M., "Lettre de M. Ampère à M. le comte Berthollet sur la détermination des proportions dans lesquelles les corps se combinent d'après le nombre et la disposition respective des molécules dont leurs parties intégrantes sont composées", *Annales de Chimie et de Physique, Paris* **90**, 43–86, April 30, 1814.
10. Prout, W., "On the Relation between the Specific Gravities of Bodies in their Gaseous State and the Weights of their Atoms", *Annals of Philosophy* **6**, 321–330, November 1815.
11. Prout, W., "Correction of a Mistake in the Essay on the Relation between the Specific Gravities of Bodies in their Gaseous State and the Weight of their Atoms", *Annals of Philosophy* **7**, 111–113, February 1816.
12. Brock, W. H. and Knight, D. M., "The Atomic Debates", in *The Atomic Debates. Brodie and the Rejection of the Atomic Theory*, pp. 1–30, Leicester University Press, Leicester, 1967.
13. Ganot, A., *Traité élémentaire de physique expérimentale et appliquée*, chez l'auteur, Paris, 1853.
14. Ganot, A., *Introductory course of natural philosophy*, A.S. Barnes & Burr, New York, 1862.
15. Tait, P. G., *Properties of matter*, Adam, Edinburgh, 1885, p. 19.
16. Nye, M. J. (editor), *The Question of the Atom from the Karlsruhe Congress to the first Solvay Conference, 1860–1911*, Tomash Publishers, Los Angeles/San Francisco, 1984.
17. Stas, J. S., "Recherches sur les rapports réciproques des poids atomiques", *Bulletins de l'Académie royale des sciences, des lettres et des beaux-arts de Belgique* pp. 208–336, session of August 4, 1860.
18. Maxwell, J. C., "Illustration of the Dynamical Theory of Gases.— Part I. On the Motions and collisions of Perfectly Elastic Spheres", *Philosophical Magazine* **19**, 19–32, January 1860.
19. Loschmidt, J. J., "Zur Grösse der Luftmolecüle", *Sitzungsberichte der Akademie der Wissenschaften in Wien* **52**, 395–413, session of October 12, 1866.
20. Nye, M. J., *Molecular reality: a perspective on the scientific work of Jean Perrin*, Macdonald, London and American Elsevier, New York, 1972.
21. Charpentier-Morize, M., *Perrin, savant et homme politique*, Belin, Paris, 1997.
22. Perrin, J., "Rayons cathodiques et rayons de Röntgen. Étude expérimentale.", *Annales de chimie et de physique, Paris* **11**, 496–554, August 1897.
23. Perrin, J., "Les hypothèses moléculaires", *Revue Scientifique* **15**, 449–61, April 13, 1901.
24. Perrin, J., "Mouvement brownien et réalité moléculaire", *Annales de Chimie et de Physique, Paris* **18**, 5–114, 1909.
25. Perrin, J., *Les Atomes*, Librairie Félix Alcan, Paris, 1913. English translation by D. Ll. Hammick: *Atoms,* Constable & Co., London, 1916.

26. Rydberg, J., "La distribution des raies spectrales", in *Congrès International de Physique*, edited by Guillaume, C.-É. and Poincaré, L., Vol. II, pp. 200–224, Gauthier-Villars, Paris, 1900.

27. Balmer, J. J., "Notiz über die Spektrallinien des Wasserstoffs", *Annalen der Physik, Leipzig* **25**, 80–87, 1885.

28. Thomson, J. J., "On the charge of electricity carried by the ions produced by Röntgen rays", *Philosophical Magazine* **46**, 528–545, December 1898.

29. Millikan, R. A., "The Isolation of an Ion, a Precision Measurement of its Charge, and the Correction of Stokes's Law", *Physical Review* **32**, 349–397, April 1911.

30. Thomson, J. J., "On the Masses of the Ions in Gases at Low Pressure", *Philosophical Magazine* **48**, 547–567, December 1899.

31. Rayleigh, John William Strutt, B., *The life of Sir J. J. Thomson*, Cambridge University Press, Cambridge, 1942.

32. Nagaoka, H., "Kinetics of a System of Particles illustrating the Line and the Band Spectrum and the Phenomena of Radioactivity", *Philosophical Magazine* **7**, 445–455, May 1904.

33. Kelvin, L., "Aepinus Atomized", *Philosophical Magazine* **3**, 257–283, March 1902.

34. Thomson, J. J., "On the Structure of the Atom : an investigation of the Stability and Periods of Oscillation of a number of Corpuscles arranged at equal intervals around circumference of a Circle; with Application to the results to the Theory of Atomic Structure", *Philosophical Magazine* **7**, 237–265, March 1904.

35. Mayer, A. M., "Floating Magnets", *Nature* **17**, 487–488, April 18, 1878, and **18**, 258–260, July 4, 1878.

36. Perrin, J., "Décharge par les rayons de Röntgen. - Rôle des surfaces frappées", *Comptes Rendus de l'Académie des Sciences* **124**, 455–458, session of March 1, 1897.

37. Sagnac, G., "Sur les propriétés des gaz traversés par les rayons X et sur les propriétés des corps luminescents ou photographiques", *Comptes Rendus de l'Académie des Sciences* **125**, 168–171, session of July 12, 1897.

38. Sagnac, G., "Sur la transformation des rayons X par les métaux", *Comptes Rendus de l'Académie des Sciences* **125**, 230–232, session of July 26, 1897.

39. Curie, P. and Sagnac, G., "Électrisation négative des rayons secondaires produits au moyen des rayons Röntgen", *Comptes Rendus de l'Académie des Sciences* **130**, 1013–1016, session of April 9, 1900.

40. Curie, P. and Sagnac, G., "Électrisation négative des rayons secondaires issus de la transformation des rayons X", *Journal de Physique* **1**, 13–21, 1902.

41. Barkla, C. G., "Secondary Radiation from Gases subject to X-Rays", *Philosophical Magazine* **5**, 685–698, June 1903.

42. Barkla, C. G., "Energy of Secondary Röntgen Radiation", *Philosophical Magazine* **7**, 543–560, May 1904.

43. Thomson, J. J., *Conduction of Electricity through Gases*, At the University Press, Cambridge, 1903. Reprint by Dover, New York 1969.

44. Thomson, J. J., "On the Number of Corpuscles in an Atom", *Philosophical Magazine* **11**, 769–781, June 1906.

45. Pais, A., *Inward Bound. Of Matter and Forces in the Physical World*, Clarendon Press/Oxford University Press, London/New York, 1986, p. 187.

46. Barkla, C. G., "On the Energy of Scattered X-radiation", *Philosophical Magazine* **21**, 648–652, May 1911.

47. Skłodowska-Curie, M., "Sur la pénétration des rayons de Becquerel non déviables par le champ magnétique", *Comptes Rendus de l'Académie des Sciences* **130**, 76–79, session of January 8, 1900.

48. Curie, P. and Curie, M., "Les nouvelles substances radioactives et les rayons qu'elles émettent", in *Travaux du Congrès international de physique réuni à Paris en 1900*, Vol. 3, pp. 79–114.

49. Caroe, G. M., *William Henry Bragg, 1862–1942 : man and scientist*, Cambridge University Press, Cambridge, New York, 1978, p. 43.

50. Bragg, W. H., "On the Absorption of α Rays, and on the Classification of the α Rays from Radium", *Philosophical Magazine* **8**, 719–725, December 1904.
51. Bragg, W. H. and Kleeman, R. D., "On the Ionization Curves of Radium", *Philosophical Magazine* **8**, 726–739, December 1904.
52. Bragg, W. H. and Kleeman, R. D., "On the α Particles of Radium, and their Loss of Range in passing through various Atoms and Molecules", *Philosophical Magazine* **10**, 318–340, September 1905.
53. McClung, R. K., "The absorption of α rays", *Philosophical Magazine* **11**, 131–142, January 1906.
54. Kučera, B. and Mašek, B., "Über die Strahlung des Radiotellurs. I", *Physikalische Zeitschrift* **7**, 337–340, May 15, 1906.
55. Kučera, B. and Mašek, B., "Über die Strahlung des Radiotellurs. II", *Physikalische Zeitschrift* **7**, 630–640, September 15, 1906.
56. Rutherford, E., "Some properties of the α Rays from Radium", *Philosophical Magazine* **10**, 163–176, July 1905.
57. Rutherford, E., "Some properties of the α Rays from Radium", *Philosophical Magazine* **11**, 166–176, January 1906.
58. Rutherford, E., "The Retardation of the Velocity of the α Particles in passing through Matter", *Philosophical Magazine* **11**, 553–554, April 1906.
59. Rutherford, E., "Retardation of the Velocity of the α Particles in passing through matter", *Philosophical Magazine* **12**, 134–146, August 1906.
60. Kučera, B. and Mašek, B., "Über die Strahlung des Radiotellurs III. Die Sekundärstrahlung der α-Strahlen", *Physikalische Zeitschrift* **7**, 650–654, October 1906.
61. Meitner, L., "Über die Zerstreuung der α-Strahlen", *Physikalische Zeitschrift* **8**, 489–491, August 1, 1907.
62. Meyer, E., "Die Absorption der α-Strahlen in Metallen", *Physikalische Zeitschrift* **8**, 425–430, July 1, 1907.
63. Feather, N., *Lord Rutherford*, 1973, p. 117.
64. Rutherford, E., "Recent Advances in Radio-activity", *Nature* **77**, 422–426, March 5, 1908. A discourse delivered at the Royal Institution on Friday, January 31, 1908.
65. Krebs, A. T., "Hans Geiger, Fiftieth anniversary of the publication of his doctoral thesis, 23 July 1906", *Science* **124**, 166, july 27, 1956.
66. Haxel, O., "Hans Geiger als Wissenschaftler und Lehrer", in *Detectors in Heavy Ion Reactions, Symposium Commemorating the 100th anniversary of Hans Geiger's birth*, edited by von Oertzen, W., pp. 1–9, Springer-Verlag, Berlin, 1982.
67. Geiger, H., *Strahlungs-, Temperatur- und Potentialmessungen in Entladungsröhren bei Starken Strömen*, Ph.D. thesis, Philosophical Faculty of the Friedrich-Aleanders-University, Erlangen, July 23, 1906.
68. Robinson, H. R., "Rutherford : Life and Work to the Year 1919, with personal Reminiscences of the Manchester Period", in *Rutherford at Manchester*, edited by Birks, J. B., pp. 53–86, Heywood and Co Ltd., London, 1962.
69. Russel, A. S., "Lord Rutherford at Manchester, 1907–19: a partial portrait", in *Rutherford at Manchester*, pp. 93–101.
70. Townsend, J. S., "The Conductivity produced in Gases by the Motion of Negatively charged ions", *Philosophical Magazine* **1**, 198–227, February 1901.
71. Townsend, J. S. and Kirkby, P. J., "Conductivity produced in Hydrogen and Carbonic Acid Gas by the Motion of Negatively Charged Ions", *Philosophical Magazine* **1**, 630–642, June 1901.
72. Townsend, J. S., "The Conductivity produced in Gases by the Aid of Ultra-Violet Light", *Philosophical Magazine* **3**, 557–576, June 1902, and **5**, 389–398, April 1903; "On Ionisation produced by the Motion of Positive and Negative Ions", *Philosophical Magazine* **6**, 358–361, September 1903.
73. Townsend, J. S., "The Genesis of Ions by the Motion of Positive Ions in a Gas and a Theory of the Sparking Potential", *Philosophical Magazine* **6**, 598–618, November 1903.

74. Kirkby, P. J., "On the Electrical Conductivities produced in Air by the Motion of Negative Ions", *Philosophical Magazine* **3**, 212–225, February 1902.

75. Rutherford, E. and Geiger, H., "A Method of Counting the Number of α-Particles from Radio-active Matter", *Memoirs of the Manchester Literary and Philosophical Society* **52**, 1–3, 1908.

76. Rutherford, E. and Geiger, H., "An Electrical Method of Counting the Number of α-Particles from Radio-active Substances", *Proceedings of the Royal Society, London* **A81**, 141–161, 1908.

77. Rutherford, E. and Geiger, H., "The Charge and Nature of the α-Particle", *Proceedings of the Royal Society, London* **A81**, 162–173, session of June 18, 1908.

78. Crookes, W., "Certain Properties of the Emanations of Radium", *Chemical News* **87**, 241, May 22, 1903.

79. Elster, J. and Geitel, H., "Über die durch radioaktive Emanation erregte szintillierende Phosphoreszenz der Sidot-Blende", *Physikalische Zeitschrift* **4**, 439–40, August 1, 1903.

80. Regener, E., "Über Zählung der α-Teilchen durch die Szintillation und die Grösse des elektrischen Elementarquantums", *Verhandlungen der deutschen physikalischen Gesellschaft zu Berlin* **10**, 78–83, session of February 7, 1908.

81. Geiger, H., "On the Scattering of the α Particles by Matter", *Proceedings of the Royal Society, London* **A81**, 174–177, session of June 18, 1908.

82. Geiger, H., "The Scattering of the α- Particles by Matter", *Proceedings of the Royal Society, London* **A83**, 492–504, session of February 17, 1910.

83. Needham, N. J. T. M. and Pagel, W. J. O. (editors), *Background to Modern Science. Ten lectures at Cambridge arranged by the History of Science Committee, 1936*, Cambridge University Press, Cambridge, 1938.

84. Rutherford, E., "Forty Years of Physics", in *Background to Modern Science*, pp. 47–74, at the University Press, Cambridge, 1938.

85. Rutherford, E., "Forty Years of Physics", in *Background of Modern Science*, p. 68.

86. Geiger, H. and Marsden, E., "On a Diffuse Reflection of the α-Particles", *Proceedings of the Royal Society, London* **A82**, 495–500, 1909.

87. Rayleigh, John William Strutt, B., *Theory of Sound*, Macmillan, London, 1877.

88. Rutherford, E., letter to B. Boltwood, dated March 18, 1912, in *Rutherford and Boltwood: letters on Radioactivity*, edited by Badash, L., p. 265, Yale University Press, New Haven, 1969.

89. Geiger, H., "Some Reminiscences of Rutherford during his time in Manchester", in *The Collected Papers of Lord Rutherford of Nelson*, edited by Chadwick, J., Vol. 2, pp. 295–298, George Allen and Unwin Ltd, London, 1963.

90. Rutherford, E., "The Scattering of α and β Rays and the Structure of the Atom", *Proceedings of the Manchester Literary and Philosophical Society* **IV, 55**, 18–20, 1911.

91. Rutherford, E., "The Scattering of α and β Particles by Matter and the Structure of the Atom", *Philosophical Magazine* **21**, 669–698, May 1911.

92. Earnshaw, S., "On the Nature of the Molecular Forces which Regulate the Constitution of the Luminiferous Ether", *Transactions of the Cambridge Philosophical Society* **7**, 97–112, 1842.

93. Rutherford, E., "The Origin of β and γ Rays from Radioactive Substances", *Philosophical Magazine* **24**, 453–462, October 1912.

94. Barkla, C. G., "Polarisation in Secondary Röntgen Radiation", *Proceedings of the Royal Society, London* **A77**, 247–255, session of February 8, 1906.

95. Barkla, C. G., "Secondary Röntgen Radiation", *Philosophical Magazine* **11**, 812–828, June 1906.

96. Barkla, C. G., "Secondary X-Rays and the Atomic Weight of Nickel", *Philosophical Magazine* **14**, 408–422, September 1907.

97. Barkla, C. G., "Note on X-Rays and Scattered X-Rays", *Philosophical Magazine* **15**, 288–296, February 1908.

98. Barkla, C. G. and Sadler, C. A., "The Absorption of Röntgen Rays", *Philosophical Magazine* **17**, 739–760, May 1909.

99. Barkla, C. G., "Typical Cases of Ionization by X-Rays", *Philosophical Magazine* **20**, 370–379, August 1910.
100. Barkla, C. G. and Ayres, T., "The Distribution of Secondary X-rays and the Electromagnetic Pulse Theory", *Philosophical Magazine* **21**, 270–278, February 1911.
101. Barkla, C. G., "The Spectra of Fluorescent Röntgen Radiation", *Philosophical Magazine* [6] **22**, 396–412, September 1911.
102. Gouy, L.-G., "Sur la réfraction des rayons X", *Comptes Rendus de l'Académie des Sciences* **122**, 1197–1198, session of May 2, 1896.
103. Haga, H. and Wind, C. H., "Die Beugung der Röntgenstralen", *Annalen der Physik, Leipzig* **10**, 305–312, 1903.
104. Walter, B. and Pohl, R., "Weitere Versuche über die Beugung der Röntgenstrahlen", *Annalen der Physik, Leipzig* **29**, 331–354, 1909.
105. von Laue, M., Friedrich, W. and Knipping, P., "Interferenz-Erscheinungen bei Röntgenstrahlen", *Sitzungsberichte der Königlich Bayerischen Akademie der Wissenschaften* pp. 303–322, session of June 8, 1912.
106. von Laue, M., "Eine quantative Prüfung der Theorie für die Interferenz-Erscheinungen bei Röntgenstrahlen", *Sitzungsberichte der Bayerishen Akademie der Wissenschaften* pp. 363–373, session of July 6, 1912.
107. Bragg, W. H. and Bragg, W. L., "The Reflection of X-rays by Crystals", *Proceedings of the Royal Society, London* **A88**, 428–438, session of April 7, 1913.
108. Bragg, W. H., "X-rays and Crystals", *Nature* **92**, 307, November 6, 1913.
109. Heilbron, J. L., *H. G. J. Moseley: the life and letters of an English physicist, 1887–1915*, University of California Press, 1974.
110. Moseley, H. G. J. and Darwin, C. G., "The Reflection of the X-rays", *Philosophical Magazine* **26**, 210–232, July 1913.
111. Moseley, H. G. J., "The High-Frequency Spectra of the Elements", *Philosophical Magazine* **26**, 1024–1034, December 1913.
112. Moseley, H. G. J., "The high frequency spectra of the elements. Part II", *Philosophical Magazine* **27**, 703–713, April 1914.
113. Rutherford, E., letter to B. Boltwood, dated September 14, 1915, in *Rutherford and Boltwood: letters on Radioactivity*, edited by Badash, L., p. 311.

Quantum Mechanics: The Unavoidable Path

Zwei Eimer sieht man ab und auf
In einem Brunnen steigen
Und schwebt der eine voll herauf,
Muß sich der andre neigen.
Sie wandern rastlos hin und her,
Und bringst du diesen an den Mund
Hängt jener in dem tiefsten Grund,
Nie können sie mit ihrer Gaben
In gleichem Augenblick dich laben.

Friedrich von Schiller

Two buckets can be seen
To rise and fall in turn,
And when one rises full,
The other always falls,
They carry on with no respite.
When one tops to your lips,
The other is rock bottom,
At one time your desire,
They cannot both fulfill.

Branching Off

In a lecture delivered on February 16, 1901, "on the molecular hypothesis," Jean Perrin explained[1]:

---

[1] See p. 52.

B. Fernandez and G. Ripka, *Unravelling the Mystery of the Atomic Nucleus:*
A Sixty Year Journey 1896 — 1956, DOI 10.1007/978-1-4614-4181-6_3,
© Springer Science+Business Media New York 2013

Let me make it clear what we mean when we say that it is into molecules that a body ends up being divided. Let us make a comparison. Assume that you see a distant white spot in the countryside and that you discern that the spot is in fact made up of separate similar but smaller white spots. The molecular hypothesis consists in saying that you are seeing a flock of sheep. It is hardly necessary to add that this does not mean that you consider the sheep to be indivisible objects. You simply mean that other means would be required to divide the sheep into their parts and that this would yield a very different picture.

Jean Perrin expresses here two important ideas. First, *the parts do not have the properties of the whole*. A water molecule is not the smallest possible quantity of water, because *the molecule is not water*. It cannot flow, it cannot freeze, it cannot evaporate. The substance we call water must be an aggregate of a very large number of molecules. Jean Perrin did not realize how right he was by stating that the division of the molecule into its parts would yield a very different picture. Indeed, to understand the internal structure of the atom, it became necessary to formulate quantum mechanics.

As soon as it became known that there was a nucleus in the center of the atom, it was clear that the known laws of mechanics and electromagnetism could not explain how electrons could form a stable structure around the nucleus. It could also not explain the identity of all atoms of a given kind. Indeed, if electrons are orbiting around the nucleus following Newton's classical mechanics, there should not be two identical atoms and each collision between atoms would modify the electron orbits.

To understand this, physicists began to wander along strange, steep, often dangerous and slippery paths, the directions of which were always dictated by experimental observations, no matter how odd they appeared. Each step forward arouse wonder, incredulity, and controversy. The outcome, namely, quantum mechanics, is probably the greatest evolution of physics in the twentieth century. Let us trace the course of this evolution, starting with the discovery of the quantum by Planck in 1900 and ending with the formulation of quantum mechanics by Dirac in 1930. Physicists will then possess the key to the understanding of atomic spectra and the structure of the atom. Nor can the structure of atomic nuclei be understood without quantum mechanics.

An Improbable Beginning

A problem concerning the observed black-body radiation causes a crack in an imposing theoretical construction. Max Planck, a conservative theorist, saves the theory, but at the price of a revolution, by postulating a quantum of action. The young Einstein spreads the revolution by postulating quanta of light, the existence of which few believed in. The effectiveness of the new ideas embarrasses physicists because the new hybrid theory lacks consistency.

The Peak of Classical Mechanics

The nineteenth century witnessed the triumph of Newtonian mechanics in numerous domains. The power of the theory was demonstrated in 1846 by Le Verrier who observed irregularities in the orbit of Uranus and predicted, to within 1°, the position of a hitherto unknown planet, Neptune [1, 2]. It is hardly surprising that physicists sought in Newtonian mechanics the *ultima ratio* which could unify all physics. Even Maxwell started there when he attempted to understand the laws of electromagnetism.

Newton thought that light was composed of "grains" which followed straight line paths. Such ideas were, however, abandoned when it was realized that light could be refracted and diffracted, causing interference, thereby displaying its wavelike nature. Maxwell showed that every perturbation of the electric or magnetic field causes the emission of electromagnetic waves which propagate at the speed of light. Light was then identified to electromagnetic waves which, as Heinrich Hertz showed, could be refracted, diffracted, and cause interference, as light does.

One disturbing feature of the wavelike theory of light was that it did not specify *what* vibrated. A vibrating violin string is a concrete mechanical phenomenon. By vibrating, the string displaces the molecules of the neighboring air, thereby provoking periodic variations of the local air pressure at the rate of 440 vibrations per second (for the open *A*-string) and the vibrations are transmitted to our ear. But when luminous or electromagnetic waves propagate, what exactly is vibrating? To answer this question, ether was invented. Ether is an indefinitely elastic, omnipresent, and massless medium which permeates everything, even the vacuum. Ether allowed one to imagine that the energy transported by an electromagnetic wave was a kind of mechanical energy, corresponding to an elastic deformation of the ether. Ether was a convenient although admittedly strange substance.

A Persistent Problem

The nineteenth century also ended with the apotheosis of classical thermodynamics, the final formulation of which was made by Clausius. The theory was based on two general principles and it explained practically all phenomena involving heat—all, except one: black-body radiation$^\diamond$. This is the radiation emanating from an ideal body which could absorb all incident radiation and which would therefore appear to be matt black. When such a body is heated, it emits a radiation which, at a certain temperature, becomes a visible radiation, as wires do in electric bulbs. The phenomenon was known for a long time. Newton concluded his treatise *Opticks* by a series of 31 questions, among which:

> Qu. 8. Do not all fix'd Bodies, when heated beyond a certain degree, emit Light and shine; and is not this Emission perform'd by the vibrating motions of their parts [3]?

The fact that the energy of the radiation increases with temperature was well known. But at what rate? In 1879, the first quantitative estimate was made by the Austrian physicist Josef Stefan, based on the available experimental data: the radiated energy seemed to grow as the fourth power of the temperature. Five years later, Boltzmann succeeded in proving this result using thermodynamics and Maxwell's electromagnetic theory. The law of Stefan concerns the total radiated energy. But *how is the radiated energy distributed among various wavelengths*? It was known that the average wavelength decreases as the temperature of the radiating body increases. As the temperature progressively rises, the spectrum starts with long wavelength infrared waves, which are invisible to the eye. Then it becomes dark red radiation at about 500–600 °C (the temperature of the resistors of our electric stoves), and finally, it becomes a whiter radiation such as that emitted, for example, by the wire inside a light bulb at about 2,600 °C. This radiation is richer in red and poorer in blue than the light emitted by the surface of the sun, which is at about 5,600 °C.[1] One may then ask whether the laws of thermodynamics are able to predict, at each temperature, how the radiation is distributed among various wavelengths. In 1893, the German physicist Wilhelm Wien finds a first answer: using the laws of thermodynamics, he shows that the curve describing the distribution of wavelengths always has the same form; it is simply shifted towards smaller wavelengths as the temperature increases.[2] This is known as Wien's law. Wien was born in 1864 in Eastern Prussia, into a family of rich landowners. His inclination towards physics led him to attend university in Göttingen, and later in Berlin, where he worked with Helmholtz. He completed his PhD thesis in 1886. In 1896, he became physics professor in Aachen and 3 years later in Giessen. In 1900, he succeeded Röntgen and became physics professor in Würtzburg. In

---

[1]Blue has a wavelength roughly two times smaller than red.

[2]He finds that the curve is a function of the product λT of the wavelength of the radiation multiplied by the temperature T.

1920, he became professor in Munich where he completed his career. He was awarded the Nobel Prize in physics in 1911 for his discoveries of the laws which govern the radiation of heated bodies.

However, Wien's law said nothing about the way the energy is distributed. It simply states what it is at one temperature, in terms of what it is at another temperature. A young physicist in Hanover, Friedrich Paschen, set out to measure it experimentally. His black body was a platinum wire, blackened by carbon, which gave a reasonable approximation to the ideal black body. To display his results, he devised a mathematical expression which fitted quite accurately the observed results. He showed his formula to Wien, who had obtained practically the same formula from a model, which described the thermodynamics of an ensemble of oscillators.[1] In Wien's model, the black body was represented by a closed container, inside which the radiation in is thermal equilibrium with the inner surface of the container. The radiation propagates inside the container, it is absorbed and reflected on its walls, which radiate in turn. Wien described the walls in terms of oscillating electric charges which act as emitting and absorbing little antennas. Following Boltzmann's kinetic theory of gases, he attributed to these oscillators the distribution in energy which molecules have in a gas [4]. Although the calculation of Wien relied more on a conjecture than on a rigorous proof, it did fit the available experimental observations, and physicists believed that this was the long sought law, even if it was not proved rigorously.

In the *Physikalisch-technische Reichsanstalt* of Berlin-Charlottenburg, probably the best equipped laboratory in the world, several young physicists develop new methods to measure the long wavelength radiation. Otto Lummer and Ernst Pringsheim measure the infrared radiation, down to wavelengths between 12 and 18 μm[2] at temperatures ranging from 300 to 1,650 K. Heinrich Rubens and Ferdinand Kurlbaum [5] measure wavelengths ranging from 30 to 60 μm at temperatures in the 300–1,500 K range. Those are real experimental feats. At 300 K, the wavelength of light ranges from 4 to 60 μm, but its energy is very difficult to measure because the temperature is close to room temperature.[3] At a temperature of 1500 K, the wavelengths are roughly 1–10 μm; only 1.25 % of the energy is in the 12–18 μm range and only 0.13 % in the 30–60 μm range.

The results of these measurements were in complete disagreement with Wien's law. The disagreement increased as the wavelength of the radiation increased. To make things worse, in June 1900, the English physicist Lord Rayleigh, while analyzing the results of the young James Jeans, noticed that the strict application of

---

[1]His exponential law was: $\rho(v, T) = \alpha v^3 e^{-\beta v/T}$, where v is the frequency of the radiation. The constants α and β are chosen so as to fit the experimental data and they are not given physical meaning.

[2]The longest wavelength of visible light is in the red and equal to about 0.8 μm. The *micrometer* (μ)m is equal to one thousandth of a millimeter.

[3]A temperature of 300 K corresponds to $300 - 273 = 27\,°C$, which is the temperature of a hot summer day.

the thermodynamics of oscillators coupled to an electromagnetic field did not lead to Wien's law but to a quite different expression [6]. For large wavelengths, the formula derived by Rayleigh gave good results, in contrast to Wien's law. However, it failed at short wavelengths where Wien's law succeeded.[1] Rayleigh's formula leads in fact to an absurd result: it predicts that the energy emitted increases indefinitely when the wavelength decreases (or, equivalently, when the frequency increases). Today, we call this an ultraviolet catastrophe. How can one resolve these difficulties?

1900: Max Planck Invents the Quantum of the Action

Max Planck, a close friend of Wilhelm Wien, had been working for years on black body radiation. He is now 42 years old and director of the *Kaiser Wilhelm Institut*. He was born on April 23, 1858, into a bourgeois family of lawyers and protestant ministers from Kiel, Germany [7–9]. He completed his PhD in Munich in 1879, and in 1885, he became extraordinary professor in Kiel.[2] When Kirchhoff dies in 1887, the University of Berlin offers him the chair of theoretical physics professor. There, he meets Heinrich Rubens, who becomes his close friend. Planck, then a specialist of classical thermodynamics, distrusts the statistical thermodynamics of Boltzmann which is based on probabilities. His honesty is recognized by all and he has an ideal. What he likes most about thermodynamics is its ability to derive universal laws starting from very general principles. And yet, up to 1900, all his attempts to derive a universal law of black body radiation had failed. On Sunday, October 7, 1900, Rubens and his wife are invited by Planck for tea [10]. Rubens brings the results of his latest measurements of long wavelength radiation. He is impatient to show them to Planck, because they definitely contradict Wien's law. After the guests leave, Planck starts working. He has to find a formula which fits the long wavelength observations of Rubens, while fitting also Wien's law at short wavelengths. This is how the celebrated formula:

$$\rho(v, T) = \frac{8\pi h v^3}{c^3} \frac{1}{e^{\frac{hv}{kT}} - 1}$$

was born, rather contrived with no theoretical foundation. The formula yields the radiated energy with a frequency v at the absolute temperature T. In the formula, c is the speed of light k and h are two constants which Planck determined so as to fit the observed radiation. Once the constants are determined, the formula describes the radiation observed at different wavelengths with astounding accuracy. He immediately writes a letter to inform Rubens, who is about to present his

---

[1] The formula of Raleigh is $\rho = \frac{8\pi v^2}{c^3} kT$, where v is the frequency, k Boltzmann's constant, and c the speed of light.

[2] Meaning that he did not hold a chair.

results to the Prussian Academy on October 19. Rubens shows there that Planck's formula explains all the experimental data available at the time, within experimental error [5]. After the Rubens' presentation, a lively discussion takes place, during which Planck presents his formula as an improvement of the Wien formula, but without a formal derivation [11].

Planck is sure he has the right formula. But how can it be justified? In his Nobel lecture, delivered 20 years later, he explained:

If the radiation formula should prove to be absolutely accurate, it would still only have, within the significance of a happily chosen interpolation formula, a strictly limited value. For this reason, I busied myself, from then on, that is, from the day of its establishment, with the task of elucidating a true physical character for the formula, and this problem led me automatically to a consideration of the connection between entropy and probability, that is, Boltzmann's trend of ideas; after some weeks of the most strenuous work of my life, light came into the darkness, and a new undreamed of perspective opened up before me [12].

In order to attain his goal, the so far conservative and cautious Planck is ready to abandon all except, of course, the two basic principles of thermodynamics. He had rejected the statistical mechanics of Boltzmann, but now, he adopts it. John Heilbron calls it a real capitulation [9]. Later, in a letter sent to Robert Wood, Planck admits:

Briefly summarized, what I did can be described simply as an act of desperation. By nature I am peacefully inclined and I reject all doubtful adventures. But by then I had been wrestling unsuccessfully for six years (since 1894) with the problem of equilibrium between radiation and matter and I knew that this problem was of fundamental importance to physics; I also knew the formula that expresses the energy distribution in normal spectra. A theoretical interpretation therefore had to be found at any cost, no matter how high [13].

Planck also used the model consisting of tiny oscillators. But in order to evaluate the probability that they emit a radiation of a given energy, he took a decisive step: instead of assuming that the energy is absorbed or emitted in a continuous fashion, he postulated that it does so in finite packets of energy, called *quanta*. The energy of one quantum of radiation is simply equal to its frequency multiplied by a universal constant h, which now bears the name of *Planck's constant*. Planck was not immediately aware of the implications of such a hypothesis, which became a building block of what was to become quantum mechanics. Planck understood that the other constant, namely, k, characterized the entropy which had been defined by Boltzmann. If the constant R of perfect gases[1] is divided by k, one obtains *Avogadro's number*! Planck actually estimated this number, which was not

---

[1] For one mole of a perfect gas, that is, for 22.4 l at a temperature of 20 °C at atmospheric pressure, the constant R, multiplied by the absolute temperature T (which is the usual measured temperature plus 273 °C), is equal to the pressure of the gas multiplied by its volume.

accurately known at the time, and he was overwhelmed with joy upon learning, a few years later, that Ernest Rutherford and Hans Geiger had obtained a similar value by counting the number of α-particles emitted by a radioactive substance.[1] In honor of Boltzmann, Planck proposed to call k the *Boltzmann constant*, the name it bears today [14, 15].

A Quantum of Action

This is how the quantum of action was born one day in December in 1900. What reality does this strange term designate, and what exactly is the *action*? It designates a somewhat abstract physical quantity. In classical mechanics, the action is a property of the trajectory of a moving body, such as a particle, a billiard ball, or a planet. More precisely, the action is the product of its mass times its velocity, summed (integrated) along its trajectory. The action has a remarkable feature, which was discovered by Maupertuis in the eighteenth century: when a particle travels from one point to another in space, following the classical (Newton's) laws of mechanics, the trajectory which it describes is the one for which the action is minimum. This is known as the *principle of least action*. Planck discovered that the action cannot acquire any value, but that only integer multiples of a fundamental (and very small) action, which he called a *quantum of action h*.

Einstein and Light Quanta

In Planck's theory, it is only the transfer of energy between matter and light which occurs in finite packages, called *quanta*. Light itself propagates through ether according to the laws of Maxwell, that is, as light waves. The success of Maxwell's equations was such that physicists were reluctant to consider even the slightest modification. Nonetheless, an unknown employee of the patent office of Bern, Switzerland, causes havoc. Born on March 14, 1879, the young Albert Einstein [16–18] is the son of an electrical engineer. In school, at the *Gymnasium* of Munich, he excels in mathematics but not in the other subjects. He resents the strict teaching style of his teachers. Business is not doing well in Germany and his parents move to Italy. Before joining his parents, he was supposed to finish school. But in fact, he leaves school without the *Abitur*, the final school diploma, and this prevents him from being admitted to University. He applies to the Zurich Polytechnic. His first application is refused, but he is admitted the following year, in 1896. He graduates in 1900 and seeks a job. Einstein begins to like physics and he applies to become assistant to several professors in Switzerland, Germany, and Holland, without success. After managing at first by giving private math lessons, he

---

[1]See p. 69.

finally obtains a job in 1902 as a technical expert at the patent office in Bern. For him, this is an ideal job. The salary is adequate and steady, the job quite interesting.

He devotes his free time to physics. He begins to work on his PhD and his interests widen. One of his favorite subjects is the statistical mechanics of Boltzmann, whom he admired all his life. On March 1905, 3 days before his 25th birthday, Einstein sends a paper to the *Annalen der Physik*, bearing the strange title "A heuristic approach to the production and transformation of light." Einstein is well aware that he will not be believed and that is why he cautiously presents his argument as "heuristic," claiming neither rigor nor justification. He invites the reader to weigh the consequences, as if it were a an intellectual game. He begins by recalling the success of the wave theory of light. Then he lets the cat out of the bag:

> It seems to me that the observations associated with blackbody radiation, fluorescence, the production of cathode rays by ultraviolet light, and other related phenomena connected with the emission or transformation of light are more readily understood if one assumes that the energy of light is discontinuously distributed in space. In accordance with the assumption to be considered here, the energy of a light ray spreading out from a point source is not continuously distributed over an increasing space but consists of a finite number of energy quanta which are localized at points in space, which move without dividing, and which can only be produced and absorbed as complete units.
>
> In the following I wish to present the line of thought and the facts which have led me to this point of view, hoping that this approach may be useful to some investigators in their research [19].

Then he goes on to show that the black body radiation behaves, as far as entropy is concerned, as a gas composed of quanta of light. He shows that this hypothesis provides a simple explanation of the photoelectric effect, as well as Stoke's law of luminescence. His model allows him to derive Planck's law of radiation without recourse to Planck's quanta. His results concerning the photoelectric effect account remarkably well for the experimental observations. Einstein assumes that a quantum of light communicates all its energy to an electron of the metal, which therefore acquires the energy of the photon minus the energy required to extract the electron from the metal. This explains why, below a certain energy (or light frequency), no electrons are emitted, whereas above this frequency, the energy of the emitted electrons is a linear function of the photon frequency. In spite of his strong arguments, his model is hard to swallow [20]. How can light, whose wavelike character is so well established, behave at the same time as if it were composed of small "particles"? The latter should follow Newton's laws of motion but how can then light display wavelike properties?

Physicists had mixed feelings concerning Einstein's hypothesis. Nobody was really convinced, certainly not Planck. Einstein was well aware of the problem. In 1909, he was invited to a meeting of the German Physical Society in Salzburg where he delivered a lecture on "The development of our conception of the nature and of the constitution of radiation" [21]. He expressed a wish that a theory could be formulated which would encompass both mechanics and light. He was visionary but somewhat optimistic. Twenty years were to pass before such a theory was constructed. In 1909, physicists did not believe in the quantum of light. Lorentz and Planck, to name a few, were very reluctant to accept the idea. Nine years later, Einstein wrote a letter to his friend Michele Besso, an engineer he met in Bern and with whom he corresponded all his life:

I have spent incalculable hours thinking about the question of quanta, naturally without making real progress. But I no longer have doubts concerning the reality of quanta in radiation, in spite of being alone with this conviction. It will remain so as long as we will not succeed in establishing a mathematical theory [22].

Nonetheless, even those who disliked the idea were forced to admit that, in the case of the photoelectric effect, all successive experiments gave results which agreed with Einstein's explanation. It also explained other data, which so far had not been explained, namely, Stoke's law of fluorescence and the ionization of gases by ultraviolet radiation. Robert Millikan, the physicist who became known for his precise measurement of the electric charge of the electron, made a systematic study of the photoelectric effect and, to his dismay, his results fitted exactly Einstein's theory. He wrote:

It was in 1905 that Einstein made the first coupling of photo effects and with any form of quantum theory, by bringing forward the bold, not to say reckless, hypothesis of an electro-magnetic light corpuscle of energy hν, which energy was transferred upon absorption to an electron [. . .].

It must be admitted that the present experiments constitute very much better justification for such an assertion than has heretofore been found, and if that equation be of general validity, then it must certainly be regarded a one of the most fundamental and far reaching of the equations of physics [. . .] Yet the semi-corpuscular theory by which Einstein arrived at his equation seems at present to be wholly untenable [23].

Millikan expresses clearly the thought shared by many. The *light quanta* caused much perplexity at the time. Millikan was awarded the Nobel Prize in 1923 for his work on the fundamental electric charge.

A few months after the publication of his 1905 paper on the photoelectric effect, Einstein writes another paper which is the foundation of the theory of relativity. From then on, his reputation follows a steady growth. In 1909, he obtains a (*extraordinarius*) chair in the University of Zurich, in 1911 a chair in the University of Prague, and finally a chair in Berlin. In 1914, he becomes member of the Prussian Academy of Science. This is where he will develop a theory of gravitation, called "the general theory of relativity."

The Specific Heat of Solids

With his "heuristic" hypothesis of light quanta, Einstein made the first application of Planck's theory to physical phenomena other than black body radiation. Towards the end of 1906, he goes further and applies this concept to a different field, namely, to the specific heat of solids. The specific heat of a body is the quantity of heat required to raise its temperature by $1°$. It measures the capacity of the body to absorb heat. In 1819, Pierre Louis Dulong and Alexis Thérèse Petit established an empirical law [24] which was accepted during 50 years. The law states that "the atoms of all simple bodies have exactly the same specific heat," about 6 cal per mole$^\diamond$. However, measurements, performed beginning 1870 at temperatures considerably lower than room temperature, indicated that the specific heat decreased at low temperature and would become zero at absolute zero temperature, that is, at $-273\,°C$. Boltzmann's theory could account for Dulong's law at normal temperature but not for the decrease of specific heat at low temperatures. This is where Einstein came in. In 1907, he argued that if heat could only be transferred to a body in indivisible *quanta*, that is, in small but finite amounts, and not in a continuous fashion, then the specific heat of the body would become zero at absolute zero temperature [25]. With this, the theory of quanta began to acquire some truth and it meant more than the simple interpolation formula which Planck suggested for black body radiation. It appeared to explain quite unrelated phenomena. Was it a reality which appeared at the atomic scale?

The First Solvay Council and the Theory of Quanta

We are now in Brussels. Ernest Solvay earned a fortune during the second industrial revolution, in the 1870s. He invented a new method to produce soda ash (anhydrous sodium carbonate) and he acquired a good part of the European and even of the world market. He was a self-taught person and his health prevented him from attending university. He became a chemist but he had far greater scientific, political, and social ambitions. He believed that matter and energy are one and the same thing and that "gravity" could explain everything [26, 27]. Politically, he considered himself progressive and to belong to a modern left wing. He was an adept of scientism and he believed that science could appease social tensions. In order to discuss his own ideas on physics, Solvay decided to bring together some of the great scientists of his time. Through Robert Goldschmidt, professor at the *Université Libre de Bruxelles*, he made contact with Walther Nernst, the great German chemist and physicist. The latter wanted to gather a *scientific council* in order to discuss what he considered a crucial question, namely, the theory of quanta, which had been invented by Planck and brilliantly applied by Einstein to explain the photoelectric effect, but which appeared to contradict the foundations of physics. The meeting, which was called the Solvay Council, was held in Brussels from October 29 to

November 4, 1911. The ideas of Solvay were not discussed but he was heartily thanked for his hospitality. Hendrik Lorentz presided the Council devoted to "*Some current questions concerning molecular and kinetic theories.*" Concerning Lorentz, Maurice de Broglie recalled in 1951:

> At the time of the Solvay Councils, he was one of the most respected masters of all physics, perfectly suited to preside not only because he was a recognized authority, but also because of his thorough knowledge of European languages and his incomparable erudition in all fields of physics. Those who were present at the Brussels meeting could only marvel at the clear and precise manner in which he could sum up most varied subjects, and also at the interventions he made concerning a wide spectrum of fields. He would address each one in his native language with both affability and precision. He was the ideal president of international meetings [28].

The first Solvay Council was a great success. Among the 23 participants, there were Hendrik Lorentz, Marie Curie, Henri Poincaré, Albert Einstein, Maurice de Broglie, Heike Kamerlingh-Onnes (specialist of low temperatures), Heinrich Rubens, Ernest Rutherford, Max Planck, Walther Nernst, Marcel Brillouin, Jean Perrin, James H. Jeans, Arnold Sommerfeld, Wilhelm Wien, and Paul Langevin. It was the élite of world physics, which, at the time, was European physics. The proceedings [29] were published in French by Paul Langevin and Maurice de Broglie under the title "*La théorie du rayonnement et les quanta.*".[1] Further Solvay Councils were held in 1913, 1921, 1924, 1927, and 1933; and after World War II. They played an important role in the development of physics during the first half of the twentieth century.

The theory of quanta, a hot subject, was passionately discussed. Everyone recognized how effective Planck's formula was, as well as Einstein's hypotheses. However, it was difficult to part with classical conceptions, and the coexistence of two different mechanics was unacceptable. In the discussion which followed his report on "The current status of the problem of specific heats," Einstein defines the problem:

> We all agree that the theory of quanta, in its present form, may be useful, but it is not a theory in the usual sense of the word, and it cannot be, at present, developed in a coherent manner. At the same time, it appears now to be well established that classical dynamics, formulated by the Lagrange and Hamilton equations, can no longer be considered as a sufficient theoretical framework for all the physical phenomena.

Poincaré sees two problems in the theory of quanta:

> What the new work seems to put in doubt, is not only the fundamental principles of mechanics, but also something which so far appeared to

---

[1] The theory of radiation and quanta.

be inseparable from a natural law. Will we still be able to express the laws in terms of differential equations? On the other hand, during our discussions, I was impressed by seeing a theory rely both on the principles of old mechanics and on hypotheses which contradict this mechanics; one must not forget that there exists no proposition which cannot easily be proved by introducing into the proof contradictory premises.

As many physicists, Planck to begin with, Nernst imagines and hopes that it will be possible to amend classical mechanics, for example, in the case of strong accelerations:

Perhaps it will be possible one day to replace the calculations performed in the theory of quanta, which is so successful, with a different conception, and thus return to continuous changes in energy in atomic oscillations; for example, by modifying pure mechanics for the extreme conditions which prevail in atomic motion.

Poincaré shares this hope:

Before admitting these discontinuities, which would lead us abandon the usual form of natural laws in terms of differential equations, it may be better to explore the path proposed by Nernst; this consists in assuming that the mass, instead of remaining constant, or of depending only on the velocity, as in electromagnetic theory, depends also on the acceleration, if the latter is large.

But when he returned to Paris, he changed his mind and he added a footnote to the proceedings:

When I returned to Paris I tried a few calculations in this direction; they lead to a negative result. The theory of quanta appears to be the only one able to explain the experimentally observed radiation, if one accepts the usual relation between the energy of the resonators and that of the ether, and if one assumes that the energy exchanges between the resonators can be caused by mechanical collisions of atoms or electrons.

Niels Bohr: The Quanta Are in the Atom

A young Danish physicist cuts the Gordian knot and solves the paradox of Rutherford's atom. He formulates a new mechanics based on the quanta of Planck and Einstein.

Niels Henrik Bohr was born on October 7, 1886, in Copenhagen [30–32]. His father, Christian Bohr, was a medical doctor and a physicist, who discovered the role played by carbon dioxide in the production of oxygen by hemoglobin. For this work, he was a nominee for the Nobel Prize in 1907 and 1908. Niels is a brilliant student in the University of Copenhagen, assiduous also in football, with his brother Harald, a year and a half younger and soon to become a great mathematician. On May 13, 1911, Niels defends his PhD thesis on "The electronic theory of metals." The Carlsberg foundation offers him a grant which can cover his expenses for a year abroad. In the footsteps of Rutherford who preceded him 16 years earlier, the young Niels chooses to go to the Cavendish laboratory which is directed by the famous J. J. Thomson, the 1906 Physics Nobel Prize winner. He arrives in Cambridge in September 1911 and is deceived to find that J. J. Thomson is not very interested in discussions. He performs a few experiments on the production of cathode rays, but deep inside, he is not an experimentalist, who needs to be a glass-blower and attentive to innumerable practical tasks. In November, Bohr visits Manchester and meets with Rutherford. He expresses his wish to work on radioactivity, for which Rutherford's laboratory in the best in the world. After a first attempt to dissuade him, in order to avoid the embarrassing situation of "abducting" a physicist from J. J. Thomson's lab, Rutherford finally accepts and Bohr comes to Manchester in January 1912.

At that time, Rutherford's model of the atom with a nucleus at its center is simply one model among others. Nobody mentions it, J. J. Thomson the least because he does not believe in it. Even Rutherford remains silent about it and he doesn't even mention it at the Solvay Council in the autumn of 1911. In the second edition of his book on radioactivity [33], in 1911, he barely mentions it. Why? Does he consider it unimportant? Does he worry about the instability of his atom, according to classical mechanics?

Bohr Introduces Quanta in the Theory of the Atom

The stumbling block of the classical model of the atom was well known. In the hydrogen atom, which was believed to consist of a nucleus and one electron, the latter would act as an *atomic vibrator*, that is, a miniature antenna, which would inevitably radiate, thereby losing energy, and eventually spiral down towards the nucleus. But another problem attracts Bohr's attention. The time required for the electron to describe an orbit around the nucleus is related, according to classical mechanics, to the radius of the orbit; the radius can take any value, and for each value, classical mechanics determines the angular velocity of the electron, just as the period of rotation of planets around the sun depends on their distance to the sun. However, the size of the atom appears to remain fixed, as well as its radiation frequencies. Later, in 1922, Bohr explained to a young German physicist, Werner Heisenberg, who entered the scene somewhat later, how he conceived his model[1]:

> By "stability" I mean that the same substances always have the same properties, that the same crystals recur, the same chemical compounds, etc. In other words, even after a host of changes due to external influences, an iron atom will always remain an iron atom, with exactly the same properties as before. This cannot be explained by the principles of classical mechanics, certainly not if the atom resembles a planetary system. Nature clearly has a tendency to produce certain forms—I use the word "forms" in the most general sense—and to recreate these forms even when they are disturbed or destroyed [34].

Bohr believes that one cannot ignore the theory of Planck and Einstein: a vibrator can only emit energy by indivisible packets, by *quanta* with energy $h\nu$. This is the key to the mystery, by means of which an atom acquires a specific size, and no other, so that all hydrogen atoms in the universe have *exactly* the same size and the same properties. The stability of matter must be secured by *Planck's constant*. On July 6, 1912, Bohr sends Rutherford a letter, later called the "Rutherford Memorandum." [35] It summarizes Bohr's thoughts on the subject. In this letter, a first draft of Bohr's ideas at that time, he analyzes Rutherford's model. He first recalls that since no static equilibrium can exist between the nucleus and the electrons, the latter must therefore rotate around the nucleus. Furthermore, no stable rotating ring exists, which contains more than seven electrons, which suggests the existence of concentric orbits.[2] In such a configuration, the interior orbits can have only little influence on the stability of the external orbits. Bohr suspects that "this might explain the periodic law of chemical properties of chemical elements (assuming that chemical properties depend only on the stability of the outermost orbit described by 'valence electrons')." The idea that the outer electrons play a privileged role in chemical properties of atoms had already been proposed by J. J Thomson. Atoms which have the same number of electrons in the outermost orbit might have similar chemical properties, as in the periodic table of Mendeleev. More on this later.

On July 24, 1912, Bohr returns to Denmark in order to perform an act unrelated to atomic physics: on the first of August, he marries Margrethe Nørlund, who is five years younger. He then obtains a position of assistant to Martin Knudsen, professor

---

[1]See p. 133.
[2]See p. 59.

at the University of Copenhagen and this interrupts his theoretical work for several months. In February, he discusses with the spectroscopist Hans Marius Hansen, who was returning from a visit to Göttingen, and who shows him the formula which Balmer had fitted to the optical rays of hydrogen. He later recalled: "As soon as I saw Balmer's formula, everything became clear to me." [36] He works frenetically because he begins to see the solution. On March 6, he finally sends a first draft of his paper to Rutherford, who raises some objections. He is skeptical about some of the radical ideas of Bohr and he even criticizes the structure of the paper:

March 20 1913

Dear Dr Bohr,

I have received your paper safely and read it with great interest, but I want to look over it again carefully when I have more leisure. Your ideas as to the mode of origin of spectrum and hydrogen are very ingenious and seem to work out well; but the mixture of Planck's ideas with the old mechanics make it very difficult to form a physical idea of what is the basis of it. There appears to me one grave difficulty in your hypothesis, which I have no doubt you fully realize, namely, how does an electron decide what frequency it is going to vibrate at when it passes from one stationary state to the other [37]?

Bohr later recalled:

I therefore felt that the only way to settle things was to go straight to Manchester and to have a thorough discussion with Rutherford. In spite of being as busy as ever, he displayed an angelic patience towards· me and, after several evening discussions during which he declared that he never imagined that I could be so obstinate, he agreed to maintain all my points, old and new, in the article [37].

The paper, dated April 5, 1913, appeared in the July issue of the *Philosophical Magazine* under the title "On the constitution of Atoms and Molecules" [38].

"On the Constitution of Atoms and Molecules"

The paper of Bohr is exceptional in many ways. His pragmatic and rigorous reasoning reflects a profound understanding of physics. The audacity of his imagination makes his paper a real classic, which any student of epistemology or of physics should study [39]. He begins by explaining the differences between the models of Thomson and Rutherford. From then on, he discusses exclusively the model of Rutherford. He considers the simplest atom, the hydrogen atom, in which a single electron is in an orbit around the nucleus. According to classical mechanics, the orbit would progressively shrink as the electron radiates. But Bohr points out an obvious fact:

A simple calculation shows that the behavior of such a system will be very different from that of an atomic system occurring in nature. [. . .]

*the actual atoms in their permanent state seem to have absolutely
fixed dimensions and frequencies"*.

He then appeals to the quanta of radiation:

*Now the essential point in Planck's theory of radiation is that the
energy radiated from an atomic system does not take place in the
continuous way, as assumed in the ordinary electrodynamics, but that
it, on the contrary, takes place in distinctly separated emissions.*

He then invites the reader to imagine how an electron, which at first is situated far
from the nucleus, becomes progressively bound to the nucleus as it radiates light.
Since it can only emit light by finite amounts, namely, *quanta*, it ends up being
blocked in the lowest energy orbit. He proceeds to calculate the energy of this orbit
and the energy loss suffered by the electron which gets trapped into this orbit. It
is the binding energy of the atom which the electron forms with the nucleus. He
somewhat miraculously finds the observed values for both the size and the binding
energy of the atom! He then states formally his assumptions:

(1) *That the dynamical equilibrium of the systems in the stationary
    states can be discussed by help of the ordinary mechanics, while
    the passing of the systems between different stationary states
    cannot be treated on that basis.*
(2) *That the latter process is followed by the emission of a homoge-
    neous radiation.*

By "homogeneous" he means "monochromatic," meaning an optical ray with a well-
defined energy and wavelength. Thus, he assumes that the electron can only describe
well-determined orbits around the nucleus. As long as it remains in the orbit, it obeys
the laws of classical mechanics, but it can only change its orbit by finite "jumps"
during which it emits a quantum of radiation. *This is the origin of spectral rays!*
The first hypothesis is easy to accept. But the second contradicts the known laws
of physics. He makes this hypothesis *because it is dictated by experimental data.*
He then applies this reasoning to the spectrum of hydrogen. Since only discrete
orbits are allowed, according to his model, he assigns each orbit an integer number
$n = 1, 2, 3, \ldots$ which is related to the radius of the orbit and therefore to the angular
velocity of the electron in this orbit. Bohr calculates the frequency of the light which
is emitted when the electron makes a transition from an orbit n to an orbit p and he
finds the enigmatic Balmer formula[1]:

$$\nu = R\left(\frac{1}{p^2} - \frac{1}{n^2}\right).$$

---

[1]See p. 56.

He even succeeds in calculating the so-called Rydberg constant R from the known values of the mass and the charge of the electron, and the constant of Planck:

$$R = \frac{2\pi^2 m e^4}{h^3},$$

where m and e represent, respectively, the electron mass and charge, and where h is the famous constant of Planck. His expression yields the value 3.1×10^{15} s^{-1} which is remarkably close to the observed value[1] 3.29×10^{15} s^{-1}.

It is this remarkable agreement which causes the success of Bohr's model. Until then, the efforts of spectroscopists, such as Rydberg, had concentrated on numerics and the constant R was an entirely empirical parameter. Now, Bohr uses physical constants, such as the mass and the charge of the electron, together with the recent Planck constant to obtain an explicit expression for R.

The revolutionary character and the strength of the arguments used by Bohr are reflected in the following story told by Max Jammer, who heard it from F. Tank [40]. During one of the meetings organized in Zurich by the University and the Institute of Technology, the paper of Bohr was presented and discussed. Max von Laue declared:

This is nonsense! Maxwell's equations apply in all circumstances, an electron in orbit must radiate!

Einstein stood up and declared:

Very remarkable! There must be something behind it. I do not believe that the derivation of the absolute value of the Rydberg constant is purely fortuitous.

This was the dilemma facing physicists. In the framework of what was known at the time, Bohr proposed indeed strange ideas, which were barely acceptable but which soon became unavoidable. Let us quote the last and most interesting remark in Bohr's paper:

The angular momentum of the electron around the nucleus in a stationary state of the system is equal to an entire multiple of a universal value, independent of the charge on the nucleus.

This universal constant is nothing but Planck's constant h divided by 2π. Bohr noticed that *the quantum of action is a quantum of angular momentum*. Recall that the action of a planet orbiting around the sun is the product of three numbers: its mass, its velocity, and the length of its orbit. Its angular momentum is the product of its mass, its velocity, and its radius (its distance from the sun). The length of its orbit is 2π times the radius of the orbit. In the case of a rotating body, the action

---

[1] With the presently known values of the mass and charge of the electron, the value of Rydberg's constant is 3.2899×10^{15} s^{-1}.

is thus equal to the angular momentum. *The quantification of the action is thus tantamount to the quantization of angular momentum.* Quantum physics began to dominate the physics of the atom. However, 20 years would pass before a coherent theory of the atom would be formulated.

Two Other Papers in Bohr's 1913 Trilogy

Two further papers were to appear. The second paper [41] extends the theory to atoms with a nucleus of arbitrary charge, which therefore has several electrons orbiting around the nucleus. Bohr tries to determine how the electrons can form concentric rings, as in the model of J. J. Thomson.

Then Bohr considers the *Pickering rays*, which had been observed in 1896 by the astronomer Edward Pickering in the star ζ *Puppis* [42]. The observed spectrum was remarkably similar to the spectrum of hydrogen. It had twice as many lines, one line out of two being almost identical to a spectral line of hydrogen. Furthermore, in 1912, the astronomer Alfred Fowler succeeded in observing the same rays in the electric discharges of a mixture of hydrogen and helium. Bohr explains these lines in terms of a "hydrogenoid" system, consisting of a helium nucleus (which has two units of positive charge) and a single electron. It appears to be an ionized helium atom. In spite of yielding spectral lines quite close to the observed ones, his calculation differs significantly from observations. The truth appears to be close, but it still manages to escape.

In his third article [43], Bohr attempts to explain, in terms of his model of the atom, how several atoms can become bound to form a molecule. He intuitively suspects the electrons to be responsible for binding several atoms. But the theory required to explain this has not been formulated yet. Bohr lays the cornerstone, which renders the rest possible.

1913–1923: Victories and Setbacks

A paradoxical theory is constructed, on shaky foundations. It relies on some know-how and a lot of physical intuition. At times it is spectacularly successful but occasionally it fails completely. The number of "quantum numbers" keeps increasing. Bohr explains the Mendeleev table but he continues to demand a real theory.

Skepticism, Enthusiasm and Adhesion

The physicists reacted to Bohr's theory with interest, enthusiasm, and circumspection [44]. Some, Sommerfeld and Paschen, for example, immediately adhered to it. Others, Max Planck and Max Born were more careful. Most of them were reluctant to accept such a radical change of classical physics. While visiting Göttingen, a high point in mathematics, Harald, the brother of Niels Bohr, a mathematician, wrote him a letter in the fall of 1913:

> People here are still exceedingly interested in your papers, but I have the impression that most of them—except Hilbert, however— and in particular, among the youngest, Born, Madelung, etc., do not dare to believe that they can be objectively right; they find the assumptions too "bold" and "fantastic". If the question of the hydrogen-helium spectrum could be definitively settled, it would have quite an overwhelming effect: all your opponents cling to the statement that, in their opinion, there is no ground whatsoever for believing that they are not hydrogen lines [45].

The hydrogen-helium spectrum, which Harald refers to, consists of the famous Pickering rays.[1] The astronomer Alfred Fowler claimed that Bohr's calculation differed systematically from the measured values. However, Bohr showed that this was due to an approximation which he had made in his first paper, where he assumed that the electron mass was negligible compared to the mass of the nucleus, which is 1836 times heavier. After correcting for his, his calculation agreed with experiment to within five digits [46]! As noted by Harald, this correction was most effective in making most physicists adhere to his theory, among whom Einstein.

---

[1]See p. 108.

Confirmation: The Experiment of Franck and Hertz

The basic assumption of Bohr, namely, the existence of "stationary" orbits with well-defined energy, was spectacularly confirmed in 1914. Ever since 1911, two young physicists in the University of Berlin, James Franck and Gustav Hertz (the nephew of Heinrich Hertz), had been bombarding various atoms with electrons of variable energy. Their apparatus was similar to a "triode", that is, a simple radio tube, which contained the gas to be studied. In 1914, they attempt to provoke an "inverse" photoelectric effect. The photoelectric effect consists in the ejection of electrons by metals exposed to light with a sufficiently short wavelength (ultra-violet light or X-rays). Einstein interpreted this phenomenon by claiming that an electron, bound to an atom, absorbs all the energy of a "quantum of light" and can thereby be liberated from the energy which binds it to the atom. After that, the electron still has a certain kinetic energy, which corresponds to a certain velocity. However, Franck and Hertz bombard mercury atoms with incident electrons, and they observe what happens when the velocity (or energy) of the electrons is varied. When an electron has exactly the energy required to excite a mercury atom from its ground state to its first excited state, at $4.9\,\mathrm{eV}$, it acquires a finite probability of being absorbed by the mercury atom. And indeed, Franck and Hertz observe a clear decrease of electrons at precisely this energy. Moreover, Franck and Herz observe the emission of light at precisely the wavelength which corresponds to the decay of the atom from its excited state to its ground state. This is convincing experimental evidence of the existence well-defined stationary electron orbits [47]. For "their discovery of the laws which govern the impact of electrons on an atom", they were awarded the Nobel Prize in 1925.

A Proliferation of Optical Lines: The Zeeman and Stark Effects

The atomic model of Bohr could explain the optical lines of the atoms of hydrogen and of ionized helium, which is like a hydrogen atom with a double electric charge. The electron can only describe orbits around the nucleus which are characterized by integer numbers $n = 1, 2, 3, \ldots$. Each orbit has a well-determined binding energy$^{\diamond}$. When the electron "jumps" from one orbit to another, it emits a quantum of light, the energy of which is equal to the difference between the energies of the two orbits. The integer n can be thought of as the *number of quanta* of the orbit. But progressively, experiments showed that it was more complicated. *There were far more optical lines than could be accounted for by Bohr's model.* Albert Michelson, who had measured the speed of light in 1887, made precision measurements of the optical lines of hydrogen and he realized that some of them at least were double lines [48, 49]. Furthermore, the young Dutch physicist Pieter Zeeman, using a spectroscope, observed in 1896 that the optical lines of lithium and sodium would split up into three very close lines when the atoms were immersed in a magnetic field and that the separation of the lines increased with the intensity of the magnetic field [50–53]. Hendrik Lorentz was able to interpret this Zeeman effect by assuming that the emission of light was due to the vibration of the electrons in the atoms. He could deduce a value of the charge to mass ratio e/m which agreed with the value found by other methods, in particular with the value found by J. J. Thomson. This also confirmed the idea that electrons were universal constituents of matter, with a well-defined charge and mass. A few months later, the Irish physicist Thomas Preston confirmed the observations of Zeeman and he also found that some lines would split up into groups of 4 or even six lines [54]. At first, this was called an "anomalous" Zeeman effect, but it was soon realized that it was in fact more frequent than the doubling of lines in the "normal" Zeeman effect [55]. Finally, in 1913, the German physicist Johannes Stark observed that when hydrogen atoms were immersed in a strong electric field, the optical lines would also split up into multiple lines, but in a manner different from the Zeeman effect [56]. This became known as the Stark effect.

Arnold Sommerfeld: Elliptic Orbits and New Quantum Numbers

These observations showed that Bohr's single quantum number n (*Quantenzahl* in German) was insufficient to define an atomic state and that each electron orbit was more like a constellation of neighboring orbits. Rather, the electron orbits appeared to form groups, each one being assigned the same quantum number n, which, however, was not sufficient to describe the variety of observed optical rays. The first to discern some order in this jungle was Arnold Sommerfeld. Born on December 5, 1868, in Königsberg, he studied there and obtained his PhD in 1891. From 1894 to 1896, he was assistant to the great mathematician Felix Klein in Göttingen, with whom he worked on the theory of the gyroscope. In 1897, he obtained a chair in mathematics at the *Bergakademie* of Clausthal and later in 1900 at the *Technische Hochschule* of Aachen. In 1906, he obtained a chair of theoretical physics in Munich, and in 1911, he began to study quanta, which became his principal field of study in 1915. Sommerfeld founded a genuine school of physics and many of his pupils played an important role: Alfred Landé, Peter Debye, and, after the First World War, Wilhelm Lenz, Adolf Kratzer, Gregor Wentzel, and Otto Laporte. Doubtlessly, his most famous pupils were Wolfgang Pauli and Werner Heisenberg. In 1951, he was killed in a car accident.

At first, Sommerfeld modifies the assumption of circular electron orbits, which Bohr had proposed for simplicity. Instead, he assumes that the electrons describe elliptic orbits, with the nucleus at one of their foci, as they would if they obeyed the laws of motion of classical mechanics (and as planets do when they describe orbits around the sun). He consequently modifies the quantum conditions upon which Bohr's model is based. He appeals to the laws of motion formulated by the Irish mathematician William Hamilton. This formulation allows Sommerfeld to generalize Bohr's quantification rule to the quantification of *each degree of freedom* of the electron orbiting around the nucleus. The "degrees of freedom" are the independent quantities in terms of which the motion is described. In the case of the hydrogen atom, Sommerfeld assumes that the orbit has three degrees of freedom, namely, its size, form, and orientation in space. He ascribes three quantum numbers (*Quantenzahl*) to each orbit:

1. The principal (or radial) quantum number n, which determines the size of the elliptic orbit, more precisely, the length of its large axis. This quantum number is similar to the one introduced by Bohr and which determines the radius of his assumed circular orbit.
2. The azimuthal quantum number n', which will later be called k, which determines the eccentricity of the elliptic orbit (related to the ratio of the principal axes). This quantum number determines the angular momentum of the electron $^\diamond$ in its orbit. Quite generally, Sommerfeld assumes that angular momentum is quantized in units of $h/2\pi$.
3. The latitude quantum number, which determines the orientation of the orbit in space, and which is also quantized.

Relativistic Corrections and the Fine Structure Constant

Sommerfeld calculates also relativistic corrections to the electron orbits. His calculation allows him to discover a further multiplication of the orbits, which occurs even in the absence of external electric and magnetic fields. When the theory of relativity is taken into account, the orbit of an electron is no longer exactly elliptic because its velocity does not remain constant as it orbits around the nucleus. This leads to a relativistic correction which, admittedly, is small, and which is due to the fact that the elliptic orbit does not exactly close upon itself. As a consequence, the orbit slowly rotates around the nucleus, instead of remaining fixed in space. The electron in fact precesses in a manner similar to the precession of the planet Mercury, the perihelia (the point of its orbit which lies closest to the sun) of which "advances" by 43 s of an arc each century. Sommerfeld shows that this relativistic correction slightly modifies the energy of the orbit by an amount which depends on the shape of the orbit, and therefore on the quantum number k. Thus, the orbits bearing a given quantum number n have, in fact, slightly different energies. This phenomenon is called the "fine structure" of the atom. The energy difference involves a universal and dimensionless physical constant formed by the electric charge e of the electron, Planck's constant h, and the speed of light c. The constant is $\alpha = e^2/\hbar c$ and it is called the *fine structure constant*.[1]

---

[1] Here $\hbar$ stands for $h/2\pi$. See below p. 128.

A Hoax!

The fine structure constant $\alpha = e^2/\hbar c$ gave rise to numerous speculations. It is a dimensionless quantity, which means that it is a pure number, equal to $e^2/\hbar c = 0.007298$. Because it is dimensionless, it does not depend on the system of units used to express e, $\hbar$, and c. Its value is very close to $1/137$ from which it differs by less than three thousandths of 1%. Why should it be so close to the inverse of an integer? Since integers had begun to invade physics, many speculated that a value so close to $1/137$ may well not be a coincidence. Such speculations prompted three young physicists, Guido Beck, Hans Bethe, and Wolfgang Riezler to write a paper, which they even succeeded in getting published in the very serious German journal *Naturwissenschaften*. In this paper, they claimed to have related the fine structure constant to the value of the absolute zero temperature, expressed in degrees Kelvin![1] Whoever claimed that physicists have neither imagination nor humor? Today's most precise measurements yield the value $\alpha = 0.007297353$, the inverse of which is 137.03599 with a possible error only in the last decimal place.

A Further Contribution of Einstein: The Interaction Between Radiation and Matter

In 1916, Einstein had just finished the most trying work in his career: he completed the formulation of his theory of gravity, called the General Theory of Relativity. His attention returns to the physics of quanta. On August 11, 1916, he writes to his friend Michele Besso:

> *A stroke of light hit me concerning the absorption and emission of radiation; you will be interested. It is a quite surprising consequence of Planck's formula, I would even say the consequence. The whole thing is completely quantal [57].*

The paper of Einstein is published at the beginning of 1917. It contains several innovations with far-reaching consequences. First, he states that when a quantum of light interacts with an atom, *it does so with all the attributes of a "real" particle*, with an energy and a momentum, at least a direction. Einstein succeeds in giving a new derivation of Planck's law of radiation, using only very general arguments. He views Planck's law *as a consequence of what might be called the "corpuscular" nature of radiation*. In Planck's original paper, radiation maintained its wavelike character and quanta appeared only in the exchange of energy between the radiation and the substance which absorbs or emits the radiation. In his 1905 work on *quanta of light*, Einstein already took a step forward by assuming that the energy of radiation could be concentrated in a small region of space, as in a particle, but he

---

[1] By purposely crazy reasoning, they "derived" the formula $T_0 = -(2/\alpha - 1)$ degrees.

did not attribute a momentum to this particle. The second innovation of Einstein is the introduction, in the case of the emission of a quantum of light by a molecule, of a probability for the event to happen during a given lapse of time. The emission is similar to a radioactive decay: it takes place at a time and in a direction to which one can only assign a probability.

The Stark Effect: A Victory of the Theory of Quanta

In the middle of the First World War, a young Russian physicist, Paul Epstein, working in Munich, and a German physicist, Karl Schwarzschild, discover independently a satisfactory explanation of the Stark effect, which is the increase in the number of optical lines when an atom in immersed in an electric field [58,59]. They appeal to the Bohr-Sommerfeld theory which is most successful: the electric field perturbs the electron orbits. The optical lines, which correspond to the emission of light by electrons which jump from one orbit to another, have thus slightly modified frequencies and therefore wavelengths. Their calculations agree with the observed effect.

In 1919, Sommerfeld publishes a treatise on atomic structure and optical lines [60]. The book has several successive editions and it becomes the "bible" of physicists. In the English edition, published in 1923, Sommerfeld writes:

> *The most beautiful and most instructive manifestation of the different elliptic orbits which belong to the same Balmer line is, however, given by Nature herself in the fine structure of space-time conditions as reflected in the fine structure of spectral lines[. . .]*
> *Therefore the observation of the fine structure unveils the whole mechanism of the intra-atomic motion until the motion of the perihelion of the elliptic orbits [61].*

Even the title of the book, *Atomic structure and spectral lines,* insists on the unique and extraordinary role played by the observed optical lines, the wavelengths of which had been painstakingly and most carefully measured by physicists since 1860, with continuous improvements and increasing precision. All of atomic physics, all which will lead to quantum mechanics, one of the deepest conceptual changes of the century, originates essentially from these observations, from the contemplation of these optical lines, endowed with rainbow colors!

There is, however, a phenomenon which remains unexplainable: in the Zeeman effect (the optical lines emitted by atoms immersed in a magnetic field), *more optical lines are observed than can be accounted for.* The situation is embarrassing. The theory explains so well phenomena which otherwise have no explanation, that one feels that the goal must be close. However, the goal keeps escaping....

The "Correspondence Principle"

In his three 1913 founding papers of atomic physics, Bohr had given deep thought to the connection between the mechanics of quanta and classical mechanics, which includes relativity. Classical mechanics applies with great precision at the usual (human) scale as well as at the larger astronomical scale. However, it ceases to describe nature correctly at the atomic scale, where Planck's quantum of action dominates. An ideal theory should be valid at all scales and it should encompass both the classical theory at large scales and the theory of quanta at the scale of the atom. The quantum of action should also be active at our scale, but its size is so small that no discontinuity is discernible during a transfer of energy at our scale. We experience a similar effect when we look at a photograph in a magazine. Seen from a normal distance, we recognize the shape of a man, a familiar object or a landscape. But if we examine the picture through a magnifying glass, or better with a microscope, all we see are blackish dots of variable size and we can no longer discern familiar shapes. Because our eye has limited resolution, from a normal distance it sees a drawing which passes continuously from black to grey and white. But it does not discern the underlying discontinuities. The key idea of Bohr is the following: since the classical theory gives precise results at large scales, and since we must be able to pass continuously from large to small scales, there must exist a correspondence between quantities involved in classical mechanics and those involved in the theory of quanta. Bohr postulates a principle: *there must exist a formal analogy between classical and quantum mechanics.* There must be a "correspondence principle" which could be used to predict some features of the theory of quanta. For the physicist, this is as much a guide as a constraint because to any feature of classical mechanics, there must exist a corresponding feature in the theory of quanta.

Bohr was invited to attend the third Solvay Council [62], the first to be held after the war. It took place in Brussels on April 1, 1921. But he did not attend because he was overworked and not in good health. It was his friend Paul Ehrenfest who delivered part of his report adding a discussion of the correspondence principle [63]. A precise formulation of the principle was published in 1923 in *Zeitschrift für Physik* [64]. It was deemed so important that an English translation was published as a supplement to the *Proceedings of the Cambridge Philosophical Society* [65]. The correspondence principle became a most useful tool to guide physicists in this unknown field. It served as a constraint which led to most efficient intuitions and finally to the formulation of quantum mechanics, which, in 1923, was soon to come. The use of the correspondence principle required both know-how and intuition, which Bohr possessed to a high degree. But it was still not a real physical theory which could be applied without further ado. Patience! A few years from now...

Kossel, Bohr and the Mendeleev Table

The theory of quanta was able to explain the structure of the hydrogen atom but not of heavier atoms. Do the electrons really form rings as Bohr had originally suggested? All calculations showed that this was not possible and that they most likely filled all the space surrounding the atom, thereby forming a kind of sphere around it. But how?

In 1916, a student of Sommerfeld, Walther Kossel, has an idea [66]. He considers the light elements, up to manganese. His reasoning is based on certain properties of the periodic table of elements: the atoms of rare gases (helium, neon, argon) are particularly stable and difficult to ionize, meaning that it is difficult to tear an electron out of them. They do not form chemical compounds. This suggests that the electrons of such atoms are on the surface of a sphere, which will shortly be called a "shell." When this shell contains a certain number of electrons, one should consider the shell as "closed" and it becomes particularly stable. It becomes difficult to extract an electron from a closed shell and also difficult to add one because the latter would have to be in another larger shell, further away and therefore less attracted to the nucleus. Kossel thus constructs a model of the atom in which the first shell contains two electrons as in the helium atom. The second shell contains eight electrons, and the corresponding atom, with a total of ten electrons, is the neon atom. The next shell has also eight electrons, and when it is filled, it forms the argon atom which has 18 electrons. The other elements have incompletely filled shells which makes it easier for them to lose an electron. Furthermore, the electrons in the outer incomplete shells are freer to form chemical bonds with neighboring atoms. Recall that J. J. Thomson's plum-pudding model[1] also has electrons forming concentric spheres.

Such a model had a lot in its favor except for one important difficulty: the problem of the rare earth atoms, which we call today the *lanthanides*. The term is constructed from the word *lanthanum*, which comes from the Greek *lanthanein* meaning "to be invisible." Only two such elements were known to Mendeleev when he conceived his classification: lanthanum and terbium$^\diamond$. The other rare earths were discovered later, mostly during the second half of the nineteenth century. Such elements have very similar but not identical chemical properties.[2] Where should one place them in the famous Mendeleev table? Many chemists thought that they should all fit into a single position but that would break the golden rule "one element in each position." Some also considered a three-dimensional table in order to offer more space to each position.

---

[1] See p. 59.

[2] There exist 15 elements which are called "rare earths" or lanthanides. Their atomic number range from 57 to 71: lanthanum, cerium, praseodymium, neodymium, promethium, samarium, europium, gadolinium, terbium, dysprosium, holmium, erbium, thulium, ytterbium, and lutetium. Promethium was the last to be identified in 1947.

Table 1 The successive electron shells in the rare gases, according to Bohr [67]

Gas	Atomic number (number of electrons)	Principal quantum number n
Helium (He)	2	1
Neon (Ne)	10	2
Argon	18	3
Krypton	36	4
Xenon	54	5
Niton (Rn)	86	6

Today, we call radon (Rn) what Bohr called niton

Bohr sent two papers to the journal *Nature*, one in February [67] and the other in September [68] 1921. Shortly afterwards, he wrote a popularized article, called "Our present knowledge of atoms," which he concluded with these words:

> We can say that it is now possible to explain qualitatively and quantitatively many individual facts and that the—partly regular and partly capricious—variations in the chemical properties as one goes from elements of lower to elements of higher atomic number, variations which are so beautifully expressed in the empirical so-called periodic system of the elements, are no longer such an incomprehensible secret as they were a few years ago [69].

How does Bohr understand the electron configuration in atoms? From a theoretical point of view, the problem seems hopeless. One has a rough understanding of the hydrogen atom which has just one electron orbiting around the nucleus, and even there, the Zeeman effect remains unexplained. It is already difficult to understand helium, which has a doubly charged nucleus of mass 4 and two electrons. How can one then understand atoms which have 10, 20, and 50 electrons? Kossel points to the way. The Mendeleev table is a precious guide. It suggests the existence of successive electron shells, the first containing two electrons, the second eight, the third eight, but that is where Kossel stops. Bohr continues and attempts to construct the whole table. He is guided by the existence of rare gases which are chemically inert, unable to combine with any other substance. Bohr assigns to each shell *a principal quantum number n*. In each successive shell, the principal quantum number n increases by one unit. This seems natural because successive shells have increasing radii and it is precisely the principal quantum number n which determines the size of the electron orbit (Table 1).

In the first paper [67], Bohr gives an overall interpretation of the manner in which the electrons are arranged, either in the same orbit or in successively larger orbits. How many electrons in each orbit? Why not put all the electrons in the lowest energy orbit, or orbits? The reasoning of Bohr is based first on an attentive, sharp, and scrupulous attention to the experimental data, that is, to the optical spectral lines (or X-ray lines). Bohr succeeds in constructing a coherent model of the electron configuration in the atom, of the way in which they form groups of orbits with one, two, three, and four quanta:

*[The correspondence principle] suggests that after the first two
electrons are bound in an orbit with one quantum, the following eight
are bound in orbits with two quanta, the following eighteen in orbits
with three quanta and the following thirty two in orbits with four
quanta [67].*

The Rare Earths

The thorough analysis of atomic spectra made by Sommerfeld and other physicists,
such as Alfred Landé, allows Bohr to discover an essential feature: in general,
successive electrons go into orbits with one, two, three, and four quanta.[1]

Bohr suggests that the filling of the orbits does not always occur in that order.
The examination of spectra and of the chemical properties of various elements
suggests that, after the filling of orbits with three quanta, a new phenomenon occurs:
the element, whose atomic number follows that of argon (18), is potassium (19).
However, there is evidence that the 19th electron goes into an orbit with four quanta,
even though the shell composed of orbits with three quanta is not yet filled. But
after adding a few electrons into the shell with four quanta, the shell with three
quanta continues to fill. This explains the properties of rare earth elements. They
have a similar electron configuration in the outermost shell, which endows them
with similar chemical properties, but they differ in the way some of their internal
shells are filled. To reach such a conclusion, Bohr relies on general arguments and on
a careful study of the optical spectra. There are no mathematical calculations in his
paper, as in most of his publications. Indeed, Bohr does not have a real and reliable
theory. The theory of quanta is still a hybrid theory, which uses both purely classical
concepts, such as Kepler's elliptic orbits, and quantum conditions which have, as
yet, no theoretical justification. This does not prevent Bohr from drawing the main
ideas upon which the future mathematical calculations will be based. Better than
anyone else, Bohr is able to read the book of Nature.

1918, 1921 and 1922: Three Nobel Prizes Attributed to Quanta

In 1918, the Nobel Prize was attributed to Max Planck "in recognition of the services
he rendered to the advancement of Physics by his discovery of energy quanta." After
long hesitations (the prize was almost awarded to him in 1908), the theory of quanta

---

[1] Recall that orbits with one, two, three... quanta are orbits whose *principal quantum number n* is
one, two, three.... The number n determines approximately the size of the orbit, and its quantum
number k the elongation (eccentricity) of the orbit. See p. 112.

was finally recognized. During the ceremony, the president of the Swedish Academy of Sciences, A. G. Ekstrand, made a speech which he concluded thus:

Planck's radiation theory is, in truth, the most significant lodestar for modern physical research, and it seems that it will be a long time before the treasures will be exhausted which have been unearthed as a result of Planck's genius.

In 1921, the Swedish Academy of Sciences attributed the Nobel Prize to Einstein "for the services rendered to theoretical physics, in particular for his discovery of the law governing the photo-electric effect." Einstein was awarded the Nobel Prize for his contribution to the theory of quanta, which was his most debated contribution, and not for this formulation of the special and general theory of relativity.

The 1922 Nobel Prize was attributed to Niels Bohr "for the services rendered in the investigation of the structure of atoms and of the radiation which they emit." In his speech, the president of the Swedish Academy of Sciences insisted on the importance of the correspondence principle[1]:

Bohr succeeded in overcoming the difficulties [of the theory of quanta] by introducing what is called a correspondence principle, which presents important perspectives. To some measure, the principle makes the new theory closer to the old one. With the correspondence principle, Bohr is able to determine, in the most important cases, the orbits of the electrons in atoms. The chemical properties of atoms depend on the outermost orbits and, on this basis, their chemical valence has been partially determined. We can have great hope for further development of this magnificent work.

Three Nobel prizes in three consecutive years! The physics of quanta is in the limelight, for a long time.

---

[1] See p. 115.

1925: Spin and the Pauli Principle

A young genius discovers a simple and general law which governs the distribution of electrons in the atom. He discovers the existence of a strange physical quantity which can only "take two values". Two young physicists show that this quantity it is nothing but the angular momentum of the electron, which behaves as a symmetric top. A strange top in fact, which appears to be spinning, when in fact no matter is rotating. It is a quantum top: the electron has "spin"!

Wolfgang Pauli

In the great *Encyclopædia of mathematical knowledge*, published by the Teubner Editions in 1921, the article which describes the theory of relativity [70] drew the attention of Einstein himself, who considered it to be one of the best presentations of the subject. The article was written by a very young man, who was barely 19 years old.

Wolfgang Pauli [71] was born on April 25, 1900. His father, Wolfgang Joseph Pascheles, from a Jewish family, was a specialist of the chemical physics of proteins. He settled in Vienna in 1892, became a Christian convert, and adopted the name of Pauli. He was a childhood friend of the great physicist Ernst Mach, physics professor in the University of Vienna, and godfather to the young Wolfgang, who was a child prodigy, mainly in mathematics. When Wolfgang graduated from school and passed his *Abitur*, he was already able to publish three papers on the general theory of relativity. Two appeared in the respectable journal *Physikalische Zeitschrift*, and the third in the prestigious *Proceedings of the German Physical Society* [72–74]. He naturally chose theoretical physics and decided to work with Arnold Sommerfeld, the great professor and master of the theory of quanta. Sommerfeld had been asked to write an article on relativity in the *Encyclopædia of Mathematics*, but he suggested that Pauli should write it instead. During all his life, Pauli admired Sommerfeld, for the quality of the students and collaborators he had assembled in Munich. It is there that Pauli will meet a student, a year younger than he and soon to become a lifelong friend, Werner Heisenberg. After defending his thesis, Pauli becomes assistant to Max Born in Göttingen.

Max Born

Max Born was born on December 11, 1882, in Breslau in Silesia.[1] His father specialized in anatomy and physiology in the University. His mother Margarete Kauffmann came from an industrial family in Silesia [75, 76]. The young Max studied in Breslau, where he was taught matrices by the mathematician Jakob Rosanes. He pursued his studies first in Zurich and then in Göttingen where he obtained his PhD in 1907. After that, he went to Cambridge and worked under the direction of J. J. Thomson. He finally returned to Breslau in 1908–1909. His rising reputation caused him to be invited to Göttingen after which he became assistant to Max Planck in Berlin in 1915. He met Einstein there and they became lifelong friends.

After the war, Born became assistant to Otto Stern in Frankfurt, and in 1921, he obtained a chair in Göttingen, at the same time as James Franck, a most talented experimentalist. Thanks to Max Born, Göttingen will become one of the most important theoretical physics laboratories and the birthplace of quantum mechanics. This is how Heisenberg described Göttingen at the time:

> Born [...] founded a school of theoretical physics in Göttingen. He held the normal courses of lectures, organized seminars, and soon succeeded in collecting a fairly large band of excellent younger physicists about him, with whom he tried to penetrate the unknown territory of the quantum theory. Göttingen was then one of the world's most important centers of modern physics. In the small university town the mathematical tradition had been carried on for more than a century by some of the most illustrious of names: Gauss, Riemann, Felix Klein and Hilbert all taught in Göttingen [77].

This was fertile soil for the elaboration of mathematical laws which govern atomic phenomena. With his experiments on electronic collisions, James Franck aroused the interest of young physicists in the strange behavior of radiating atoms. Born and his disciples attempted to discern the laws which governed the phenomena. But what Heisenberg noted most was the warm and stimulating atmosphere created by Max Born and which was probably the key to the forthcoming success of the team. There prevailed an intellectual atmosphere in which the behavior of electrons aroused more interest than political discussions:

> Born and his wife Hedwig took care scientifically and humanly of these physicists, most of whom were barely 25 years old. Born's house was always open for social gatherings with young people, and anyone who happened to meet these youngsters in the university canteen or on the ski-slopes of the Harz Mountains, may well have wondered how the academic staff succeeded in focusing their interest so exclusively on such a difficult and abstract science [78].

---

[1] Today, the town is in Poland and it is called Wrocław.

In 1922, Max Born invited Niels Bohr to give a series of lectures on the structure of atoms [79]. The lectures took place on June 12–22. The first three lectures were devoted to the theory of quanta and its application to the hydrogen atom. The other atoms were discussed in the following lectures. The event was later called the "Bohr festival." It became famous. In is on that occasion that Pauli met Bohr for the first time. He recalls:

Bohr came to me one day [. . .] and asked me whether I could come to him in Copenhagen for a year. He needed a collaborator for the editing of his works, which he wanted to publish in German. I was much surprised, and after considering for a little while I answered with that certainty of which only a young man is capable: "I hardly think that the scientific demands which you will make on me will cause me any difficulty, but the learning of a foreign tongue like Danish far exceeds my abilities." The result was a hearty burst of laughter from Bohr [. . .] and I went to Copenhagen in the fall of 1922, where both of my contentions were shown to be wrong [80].

Indeed, Pauli managed to learn some Danish, but notwithstanding his hard work, he was unable to elucidate the anomalous Zeeman effect:

A closer investigation of this problem left me with the feeling that it was even more unapproachable. A colleague who met me strolling rather aimlessly in the beautiful streets of Copenhagen said to me in a friendly manner, "You look very unhappy"; whereupon I answered fiercely, "How can anyone look happy when he is thinking about the anomalous Zeeman effect?" I could not find a satisfactory solution at that time.

After a year in Copenhagen, Pauli went to Hamburg where he became *Privatdozent* and later professor. He soon became known in all Europe for being outspoken, for his keen critical sense, for his often deadly humor, but also for his kindness and warmth. People would approach him as a judge, with both fear and hope, to discuss a new idea or a paper. Heisenberg would regularly consult him for his important papers. In Hamburg, Pauli became a friend of Otto Stern, who surprised everyone, that is, the atomic physicists, with a most puzzling experiment.

The Stern and Gerlach Experiment

Otto Stern was born on February 17, 1888. His family owned a rich cereal and flour mill business and he enjoyed a financial independence which allowed him to choose the laboratories where he wished to work. He first worked as a theoretician with Einstein, Max Born, Ehrenfest, and Max von Laue. He then attempted to measure directly the velocity of gas molecules. Theory had predicted the distribution of velocities in 1850, but no one had measured it directly. He devised a way to produce almost parallel jets of gas which became real "beams" of gas molecules. If a gas is made to penetrate through a small hole in an evacuated vessel, the gas molecules will propagate in straight lines as light rays do. Stern measured the velocities of the molecules and he found the expected values. But he then realized that he could use his apparatus to settle an important issue in quantum mechanics. According to the calculations of Sommerfeld, for example, the atoms of certain metals should behave as small magnets. Sommerfeld had calculated the magnetic moments of such atoms and he found that they were quantized in units of the *Bohr magneton*, a term coined by Pauli [81]. When a magnet is placed in a uniform magnetic field, it becomes oriented but it does not get displaced. This is what happens to the magnetized needle of a compass in the magnetic field of the earth. But if the magnetic field is inhomogeneous, that is, if it is has different magnitude in different points in space, it exerts a force on the magnet and displaces it. Stern therefore argued that if silver molecules were made to pass through the gap between the poles of a magnet, where the magnetic field is very inhomogeneous, they should be deviated. In the absence of the magnetic field, the molecules impinging on a photographic plate form a black mark. But what if they pass through an inhomogeneous magnetic field? Stern performed the experiment with one of his colleagues, Walther Gerlach. The result was spectacular! They observed two distinct black marks! It was difficult to understand this result. Einstein, together with Ehrenfest, published a paper, in which they made an unsuccessful attempt to explain the result in terms of existing theories:

> The difficulties which we encountered show that the two present attempts to explain the results obtained by Stern and Gerlach are both unsatisfactory [82].

Atoms appeared more and more to behave differently than material bodies do at our scale. A new theory was really called for.

The Compton Effect

Soon after, a paper by Arthur Compton appeared in *The Physical Review*. Born in 1892, Arthur Compton obtained his doctorate at Princeton in 1916. He spent 2 years as electrical engineer at Westinghouse, and in 1919, he spent 1 year at the Cavendish laboratory, under Rutherford. He returned in 1923 to the United States and became professor of physics at the University of Chicago. Compton had been studying for several years the interaction between X-rays and the electrons of atoms [83]. He discovered that, when X-rays pass through a gas, some are deviated at various angles while *changing their energy* (i.e., their frequency or wavelength). The only way to understand this was to assume that X-rays behave as particles, quite similar to the *light quanta* of Einstein.[1] A quantum of light which collides with an electron can transfer a certain velocity to the electron while losing some of

---

[1] See p. 96.

its energy; its trajectory is thus deviated and its frequency diminished. The change in energy is related to the change in frequency by Planck's law. Compton showed that the change in frequency did indeed correspond, for each angle of deviation, to the change in energy of a light quantum, considered as a particle, in a collision with an electron. This became a direct confirmation of the theory of Einstein,[1] who a year later, received the Nobel Prize for his theory of... *light quanta*. A new theory was really required. How could light behave both as a particle and a wave propagating through space?

A Strange Explanation of the Zeeman Effect

In spite of Bohr's magnificent feat, by which he explained Mendeleev's periodic table of elements using an empiric and intuitive mixture of the correspondence principle and symmetry considerations, the structure of the atom retained most of its mystery. Why did the electrons obediently fill successive orbits instead of all occupying the same orbit? Rydberg also noticed that they corresponded to the squares of the integers 1, 2, 3, 4 (i.e., to 1, 4, 9, 16) multiplied by 2, *that is*, 2, 8, 18, 32.... An incomprehensible numerology!

Among the physicists of the time, Alfred Landé was the one most likely to imagine empirical and practical models which could explain the spectra. Born in 1888 in Ebenfeld in Rhineland, into a cultured family, he loved music and studied the piano and composition until the age 18. He excelled in mathematics and physics and began his university studies without really knowing which subject to choose. He had a go at experimental physics but he soon discovered it was not his cup of tea. After a brief stay in Göttingen, he came to Munich in 1912 where he became fascinated by the lectures of Sommerfeld and by the group of students, which included Debye, Epstein, Ewald, and von Laue. He defended his thesis in 1914, a fortnight before the outset of the First World War.

After the war, Landé becomes *Privatdozent* in Frankfurt. He studies the spectrum of helium, which nobody could understand and, in 1920, he begins to study the "anomalous" Zeeman effect. He is quite familiar with the Bohr-Sommerfeld theory of the atom and he knows that it does not explain the "anomalous" Zeeman effect. The basic idea of the theory is that a single electron is involved, namely, the outermost one. Its rotation around the atom produces, as an electric current does, a magnetic field which is perpendicular to the plane of its orbit. Thus, a rotating electric charge, endowed with an angular momentum, behaves as a magnet, the north–south direction of which points in the direction of its rotation axis. When an atom is immersed in a magnetic field, its little internal magnet is perturbed, and this in turn modifies the binding energy of its orbit; the modification depends on the position of the orbit, so that the number of distinct spectral rays increases. The model

---

[1] However, the name of Einstein was not mentioned in Compton's paper.

predicts that each spectral ray splits up into three rays, when, in fact, experiment shows that it can split up into four or even five rays! At first, Landé imagines that the other electrons, which are orbiting closer to the nucleus, and which are not considered in the model, might acquire a collective angular momentum and produce a magnetic field. It was known how to calculate the magnetic moment of an electron which has a given angular momentum. In units of elementary angular momentum, which is Planck's constant h divided by 2π, the magnetic moment is obtained by multiplying the angular momentum by the so-called *Bohr magneton* [81], which is the smallest possible magnetic moment.

Landé discovers an amazing an incomprehensible thing: the anomalous Zeeman effect can be perfectly explained if one assumes that the internal electrons, which he calls the "core" electrons, have a *half-integer* angular momentum! This can explain the number of observed rays, but their splitting is too small. Landé then introduces a purely empirical factor g^1 with which spectral rays can all be explained. One can wonder at Landé's formula: it assumes that the "internal" or "core" electrons have a total angular momentum equal to a *half-integer* and the value of the magnetic moment is obtained by multiplying by a mysterious number, invented just for this purpose!

Pauli's Exclusion Principle

This is where Pauli comes in. How can one accept the fact that an assembly of electrons can acquire a *half-integer* angular momentum, which is half of the smallest possible angular momentum which a particle can acquire? No matter how we add up the integer angular momenta of the electrons, we never obtain a half-integer result! Pauli thinks that Landé cannot be right, but his formula gives excellent results, and it must therefore be close to the truth. Instead of attributing the half-integer angular momentum to the "core" electrons, Pauli thinks it is better to attribute *to the external electron* a *hitherto unknown property* which gives rise to a *fourth quantum number*. This assumption leads to Landé's successful formula, if one assumes that the new quantum number can have only two values [84].

Pauli then has an extraordinary idea, worthy of a genius: in the atom of Bohr and Sommerfeld, each electron orbit is labelled by a set of quantum numbers. If we ignore the original notation of Sommerfeld, which is more complicated but which leads to the same results, the quantum numbers of the electron orbits are:

- The principal (or radial) quantum number n which has integer values 1,2,3,... and which determines the size of the orbit.

---

[1]The factor g is still used today. It is no longer a mystery and it still called the *Landé factor*.

- The quantum number k which determines the shape of the orbit and which takes the values $k = 1, 2, \ldots, n$. When $n = 1$, it can only acquire the value $k = 1$; when $n = 2$, it can only acquire the values $k = 1, 2$, and so forth.
- The quantum number m which acquires integer values between $-k + 1$ and $k - 1$. This quantum number determines the orientation of the orbit. It is limited to the single value $m = 0$ when $k = 1$, and to three values ($m = -1, 0, +1$) when $k = 2$, and so forth.
- The additional quantum number m_2 which can take two values.

Pauli then postulates the following general principle:

Two electrons cannot have their four quantum numbers equal [85].

This explains why electrons occupy successive shells (orbits) of the atoms. It is as if the atom presented a certain number of boxes (orbits), each one defined by the set of four quantum numbers and differing by at least one quantum number. Each box can contain at most one electron. It follows that:

- For $n = 1$, there are only two orbits, because k and m acquire each the single value $k = 1$ and $m = 0$ and the additional quantum number m_2 can acquire two values.
- For $n = 2$, there are eight orbits.[1]
- For $n = 3$, there are 18 possible orbits.

This is how successive electron "shells" are generated. The first shell can contain two electrons, in which case it is filled and said to be *saturated*. The second shell can contain up to eight electrons, the third up to 18 electrons, and so forth. The construction of atoms becomes simple and obvious. The principle, postulated by Pauli is known as *Pauli's exclusion principle*.

Let us finally note an evolution in Pauli's wording: he no longer discusses electron orbits. He simply refers to electrons. Each electron is associated to *four quantum numbers*, one of which has no classical counterpart. In fact, he identifies the electron to this set of quantum numbers and this appeared to be quite abstract at the time.

The "Spin" of the Electron

Things now begin to evolve fast. As in a good detective story, this is a sign that we are reaching a conclusion. Pauli sent his paper in January 1925. In November of that year, a short paper is published in *Naturwissenschaften*, signed by two

---

[1]When $n = 2$, k can acquire the two values $k = 1, 2$; when $k = 1$, only the value $m = 0$ is allowed; when $k = 2$, the three values $m = -1$, $m = 0$ and $m = 1$ are allowed; all together, when $n = 2$, there are four sets of allowed values for k and m and for each set, two values of m_2, which amounts to a total of eight.

young Dutch physicists, George Uhlenbeck and Samuel Goudsmit. They propose a physical meaning to the two-valued quantum number, which was introduced by Pauli and which they call R:

For us, another path appears to be still open [...] The 4 quantum numbers attributed to each electron have lost their original meaning, which was linked to the representation of Landé. We therefore attribute 4 degrees of freedom to each electron, which is characterized by 4 quantum numbers. One can then, for example, give the following meaning to the quantum numbers:

- *n and k remain as before the principal and azimuthal quantum numbers of the electron orbit*
- *Now R is linked to a "intrinsic" rotation of the electron*

The other quantum numbers keep their old meaning [86].

The quantity R is "associated to an intrinsic rotation of the electron" and it will soon be called *spin*. The electron spins like a top. But this is where the classical analogy fails, because *nothing is actually rotating in the electron!* The idea that something must be turning in the electron leads to insurmountable difficulties: according to what was believed to be the size of the electron, it would imply that the external parts of the electron would have velocities higher than the velocity of light.

Spin is a quantity which behaves as an angular momentum, but it is not associated the rotation of a material object. It also has a strange value, namely, one half of the elementary angular momentum $h/2\pi$, which was soon to be denoted as $\hbar$. Thus, the electron has an intrinsic angular momentum equal to $\frac{1}{2}\hbar$.

It is understandable that this paper caused quite a stir among physicists. Was this a relevant interpretation of Pauli's double-valued quantum number? Bohr was due to come to Holland on December 11 to attend the celebration of the fiftieth anniversary of the PhD of Lorentz.[1] Ehrenfest, who was then professor at the University of Leyden [87], took this occasion to invite Bohr and Einstein to stay at his house in order to allow them to discuss quietly without being disturbed. Bohr would forge his convictions during discussions. He plans to stop in Hamburg, to discuss with Pauli, but he does not leave Copenhagen before December 9 and spends only a couple of hours there. He meets Pauli and Stern at the railway station. They are both hostile to the idea of a rotating electron.

Bohr arrives in Leyden.

In a letter he wrote later to Ralph de Laer Kronig, Bohr tells:

Einstein asked the very first moment I saw him what I believed about the spinning electron. Upon my question about the cause of the necessary mutual coupling between the spin axis and the orbital motion, he explained that this coupling was an immediate consequence of the theory of relativity. This remark acted as a complete revelation

---

[1] The following account is, for the most part, taken from the introduction, by Klaus Stolzenburg, to the Volume 5 of the Collected Works of Niels Bohr.

to me and I have never faltered in my conviction that we are at least at the end of our sorrows. As an apostle of this faith, I have since had quite a difficult time in trying to persuade Pauli and Heisenberg, who were so deep in the spell of the magical duality that they were most unwilling to greet any outway of the sort [88].

The "magical duality," which Bohr refers to, is obviously Pauli's theory of a double valued abstract quantity, with no classical equivalent. They were most unwilling to give up this abstract beauty in favor of an "intrinsic rotation," of this spin of the electron, which had only the appearance of being concrete and classical! On his way back, Bohr stopped in Göttingen, where he met and apparently convinced Heisenberg. He then stopped in Berlin, where Pauli purposely came from Hamburg to see him, without, however, succeeding in convincing him. Prompted by Ehrenfest, Uhlenbeck and Goudsmit published a somewhat more detailed paper in *Nature* [89]. The paper had been discussed at length with Bohr, who added a comment. The battle concerning the electron spin was won.

As so often in physics, it was incomprehensible experimental evidence which led physicists to imagine and formulate the most abstract theories. In this case, it was mostly the observed anomalous Zeeman effect which forced physicists to invent the spin of the electron. The unfolding of discoveries does not always follow a rational path. The spin of the electron had been proposed, a year earlier, by Ralph de Laer Kronig, a young American physicist, born in Germany. He had proposed the idea to Pauli and to Heisenberg, who both rejected it. He gave up the idea of publishing it, so that it was Uhlenbeck and Goudsmit to whom history usually attributes the discovery of the electron spin.

Quantum Mechanics

In the span of 5 years, a handful of physicists, de Broglie, Heisenberg, Born, Schrödinger, and Dirac, lay the foundations of quantum mechanics. It is such an astonishing theory that Einstein and Planck refuse to accept some of its consequences. It sets a limit to what man can observe in nature. But, when applied, the theory yields spectacular results.

Louis de Broglie

The next important step is taken in 1923 by a 31-year-old French physicist, Louis de Broglie. Born in 1892 in Paris, he began to study history, but later switched to physics. He obtained his B.Sc. in 1913. He then began to work with his older brother Maurice de Broglie, an renown X-ray specialist, who used his personal fortune to pursue physics, his passion. Maurice de Broglie had been one of the two secretaries of the first Solvay Council in 1911, the other being Paul Langevin.[1] Drafted during the 1914-1918 war, Louis de Broglie worked at the radio station on top of the Eiffel Tower, and, after the war, he returned to physics, mainly theoretical physics. In 1963, in the foreword to the second edition of his thesis, he recalls:

I can remember those short years where my thoughts, fed by innumerable books concerning the most varied subjects, kept returning to the serious problem of the double nature, granular and wavelike, of light, which Einstein had uncovered, some twenty years earlier, in his brilliant theory of light quanta. After long and solitary meditations, the idea suddenly occurred to me that one should generalize Einstein's 1905 discovery to all material particles, in particular to electrons. I then discovered the relations, which generalize those which applied to light quanta, and which establish the same relation between a material particle and an associated wave as that which Einstein had established between the electromagnetic wave and what today we call the photon [90].

The starting point of de Broglie is the irritating double nature of light which sometimes appears in the form of a wave, and another time in the form of a particle. He has a far reaching idea, which he reveals in two short notes [91, 92] submitted to the *Académie des Sciences,* on September 10 and 24, 1923. He first proposes

---

[1]See p. 100.

to associate to an "atom of light," which today we call a photon,[1] a wave which propagates in the same direction as the particle. *He then extends this idea to material particles, in particular to electrons*, which describe closed orbits as in the atom of Bohr and Rutherford. He shows that Bohr's condition for an orbit to be stable is identical to the condition required for the particle *to remain in phase with the associated wave*. The condition is similar to the one of a surfer who needs to move forwards at the same speed as the wave, to avoid sinking. The condition associates to every particle "a matter wave," the wavelength of which is obtained by dividing Planck's constant h by the momentum$^\diamond$ of the particle. For Louis de Broglie, the wave was a kind of "pilot wave" and he thought this could be a possible unification of all physics. A year later, he defended his thesis on this subject. His thesis advisor was no less than Paul Langevin, who sent a copy of the thesis to Einstein, asking his opinion. Einstein was interested and replied:

> *The work of Louis de Broglie left a strong impression on me. He has uncovered a part of the veil [94].*

In a letter to Lorentz, he writes:

> *A younger brother of the de Broglie whom we know has made an interesting attempt to interpret Bohr-Sommerfeld rule (thesis defended in Paris, 1924). I believe that it sheds the first feeble light on one of our worst enigmas in physics. I have also found a couple of things which speak in favor of his construction [95].*

The "de Broglie whom we know" is, of course, Maurice de Broglie. The thesis of Louis de Broglie was defended on November 25, 1924. The jury consisted of Jean Perrin, Élie Cartan, Charles Mauguin, and Paul Langevin. It was published in the *Annales de physique* [96] in 1925.

Two years later, two independent experiments confirmed the wavelike nature of electrons, by revealing diffraction patterns of low energy electrons passing through a crystal. They were the experiments of Clinton Davisson and Lester Germer, in the Bell laboratories in New York [97] on the one hand, and those of George Paget Thomson (the son of J. J. Thomson) in London [98, 99] on the other. Furthermore, the measured wavelength was, within the experimental accuracy, equal to the one predicted by de Broglie.

Matter waves were a reality and electrons could also appear as waves. The paradox of the double aspect, particle and wavelike, now extended to all matter! However, in his thesis, Louis de Broglie wrote:

> *One must admit that the real structure of luminous energy remains a mystery.*

---

[1] The word *photon* was proposed in 1926 by the American physicist Gilbert Lewis [93]. It was quickly adopted by physicists.

Heisenberg and Matrix Mechanics

At about the same time, a 23-year-old German physicist tackles the problem from a quite different angle. Werner Heisenberg was born on December 5, 1901, in Würzburg, Germany [100–102]. After studying in the university of Munich, where his father was a professor of Greek, he prepared his thesis under the direction of Arnold Sommerfeld. One of his fellow students was Wolfgang Pauli, who became his friend. In June 1922, Max Born invited Niels Bohr to deliver seven lectures in Göttingen on the structure of the atom. This was the "Bohr festival" which we mentioned in connection with Wolfgang Pauli.[1] Sommerfeld, who was also invited, took the young Heisenberg along with him. Heisenberg made a critical remark during one of the lectures of Bohr and Bohr invited him to join him in a walk on the Hainberg hill, to discuss physics. This was a distinctive feature of Niels Bohr: he found this young man, this student, interesting because he had been criticized by him, and he wanted to know more about him. For Heisenberg, this was a real shock:

This walk was to have profound repercussions on my scientific career, or perhaps it is more correct to say that my real scientific career only began that afternoon [103].

Bohr quickly distinguished whom he beheld and invited him to work with him in Copenhagen. However, Heisenberg had not yet completed his studies. During the 1922–1923 academic year, Sommerfeld was invited for a sabbatical in the university of Wisconsin, in Madison, USA. He suggested that Heisenberg should go to Göttingen and work with Max Born during that time. Born needed an assistant to replace Wolfgang Pauli, who had just left and who recommended... Werner Heisenberg. Later, Born described his first contact with Heisenberg as follows:

Sommerfeld advised him to accept my offer in order to breathe a different scientific atmosphere. When he arrived (it must have been October 1923) he looked like a simple peasant boy, with short, fair hair, clear bright eyes and a charming expression. He took his duties as an assistant more seriously than Pauli and was a great help to me. His incredible quickness and acuteness of apprehension enabled him to do a colossal amount of work without much effort; he finished his hydrodynamic thesis, worked on atomic problems partly alone, partly in collaboration with me and helped me to direct my research students [104].

Finally, Heisenberg had to finish his thesis. Sommerfeld had chosen for him a subject which was more classical than the physics of the atom: it consisted in calculating how a liquid passes from a laminar to a turbulent flow. It was a mathematically difficult calculation and Heisenberg succeeded in obtaining an approximate solution, which Sommerfeld praised highly when the thesis was defended on July 10, 1923, in Munich [105].

---

[1]See p. 123.

However, Wilhelm Wien, his professor of experimental physics was not happy with Heisenberg, possibly because he had somewhat neglected his lectures. He asked Heisenberg some difficult questions concerning experimental techniques and Heisenberg was unable to answer. Wien became furious and even expressed a wish to refuse the thesis. Sommerfeld was obviously most embarrassed, because he considered Heisenberg to be the best student he ever had. A compromise was found: the thesis of Heisenberg was accepted with the mention *rite*, that is, with minimum honors, instead of *cum laude, magna cum laude*, or *summa cum laude*. That was defamatory and could ruin his career. The following day, Heisenberg returned humbly and peevishly to Max Born and told him what happened. But after hearing what questions Wien has asked, Born decided to admit him as an assistant. That same year, Heisenberg obtained in Göttingen his *venia legendi*, which enabled him to teach in the University. Finally, at Easter 1924, his dream became true: he went to Copenhagen to work with Bohr. He spent a year there, thanks to a scholarship delivered by the Rockefeller foundation, and in July 1925, he returned to Göttingen as an assistant of Max Born.

For some time, Heisenberg had been wondering whether notions such as positions, velocities, and trajectories were actually relevant to electrons in atoms. He shared some doubts with Pauli, who did not even mention trajectories in his latest work on the exclusion principle. But what can replace such quantities which appear to be so obvious? Certainly, an electron must be at some point at a given time and it must move with a given velocity and describe a given trajectory. Heisenberg makes a thorough critique and banishes every unnecessary notion: what does one actually know about the electrons in the atom? The only things which are observed are the spectral rays which bear a mark on a photographic plate in a spectroscope. Heisenberg does admit, however, Bohr's hypothesis, according to which each spectral ray corresponds to the emission of light which is produced when an electron "jumps" from one stationary state to another. In Newton's mechanics, the simplest formulation consists in saying that a given force (which has an intensity and a direction, as a vector) acts on the electron and changes its velocity at a rate which we call its acceleration and which is proportional to the force.[1] Heisenberg proposes to represent every quantity which concerns the electron (its position, velocity, acceleration,...) not simply by a number, but by a set of numbers cast into a two-dimensional array: each line and each column corresponds to a stationary state and the number situated at the intersection of a given line and a given column represents, for example, the intensity of the spectral ray. But how can one formulate mechanics with such an array? What calculations can one perform with it? Progressively, Heisenberg discovers how to make mathematical operations, such as addition and multiplication of such arrays. He constructs an empirical *algebra*, consisting of rules stating how the arrays should be added, subtracted, and multiplied. It is in the course of a solitary stay on the Heligoland island, where

---

[1]That is the content of Newton's famous equation $f = ma$ where f is the force acting on the electron with a mass m. The electron undergoes an acceleration a.

Heisenberg was trying to placate a violent hay fever, that he succeeded in setting the basis of a new mechanics:

Apart from daily walks and long swims, there was nothing in Heligoland to distract me from my problem, and so I made much swifter progress than I would have done in Göttingen. A few days were enough to jettison all the mathematical ballast that invariably encumbers the beginning of such attempts, and to arrive at a simple formulation of my problem. Within a few days more, it had become clear to me what precisely had to take the place of the Bohr-Sommerfeld quantum conditions in an atomic physics working with none but observable magnitudes [106].

Heisenberg had a lightning intuition, but he had not completed his task. He wanted to check that his new mechanics is really coherent and that it satisfies the principle of energy conservation, failing which it would be meaningless. A few days later, he succeeds and sends a manuscript to Max Born, his boss, to seek his opinion. In a letter sent to Einstein on July 15, 1925, Born writes:

Heisenberg's latest paper, soon to be published, appears rather mystifying but is certainly true and profound [107].

Born quickly realizes that the arrays introduced by Heisenberg are nothing but *matrices*, which were well known to mathematicians at the time, but not to Heisenberg. Together with another assistant, Pascual Jordan, Born soon publishes an essential paper which completes the work of Heisenberg [108] and which forms a basis, with a firm mathematical foundation, of what will for some time be called *matrix mechanics* [109]. A few months later, the paper is followed by another one this time signed by Born, Heisenberg, and Jordan [110]. The aim is to formulate a new mechanics, based, as was Newton's mechanics, on some general principles, from which everything can be derived; in other words, a *quantum mechanics* which would be exempt from the contradictions inherent in the previously formulated mechanics of quanta. Bohr considered that it was a step in the right direction, but he noted that it had not yet led to any concrete calculation which would explain experimental results [111].

New Physics

The papers of Heisenberg, Born, and Jordan project us into a new era of contemporary physics: their formalism makes it difficult to describe quantum phenomena in terms of everyday language. This has been discussed in depth in the enlightening essay *Quantum philosophy: understanding and interpreting contemporary science* of Roland Omnès [112]. The new quantum mechanics is in fact quite similar to classical mechanics. One "simply" replaces ordinary numbers, representing positions, velocities..., by matrices, but the equations are similar. Einstein's reaction confirms this. In a letter to his friend Besso, he writes:

The most interesting recent contribution is the Heisenberg-Born-Jordan theory of quantum states. It is a devilish calculation which involves infinite determinants (matrices) instead of cartesian coordinates. It is ingenious and sufficiently protected, by its complexity, from being proven wrong [113].

He expresses a similar reaction in a letter sent to Lorentz, March 13, 1926:

I have studied Born-Heisenberg unrelentingly. In spite of the admiration I have for this work, I react instinctively against this conception [114].

Not that the formalism should *hide* anything. What shatters physicists, and others, is the emergence of a description of infinitely small systems which is quite incompatible with our common sense, with the intuition which we have developed, as Einstein says, from experience at our scale. Atoms and electrons cannot be described as small objects, such as sand grains, which are simply very small. Their nature appears to be radically different and irreducible to concepts derived from our everyday experience.

Pauli Applies the New Mechanics to the Spectrum of Hydrogen

In January 1926, Pauli submits to *Zeitschrift für Physik* a paper which contains the first calculation which uses the new quantum mechanics of Heisenberg, Born, and Jordan [115]. He derives the Balmer formula, which Bohr had calculated in 1913 using audacious intuition. This answers Bohr's initial objections. The theory gives correct results and has a solid theoretical foundation.

The Schrödinger Equation

Erwin Schrödinger was born on August 12, 1887, in Vienna into a well-to-do family [116]. After brilliant studies in school, he attends university in Vienna in 1906 and he defends his Ph.D. thesis in 1910. After several positions in Germany, he obtains in late 1921 a chair as professor in Zurich, where he replaces Max von Laue, who moved to Berlin. There he meets Hermann Weyl and Peter Debye. In 1921, he studies atomic physics, and he turns to quantum statistics in 1924. Felix Bloch, a young student at that time, remembers a conversation between Debye and Schrödinger, in the fall 1925:

Once at the end of a colloquium I heard Debye saying something like: "Schrödinger, you are not working right now on very important problems anyway. Why don't you tell us some time about that thesis of de Broglie, which seems to have attracted some attention" [117].

Indeed, a paper by Einstein [118] brought the thesis of Louis de Broglie to the attention of physicists in the University of Zürich.

On November 3, Schrödinger writes to Einstein:

A few days ago I read with the greatest interest the ingenious thesis of Louis de Broglie, which finally I got hold of [119].

The idea published by de Broglie acted as a revelation. The famous integers, "quantum numbers" measuring the "number of quanta" could thereby be naturally introduced without imposing, more or less arbitrarily, "quantum conditions". At the end of a colloquium which was devoted to de Broglie's thesis, and which probably took place on November 23, 1925, Felix Bloch recollected:

Schrödinger gave a beautifully clear account of how de Broglie associated a wave with a particle and how he could obtain the quantization rules of Niels Bohr and Sommerfeld by demanding that an integer number of waves should be fitted along a stationary orbit. When he had finished, Debye casually remarked that he thought this way of talking was rather childish. As a student of Sommerfeld he had learned that, to deal properly with waves, one had to have a wave equation [117].

Schrödinger spends the Christmas holidays at the villa Herwig, at Arosa, a resort in the Swiss Alps, near the Austrian border. He has invited an old girlfriend from Vienna, who has never been identified. It is there that he finds the wave equation he has been working on [120]. He returns to Zurich on January 9, 1926, and after a few days, he sends a paper to *Annalen der Physik*:

In this paper I wish to consider, first, the simple case of the hydrogen atom (non-relativistic and unperturbed), and show that the customary quantum conditions can be replaced by another postulate, in which the notion of "whole numbers", merely as such, is not introduced. Rather when integralness does appear, it arises in the same natural way as it does in the case of the node-numbers of a vibrating string. The new conception is capable of generalization, and strikes, I believe, very deeply at the true nature of the quantum rules.
[. . .]
Above all, I wish to mention that I was led to these deliberations in the first place by the suggestive papers of M. Louis de Broglie, and by reflecting over the space distribution of those "phase waves", of which he has shown that there is always a whole number, measured along the path, present on each period or quasi-period of the electron [121].

Indeed, integer numbers often appear in physics in connection with *resonant* phenomena, such as the vibration of a violin string. The tone of the violin results from a superposition of the basic frequency (equal to 440 vibrations per second for an open A string) and of "harmonics," which are integer multiples (880,1320,...) of the basic frequency and which correspond to vibrations of strings which have lengths equal to 1/2, 1/3,... of the full length of the string. A violin string cannot vibrate at an arbitrary frequency but only at frequencies which are integer multiples of its basic frequency. In a paper, published in two parts in *Annalen der Physik*, Schrödinger was inspired by such an analogy on order to find an equation which describes the propagation of the "matter waves."

In this paper, Schrödinger constructs an equation which determines the behavior of an electron on its orbit and which is inspired by the equation of a vibrating string. His equation leads to the quantum conditions of Bohr! The equation was to become the famous *Schrödinger equation*. It gives rise to the first appearance of what will soon be called the *wavefunction* and which he denotes by ψ (Einstein will never refer to it otherwise than "the ψ function"). In the case of a vibrating string, the wavefunction represents the distance from which a point on the string is displaced from its static equilibrium position. But what meaning does it have in an atom? This point remained mysterious! Nonetheless, it had an operational value: whenever it could be determined, it would yield the energies of the various "stationary states," and therefore the famous spectral rays. *The main point of the Schrödinger equation was that it allowed one to calculate the spectrum of hydrogen, without recourse to additional hypotheses.* The quantum numbers were a natural consequence of the equation.

Planck and Einstein reacted most favorably. Planck writes to Schrödinger on April 2, 1926:

> *Many thanks for the off-print. I am reading your paper as a curious child who would listen with the greatest attention to the solution of an enigma which had worried him for a long time, and I am delighted by the beauties which unfold under my eyes and which I must still study in detail in order to understand them completely [122].*

And on April 16, 1926, Einstein wrote to Schrödinger:

> *Mr. Planck presented to me your theory with great enthusiasm, and I have studied it very carefully [122].*

Einstein adds a marginal *post-scriptum*:

> *The basic idea of your work is worthy of a genius.*

The immediate success of Schrödinger's equation is also due to the fact that it leads to considerably simpler calculations than those of matrix mechanics. It also revived the hope, to physicists such as Planck and Einstein, to find a theory which would be closer to what they considered a proper theory should be. It appeared to stand on more solid ground. Schrödinger hoped that his equation would resolve all the contradictions of the theory of quanta, and that the concept of particles would be replaced by that of waves which would be material analogs of Maxwell's electromagnetic waves. He believed he could do without the famous "quantum jumps" which the electrons were claimed to perform in order to emit quanta of light.

In the December 1926 issue of *Physical Review*, Schrödinger published a summary of his original papers [123].

Heisenberg and Schrödinger, Two Sides of the Same Coin

Oddly, identical results were always obtained, whenever the same problem was solved using either Schrödinger's or Heisenberg's methods. Schrödinger soon discovers why. In 1926, he publishes a paper which begins thus:

Considering the extraordinary differences between the starting-points and the concepts of Heisenberg's quantum mechanics and the theory which has been designated "undulatory" or "physical" mechanics, and has lately been described here, it is very strange that these two new theories agree with one another with regard to the known facts, where they differ from the old quantum theory[. . .] That is really very remarkable, because starting-points, presentations, methods, and in fact the whole mathematical apparatus, seem fundamentally different[. . .]

[Heisenberg, Born and Jordan] themselves describe the theory as a "true theory of a discontinuum". On the other hand, wave mechanics shows just the reverse tendency; it is a step from classical point-mechanics towards a continuum-theory [124].

Schrödinger then announces what he discovered:

In what follows the very intimate inner connections between Heisenberg's quantum mechanics and my wave mechanics will be disclosed. From the formal mathematical standpoint, one might well speak of the identity of the two theories.

The loop is closed. The two theories are simply two representations of one same thing.

The Probabilistic Interpretation of Max Born and the End of Determinism

Schrödinger's equation involves a "wavefunction" ψ and nobody really knew what it meant. The equation made it possible to calculate, at least in simple cases, the successive stationary states and the corresponding optical rays of hydrogen, for example. The calculation is quite similar to that which an acoustic engineer would perform to calculate the sound waves in a box, with one important difference: whereas the amplitude of the sound wave represents a variation in the pressure of the air, Schrödinger's wavefunction ψ did not seem to represent any physical quantity.

This is where Max Born comes in with two successive papers, in which he studies, with the new wave mechanics, not the structure of the atom, but the scattering of an electrically charged particle, such as an electron, by a nucleus, assumed to be a static charge [125, 126]. His calculation makes use of an approximation, which consists in assuming that, before and after the collision, the electron propagates in a straight line trajectory and that its interaction with the nucleus, simply modifies the incident straight line trajectory. This approximation still often used today is called the *Born approximation*. Born explains his calculation and he makes the following far reaching statement:

> Should one wish to analyze this result from the point of view of the particle, only one interpretation is possible: [the wavefunction ψ] determines the probability that the electron [...] is deviated [to a given direction...] Schrödinger's quantum mechanics gives a well determined answer for the result of a collision, but the answer is not causal. It does not answer the question "what is the state of the electron after the collision?" but only "what is the probability that the collision yields a given state?" [125]

Upon correcting the proofs, Born adds a note which makes things more precise:

> A deeper thought shows that the probability is proportional to the square of the quantity ψ.

Probability! The word is uttered. Born notes that the only thing which we can determine, the only thing accessible to our knowledge, is the wavefunction which, in turn, allows us to calculate *the probability of one or another outcome* of the collision, such as, for example, that the electron should be deviated by a given angle.
Born adds:

> This raises the question of determinism. The wave mechanics does not yield a quantity which, in a given case, determines causally the result of a collision; but so far, experiment also fails to yield the slightest cue that there should exist some internal properties of the atom which would determine the result of a collision. Should we hope to discover such properties in the future... and to determine them in each case? Or should we believe that such an agreement between theory and experiment, that this inability of both the theory and the experiment to unveil conditions for a causal evolution, results from some pre-established harmony which is based on the inexistence of such conditions? As far as I am concerned, I would be inclined to think that, as far as the atom is concerned, determinism should be given up. This is however a philosophical question which physical arguments alone cannot resolve.

One should keep in mind that the initial argument of Heisenberg, repeated by Born and Jordan in order to obtain quantum mechanics, consisted in ignoring the very concept of a trajectory. One should therefore not be too surprised to see that notions, such as the precise position of a particle, do not occur in quantum mechanics. But, from a philosophical point of view, this is a great leap: Born claims to be willing to abandon determinism which is the basis of classical mechanics! One should, however, read carefully what is stated: *if one wishes to interpret the results from a particle point of view,* determinism must be given up. Only probabilities can be calculated. However, the time evolution of the wavefunction, and therefore of the probabilities, is perfectly deterministic.

The probabilistic interpretation was rapidly accepted by Bohr, as well as by Heisenberg and Pauli, who both made regular visits to Copenhagen. This is why it became known as the "Copenhagen interpretation" when in fact, it could equally have been called the "Göttingen interpretation". However, Schrödinger and Einstein fought against it, in spite of recognizing its operational value. They did not believe that a complete theory would only yield probabilities. The hot debate ended in favor of Copenhagen but it leaves some unsatisfied even to this day. Einstein considered quantum mechanics to be an "incomplete" theory and all his life he sought a deterministic theory, without success.

The Pauli Matrices

Both the Heisenberg and the Schrödinger formulations of quantum mechanics suffered from one drawback. The new mechanics said nothing about *spin*, this strange quantity which appeared in October 1925 and which in fact was an effect which could only be understood in terms of quantum mechanics since it was an angular momentum$^{\diamond}$ which not only did not correspond to the rotation of a particle but which acquired half-integer values. In 1927, Wolfgang Pauli proposed a formalism which could take into account the half-integer spin. In a paper sent to *Zeitschrift für Physik* in May, he uses Schrödinger's quantum mechanics and he adds the double-valued spin degree of freedom [127]. He achieves this by using two wavefunctions and the spin is represented, as in Heisenberg's formulation, by a matrix, which is a simple two-dimensional matrix with two rows and two columns. The three components of the spin angular momentum are represented by three matrices which have been called the *Pauli matrices* ever since. There is, however, more to come, soon.

Indistinguishable Particles: Bose-Einstein "Statistics"

Two electrons are considered to be identical particles, as well as two protons. They have the same mass, the same charge, and the same "spin." Nobody doubts that all the photons which have the same wavelength, or, equivalently, the same frequency, are identical. Let us observe a beam of protons emerging from an accelerator and suppose that the beam passes through a thin strip of organic matter, containing hydrogen. From time to time, a proton of the beam will interact, "collide" as physicists say, with a proton of the thin strip. In this case, two protons emerge on the opposite side of the strip and they can both be detected. However, which is which? In classical mechanics, the answer is trivial because, in principle at least, the trajectory of each proton can be traced as a function of time. In other words, each proton can be tagged and identified. This is how Boltzmann argued. But in quantum mechanics, *the very notion of a trajectory is absent.* All that is known, *all that can be known*, is that two protons existed before the collision and that two exist after. If the two protons are indistinguishable, the two possibilities will mix as waves do, the waves will interfere[◇] and the resulting observation will differ from the one predicted by classical mechanics. But how will the waves mix?

In June 1924, a practically unknown physicist from India, Satyendra Nath Bose [128], sends a letter to Einstein. He encloses a paper, which is written in English and which had been rejected by the *Philosophical Magazine*. In the letter, Bose asks Einstein whether he thinks that his paper deserves to be published and whether he would help him to get it published in *Zeitschrift für Physik*. His short paper bears the simple title "Planck's law and the hypothesis of light quanta." He proposes a new derivation of Planck's law, which is not based on the theory of electromagnetism. Could a new derivation, following that of Planck himself and the two of Einstein, be of any interest? Einstein at least thought so. He translated himself the paper into German and sent it to *Zeitschrift für Physik* [129] with the following note:

> Remark by the translator: the proof of Planck's law by Bose is, in my opinion, an important step forward. The method which is used also leads to the quantum theory of perfect gases, as I will show elsewhere [129].

To derive his law, Planck had to evaluate the number of waves for each frequency, at least for each frequency interval. He considered stationary electromagnetic waves in a cavity, just like acoustical waves in a room which resonate at certain frequencies. Bose starts by assuming the existence of light quanta and he considers, just as Einstein did, that the light quanta should behave as the molecules of a gas. He estimates the different ways which the molecules would be distributed to achieve a given temperature. This allows him to obtain Planck's law.

The "perfect gas" to which Einstein refers to is an ideal gas in which the distance travelled by a molecule before colliding with another is very large, so that the molecules rebound on the wall of the cavity without colliding with others. In this sense, the "photon gas" assumed by Bose is a perfect gas[◇] since he implicitly assumes that there are no interactions between the photons. The interesting point of Bose's derivation of a physical law, which had been famous for 24 years, is that Bose makes, with apparent innocence, a fundamental assumption, namely, that *light quanta are indistinguishable*. He counts the number of light quanta in a given stationary state. However, two configurations in which each stationary state contains the same number of quanta, are considered identical and are only counted once. He considers this to be obvious. He considers that it does not make sense to exchange two identical photons. He derives Planck's law, thereby validating his assumptions.

Einstein immediately perceives the consequences of applying Bose's method to a gas, especially at low temperature. He soon publishes three papers on the quantum theory of perfect gases [118, 130, 131], in which he predicts a special phenomenon, which today is still called "Bose–Einstein condensation." It occurs at very low temperature, at which, for example, "normal" liquid helium makes a phase transition to superfluid helium. The transition in helium was not observed before 1928 and it is Fritz London who suggested that it might be due to Bose–Einstein condensation. Not until 1995 was Bose–Einstein condensation directly and unambiguously observed by an American group in Colorado, directed by Eric Cornell and Carl Wieman [132]. "Bose–Einstein statistics" was born, according to which any number of indistinguishable particles could be in a given single-particle state. The state of a gas is determined by the number of indistinguishable particles which occupy each given single particle state.

Enrico Fermi: A New "Statistics"

At first, Einstein believed that electrons also obey Bose–Einstein statistics. A very young Italian physicist will realize that this is not the case. Enrico Fermi was born on September 21, 1901. His father was a train worker. Enrico had a sister, Maria, 2 years older, and a brother Giulio, a year older. He was brilliant in school and he displayed a special talent in mathematics as well as a prodigious memory. His best friend is his brother with whom he plays all sorts of games, such as constructing electrical motors. Unfortunately, Giulio dies suddenly after an operation. It is tragic for Enrico who looses his best friend. He then begins an active study of mathematics and physics. A friend of his father, Adolfo Amidei, a train engineer, lends him a few books which the young Fermi digests in a very short time. After the *liceo* (high school), he should have naturally continued to study in the University of Rome. Instead, following the advice of Amidei and with his parents, approval, he attempts on November 14, 1918, the entrance examination to the *Scuola Normale* of Pisa. He flabbergasts the physics professor who questions him. He spends the following 4 years in Pisa, where he obtains his PhD in 1922. By then, he has already published four papers on electromagnetism and relativity. In Italy, he is considered an expert on relativity.

He obtains a postdoc grant which enables him to go to Göttingen in 1923 where he meets Max Born. He then returns to Italy where he obtains a temporary position to work with Orso Mario Corbino, one of the best Italian physicists at the time. Corbino, born in 1876, is 25 years older than Fermi. He was an excellent physicist, as well as a senator. He regretted not to have succeeded in his scientific career because of the archaic structures which prevailed in the Italian universities. His dream was to favor the emergence of a new generation of physicists which would give Italy the same luster it had at the time of Galileo. It is on Fermi that he lays his hopes. Fermi then obtains a position in Florence where he finds his friend Franco Rasetti. They had met in Pisa, while Fermi was at the *Scuola Normale* and Rasetti was studying physics at the University. They remained close friends. Rasetti had a great talent as an experimentalist. He was also interested in literature, botanics, and alpinism [133]. Together, the two friends begin to make experiments. It is in

Florence that Fermi discovers the paper of Pauli on the exclusion principle. For some time, he is concerned with the theory of perfect gases, especially at low temperature to which Einstein applied the Bose–Einstein statistics. The latter, however, gave poor results, when applied to a gas of electrons, in a metal, for example, in which the electrons propagate freely, as molecules do in a gas. In a book written later, his wife Laura recalls:

> The precise law which such a gas obeys had baffled him for some time. Some factor that would bring full comprehension was missing, and he could not figure out what it was [134].

The exclusion principle was missing! He soon publishes a paper on "The quantization of a perfect monoatomic gas." In the introduction he announces:

> The aim of this work is to present a method to quantize a perfect gas which would be, in our opinion, as independent as possible of unjustified hypotheses concerning the statistical behavior of the gas molecules [135].

After mentioning Pauli's exclusion principle, he formulates his hypothesis:

> We now propose to see if a similar hypothesis couldn't give equally good results in the case of the quantization of a perfect gas: we will admit that, in our gas, there can be at most one molecule, whose motion is characterized by certain quantum numbers, and we will show that this hypothesis leads to a perfectly consistent quantification of perfect gases, and that in particular, it accounts for the exact lowering of specific heat predicted at low temperature, and that it leads to the correct value of the entropy constant of perfect gases.

The absolute zero temperature corresponds to the lowest energy state. In Bose–Einstein statistics, all the particles are in a zero energy state. But Fermi claims that in an electron gas, if Pauli's exclusion principle is applied, *there can only be one electron in this lowest energy state*, one electron in the next state, and so on. Thus, the lowest energy state for the system of electrons does not have zero energy! Where does such an electron gas occur? In metals whose electrical conduction properties are described in terms of freely moving conduction electrons which behave *as a gas*. Fermi explains thus the behavior of metals at very low temperature.

Paul Adrien Maurice Dirac

Meanwhile, a young English physicist was becoming known in the small community of physicists who were in the process of building the new quantum mechanics. Born on August 8, 1902 in Bristol [136, 137], from a Swiss father and an English mother, Paul Adrien Maurice Dirac studied in the university of Bristol, where he obtained a diploma for electrical engineering in 1921. It was during his studies, in 1919, just after the First World War, that he discovered and was most impressed by relativity:

At that time, a wonderful thing happened. Relativity burst upon the world, with a tremendous impact. Suddenly everyone was talking about relativity. The newspapers were full of it. The magazines also contained articles written by various people on relativity, not always for relativity but sometimes against it. [. . .] The impact of relativity involved simultaneously the special theory and the general theory. Now, the special theory was actually very much older, dating from 1905, but no one knew anything about it except a few specialists in the universities. The ordinary person had never heard of Einstein. Suddenly Einstein was on everyone's lips [. . .] [138].

Dirac had a passion for mathematics, especially geometry. He continued his studies in Bristol and in 1923 he began research at St. John's College, in Cambridge, under the direction of Ralph Fowler, no other than the son in law of Rutherford. It is there that Fowler, in the summer of 1925, sends him a copy of Heisenberg's first paper on the new quantum mechanics, with the question: "What do you think of this?" Dirac recalls what followed:

I received it either the end of August or the beginning of September, I am not sure of the date, and of course I read it. At first I was not much impressed by it. It seemed to me to be too complicated. I just did not see the main point of it, and in particular his derivation of quantum conditions seemed to me too far fetched, so I just put it aside, as being of no interest. However, a week or ten days later I returned to this paper of Heisenberg and studied it more closely. And I suddenly realized that it did provide the key to the whole solution of the difficulties which we were concerned with.

Within a few weeks, Dirac formulates in his own manner the matrix mechanics of Heisenberg, which, as seen above, replaces the numbers, representing the position and the momentum of an electron, by matrices which can be added, subtracted, and multiplied [139]. There is however one important difference: the product qp of the position matrix q by the momentum matrix p is not equal to the product pq of the momentum matrix p by the position matrix q. The difference is equal to the *imaginary number*:

$$pq - qp = \frac{h}{2\pi i},$$

where i represents the square root of -1.

Dirac was impressed by the similarity of this result with that of "Poisson brackets" in classical mechanics. He made the relation above the very foundation of his formulation of quantum mechanics. His formulation had the elegance of much of his later work and his formulation of quantum mechanics was simpler to use than that of Heisenberg [140]. Dirac was then 24 years old and had not yet obtained his Ph.D.

From then on, Dirac sent regular communications to the *Royal Society*. In January, a second paper presented a "preliminary" calculation of the hydrogen spectrum, using his formulation of quantum mechanics [141]. He introduced what he called q-numbers, which had the same properties as ordinary numbers, except for the fact that the products $x \times y$ and $y \times x$ are not the same when x and y are q-numbers. His q-numbers had properties similar to matrices. He listed the rules required to make calculations with the q-numbers as well as with ordinary numbers, which he called c-numbers.

In August 1926, Dirac shows how to treat systems composed of several electrons [142]. He uses the new equation of Schrödinger, in which the system is represented by a "wavefunction" ψ. Dirac discusses how to express mathematically the indistinguishability of the electrons. He claims that there are two possibilities. The wavefunction must be either symmetric or antisymmetric with respect to the exchange of the electron quantum numbers. This means that, when the quantum numbers, such as the position and the spin, of two electrons are exchanged, the wavefunction must either remain the same, in which case it is said to be *symmetric*, or simply change sign, in which case it is said to be *antisymmetric*. This has a far reaching consequence regarding the statistics of the particles. *A symmetric wavefunction corresponds to Bose-Einstein statistics*, whereas *an antisymmetric wavefunction corresponds to* what was soon to be called *Fermi-Dirac statistics*. Fermi and Dirac came to similar conclusions concerning the electron gas.

But Dirac had more in store. So far, the quantum mechanics of Heisenberg, Born and Jordan, as well as the quantum mechanics of Schrödinger, were quantum analogs of *nonrelativistic* mechanics. Several attempts to find a relativistic equation had failed. Schrödinger tried at first and failed. In 1928, Dirac publishes a paper called "The quantum theory of the electron." In the introduction, he explains:

> The new quantum mechanics, when applied to the problem of the structure of the atom with point-charge electrons, does not give results in agreement with experiment. The discrepancies consist of "duplexity" phenomena, the observed number of stationary states for an electron in an atom being twice the number given by the theory. To meet the difficulty, Goudsmit and Uhlenbeck have introduced the idea of an electron with a spin angular momentum of half a quantum and a magnetic moment of one Bohr magneton [...].
>
> The question remains as to why Nature should have chosen this particular model for the electron instead of being satisfied with the point-charge. One would like to find some incompleteness in the previous methods of applying quantum mechanics to the point-charge electron such that, when removed, the whole of the duplexity phenomena follows without arbitrary assumptions. In the present paper it is shown that this is the case, the incompleteness of the previous theories lying in their disagreement with relativity [143].

Walter Gordon [144] and Oskar Klein [145] had established a relativistic equation for the electron, but their equation had defects which Dirac discusses before proposing his own solution. The equation he proposes is not only better, but really miraculous! Indeed, with simply requiring that his equation should obey the theory of relativity, and that it should be first order in space-time derivatives, Dirac's equation predicts that the electron has a half-integer spin which can be described by the Pauli matrices. A short time later Dirac deduced further properties from his equation, namely, the correct magnetic moment of the electron and the existence of positrons, that is, of antiparticles.

Quantum mechanics appears now to yield correct equations, even if they can't always be solved. The "statistics" which should be assigned to other elementary particles, such as the proton and the neutron, remained to be determined.

"Bosons" and "Fermions"

Two kinds of particles exist: those which obey Bose-Einstein statistics (any number of particles can bear the same set of quantum numbers) and those which obey Fermi-Dirac statistics (at most one particle can bear a given set of quantum numbers). They obey, respectively, Bose–Einstein or Fermi-Dirac statistics, and they are called *bosons* and *fermions*. So far, only photons were known to be bosons and only electrons to be fermions.

The statistics also apply to the *interacting* electrons of an atom, of a molecule, or of a metal. However, when the particles are far apart, the statistics applied to them has no observable effect.

But what statistics should one apply to protons and to neutrons? Are they bosons of fermions? The problem is more general: should the statistics be applied (and if so, which one) to a nucleus composed of neutrons and protons, to a molecule? Consider liquid helium: all the helium atoms are identical and indistinguishable. What is the symmetry of the liquid helium wavefunction? Nitrogen molecules are also identical. What statistics should be applied to them?

The fifth Solvay Council was held in Brussels on October 24–29, 1927. It was devoted to quantum mechanics and we shall return to it shortly. The question of "statistics" was raised during the discussion which followed the talk delivered by Bohr. To a question of Langevin, Heisenberg gave the following answer:

> There is no reason, in quantum mechanics, to prefer one statistic to another [. . .] We feel however, that the statistics of Einstein - Bose might be more suitable to light quanta, the statistics of Fermi-Dirac to positive and negative electrons. The difference might be related to the difference between matter and radiation, as noted by Bohr [146].

A bit later, he added:

> According to the experiments, protons and electrons have both rotational momentum and they obey Fermi-Dirac statistics; the two properties seem connected. The He[lium] atom does not have rotational momentum and an assembly of He atoms obeys the statistics of Bose–Einstein.

What Heisenberg call "rotational momentum" is what we have called *spin*, or intrinsic angular momentum. What about the proton? Heisenberg thinks that it is a fermion and he bases his argument on a recent paper of the American physicist David Dennison [147], who interpreted experimental results of the Japanese physicist Takeo Hori [148], adding the comment: *"these results are exactly what one would expect if the nuclei of hydrogen, protons, had a spin 1/2 and that they are fermions."*

Between 1925 and 1928, physicists did not have a clear idea about which statistics to apply. In the remark quoted above, Heisenberg suggests a basis for the rules which will be verified empirically before Pauli actually proves them mathematically [149] in 1940, namely, that particles of integer spin (0, 1, 2,...) are bosons and particles of half-integer spin (1/2, 3/2,...) are fermions.

Thus, for example, the helium atom, which is the isotope 4 of helium, composed of two protons, two neutrons, and two electrons, is a boson because the helium-4 atom has a total spin equal to zero. On the other hand, the helium-3 atom is composed of two protons, one neutron, and two electrons. It has a spin equal to 1/2 and it behaves as a fermion. Indeed, liquid helium-4 and liquid helium-3 have different properties.

The Uncertainty Relations of Heisenberg

In the spring of 1926, Heisenberg is invited to give a talk on the new quantum mechanics at the University of Berlin, the most notable institution of German physics, in the presence of Planck, Einstein, von Laue, Nernst, and Rubens. This is where he meets Einstein who invites him home in order to discuss. Heisenberg notes that the main question raised by Einstein was the following:

> *What you have told us sounds extremely strange. You assume the existence of electrons inside the atom, and you are probably quite right to do so. But you refuse to consider their orbits, even though we can observe electron tracks in a cloud chamber[1]. I should very much like to hear more about your reasons for making such strange assumptions.*

Heisenberg explained the reasons which led him to ignore the electron orbits and to introduce only directly observable quantities such as the position or the intensity of an optical ray:

> *Now, since a good theory must be based on directly observable magnitudes, I thought it more fitting to restrict myself to these,*

---

[1] We will describe later (p. 222) the Wilson chamber, in which an electrically charged particle, such as an electron, induces the condensation of small water droplets, a mist, which gives a material and visible form to the particle trajectory, in a fashion similar to the one of the white trail which is often observed behind high altitude jet planes in the sky.

treating them, as it were, as representatives of the electron orbits."
"But you don't seriously believe," Einstein protested, "that none but
observable magnitudes must go into a physical theory?"
"Isn't that precisely what you have done with relativity?" I asked in
some surprise. "After all, you did stress the fact that it is impermissible
to speak of absolute time, simply because absolute time cannot be
observed; that only clock readings, be it in the moving reference
system or the system at rest, are relevant to the determination of
time [150].

Indeed, referring to idealized observers with clocks and moving one relative to another, Einstein made a radical criticism of the notion of simultaneity, showing that paradoxes occur when the signals used by the observers to communicate with one another cannot travel faster than light. Two events may appear simultaneous to one observer, but not to another. Heisenberg had been most impressed by Einstein's demonstration and it remained an inspiration for him. However, Einstein replied:

I may have used such a philosophy, but nonetheless it is absurd. I
might be more careful by saying that, from a heuristic point of view,
it can be useful to keep in mind what one actually observes. But is
quite wrong to base a theory, as a matter of principle, exclusively on
observable quantities.

Einstein explains that indeed a physical measurement is a complex procedure during which one is led to interpret many physical phenomena before achieving a measurement. The measurement is thus the result of an interpretation of numerous signals delivered by a complex apparatus and the interpretation is necessarily guided by the theory. This makes Einstein a perfect Cartesian. He is very close to the second *Metaphysical Meditation* of Descartes:

If I happen to look through the window at men walking in the street,
I will not fail to say that I see men [. . .] and yet, what do I really
see through the window? Hats, coats which could cover either ghosts
or imitations of men, or artificial men which are activated by springs.
However, I consider that they are real men; and I understand thus,
using only the judgement which lies in my mind, what I see with my
eyes [151].

Anyway, Einstein raised a delicate question, which Heisenberg was well aware of: how could one reconcile the negation of electron trajectories in an atom with the fact that one can observe what appears to be electron trajectories in a Wilson chamber?

In May, 1926, Niels Bohr offered Heisenberg a position as lecturer in the University of Copenhagen, and he invited him to become his assistant, a position previously held by Hendrik Kramers. Heisenberg was happy to accept. He lectured in the University in Danish and continued to work with Bohr. In September 1926, Bohr invited Schrödinger to Copenhagen in order to explain his theory but also to discuss with him, as he always did. He invited him to his house and the conversations would begin in the morning and end late at night. Schrödinger was exhausted and bedridden with a fever. He was cared for by Margrethe Bohr who served him cakes and tea. Heisenberg took an active part in the discussions but no agreement could be reached. Schrödinger refused to admit the existence of the famous "quantum jumps" and Bohr explained that they could not be avoided. In fact, Schrödinger entertained the hope that his "wave mechanics" would be able to suppress all jumps and discontinuities, and that a continuous physics could be formulated in which particles would be replaced by waves. The same hope had animated Planck, namely, to reconcile continuous physics with quanta. Neither Bohr nor Heisenberg adhered to this view. Heisenberg recalls that at one point Schrödinger exclaimed:

> *If we are going to stick to this damned quantum-jumping, then I regret I ever had anything to do with quantum theory [152].*

To which Bohr replied, as Heisenberg recalls:

> *But the rest of us are thankful that you did, because you have contributed so much to the clarification of the quantum theory.*

During the following months, Heisenberg and Bohr had intense discussions. Bohr needed to talk, to discuss, in order to make progress in his mind, and he was particularly happy to be faced with someone who could answer and contradict him. In return, Bohr did not fail to exert a profound influence on Heisenberg, who was only 25 years old.

In February 1927, Bohr was tired by intense work and decided to take a skiing vacation in Norway. Before he left, he had long discussions with Heisenberg about the paradox of the electron trajectories: whereas quantum mechanics forbade the existence of electron trajectories, one could observe them with the naked eye in a Wilson chamber![1] While Bohr was away, Heisenberg remained in Copenhagen and decided to think more precisely about the problem of the position and the trajectory of an electron. Does it make sense to speak of its position, its orbit, and its velocity? The complete and abstract formalism constructed by Heisenberg, Born, and Jordan, and later by Dirac, had already led Jordan to make a surprising remark concerning "conjugate variables," such as the position and the velocity: when one acquired a well-determined value, the other could take any value!

Heisenberg recalls:

> *It must have been one evening after midnight when I suddenly remembered my conversation with Einstein and particularly his statement, "It is the theory which decides what we can observe." I was immediately convinced that the key to the gate that had been closed for so long must be sought right here. I decided to go on a nocturnal walk through Fælledpark to think further about the matter. We had always said so glibly that the path of the electron in the cloud chamber could be observed. But perhaps what we really observed was something much*

---

[1] See the footnote (p. 148) and the description of the Wilson cloud chamber(p. 222).

less. Perhaps we merely saw a series of discrete and ill-defined spots through which the electron had passed. In fact, all we do see in the cloud chamber are individual water droplets which must certainly be much larger than the electron [153].

Heisenberg has the hunch that this was the key to the paradox: quantum mechanics prevents an exact localization of the electron at a given time and this is what prevents the electron trajectory from being precisely defined as in classical mechanics. It does however specify, at least roughly, where the electron is at successive times, and this defines what appears to be a continuous trajectory, when it is viewed at our scale. When he returns to the lab, his calculations confirm the idea. If you ask "What is the position of the electron?", quantum mechanics gives an answer which bears a certain imprecision, as it does if you ask "What is the velocity of the electron?". The two imprecisions are related. The more precision you require for the position, the less the velocity is determined. In the following days, Heisenberg acquires a deeper understanding of this and he prepares a paper, which he submits to Bohr when the latter returns from Norway. Bohr welcomes the idea at first but then he expresses some objections. The paper is nonetheless submitted to *Zeitschrift für Physik*, which accepts it on March 23, 1927. Meanwhile, Heisenberg had written to his friend Pauli who replied most favorably. In his paper [154], Heisenberg begins by stating that the general interpretation of quantum mechanics is always flawed by the confrontation of the corpuscular and wave representations of the theory. He adds:

This fact alone would lead us to conclude that an interpretation of quantum mechanics is in any case impossible in terms of the usual kinematic and mechanical concepts.

He then takes a radically different course: he redefines what is meant by "position", "velocity", "orbit" of an object such as an electron:

If we want to make it clear what we mean by the words 'the position of an object', such as an electron, [. . .] we must define the experiments which allow us to measure the 'position of the electron'; failing this the words mean nothing. Such experiments, which, in principle allow us to measure 'the position of the electron', are far from lacking. For example, one can shed light on the electron and observe it with a microscope. It is essentially the wavelength of the light which determines the highest possible accuracy of the measurement. In principle, one should construct a gamma-ray microscope to perform the measurement with a required accuracy.

Indeed, photographers and astronomers know only too well that the precision is limited by the wavelength of light. Details smaller than the wavelength cannot be resolved. The wavelengths of visible light range between 0.4 and 0.8 μm, which is 10 000 times larger than the size of an atom. To observe an electron orbiting in an atom, it would need to be illuminated by light waves with a wavelength 100 000 times shorter than visible light. To illuminate the electron, the corresponding light quantum would have to collide with the electron, thereby disturbing its orbit by modifying its velocity. The better the orbit is to be resolved, the shorter the wavelength of the light needs to be and therefore the more the electron orbit will be disturbed. Heisenberg proves that quantum mechanics predicts that the precision Δx with which the position x of the electron is measured, and the precision Δp with which the momentum p^1 of the electron can be measured, are related by a simple formula: their product cannot be smaller than Planck's constant h. Heisenberg's rule is expressed by the famous expression:

$$\Delta x \, \Delta p \simeq h$$

The symbol $\simeq$ means "approximately equal to". Heisenberg also shows that a similar relation exists between the precision ΔE, with which the energy of a stationary state can be determined during a time interval Δt:

$$\Delta E \Delta t \simeq h$$

It follows that a system which lives only for a short time does not have a well defined energy. The precision with which the energy is determined is greater if the system lives for a long time. The equations above are usually referred to as Heisenberg's "uncertainty relations". This is a somewhat unfortunate wording. The uncertainty relations simply relate the unavoidable uncertainty of a given observable to that of another related observable. The uncertainty in the position of a particle is inversely proportional to the uncertainty of its velocity. Such uncertainties are familiar in wave phenomena. For example, the A-string of a violin vibrates at a frequency of 440 Hz, meaning that it comes and goes 440 times every second. Consider next a lower A which is five octaves lower. Its frequency is approximately 18 Hz. If our ear is to identify this note, it must be able to discern the frequency of the vibration. It must therefore perceive at least two or three successive vibrations. The sound must therefore last for at least 1/4 of a second. If the sound is emitted in a shorter time interval, one cannot attribute to the sound the frequency required to identify the note. This shows that a relations exists between the duration of a sound and the precision with which one can identify its frequency, which, in quantum mechanics, is energy.

Nobel Acknowledgments

The 1918, 1921 and 1922 Nobel prizes had honored Planck, Einstein and Bohr for their work on the theory of quanta. They were later attributed to the founders of the new wave mechanics: Louis de Broglie received the Nobel prize in 1929 "for his discovery of the wavelike nature of electrons", followed by Heisenberg in 1932

---

[1] The momentum p of the electron is equal to the product mv of its mass m by it velocity v.

"for the creation of quantum mechanics, the application of which has, *inter alia,* led to the discovery of the allotropic forms of hydrogen", Erwin Schrödinger and Paul Dirac in 1933 "for the discovery of new productive forms of atomic theory", Wolfgang Pauli in 1945 "for the discovery of the exclusion principle, also called the Pauli principle", Max Born in 1954 "for his fundamental research in quantum mechanics, especially for his statistical interpretation of the wavefunction".

In 1930, quantum mechanics is still developing. However, the foundations of quantum mechanics will no longer change. The physics of the atom has a solid foundation. The physics of the atomic nucleus is however still in its infancy, but the tool to tackle it with is ready.

The Fifth Solvay Council: An Assessment of the New Mechanics

The fifth Solvay council was held in Brussels from October 24–29, 1927. It was the last one to be presided by Hendrik Lorentz, who died shortly after, in February 1928. Initially it was dedicated to "Electrons and Photons" but in fact all the new quantum mechanics was discussed [155]. Traditionally, the invited participants were carefully selected: Marie Curie, Niels Bohr, Max Born, William Lawrence Bragg, Léon Brillouin, Arthur Compton, Louis de Broglie, Peter Debye, Paul Dirac, Paul Ehrenfest, Albert Einstein, Ralph Fowler, Werner Heisenberg, Martin Knudsen, Hendrik Kramers, Paul Langevin, Wolfgang Pauli, Max Planck, Owen Richardson, and Charles Wilson.

The discussions were very vivid, particularly between Bohr and Einstein, who made a point of inventing so-called *Gedankenexperiments*, which are thought or virtual experiments, designed to show flaws and internal incoherences of Bohr's interpretation, which soon became known as that of the "Copenhagen school". However, each time, Bohr found a satisfactory answer.

Bohr presented what he called the "*quantum principle*" and he discussed its inevitable consequences:

> The meaning of the theory can be expressed in terms of what is called the quantum postulate, according to which any atomic process involves a discontinuity or rather an individuality which is totally absent in the classical theories and which is characterized by Planck's quantum of action. This postulate forces us to renounce to a causal description of atomic phenomena in space and time.

Why does quantum mechanics necessarily lead us to abandon causality? Because, as discussed in connection with Heisenberg's uncertainty relations, it is not possible to observe a phenomenon without disturbing it. In other words, it is not possible to dissociate the measured quantity from the measuring apparatus. Bohr expresses ideas which he had doubtlessly amply discussed with Heisenberg, before as well as after the publication of his paper on the "uncertainty relations". Bohr insists on the fact that *it is impossible to observe a microscopic system without*

perturbing it in an important and unpredictable fashion, and this destroys causality.
In his analysis of the use of light, Bohr shows that one must either give up causality
or the description of events occurring in space and time. He makes a philosophical
conclusion, which is still valid today:

> We find ourselves here on the path, followed by Einstein, consisting in
> adapting our intuition, borrowed from our sensory perceptions, to a
> continuously deeper knowledge of the laws of nature. The obstacles,
> which we face on this path, are mainly due to the fact that, in
> some fashion, every term of our language is related to the forms of
> our representations. In the theory of quanta, this difficulty becomes
> immediately apparent in the question concerning the impossibility of
> avoiding indeterminism and this is inherent in the quantum postulate.

Note that we must give up a "causal link" only when one describes an individual
event, such as a single collision of an electron with an atom. But if a million
electrons impinge of a sample of matter, quantum mechanics does predict the
fraction of electrons which are deviated to a given direction. *The wavefunction
evolves in a completely deterministic manner.* However it allows us only to know the
probability that an electron should scatter at this or that angle, or, more precisely,
the probability of detecting an electron deviated at a given angle.

At this point, we interrupt the story, because the essential part of what is to
become the tool of the nuclear physicist, is in place.

The German Language, the Language of Quantum Mechanics

One last thing. From the outset, quantum physics is effectively a German lan-
guage physics. Planck, Einstein, Sommerfeld, Landé, Pauli, Heisenberg, Born,
Schrödinger, as well as the experimentalists Paschen, Lummer, Pringsheim, Rubens,
Kurlbaum, Stern, Gerlach, Franck and Hertz were all German speaking physicists.
One should add that physicists, such as Uhlenbeck and Goudsmit, who were Dutch,
or Fermi and Majorana, who were Italian, all published their important papers in
German. This illustrates the fact that the development of quantum theory, during the
first half of the twentieth century, was mostly formulated in the German language
even if one takes into account the French contribution (of de Broglie who invented
matter waves, although it is Schrödinger who wrote down the wave equation) and
the notable English contribution, due, for the most part, to Paul Dirac.

This vitality survived the disaster of the 1914–1918 war, the ensuing economic
crisis and the boycott of German physicists after the war. When young physicists,
from Italy, the United States, or Japan, wanted to learn "modern physics", that is,
quantum physics, they went mainly to Germany. Such was the case of Fermi, Majorana, Oppenheimer, Millikan and Nishina. We mentioned the German *language*,
because the use of the German language as a common scientific language spread
well beyond the frontiers of Germany. It naturally included all of what had been

the Austro-Hungarian empire, as well as Holland, Poland and Italy (until 1933). However, when Nazism hunted down and eliminated so many Jewish physicists, it decapitated German science.[1] Beginning 1933, the use of the German language, which had been an important language in international scientific communication, began to decline and, by 1945, it had become replaced by English.

A Brief Bibliography

Several books on the history of quantum mechanics exist for non specialists:

- George Gamow, *Mr. Tomkins in Wonderland,* Cambridge, University Press, Cambridge, University Press, 1939.
- George Gamow, *Mr. Tompkins explores the atom,* Cambridge University Press, 1945. The books involving M. Tompkins became very famous and have been translated into many languages.
- George Gamow, *Thirty Years that Shook Physics*, Garden City, N.Y., Anchor Books, 1966, illustrated by the author.
- Banesh Hoffmann, *The strange story of the quantum, an account for the general reader of the growth of the ideas underlying our present atomic knowledge,* Harper & Brothers, New York, 1947; Dover, New York, 1959.
- Roland Omnès, *Quantum philosophy: understanding and interpreting contemporary science,* Princeton University Press, Princeton, 1999.

Among books aimed at specialists, let us quote:

- B. L. van der Waerden (ed.), *Sources of quantum mechanics*, North Holland, Amsterdam, 1967. A reprint of 17 fundamental articles on quantum mechanics, translated into English.
- F. Hund, *The history of quantum mechanics*, London, Harrap, 1974. The history of the development of quantum mechanics by one of the contemporary actors.
- J. Mehra and H. Rechenberg, *The Historical Development of Quantum Mechanics*, Heidelberg, Berlin, 1982. This monumental 7-volume book is well documented. It covers the period 1925-1932, which marks the development of quantum theory in its modern form.
- Abraham Pais, *Inward Bound,* Oxford, Clarendon Press, 1986. This is an excellent history of particle physics and it covers the history of quantum mechanics at a rather technical level.
- Abraham Pais, *Niels Bohr Times,* Oxford University Press, 1991. This is a scientific biography of Niels Bohr which is intimately related to the history of quantum mechanics.

---

[1] See p. 327–331.

- Sin-itiro Tomonaga, *The Story of Spin,* translated from Japanese by Okeshi Oka, University of Chicago Press, Chicago and London, 1997. A wonderful description of the crucial rôle of spin in the development of quantum mechanics. This mathematically precise book is aimed at physicists.
- Olivier Darrigol, *From c-Numbers to q-Numbers: The Classical Analogy in the History of Quantum Theory,* Berkeley, University of California Press, 1992. http://ark.cdlib.org/ark:/13030/ft4t1nb2gv/. A mathematically precise analysis of the structure and development of analogies between quantum mechanics and classical mechanics in three cases: Planck's radiation theory, Bohr's atomic theory, and Dirac's quantum mechanics.
- Helmut Rechenberg, *Quanta and quantum mechanics*, the third chapter of *Twentieth Century Physics*, published under the direction of Laurie M. Brown, Abraham Pais et Sir Brian Pippard, by the Institute of Physics Publishing, Bristol and Philadelphia and the American Institute of Physics, New York, 1995 (three volumes).

References

1. Le Verrier, U. J. J., "Recherches sur les mouvements d'Uranus", *Comptes Rendus de l'Académie des Sciences* **22**, 907–918, session of June 1, 1846.
2. Le Verrier, U. J. J., "Sur la planète qui produit les anomalies observées dans le mouvement d'Uranus", *Comptes Rendus de l'Académie des Sciences* **43**, 428–438, session of August 31, 1846.
3. Newton, I., *Opticks*, 1704, p. 314.
4. Wien, W., "Über die Energievertheilung im Emissionsspektrum eines schwarzen Körpers", *Annalen der Physik, Leipzig* **58**, 662–669, 1896.
5. Rubens, H. and Kurlbaum, F., "Über die Emission langwelliger Wärmestrahlen durch den schawarzen Körper bei verschiedenen Temperaturen", *Sitzungsberichte der Kaiserliche Preussischen Akademie der Wissenschaften zu Berlin* , No. 2, 929–941, session of October 25, 1900.
6. Rayleigh, John William Strutt, Baron, "Remarks upon the Law of Complete Radiation", *Philosophical Magazine* **49**, 539–540, June 1900.
7. Planck, M., *Scientific Autobiography and other papers,* Williams & Norgate, London, 1950.
8. Hermann, A., *Max Planck in Selbstzeugnissen und Bilddokumenten,* Rowohlt, Reinbek bei Hamburg, 1973.
9. Heilbron, J. L., *The dilemmas of an upright man: Max Planck as spokesman for German science,* University of California Press, Berkeley, 1986.
10. Hermann, A., *The Genesis of Quantum Theory (1899-1913),* M.I.T. Press, Cambridge, 1971.
11. Planck, M., "Über eine Verbesserung der Wien'schen Spektralgleichung", *Verhandlungen der deutschen physikalischen Gesellschaft* **2**, 202–204, session of October 19, 1900.
12. Planck, M., "Die Entstehung und bisherige Entwickelung der Quantentheorie", in *Les prix Nobel en 1919 et 1920 (published in 1922),* pp. 1–14, Imprimerie Royale, P. A. Norsted & Söner, Stockholm, 1920.
13. Planck, M., "letter to Robert Wood, dated October 7, 1931", cited by A. Hermann, *The Genesis of Quantum Theory*, Cambridge, The M.I.T. Press, 1971, p. 23.
14. Planck, M., "Zur Theorie des Gesetzes der Energieverteilung im Normalspectrum", *Verhandlungen der Deutsche physikalischen Gesellschaft* **2**, 237–245, session of December 14, 1900.
15. Planck, M., "Über das Gesetz der Energieverteilung im Normalspektrum", *Annalen der Physik* **4**, 553–563, 1901.
16. Frank, P., *Einstein, his life and times*, Jonathan Cape, London, 1948.
17. Pais, A., *Subtle is the Lord. . . . The Science and the Life of Albert Einstein*, Oxford University Press, Oxford, 1982.
18. Fölsing, A., *Albert Einstein: a biography,* Viking, New York, 1997.
19. Einstein, A., "Über einen die Erzeugung und Verwandlung des Lichtes betreffenden heuristischen Gesichtspunkt", *Annalen der Physik, Leipzig* **17**, 132–148, 1905.
20. Soler, L., "Les quanta de lumière d'Einstein en 1905, comme point focal d'un réseau argumentatif complexe", *Philosophia Scientiæ* **3**, 107–144, 1999.
21. Einstein, A., "Über die Entwickelung unserer Anschauungen über das Wesen und die Konstitution der Strahlung", *Verhandlungen der deutschen physikalischen Gesellschaft zu Berlin* **7**, 482–500, 1909.
22. Einstein, A., letter to Michele Besso, dated July 29, 1918, in *Albert Einstein/Michele Besso: Correspondance 1903-1955*, edited by Speziali, P., Hermann, Paris, p. 131.
23. Millikan, R. A., "A Direct Photoelectric Determination of Planck's Constant "*h*"", *Physical Review* **7**, 355–388, March 1916.
24. Dulong, P. L. and Petit, A. T., "Sur quelques points importants de la théorie de la chaleur", *Annales de Chimie et de Physique, Paris* **10**, 395–413, 1819.
25. Einstein, A., "Die Plancksche Theorie der Strahlung und die Theorie der spezifischen Wärme", *Annalen der Physik, Leipzig* **22**, 180–190, 1907.

26. Mehra, J., *The Solvay Conferences on physics : aspects of the development of physics since 1911*, D. Reidel, Dordrecht, 1975.
27. Marage, P. and Wallenborn, G. (editors), *Les Conseils Solvay et les débuts de la physique moderne*, Université Libre de Bruxelles, Bruxelles, 1995.
28. de Broglie, M., *Les premiers conseils de physique Solvay et l'orientation de la physique depuis 1911*, Albin Michel, Paris, 1951.
29. Langevin, P. and de Broglie, M. (editors), *La théorie du rayonnement et les quanta, rapports et discussions de la Réunion tenue à Bruxelles du 30 octobre au 3 novembre 1911*, Gauthier-Villars, Paris, 1912.
30. Rozental, S. (editor), *Niels Bohr; his life and work as seen by his friends and colleagues*, North-Holland Pub. Co., Amsterdam, 1967.
31. Pais, A., *Niels Bohr Times*, Clarendon Press, Oxford, 1991.
32. Rozental, S., *Niels Bohr: memoirs of a working relationship*, Christian Ejlers, Copenhagen, 1998.
33. Rutherford, E., *Radioactive substances and their radiations*, Cambridge University Press, Cambridge, 1913.
34. Heisenberg, W., *Physics and Beyond*, Harper & Row, New York and London, 1971, p. 39.
35. Heilbron, J. L. and Kuhn, T., "The Genesis of the Bohr Atom", *Historical Studies in the Physical Sciences* **1**, 211–290, 1969.
36. *Niels Bohr, Collected Works. Volume 2*, p. 110.
37. Bohr, N., "Reminiscences of the founder of nuclear science and some developments based on his work", in *Rutherford at Manchester*, pp. 114–167.
38. Bohr, N., "On the Constitution of Atoms and Molecules", *Philosophical Magazine* **26**, 1–25, July 1913.
39. Bohr, N., *Collected Works. Volume 2*, p. 103.
40. Jammer, M., *The Conceptual Development of Quantum Mechanics*, McGraw Hill, New York, 1996, p. 85.
41. Bohr, N., "On the Constitution of Atoms and Molecules. Part II : Systems containing only a single nucleus", *Philosophical Magazine* **26**, 476–502, 1913.
42. Pickering, E. C., "Stars Having Peculiar Spectra : New Variable Stars in Crux and Cygnus", *Astrophysical Journal* **4**, 369–370, December 1896.
43. Bohr, N., "On the Constitution of Atoms and Molecules. Part III : Systems containing several nuclei", *Philosophical Magazine* **26**, 857–875, November 1913.
44. Pais, A., *Niels Bohr Times*, 1991, p. 152–155.
45. Bohr, H., letter to Niels Bohr, fall 1913, in *Niels Bohr collected works*, Vol. 1, p. 567.
46. Bohr, N., "The spectra of helium and hydrogen", *Nature* **92**, 231–232, October 23, 1913.
47. Franck, J. and Hertz, G. L., "Über Zusammenstöße zwischen Elektronen und den Molekülen des Quecksilberdampfes und die Ionisierungsspannung desselben", *Verhandlungen der deutschen physikalischen Gesellschaft zu Berlin* **16**, 457–467, session of April 24, published May 30, 1914.
48. Michelson, A. A., "On the Application of interference methods to spectroscopic measurements", *Philosophical Magazine* **31**, 338–346, April 1891.
49. Michelson, A. A., "On the Application of interference methods to spectroscopic measurements", *Philosophical Magazine* **34**, 280–299, September 1892.
50. Zeeman, P., "Over den invloed eener magnetisatie op den aard van het door een stof uitgezonden licht", *Verslagen van de gewone vergaderingen der Wis- en Natuurkundige der Koninklijke Akademie van Wetenschappen te Amsterdam* **5**, 181–184, Session of October 31, 1896.
51. Zeeman, P., "On the Influence of Magnetism on the Nature of the Light emitted by a Substance", *Philosophical Magazine* **43**, 226–239, March 1897.
52. Zeeman, P., "Doublets and Triplets in the Spectrum produced by External Magnetic Forces", *Philosophical Magazine* **44**, 55–60, July 1897.
53. Zeeman, P., "The Effect of Magnetisation on the Nature of Light Emitted by a Substance", *Nature* **55**, 347, February 11, 1897.

54. Preston, T., "Radiation phenomena in a strong magnetic field", *Scientific Transactions of the Royal Dublin Society* **6**, 385–391, 1898. Read December 22, 1897.

55. Paschen, F. and Back, E., "Normale und anomale Zeemaneffekte", *Annalen der Physik, Leipzig* **39**, 897–932, 1912.

56. Stark, J., "Beobachtungen über den Effekt des elektrischen Feldes auf Spektrallinien. I. Quereffelt", *Annalen der Physik, Leipzig* **43**, 965–982, 1914.

57. Einstein, A., letter to Michele Besso, dated August 11, 1916, in *Albert Einstein/Michele Besso : Correspondance 1903-1955*, p. 78.

58. Epstein, P. S., "Zur Theorie des Starkeffekts", *Physikalische Zeitschrift* **17**, 148–150, April 15, 1916.

59. Schwarzschild, K., "Zur Quantentheorie", *Sitzungsberichte der Preussische Akademie der Wissenschaften* pp. 548–568, 1916.

60. Sommerfeld, A., *Atombau und Spektrallinien*, Vieweg, Braunschweig, 1919.

61. Sommerfeld, A., *Atomic structure and spectral lines*, Methuen, London, 1923. Translated from the 3rd German edition, 1922.

62. *Atomes et électrons, rapports et discussions du Conseil de Physique Solvay tenu à Bruxelles du 1er au 6 avril 1921*, Gauthier-Villars, Paris, 1923.

63. Bohr, N., "Le principe de correspondance", in *Atomes et électrons, rapports et discussions du Conseil de Physique Solvay tenu à Bruxelles du 1er au 6 avril 1921*, pp. 248–254.

64. Bohr, N., "Über die Anwendung der Quantentheorie auf den Atombau. I. Die Grundpostulaten der Qnantentheorie", *Zeitschrift für Physik* **13**, 117–165, 1923.

65. Bohr, N., "On the application of the quantum theory to atomic structure. I. The fundamental postulates", *Proceedings of the Cambridge Philosophical Society. Supplement*, 1924.

66. Kossel, W. L., "Über Molekülbildung als Frage des Atombaus", *Annalen der Physik, Leipzig* **49**, 229–362, 1916.

67. Bohr, N., "Atomic Structure", *Nature* **107**, 104–107, March 24, 1921.

68. Bohr, N., "Atomic structure", *Nature* **108**, 208–209, October 13, 1921.

69. Bohr, N., "Unsere heutige Kenntnis vom Atom", *Die Umschau* **25**, 229–234, 1921. English translation in *Niels Bohr Collected Works*, vol. 2, p. 84.

70. Pauli, W., "Relativitätstheorie", in *Encyklopädie der mathematischen Wissenschaften*, Vol. 5, part 2, pp. 539–775, B. G. Teubner, Leipzig, 1921.

71. Enz, C. P., *No Time to Be Brief: A Scientific Biography of Wolfgang Pauli*, Oxford University Press, Oxford, 2002.

72. Pauli, W., "Zur Theorie der Gravitation und der Elektrizität von Hermann Weyl", *Physikalische Zeitschrift* **20**, 457–67, October 15, 1919.

73. Pauli, W., "Über die Energiekomponenten des Gravitationsfeldes", *Physikalische Zeitschrift* **20**, 25–27, January 15, 1919.

74. Pauli, W., "Mercurperihelbewegung und Strahlenablenkung in Weyls Gravitationstheorie", *Verhandlungen der deutschen physikalischen Gesellschaft zu Berlin* **21**, 742–750, session of December 5, 1919.

75. Born, M., *My Life and my Views*, Charles Scribner's Sons, New York, 1968.

76. Born, M., *My life: recollections of a Nobel Laureate*, Taylor and Francis, London, 1978. Translation of 'Mein Leben: Erinnerungen des Nobelpreisträgers', Nymphenburger, München, 1975.

77. Heisenberg, W., Foreword to *The Born Einstein Letters : correspondence between Albert Einstein and Max and Hedwig Born from 1916 to 1955 with commentaries by Max Born.*, Macmillan, London, Basingstoke, 1971, p. viii.

78. *Ibid.*, p. ix.

79. *Niels Bohr Collected Works. Volume 2*, p. 341.

80. Pauli, W., "Remarks on the History of the Exclusion Principle", *Science* **103**, 213–215, February 22, 1946.

81. Pauli, W., "Quantentheorie und Magneton", *Physikalische Zeitschrift* **21**, 615–617, 1920.

82. Einstein, A. and Ehrenfest, P., "Quantentheoretische Bemerkungen zum Experiment von Stern und Gerlach", *Zeitschrift für Physik* **11**, 31–34, 1922.

83. Compton, A. H., "A Quantum Theory of the Scattering of X-Rays by Light Elements", *Physical Review* **21**, 483–502, May 1923.

84. Pauli, W., "Über den Einfluss der Geschwindigkeitsabhängigkeit der Elektronenenmasse auf den Zeemaneffekt", *Zeitschrift für Physik* **31**, 373–385, 1925.

85. Pauli, W., "Über den Zusammenhang des Abschlusses der Elektronengruppen in Atom mit der Komplexstruktur der Spektren", *Zeitschrift für Physik* **31**, 765–783, 1925.

86. Uhlenbeck, G. E. and Goudsmit, S. A., "Ersetzung der Hypothese vom unmechanischen Zwang durch eine Forderung bezüglich des inneren Verhaltens jeden einzelnen Elektrons", *Naturwissenschaften* **13**, 353–354, November 20, 1925.

87. Klein, M. J., *Paul Ehrenfest*, North-Holland, Amsterdam & New York, 1985.

88. *Niels Bohr Collected Works. Volume 5*, p. 234.

89. Uhlenbeck, G. E. and Goudsmit, S. A., "Spinning Electrons and the Structure of Spectra", *Nature* **117**, 264–265, February 20, 1926.

90. de Broglie, L., *Recherches sur la théorie des quanta*, Masson, Paris, 1963, p. 4.

91. de Broglie, L., "Ondes et quanta", *Comptes Rendus de l'Académie des Sciences* **177**, 507–510, session of September 10, 1923.

92. de Broglie, L., "Quanta de lumière, diffraction et interférences", *Comptes Rendus de l'Académie des Sciences* **177**, 548–550, session of September 24, 1923.

93. Lewis, G. N., "The conservation of photons", *Nature* **118**, 874–875, December 18, 1926.

94. Einstein, A., *letter to Paul Langevin, dated December 16, 1924*, in *Œuvres choisies. Vol. 4: Correspon- dances françaises*, edited by Biezunski, M., Éditions du Seuil/Éditions du CNRS, 1989, p. 172.

95. Einstein, A., letter to H. A. Lorentz, dated December 16, 1924, in *The Scientific Correspondence of H. A. Lorentz*, edited by Kox, A. J., p. 568, Springer, New York, 2008.

96. de Broglie, L., "Recherches sur la théorie des quanta", *Annales de Physique (Paris)* **3**, 22–128, 1925.

97. Davisson, C. J. and Germer, L. H., "The scattering of electrons by a single crystal of nickel", *Nature* **119**, 558–560, April 16, 1927.

98. Thomson, G. P., "Diffraction of a cathode ray by a thin film", *Nature* **119**, 890, June 18, 1927.

99. Thomson, G. P., "Experiments on the Diffraction of Cathode Rays", *Proceedings of the Royal Society, London* **A117**, 600–609, 1928.

100. Hermann, A., *Werner Heisenberg, 1901-1976*, Inter Nationes, Bonn, 1976. Translated from German by Timothy Nevill. Original publication : *Heisenberg*, Rowolt, 1976.

101. Heisenberg, E., *Inner exile : recollections of a life with Werner Heisenberg*, Birkhäuser, Boston, 1984.

102. Cassidy, D. C., *Uncertainty : the life and science of Werner Heisenberg*, Freeman, New York, 1992. Heisenberg-1971p38,Born-1978p212

103. Heisenberg, W., *Physics and Beyond*, p. 38.

104. Born, M., *My life*, p. 212.

105. Heisenberg, W., *Stabilität und Turbulenz von Flüssigkeitsströmen*, Ph.D. thesis, Ludwig-Maximilian, 1923.

106. Heisenberg, W., *Physics and Beyond*, p. 61.

107. Born, M., letter to Albert Einstein, dated July 15, 1925, in *The Born Einstein Letters: correspondence between Albert Einstein and Max and Hedwig Born from 1916 to 1955 with commentaries by Max Born.*, p. 84, Macmillan press, London, Basingstoke, 1971. Translated by Irene Born. Original edition: *Albert Einstein/Hedwig and Max Born: Briefwechsel 1916–1955*, Nymphenburger, München, 1969.

108. Heisenberg, W., "Über die quantentheoretische Umdeutung kinematischer und mechanischer Beziehungen", *Zeitschrift für Physik* **33**, 879–893, 1925.

109. Born, M. and Jordan, E. P., "Zur Quantenmechanik", *Zeitschrift für Physik* **34**, 858–888, 1925.

110. Born, M., Heisenberg, W. and Jordan, E. P., "Zur Quantenmechanik. II", *Zeitschrift für Physik* **35**, 557–615, 1926.

111. Bohr, N., "On the law of conservation of energy", *Nature* **116**, 262, August 25, 1925.

112. Omnès, R., *Quantum philosophy: understanding and interpreting contemporary science*, Princeton University Press, Woodstock, 1999.

113. Einstein, A., letter to Michele Besso, dated December 25, 1925, in *Albert Einstein/Michele Besso: Correspondance 1903-1955*, p. 128.

114. Einstein, A., letter to H. A. Lorentz, dated March 13, 1926, in *The Scientific Correspondence of H. A. Lorentz*, p. 592.

115. Pauli, W., "Über das Wasserstoffspektrum vom Standpunkt der neuen Quantenmechanik", *Zeitschrift für Physik* **36**, 336–363, 1926.

116. Moore, W. J., *Schrödinger: life and thought*, Cambridge University Press, Cambridge, 1989.

117. Bloch, F., "Heisenberg and the early days of quantum mechanics", *Physics Today*, December 1976.

118. Einstein, A., "Quantentheorie des einatomigen idealen Gases. Zweite Abhandlung", *Sitzungsberichte der Preussischen Akademie der Wissenschaften* pp. 3–14, 1925.

119. Schrödinger, E., letter to Einstein, dated November 3, 1925, in Walter Moore, *Schrödinger, life and thought*, p. 192.

120. Moore, W. J., *Schrödinger: life and thought*, p. 194.

121. Schrödinger, E., "Quantisierung als Eigenwertproblem. Erste Mitteilung", *Annalen der Physik* **79**, 361–376, 1926. English translation in E. Schrödinger, *Collected papers on wave mechanics*, London: Blackie, 1928.

122. Przibram, K. (editor), *Schrödinger, Planck, Einstein, Lorentz : Briefe zur Wellenmechanik*, Springer Verlag, Vienne, 1963.

123. Schrödinger, E., "An Undulatory Theory of the Mechanics of Atoms and Molecules", *Physical Review* **28**, 1049–1070, December 1926.

124. Schrödinger, E., "Über das Verhältnis der Heisenberg-Born-Jordanschen Quantenmechanik zu der meinen", *Annalen der Physik* **79**, 734–756, 1926. English translation in E. Schrödinger, *Collected papers on wave mechanics*, London: Blackie, 1928.

125. Born, M., "Zur Quantenmechanik der Stoßvorgänge (Vorläufige Mitteilung)", *Zeitschrift für Physik* **37**, 863–867, 1926.

126. Born, M., "Zur Quantenmechanik der Stoßvorgänge", *Zeitschrift für Physik* **38**, 807–827, 1926.

127. Pauli, W., "Zur Quantenmechanik des magnetisches Elektrons", *Zeitschrift für Physik* **43**, 601–623, 1927.

128. Wali, K. C. (editor), *Satyendra Nath Bose, his life and times. Selected works (with commentary)*, World Scientific, Singapore, 2009.

129. Bose, S., "Plancks Gesetz und Lichtquantenhypothese", *Zeitschrift für Physik* **26**, 178–181, 1924.

130. Einstein, A., "Quantentheorie des einatomigen idealen Gases", *Sitzungsberichte der Preussischen Akademie der Wissenschaften* pp. 261–267, 1924.

131. Einstein, A., "Zur Quantentheorie des idealen Gases", *Sitzungsberichte der Preussischen Akademie der Wissenschaften* pp. 18–25, 1925.

132. Jin, D. S., Ensher, J. R., Matthews, M. R., Wieman, C. E. and Cornell, E. A., "Collective Excitations of a Bose-Einstein Condensate in a Dilute Gas", *Physical Review* **77**, 420–423, July 15, 1996.

133. Ouellet, D., *Franco Rasetti, physicien et naturaliste*, Guérin, Montréal, 2000.

134. Fermi, L., *Atoms in the family. My life with Enrico Fermi*, The University of Chicago Press, Chicago, 1954, p. 36.

135. Fermi, E., "Zur Quantelung des idealen einatomigen Gases", *Zeitschrift für Physik* **36**, 902–912, 1926.

136. Kursunoglu, B. N. and Wigner, E. P. (editors), *Reminiscences about a great physicist: Paul Adrien Maurice Dirac*, Cambridge University Press, Cambridge, Mass., 1987.

137. Farmelo, G., *The Strangest Man. The hidden life of Paul Dirac, quantum genius*, Faber and Faber, 2009.

138. Dirac, P. A. M., "Recollections of an exciting era", in *History of Twentieth Century Physics, International School of Physics 'Enrico Fermi' LVII, Varenna on Lake Como*, pp. 109–146, Academic Press, London et New York, 1977.

139. Darrigol, O., *From c-Numbers to q-Numbers: The Classical Analogy in the History of Quantum Theory*, University of California Press, Berkeley, 1992. http://ark.cdlib.org/ark:/ 13030/ft4t1nb2gv/.

140. Dirac, P. A. M., "The fundamental equations of quantum mechanics", *Proceedings of the Royal Society, London* **A109**, 642–53, 1925.

141. Dirac, P. A. M., "Quantum mechanics and a preliminary investigation of the hydrogen atom", *Proceedings of the Royal Society, London* **A110**, 561–79, 1926.

142. Dirac, P. A. M., "On the Theory of Quantum Mechanics", *Proceedings of the Royal Society, London* **A112**, 661–677, Received August 26, 1926.

143. Dirac, P. A. M., "The Quantum Theory of the Electron", *Proceedings of the Royal Society, London* **A117**, 610–624, 1928.

144. Gordon, W., "Der Comptoneffekt nach der Schrödingerschen Theorie", *Zeitschrift für Physik* **40**, 117–133, 1926.

145. Klein, O., "Elektrodynamik und Wellenmechanik vom Standpunkt des Korrespondenzprinzips", *Zeitschrift für Physik* **41**, 407–442, 1927.

146. Heisenberg, W., in *Électrons et photons, rapports et discussions du cinquième conseil de physique Solvay tenu à Bruxelles du 24 au 29 octobre 1927*, p. 269.

147. Dennison, D. M., "A note on the Specific Heat of the Hydrogen Molecule", *Proceedings of the Royal Society, London* **A115**, 483–486, 1927.

148. Hori, T., "Über die Analyse des Wasserstoffbandenspektrums im äußersten Ultraviolet", *Zeitschrift für Physik* **44**, 834–854, 1927.

149. Pauli, W., "The Connexion between Spin and Statistics", *Physical Review* **58**, 716–722, October 1940.

150. Heisenberg, W., *Physics and Beyond*, p. 63.

151. Descartes, R., *Meditations and other metaphysical writings*, Translated with an introduction by Desmond M. Clarke. Penguin, London, 1998. Original edition: *Meditationes de prima philosophia*, L. & D. Elzevirios, Amsterdam, 1641.

152. Heisenberg, W., "The Development of the Interpretation of the Quantum Theory", in *Niels Bohr and the development of physics*, edited by Pauli, W., pp. 12–29, Pergamon Press, London, 1955.

153. Heisenberg, W., *Physics and Beyond*, p. 77.

154. Heisenberg, W., "Über den anschaulichen Inhalt der quantentheoretischen Kinematik und Mechanik", *Zeitschrift für Physik* **43**, 172–98, 1927.

155. *Électrons et photons, rapports et discussions du cinquième conseil de physique Solvay tenu à Bruxelles du 24 au 29 octobre 1927*, Gauthier-Villars, Paris, 1928.

A Timid Infancy

For the Snark's a peculiar creature, that won't
Be caught in a commonplace way.
Do all you know, and try all that you don't:
Not a chance must be wasted to-day.

Lewis Carroll, *The Hunting of the Snark.*

The Atomic Nucleus in 1913

In the second of the three famous papers,[1] which Niels Bohr wrote in 1913, and in which he described the properties of atoms in terms of quantum physics, he concluded that radioactive phenomena most likely occur in the atomic nucleus:

> *A necessary consequence of Rutherford's theory of the structure of the atom is that α-particles originate in the nucleus. Similarly, in our theory, it seems necessary to assume that the high velocity β-particles are emitted from the nucleus [1].*

What could be the composition of this incredibly small nucleus, as Rutherford qualified it, which is ten or a 100 000 times smaller than the atom, itself only one ten-millionth of a millimeter?[2] In his paper, Rutherford does not address the problem. It is however addressed at the second Solvay Council, held in Brussels on October 27–31, 1913. In the discussion of the report, given by J. J. Thomson, it appears clear to both Marie Curie and Rutherford, that radioactive phenomena, β radioactivity in particular, in which electrons are emitted, originate in the nucleus, *which therefore must contain electrons.* However, Paul Langevin asks what determines the number

---

[1] See p. 108.

[2] See p. 51.

B. Fernandez and G. Ripka, *Unravelling the Mystery of the Atomic Nucleus: A Sixty Year Journey 1896 — 1956*, DOI 10.1007/978-1-4614-4181-6_4, © Springer Science+Business Media New York 2013

of electrons which are detected by Barkla?[1] Marie Curie thinks that they are *peripheral* electrons orbiting around the nucleus and totally separated from the electrons which lie inside the nucleus and which occasionally escape, giving rise to β-rays, because:

> Such an electron is characterized by the fact that it cannot be separated from the atom which contains it without necessarily destroying the atom [2]..

On December 6, 1913, Rutherford sends a short note to *Nature* in which he describes his view of the structure of the nucleus:

> There appears to me no doubt that the α particle does arise from the nucleus, and I have thought for some time that the evidence points to the conclusion that the β particle has a similar origin. This point has been discussed in some detail in a recent paper by Bohr (Phil. Mag., September, 1913) [3].

At this point, everybody agrees that:

1. Radioactive phenomena stem from the nucleus.
2. The nucleus probably contains α-particles because they are emitted at great speed in the so-called α radioactivity. It must also contain electrons since it emits them violently during the so-called β radioactivity.

However, the electrons, which Marie Curie proposed to qualify as *essential*, are indeed strange: except for being confined in the nucleus, they appear to play no role in the atom; they do not mix with the peripheral electrons which are orbiting at distances about 10 000–100 000 times larger than the size of the nucleus. Furthermore, the electrons emitted during β-decay have much higher velocities than the peripheral electrons. One is led to conclude, with Marie Curie, that the electrons inside the nucleus do not interact with X-rays and that they are not counted in the X-ray measurements of Barkla, which concern only peripheral electrons. This is really odd.

What else is known about the atomic nucleus? Precious nothing, only the upper limit of its size (about 10^{-12} cm), which Rutherford had deduced from the experiments of Geiger and Marsden.[2] And that this pinhead-sized nucleus contained 99.95 % of the mass of the atom! Its internal structure remained totally unknown.

The following 20 years belong to atomic physics. The greatest minds of the century tackle the task of understanding the structure of the atom. Quantum mechanics is created for this purpose. Only a handful of physicists, among whom Rutherford, will pursue the study of the atomic nucleus, thereby opening a new field: nuclear physics.

---

[1] See p. 60.
[2] See p. 72 and 74.

The Discovery of Isotopes and the Measurement of Masses of Nuclei

The chemist Frederick Soddy is led to assume that atoms of a given element do not all have the same mass. He calls "isotopes" atoms of the same element which differ in mass. The chemist Francis Aston becomes a physicist and learns to measure the masses of atoms, thereby discovering dozens of new isotopes. It becomes apparent that the masses of atoms can be expressed in integer units of a given mass, just like their electric charge. But it turns out, finally, that the masses are not exactly integer units of the given mass.

The Chemistry of Radioactive Products

By about 1910, the radioactive families included numerous radioactive particles with short half-lives. They were endowed with rather barbaric names such as, in the case of thorium, "MsTh1 (mesothorium 1), MsTh 2 (mesothorium 2), RaTh (radiothorium), ThX (thorium X), Tn (thoron), ThA, ThB, ThC, ThC', ThD, ThC." The names all involve "thorium" simply to indicate that they originate from the radioactive decay of thorium, with no reference to their chemical properties, by which they differ from thorium.

The study of their chemical properties was difficult because most of them decayed in a short time and only small amounts would accumulate in the samples of uranium, thorium, or radium. Marie Curie already had a hard time identifying and isolating radium, in spite of its half-life of 1600 years. Furthermore, thorium decays first into "mesothorium 1," which itself decays into "mesothorium 2," which in turn decays into radiothorium, and so on. A given sample therefore always consists of a mixture of the decay products. However, chemists stubbornly tackled the problem, and they gradually succeeded in determining the chemical properties of a large number of the decay products.

One fact became gradually clear: some of the products had very similar, or even identical, chemical properties. For example, it was impossible to distinguish the chemical properties of thorium and those of so-called ionium, which was produced during the radioactive decay of uranium.[1] There seemed to be no difference between the chemical properties of radiothorium 1 and lead.

---

[1] The name "ionium" was coined by Bertram Boltwood who identified it as the parent of radium [4]. It is a decay product of uranium 238 and was later shown to be thorium 230.

Frederick Soddy

During the years 1900–1902, Rutherford, who was then professor at the University of McGill in Montreal (Canada), succeeded in showing that radioactivity was a transformation of atoms, a decay in fact, even a succession of decays. He worked this out in collaboration with a young chemist, Frederick Soddy,[1] who had come from England. Born in Sussex, on September 2, 1877, Soddy was the son of a London merchant [5]. In 1898, he defended his PhD thesis in chemistry in the University of Oxford. After two years of research, he obtained the position of assistant demonstrator in chemistry at the University of McGill, where Rutherford enlisted him for a brief but fruitful collaboration. In the spring of 1903, Soddy accepted a position offered to him by William Ramsay in the University College of the University of London. Ramsay and Soddy showed that helium was produced during the decay of radium.[2]

From 1904 to 1914, Soddy was *lecturer* in the University of Glasgow. People remembered him there as a man of principle, friendly with the students, and often caustic with his colleagues. In Glasgow, he undertook a detailed study of all the short-lived radioactive substances, which he and Rutherford proposed to call *metabolons*.

Isotopes

Beginning 1904, the *Chemical Society* began publishing annual reports on the progress in chemistry, and the chapter devoted to "Radioactivity" was entrusted to Frederick Soddy, who thereafter signed the *Annual Progress Reports on Radioactivity* every year from 1905 to 1908, from 1910 to 1915, and then every 2 years until 1921. These reports have been reedited in facsimile, under the title *Radioactivity and Atomic Theory*, together with a well-documented introduction by Thaddeus J. Trenn [6]. They form a precious record of the evolution of experiments and ideas during this period, viewed by a great physicist and chemist. In his 1911 report, Soddy assesses the progress made in 1910. He begins by describing the methods which were used to determine the chemical properties of the *metabolons*:

> A method of determining the chemical nature of a member of a disintegration series by isomorphism consists in adding varying salts to the solution, allowing them partly to crystallize out, and determining which kinds of salts crystallize with the active material.

After describing other examples, he is led to an inevitable conclusion, which is so surprising that he takes great care to state it only after a rigorous reasoning:

---

[1] See p. 24.
[2] See p. 35.

> *These regularities may prove to be the beginning of some embracing generalization, which will throw light, not only on radioactive processes, but on the elements in general and the Periodic Law [. . .] The complete identity of ionium, thorium, and radiothorium, of radium and mesothorium-1, and of lead and radium-D, may be considered thoroughly well established. Indeed, when it is considered what a powerful means radioactive methods of measurement afford for detecting the least change in the concentration of a pair of active substances, and the completeness and persistence of some of the attempts at separation which have been made, the conclusion is scarcely to be resisted that we have in these examples no mere chemical analogues, but chemical identities [7].*

A daring idea, the so-called radium D, the residue of the decay of radium, has all the chemical properties of lead, and it differs only by its atomic mass and by its radioactivity; *it is therefore chemically identical to lead.* Soddy thinks that this property of matter may also apply to nonradioactive elements. Such examples simply had not been observed in the absence of radioactivity, in which case they would have been identified.

In a 1911 paper, Soddy finds that the substances called *mesothorium X* and *thorium X* are chemically equivalent to radium:

> *It appears that chemistry has to consider cases, in direct opposition to the principle of the Periodic Law, of complete chemical identity between elements presumably of different atomic weight[1], and no doubt some profound general law underlies these new relationships [8].*

What Soddy calls the "periodic law" is the Mendeleev's table, or "periodic table of elements," which lists the elements in order of increasing mass. It was established in 1869 by the Russian chemist Dmitri Mendeleev and the German chemist Lothar Meyer, who observed that, if the elements are ordered by increasing mass, elements with similar chemical properties reappear regularly [9–12]. The Mendeleev table assumes implicitly that a single element corresponds to a given mass.

However, the idea that all atoms of a given element may not necessarily have the same mass, had been formulated some 25 years earlier by the famous chemist William Crookes, in a talk presented to the Chemistry Section of the British Association[2] in Birmingham:

> *I conceive, therefore, that when we say the atomic weight of, for instance, calcium is 40, we really express the fact that, while the majority of calcium atoms have an actual atomic weight of 40, there*

---

[1]The term *atomic weight* was used at that time. Today, we prefer to use the term *atomic mass*. See these terms in the Glossary.

[2]The British Association for the Advancement of Science (formerly known as the BA) was founded in 1831. It is a learned society with the object of promoting science, directing general attention to scientific matters.

*are not a few which are represented by 39 or 41, a less number by 38
or 42, and so on [13].*

But whereas Crookes was simply speculating on this possibility, Soddy showed
that it was absolutely necessary.

Two years later, Soddy proposed a name for the different varieties of a given
chemical element [14]. Since they have identical chemical properties, they occupy
the same position in the periodic table of Mendeleev, the bible of chemists. Soddy
proposed to call them "isotopes," from the Greek words *isos,* meaning "equal,"
and *topos*, meaning "location." All the isotopes of a given element belong to the
same position in the Mendeleev classification because they are varieties of the same
element.

Indeed, the existence of isotopes was also a consequence of the *displacement
law*, which was formulated almost simultaneously by the Polish chemist Kasimir
Fajans [15–20] and Frederick Soddy [21]: "the expulsion of an α particle causes
the element to shift its position in the Periodic Table by two places in the direction
of diminishing atomic number (*i. e.* of the charge of the nucleus). The expulsion
of a β particle causes a shift of its position of one place in the opposite direction."
Therefore, the new elements which result from radioactivity (the "metabolons" of
Rutherford), are assigned a position in the periodic table *which is already occupied.*
In the language of Soddy, they are "isotopes" of this element.

Soddy became professor of chemistry in the University of Oxford in 1919, and
he remained there until he retired in 1937. In 1921, was awarded the Nobel Prize
in chemistry "for his contributions to our knowledge of the chemistry of radioactive
substances and for his research on the origin and the nature of isotopes." He died in
Brighton on September 22, 1956.

The Revival of Positively Charged "Canal Rays"

In 1886, the German physicist Eugen Goldstein was studying cathode rays. He
noticed that in his Crookes tube, some rays propagated in a direction opposite to
that of cathode rays. If one bores a hole (a "channel" or a "canal") which crosses the
cathode perpendicularly to its surface, some "rays" pass through this canal and they
form, when they emerge on the opposite side, a luminous pencil of light, the color
of which depends on the residual gas. Since they travel in the opposite direction to
the cathode rays, they carry a positive electric charge. In due time, the study of such
"canal rays" will become increasingly important.

The First Physical Measurements of Atomic Masses

In 1907, J. J. Thomson undertakes the study of canal rays which he proposes to call
simply "positive rays" [22, 23]. He studies their nature, their charge, and their mass.

The positive charge rays are produced by an electric discharge in a tube, similar to the one used to produce cathode rays. The rays pass through a narrow canal, and they are then subject to electric and magnetic fields which deviate them in perpendicular directions. A simple calculation shows that ions of given mass and electric charge[1] will produce a spot with a parabolic shape on a fluorescent screen,[2] different points of the parabola corresponding to different velocities of the ions. This so-called *parabola method* became quite famous [24]. The curvature of the parabola depends on the mass of the ions and permits to distinguish ions of different mass. J. J. Thomson perfects his apparatus and decides to record the parabolas on photographic plates. The method is not very precise (one cannot distinguish masses which differ by less than 10 %) but it is the first direct measurement of an ion mass. If isotopes of normal nonradioactive ions exist, this is a way to detect them. In fact, this is just what J. J. Thomson observed in 1913: a parabola was formed by neon ions which appeared to be thicker than the other parabolas, as if, in addition to the "normal" neon ion with atomic mass close to 20, there would be another hitherto unknown ion with atomic mass of about 22. He suspected the existence of an isotope of neon, which would be the first nonradioactive isotope observed so far and which would explain why the atomic mass of neon, which had been measured 3 years earlier,[3] was found equal to 20.2: neon could be a mixture of isotopes with masses 20 and 22, with, respectively, 80 % and 20 % concentrations. However, the isotope 22 still remained to be identified. On January 17, 1913, J. J. Thomson presents his results to the Royal Institution and he describes a faint parabola, which occurs close to the neon parabola, as being due to ions with mass 22. He is however unable to conclude [26]. Shortly afterwards, he publishes a book on positively charged rays [27], which will remain a reference for long time. During the following 2 years, his assistant, the young Francis Aston, will attempt to separate the neon isotopes, using first fractional distillation and later gas diffusion, however without success.[4]

Francis Aston and the First Mass Spectrometer

Francis Aston was born in Birmingham in 1877. After graduating in chemistry from the University of Birmingham, he spent 3 years working in the laboratory

---

[1]In such experiments, the electric charge is always the same, namely, one unit of positive elementary charge, because one electron is torn out during the ionization of any atom.

[2]The screen was made of willemite, a zinc silicate crystal.

[3]The English chemist Herbert Watson had measured the density of neon [25] and this was sufficient to deduce its mass. Indeed, according the hypothesis of Avogadro, two identical volumes of any gas, under the same pressure, contain the same number of molecules. The masses of molecules are therefore in the same ratio as their densities.

[4]The time taken by a gas to pass through a porous membrane is shorter for a light gas than for a heavy gas. The mixture of different gases will therefore be different after it passes through a porous membrane. Aston attempted to use this property to isolate the isotope 22 of neon.

of a brewery. However, his interest in physics and the talent which he displayed in constructing vacuum pumps, led him to physics. In 1903, he obtained a grant to work in the University of Birmingham on discharges in Crookes tubes. He soon discovered a new phenomenon (the "dark space of Aston"). At the end of 1909, J. J. Thomson invited him to become his assistant at the Cavendish and this is where we find him, in 1912–1914, trying to separate neon isotopes. Unfortunately, the research is interrupted during the First World War, during which he studies various coatings which could strengthen the resistance of aeroplane cloth to atmospheric conditions.

In 1919, after the end of the war, Aston returns to the Cavendish, where he again tries to separate neon isotopes. After his previous unsuccessful attempts, he considers first using magnetic fields, but the method requires a far greater separating power than that which had been achieved by J. J. Thomson. Aston does not see how to improve the apparatus without making drastic changes. The problem is that, no matter how fine the "channel" is, the particles penetrate it in slightly different directions, thereby blurring the produced spot. One cannot reduce the diameter of the channel indefinitely, because then too few particles pass through and it takes far too long to expose the photographic plate. The same problem was encountered with the pinhole camera, which consisted of a chamber with a small hole at one end and a photographic plate at the other. The rays, emitted from a point of the subject which is being photographed, pass through the pinhole and produce a small spot on the photographic plate. The resolution of the image depends on the diameter of the small pinhole. Besides being simple, the pinhole camera presents several advantages: a large depth of field and no distortion. But it also has disadvantages: the pictures are never very sharp and the exposure times are very long since little light passes through the small pinhole. In cameras, this problem is solved by the use of lenses which focus *all* the rays, emanating from a point on the subject to a single point on the photographic plate. The lens aperture can then be considerably larger than a pinhole, and pictures can be recorded much faster.

Aston discovers that a result, similar to that of the lens of a camera, can be obtained with the help of magnetic fields. Previously, electric and magnetic fields were only used to deviate the particles in different directions in order to separate particles which have the same mass and charge, but different velocities. Particles with a given velocity were made to pass through a small slit, exactly like light rays through a pinhole. The apparatus shared the disadvantages of the pinhole camera. However, a magnetic field, if judiciously placed, produces an unexpected effect: it focuses onto one point (in fact a small but finite-sized point, as in optics) all the particles with a given mass, charge and velocity *which penetrate the apparatus at different angles*. When the photographic plate is placed in the "focal plane" (which is the position where the spot has the smallest size), ions with different masses appear as distinct spots. The image thus formed can distinguish particles with different masses. Aston succeeds in building this apparatus which he calls a "mass spectrometer" [28]. It is capable of resolving mass differences as small as 1 %. Somewhat later, in a collaboration with Ralph Fowler, he made a detailed mathematical study of his mass spectrometer [29]. His first application was naturally devoted to neon. He was able to prove unambiguously what Thomson had suspected, namely, that neon is composed of two isotopes, with masses 20 and 22[1]:

> The first suggestion that it could be a mixture [of isotopes] was the observation, made in 1912 by J. J. Thomson, that a faint but unquestionable parabola at a position, corresponding to an atomic mass of roughly 22, [...] would appear each time neon was present in the discharge tube. The measurements [...] conclusively show that neon contains two isotopes, with atomic masses 20.00 and 22.00, within a precision of one tenth of a percent [30].

From then on, Aston explores many other elements and within a few years, he discovers several elements, which are composed of two or more isotopes. In a publication submitted to the *Philosophical Magazine* in 1920, he lists 11 such elements. In addition to neon, he reports two or more isotopes of chlorine, argon, krypton, xenon, and mercury. He discovered in fact six isotopes of xenon [31]! The elements studied were chosen mostly for technical reasons. Hydrogen, helium, nitrogen, oxygen, neon, chlorine, argon, krypton, and xenon are gases at room temperature, and others form gaseous compounds (carbon forms carbon dioxide). It is also easy to form mercury vapor. The point is that, in a discharge tube, it is far easier to form positive ions from gases than from solids.

In 1922, Aston publishes his first book, *Isotopes* [32], which contains an impressive list of 48 identified isotopes measured in 27 elements. They include isotopes of sodium, magnesium (with three isotopes), silicon, phosphorus, sulfur, chlorine (at least two isotopes), and so on.

The "Whole Number Law" and the Old Hypothesis of William Prout

There was one fact which Aston, as well as all those who examined the masses of isotopes, could not fail to notice:

> By far the most important result of the measurements detailed in the foregoing chapters is that, with the exception of hydrogen, the weights of the atoms of all the elements measured, and therefore almost certainly of all elements, are whole numbers to the accuracy of experiment, in most cases about one part in a thousand. Of course, the error expressed in fractions of a unit increases with the weight measured, but with the lighter elements the divergence from the whole number rule is extremely small.

---

[1]Recall that, at that time, the unit of mass was 1/16 of the mass of oxygen. A mass 20 means 20/16 of the mass of oxygen.

This enables the most sweeping simplifications to be made in our ideas of mass, and removes the only serious objection to a unitary theory of matter [33].

Aston then summarizes the theory of William Prout, mentioned above[1]:

The first definite theory of the constitution of the atoms of the elements out of atoms of a primordial element (Protyle, Urstoff, etc.) was made by Prout in 1815. Prout's Hypothesis was that the atoms of the elements were different aggregations of atoms of hydrogen. On this view it is obvious that the atomic weights should all be expressed by whole numbers when the atomic weight of hydrogen itself is taken as unity.

Recall that the hypothesis of Prout was given up when it became known that numerous atomic masses did not have integer values. Chlorine, for example, had a mass equal to 35.46. However, all changed when isotopes were discovered, because *the masses of isotopes were found to be integer numbers.* Indeed, the existence of isotopes provided a simple explanation of the fact that some atomic masses have close to integer values (4 for helium, 12 for carbon, 16 for oxygen) and that others differ significantly from integer values, as in the case of chlorine. Thus, carbon is a mixture of two isotopes with masses equal to 12 and 13. In nature, the abundance of the two isotopes is 98.9 % and 1.1 %, respectively. This is why the atomic mass of carbon is so close to 12. Similarly, oxygen is a mixture of three isotopes, of masses 16, 17, and 18. The respective abundances are 99.76 %, 0.04 %, and 0.2 % so that the oxygen in nature has a mass very close to 16. However, chlorine is a mixture of two isotopes with masses equal to 35 and 37. They occur with abundances which are, respectively, 76 % and 24 % so that the measured mass of chlorine in nature is equal to 35.46. Aston concludes:

The only serious obstacle, the fractional atomic weights, has now been removed so that there is nothing to prevent us accepting the simple and fundamental conclusion:—The atoms of the elements are aggregations of atoms of positive and negative electricity.

This is what Aston calls the "unified theory of matter." It was indeed very tempting to assume that all atomic nuclei are composed of a single *elementary particle* which could only be the nucleus of hydrogen and which was therefore called the *H-particle*. In a meeting of the *British Association*, held in Cardiff in 1920, Rutherford proposed to call this particle either a *proton* or a *prouton*, in recognition of William Prout [34]. The term *proton* survived and was universally adopted.

---

[1] See p. 50.

The Exceptional Mass of the Hydrogen Atom

Recall that chemists were only able to measure ratios of masses of various elements. Therefore, they had to choose a reference element. They chose the element oxygen, which was attributed a mass equal to 16 because, this way, hydrogen had a mass close to 1 and many elements had almost integer masses. When one says that an element has a mass equal to 32, for example, it means that it is twice heavier than oxygen.

The mass of hydrogen was however too large, equal to 1.008. That may appear to be very close to 1, but the mass of hydrogen was measured with an accuracy of plus or minus 0.001 so that it could not have an value equal to 1. What did that signify? It seemed clear at the time that the weight of several hydrogen nuclei, assembled to form a nucleus of a heavier atom, was less than the weight they would have if they were separated. The explanation proposed by Aston in 1922 was based on the "electromagnetic theory of mass," a theory given up today [35]. The idea was first proposed by J. J. Thomson in 1881 and taken up by several physicists, in particular by Max Abraham, who developed it with the hope of deducing from it all of physics, at least the laws of mechanics, and of electromagnetism. This is an age-old dream of all physicists: to formulate a unified theory of matter, mechanics and electromagnetism. Abraham reasoned as follows: if I wish to make an electrically charged particle move, I need to accelerate it. It will then necessarily radiate an electromagnetic wave which has a certain energy. I must furnish this energy and therefore the particle will try to resist the imparted change of velocity, and the resistance will be the same as that of a mass. This is the phenomenon called induction, well known to electricians. Thus, in addition to its usual mass, the charged particle displays an inertia due to its electric charge, and this inertia behaves exactly as a mass. It could be considered to be an *electromagnetic mass*. This theory seduced many physicists, among whom Wilhelm Wien and Hendrik Lorentz. And what if this was the origin of all mass? This is what Abraham thought. In 1902, in a meeting held in Karlsbad, he declared:

The mass of the electron is of purely electromagnetic nature [36].

In this theory, a relation existed between the mass and the size of a particle, assumed to be a uniformly charged little sphere. The mass became larger as the radius of the particle decreased. The theory assigned a radius to the electron, which we still call today the "classical radius of the electron".[1] The nucleus of the hydrogen atom, which is 1,836 times heavier, should, according to this theory, be much smaller. At the time, it was believed that the hydrogen nuclei were very closely packed in nuclei and that this packing diminished the mass of the nucleus. Aston called this the *packing effect*:

---

[1] The classical radius of the electron is equal to 2.82×10^{-13} cm, which is comparable to the size of an atomic nucleus. But today it is no longer believed to be the size of the electron, which is considered to be a point particle.

In the nuclei of normal atoms the packing of the electrons and protons is so close that the additive law of mass will not hold and the mass of the nucleus will be less than the sum of the masses of its constituent charges [37].

However, another explanation appeared in 1905. In a three-page paper published 4 months after the paper in which he formulated the theory of relativity [38], Einstein raised the question: "Does the mass of a body depend on the energy which it contains?" He showed that the equations of the theory of relativity had an important and unexpected consequence:

If a body emits an energy L in the form of radiation, then its mass diminishes by L/v^2 [. . .] the mass of a body is a measure of its energy content; if the energy changes by a quantity L, then its mass changes in the same direction by $L/9.10^{20}$, when the energy is expressed in ergs and the mass in grams. One cannot exclude that bodies, the energy content of which undergo a large variation (radium salts for example) could test the theory. If the theory explains facts correctly, then radiation transports mass from the emitting body to the absorbing body [39].

To the question expressed in the title of the paper, Einstein replies positively: *energy is equivalent to mass.* Today, this is expressed by the most famous formula $E = mc^2$. We see that from the outset, Einstein attempted to confront his theory to experiment. This required the measurement of the variation of the mass of a body before and after it had emitted a radiation. In usual phenomena, such as chemical reactions or the cooling of a body heated to a high temperature, the energy loss corresponds to an almost infinitesimally small variation of mass which there is no hope to measure even by the most delicate scale. Indeed, as Einstein notes, one has to divide the energy loss by 9×10^{20}, which is 9 followed by 20 zeroes! That is precisely why Einstein mentions "radium salts," more generally radioactivity, a phenomenon which involves energies about a million times larger than chemical reactions. But even in this case, the variation of the mass is extremely small.

In 1913, Paul Langevin, one of the first in France to have understood the importance of relativity, published a paper on the subject. He showed that the variation of the mass of a body was always extremely small except for the case of radioactive transformations, and therefore of nuclei [40]. In the paper, written in 1921 for the *Encyclopedia of Mathematics*,[1] Pauli came to the same conclusion:

Perhaps we shall be able, in the future, to verify the principle of the inertia of energy by observing the stability of nuclei.

The stability, which Pauli refers to, measures the energy required to break up a nucleus, in order to separate its constituents. This is what we call its *binding energy*.

---

[1]See p. 121.

A very stable nucleus, composed of given constituents, must therefore be somewhat lighter than an unstable nucleus.

Was it so difficult for theories, mainly disturbing theories, to cross the English Channel? Or did an experimentalist, such as Aston, consider that such theories could only be of interest to a handful of theoreticians, without realizing what practical consequences they might have, in particular concerning nuclear masses? One must recall that the electromagnetic theory of matter offered another explanation of the loss of mass, an explanation which was well rooted in many minds and appeared to be simple, natural, and almost inevitable. Furthermore, in 1906, certain experiments appeared to favor the theory of Abraham rather than that of Einstein [41]. But relativity definitely won in the years 1915–1920, in spite of the fact that Aston had still not adopted it in 1926.

A Nobel Prize for the "Whole-Number Rule"

Although pleasant in his manner, Francis Aston was reserved and even timid. This made him a rather bad teacher. He loved music, and he played the piano, the violin, and the cello. He was proficient in skiing, mountain climbing, tennis, and swimming. He died in Cambridge in 1945. In 1922, he was awarded the Nobel Prize in chemistry "for his discovery, by means of his mass spectrograph, of isotopes, in a large number of nonradioactive elements, and for his enunciation of the whole-number rule." This "whole-number rule" was indeed extraordinary. It explained why certain atomic masses are almost integer, whereas other are not at all. In his introductory speech at the Nobel prize ceremony, H. G. Söderbaum, a member of the Chemistry Nobel Prize Committee and of the Academy of Sciences, expressed it thus:

At the present standpoint of science, the simplest small parts of matter must be conceived as consisting of two essentially different kinds, namely of positively and negatively charged small particles, protons and electrons.

The broken numbers in the atomic weights of certain fundamental substances, in fact, now appear simply as statistical effects of the internal quantitative relations of their isotopic constituents.

Thanks to the measurements of Aston, people were now convinced that atomic nuclei were composed of identical particles with a mass close to the mass of hydrogen. It was natural to assume that the constituents were in fact hydrogen nuclei. There was however a problem. The electric charges of nuclei were known from the measurements of Moseley. Their charge was equal to the atomic number in the Mendeleev classification. However, the number of hydrogen nuclei required to yield the observed mass yielded an electric charge which was roughly twice the observed charge for light elements. The nucleus must therefore contain a certain number of negative electric charges in order to neutralize the charge excess. It was natural to assume that electrons could do the job. The fact that the β-rays emitted by some radioactive substances were in fact electrons, seemed to confirm this view. The electron mass is 1836 times smaller than the mass of the hydrogen nucleus, and it could therefore be neglected. A clear picture of the nucleus appeared to emerge.

At the *Ryerson Physical Laboratory*, in Chicago, Arthur Dempster constructed a spectrometer similar to that of Aston, but with some differences [42]: the particles were made to describe a semicircle trajectory in a magnetic field, and they were detected in an electrometer instead of impinging on a photographic plate. The method had the advantage of being quantitative. It allowed one to measure the relative abundance of the different isotopes of a given element. The performance of this spectrometer was similar to that of Aston (it had a precision$^\diamond$ of one part in a 1000). Dempster published a series of papers on his measurements of lithium, magnesium, zinc, calcium, and potassium [43, 44]. In the *Laboratoire de Chimie Physique* of Jean Perrin in Paris, J.-L. Costa constructed a spectrometer which was better than the one of Aston (with a precision of three parts in 10 000). This enabled him to measure the masses of the isotopes 6 and 7 of lithium. These measurements remained a reference for a long time because it was particularly difficult to produce positive ions of lithium [45, 46]. However, Aston did not remain inactive. Already in 1921 he was thinking of ways to improve on the precision, and he decided to build a new spectrometer. In 1925, he dismounted his first spectrometer and replaced it with a new one which had a precision of two parts in a 1000 and which could determine atomic masses with a precision of one part in 10 000. Aston decided to make new measurements of all the possible atomic masses with his new spectrometer.

The Atomic Masses Known in 1932: The Binding Energy of Nuclei

In June 1927, Aston was invited to deliver the prestigious Bakerian lecture at the Royal Society in London.[1] This provided him with an occasion to present his new spectrometer and the results which it yielded [47]. Six years later, the second edition of his book *Mass Spectra and Isotopes* [48] allows one to assess the progress made. *None of the masses are found to be integer numbers.* Although some come close to integer numbers, each atomic mass deviates from an integer value by an unquestionable amount. By 1927, the theory of relativity had permeated the minds of physicists, so that the mass defect, which measures the difference between the mass of a nucleus and the sum of the masses of its constituents, assumed to be hydrogen nuclei, was understood as meaning the binding energy$^\diamond$ of the nucleus, that is, the energy which binds the constituents of the nucleus. This energy was immensely large. Aston notes this in his book:

---

[1] See p. 35.

Two results which were first demonstrated by the mass spectrograph, firstly the whole-number rule, which showed the high probability of all atoms being composed of the same ultimate units, and secondly the fact that one helium atom does actually weigh less that four hydrogen atoms, have profound and far-reaching theoretical implications.

We know from Einstein's Theory of Relativity that mass and energy are interchangeable [...] Even in the case of the smallest mass this energy is enormous [...]

Take the case of one gramme atom of hydrogen[...] If this is entirely transmuted into helium the energy liberated will amount to about 6.5×10^{18} ergs, or 200,000 kW h. This transformation of mass into radiation by the partial or complete annihilation of matter is the so-called "atomic energy" believed to be the source of the heat of the stars and which, it was predicted, might be tapped and used when a means of artificial transmutation of elements was discovered [49].

The vision of Aston was correct. Nuclear fusion is indeed the source of solar energy. However, so far, nuclear fusion has only been used destructively in the hydrogen bomb, the terrifying "H-bomb."

An Enquiry Full of Surprises: β Radioactivity

We meet a German chemist, an Austrian physicist and two English physicists. At first, β radioactivity appeared to be similar to α radioactivity. But the β spectrum soon turned out to be a continuous spectrum. This seemed be unbelievable at first, incomprehensible and even scandalous on second thoughts. The great magician Pauli pulls out of his hat an explanation of what was incomprehensible, namely an unobservable particle [50].

Rutherford was the first to distinguish α- and β-rays, the former being later identified to nuclei of the helium atom and the latter to high-speed electrons.[1] But exactly at what speed are the electrons ejected from the nucleus? How are they slowed down as they pass through matter? In his first observations of the radiation of uranium, Rutherford measured the ionization of air exposed to radiation.[2] The ionization is produced by electrons ejected from the atoms by the radiation, thereby producing ions which in turn make air become an electric conductor. The electric current, passing through the air, is a measure of the intensity of the radioactivity.

By measuring the current produced by the β-rays after they had passed through aluminum sheets of various thicknesses, Rutherford measured how the intensity of β-rays decreased progressively as they passed through matter:

The β radiation passes through all the substances tried so far with far greater facility than the α radiation. For example, a plate of thin cover glass placed over the uranium reduced the rate of leak to one-thirtieth of its value; the β radiation, however, passed through with hardly any loss of intensity. Some experiments with different thicknesses of aluminum seem to show, as far as the results go, that the β radiation is of an approximately homogeneous character [51].

He adds:

The intensity of the radiation diminishes with the thickness of metal traversed according to the ordinary absorption law.

This "ordinary absorption law" is the so-called *exponential* law, sometimes also called the *geometric* law. It had been stated by Pierre Bouguer in 1760 in his *Traité d'optique sur la gradation de la lumière*.[3] Bouguer had studied the attenuation of light which passed through tinted glass of various thickness:

---

[1] See p. 19.

[2] See p. 18.

[3] Optics treatise on the gradation of light.

If a given thickness intercepts half of light, a second equal thickness will not intercept the other half, but only half of the other half. It will reduce the intensity to a quarter: all the other layers will destroy similar parts, it follows that the light will diminish as a geometrical progression [52].

For Rutherford as well as for the other physicists of that time, this was the ordinary law of the attenuation of *radiation*. What is meant by "homogeneous radiation" is still quite vague in 1899. What Rutherford probably means is that the β radiation does not display a complex nature, which could, for example, suggest a mixture of several different radiations.

The Velocity of the β Electrons

In 1904, W. H. Bragg showed that all the α-particles, emitted by a given radioactive substance, had the same velocity[1] and that this velocity appeared to characterize the emitting radioactive substance. As they passed thorough matter, the β electrons appeared to follow, at least approximately, the exponential law. This was compatible with the assumption that they were all emitted by a given radioactive substance with a fixed velocity. Between 1905 and 1907, several works seem to indicate this, but none is really conclusive. In his 1908 annual report on radioactivity, Frederick Soddy does not commit himself:

The whole question of the nature of the beta rays, whether homogenous or heterogeneous as regards velocity, and the exact meaning to be attached to their "absorption" in passage through matter is, in spite of numerous researches, still in a highly controversial state and would scarcely repay very detailed discussion at the present stage [53].

In Berlin, the chemist Otto Hahn and a young Austrian physicist, Lise Meitner, tackle the problem.

Otto Hahn

Otto Hahn was born on March 8, 1879, in Frankfurt. He spent his youth there [54–56]. His father was a glazier and succeeded in creating quite a prosperous business. Otto went to the University of Marburg with the intention of becoming a chemist and getting a job in one of Germany's great chemistry companies, which were developing at the time. In 1901, he completed his PhD in organic chemistry, did his military service, and became assistant to his previous thesis advisor,

---

[1] See p. 63.

Theodor Zincke, in Marburg. Two years later, he decides to go to England, to perfect his English. He obtains a position in the laboratory of the chemist Sir William Ramsay and begins to work with him in the fall of 1904. Ramsay suggests that he should separate the radium contained in a solution of barium chloride, which he thought would contain about 10 mg of radium. While attempting to do so, Hahn discovers that the radioactive substance is not radium but an unknown substance which he called *radiothorium* [57], because the disintegration produces the same radiation as thorium. It was reasonable to assume that it was one of the substances produced in the disintegration chain of thorium. At first, Rutherford, at McGill in Montreal at the time, is quite skeptical. He expresses his reservations to his friend Boltwood who answered:

I bet that the substance he obtained is some Th-X mixed with some radium [58].

and somewhat later:

I think that is only a new compound of Thorium-X and stupidity [59].

In the summer of 1906, Hahn has the intention of returning to Germany, hoping to get a job as a chemistry engineer. Ramsay, who can appreciate the qualities of this modest, talented, tenacious, and rigorous researcher, tries to dissuade him from leaving. He advises him to seek an academic career in Berlin. He sends a very favorable letter of recommendation to Emil Fischer, the director of the Chemistry Institute in the University of Berlin. Fisher is willing to help him, but at the time, radiochemistry is neither taught nor investigated in Berlin. On the other hand, Hahn thinks that, in order to make a real research career in radioactivity, he needs to learn more about the subject. He writes to Rutherford suggesting to work in his laboratory during "one winter." Rutherford accepts, and Hahn goes to Montreal and shows him his results. Rutherford rapidly changes his mind about Hahn. In a letter, dated October 10, he writes to Boltwood:

Hahn has arrived and he began to work; he seems to be a subtle man [. . .] who lacks knowledge in physics but I hope to be able to change that. According to what he showed me, there is no doubt that he has separated a very active and rather permanent constituent of thorium [60].

Thus, radiothorium does indeed exist. It is a product appearing in the decay chain of thorium. It appears before thorium X, which Rutherford and Soddy had discovered in 1902. Hahn remains in Montreal until the summer of 1906. He makes friends there, among whom Rutherford whom he will always respect and admire.

He then obtains a position in the Chemistry Institute of the University in Berlin, directed by Emil Fischer. Since radiochemistry was not an official research subject in the Institute, Hahn cannot, at first, become officially the assistant of Fischer. Nonetheless, he is allotted a laboratory. He succeeds in obtaining radioactive sources, and he gets to work. He soon discovers a new radioactive substance, which he calls *mesothorium*, because it is also created in an intermediate step of the disintegration of thorium, before the generation of radiothorium. He actually discovers two such substances, which he calls *mesothorium I* and *mesothorium II* [61, 62]. In the spring of 1907, he becomes *Privatdozent* in chemistry, an obligatory stage before becoming professor. He feels quite isolated among the chemists, mostly because the chemistry he practices has little in common with that of the other chemists. He is better appreciated by physicists who are concerned with similar problems to his. He attends regularly the *Colloquium* of the Physics Institute, directed by Heinrich Rubens. In the German tradition, the *Colloquium* was a usually weekly talk on subjects covering a wider area of subjects than those of the ongoing research in the physics department. This is where, on September 28, 1907, he meets a young woman who had recently come from Vienna in order to follow the theoretical physics lectures of Max Planck. She is called Lise Meitner. The lecture schedule leaves her free time to do some experimental work, and Rubens suggests that she should work with Hahn. A 30-year-long collaboration begins.

Lise Meitner

Lise Meitner was born on November 7, 1878, in Vienna. Her family was Jewish but practiced religion only sparingly [63, 64]. Her father, Philipp Meitner, was a lawyer, and he belonged to the first generation of Jews who were allowed to practice law in the Austrian state. Without being rich, he could comfortably keep up his family, send his children to university, and offer them music lessons, as became in Vienna. From the outset, Lise was attracted to mathematics and physics. Unfortunately, she could not attend the *Gymnasium*, which was reserved for boys, and she could therefore not obtain the *Matura*, which was required to attend university. Girls usually left school at the age of 14 and would then devote themselves to domestic tasks and seek a husband. But in 1897, Austria finally admitted women to university, provided they obtain the *Matura*, without however giving them the possibility to prepare this exam in the *Gymnasium*, reserved for boys.

In the footsteps of her elder sister, Lise prepares the *Matura* with private lessons, passes the exam, and enters the university in the fall of 1901. In the first year, she studies mathematics and physics. After that, she chooses physics and follows the lectures of the great physicist and professor Ludwig Boltzmann. Lise Meitner begins to work on her thesis under the direction of Franz Exner, an assistant of Boltzmann. On February 1, 1906, she obtains her PhD with *summa cum laude*. The course of her life is determined. After having worked on optics, she decides to study radioactivity. She follows the seminar of Egon von Schweidler. Stefan Meyer, an assistant of Boltzmann, suggests that she should study the absorption of α- and β-rays passing through thin layers of various metals. She publishes her first work on radioactivity [65] in June 1906. At that time, she feels so unsure about pursuing research in science, that she passes an exam allowing her to teach in a girls school. She obtains the job which keeps her busy the day, and she does research at night [66]. The chances of obtaining a position in the university were practically nil for women in Vienna at the time. But her first two studies in physics prompt her to ask her father to pay for a stay of several months (she suggested 6 months) in Berlin in order to attend the theoretical physics lectures of the great Max Planck. She arrives in Berlin in 1907, at the age of 29. In spite of her dawning experience, she is timid, modest, and lacks self-assurance. But her master Boltzmann had endowed her with a real passion for physics. Her second passion is music. She does not play an instrument, although she had played the piano, but she attends many concerts, often with the score in hand. She has a warm and reserved manner, and she has the knack making faithful friends. She signs up for the course of Planck, who accepts her kindly (in spite of his opinions against intellectual work of women). His lectures give her ample free time which she wants to make the best of:

> I wanted to do some experimental work and approached Professor Rubens, head of the department of experimental physics in Berlin. He told me the only space he had was in his own laboratory, where I could work under his direction, that is, to a certain extent with him [66].

She is not seduced by the idea: she is timid and very impressed by Professor Rubens. An unexpected event arises:

> Now it was quite clear to me then, as a beginner, how important it would be for me to be able to ask about anything I did not understand, and it was no less clear to me that I should not have the courage to ask Professor Rubens. While I was still considering how I could answer without giving offense, Rubens added that Dr. Otto Hahn had indicated that he would be interested in collaborating with me, and Hahn himself came in a few minutes later. Hahn was of the same age as myself and very informal in manner, and I had the feeling that I would have no hesitation in asking him all I needed to know.

As it was in Vienna, a university career was far from obvious for a woman in Berlin in 1907. Otto Hahn worked in the institute directed by Emil Fischer, a wine and music amateur and a great chemist who was awarded the Nobel Prize in 1902, but who did not allow women to attend his lectures nor to penetrate into his laboratories. Nonetheless, Hahn asked him if Lise Meitner could work with him. She continues to recollect:

> I went to Fischer to hear his decision, he told me that his reluctance to accept women students stemmed from his constant worry with a Russian student, lest her rather exotic hairstyle should catch fire on the Bunsen burner. He finally agreed to my working with Hahn, if I promised not to go into the chemistry department where the male students worked and where Hahn conducted his chemical experiments. Our work was to be confined to a small room originally planned as a carpenter's workshop; Hahn had fitted it out as a room for measuring radiation. For the first few years I was naturally restricted to this work and could not learn any radiochemistry. But when women's education became officially regulated in Germany in 1909, Fischer at once gave me permission to enter the chemistry department.

One may wonder why Fischer chose the ridiculous pretext of the Russian student to forbid Lise to enter his laboratory. A strange pretext indeed!

Hahn, Meitner and β Radioactivity

Otto Hahn and Lise Meitner begin to work together in October 1907. They study the absorption of β-particles using several β emitters. They place thin strips of metal, usually aluminum, between the radioactive source and the electrometer with which they measure the current. In almost all cases, they find that the current diminishes exponentially when the number of strips is increased. Whenever such an attenuation of the current is not observed, they believe that it means that the source is a mixture of two sources. They publish their first paper in *Physikalische Zeitschrift* [67]. After this, they study the radioactivity of actinium and, since the absorption of β-rays is not observed to be exponential, they suspect and soon discover that their actinium sample contains a new substance which they call actinium C.

From the outset, the collaboration of Lise Meitner and Otto Hahn is extremely fruitful. The mathematical and theoretical physics knowledge of Lise Meitner complements the rigor of Otto Hahn in chemistry. They publish two papers in 1908 and six in 1909. They assume in their work, and this reflects the opinion of Lise Meitner, that laws of nature must be simple, and that the electrons emitted by a given substance must all have the same velocity:

> Our results suggest that substances which emit only β particles, emit only β particles of a single type, as in the case of α rays [67].

If this postulate turns out to be true, it provides for a simple way to detect unknown radioactive substances which are admixed to known sources. When applied to actinium, this idea appears to give encouraging results:

> The hypotheses which we made previously, namely that pure sub-stances emit unique β rays and that their absorption in aluminum obeys an exponential law, is entirely confirmed also as a working hypothesis for actinium and it has led to the discovery of new groups of β rays [68].

They proceed to tackle radium, and complications appear to arise because the absorption law no longer appears to be exponential. That does not lead them to doubt their hypothesis. On the contrary, they reaffirm it and they conclude that radium has a complex nature:

> In view of our hypothesis, according to which a complex radiation corresponds to a complex substance, we are led to conclude that the nature of radium is complex.

Let us keep this in mind.

The First "β Spectrometer"

Until then, Otto Hahn and Lise Meitner tackled the problem by measuring the absorption of β-rays. However, another and more direct method existed to check whether all the β-rays had the same velocity. It consisted measuring their deviation in a magnetic field. The method had been used by several physicists in the beginning of the century. Indeed, when a charged particle, such as an electron, passes between the poles of a magnet, through a zone where a magnetic field is present, and when the magnetic field points in a direction which is perpendicular to its trajectory, the electron no longer travels in a straight line. Its trajectory becomes a circle, the radius of which increases with the velocity of the electron, provided that at all points of the trajectory, the magnetic field maintains a constant value and direction. This allows one to measure the velocity of the electron, if its mass is known, as was the case. Otto Hahn had participated in an experiment with Rutherford, in which α-particles, emitted by thorium, were deviated by a magnetic field. The experiment showed that the α-particles were all emitted by a given substance with the same velocity [69]. Otto Hahn and Lise Meitner are inspired by this experiment, and they construct a first β spectrometer, an apparatus which made it possible to measure the radius of the circle described by the electrons, in a manner similar to the optical spectrometers which allow us to measure the wavelengths of light waves:

> We thought, Lise Meitner and I, that the results of this research on α rays would apply to β radiation. Some further studies of β rays did not yield conclusive results but the year 1910 brought a real progress. The carpenter's shed was little adapted to the kind of work we wanted to pursue, so that we joined forces with Otto von Baeyer from the Physics Institute of the University. We turned our attention to the study which Rutherford had made of the deviation of α rays and we constructed a similar apparatus to that which had been used by Rutherford [70].

The first paper which, oddly, is not signed by Lise Meitner [71] shows unambiguously two very clear lines which could be understood as due to two groups of electrons emitted at different velocities. The paper is soon followed by another, this time jointly signed by Otto von Baeyer, Otto Hahn, and Lise Meitner [72]. Their results appear to confirm the hypothesis of homogeneous velocities, the presence of several lines indicating the presence of several emitters.

In his *Annual report on radioactivity*[1] which Soddy wrote for the *Chemical Society* in 1910, Soddy acknowledges the result of Lise Meitner and Otto Hahn:

> Foremost in the work on β rays must be placed a notable advance, bearing out the working theory before alluded to, that in any single disintegration only one type of β radiation is expelled, which, like the α radiation, is homogeneous as regards initial velocity of expulsion, and is exponentially absorbed by matter [73].

---

[1] See p. 166.

In 1910, after 10 years of study of β-rays, the conclusion appeared to be that, just like α-particles, β-particles, emitted by a given substance, all have the same velocity and therefore the same energy. The velocity characterizes the radioactive substance. There is however one shadow of doubt: experiments performed by William Wilson [74] in the Manchester laboratory directed by Rutherford, contradict the hypothesis of Lise Meitner and Otto Hahn [75]. They become the subject of some controversy, but they are not yet regarded as conclusive.

The Kaiser Wilhelm Institut

Shortly before 1910, the emperor *Kaiser Wilhelm* was convinced that it had become necessary to create "institutes" devoted to fundamental science and independent of the universities, while maintaining close contact with them and with industry. He recognized that scientific research played a role in maintaining Germany influential and important. He created the *Kaiser Wilhelm Gesellschaft*, which was to include institutes in physics, chemistry, biology, medicine, etc. These institutes would be funded by the state, but also by industry and individual patrons. The first institute to be created was the *Kaiser Wilhelm Institut für Chemie*, which was inaugurated during a great ceremony by the emperor himself on October 23, 1912. The first director, Ernest Beckmann, offered Otto Hahn the position of director of a small laboratory devoted to radioactivity. Lise Meitner obtained the position of "visiting physicist," which meant that she was allowed to work there without a salary (a real favor!). So far, she had been living on a small allowance from her family (her father died in 1910). It was not until 1912 that Max Planck offered her an assistant position, her first paid job. She is then 34 years old.

Clouds Are Gathering

Certain problems remained which were not resolved by the hypotheses made by Otto Hahn and Lise Meitner. They assumed that a given radioactive substance displayed one kind of radioactivity, α or β, and that it emitted particles which all had the same energy. However, in the "β spectra," *more lines were always observed than could be predicted*. As the experiments progressed, the spectrometers improved and, logically, increasingly fine lines should have been observed. But such was not the case. Numerous diffuse lines were observed.

In Paris, Jean Danysz, a physicist of Polish origin working in the laboratory of Marie Curie, is also measuring the deviations of β-rays and he discovers far more lines than Otto Hahn and Lise Meitner, using a sample containing radium B and radium D. His spectrometer is more precise than the one in Berlin, because the electrons are made to follow three quarters of a circle (270°), which makes it more sensitive to variations of its radius. After two communications to the *Académie des Sciences* [76, 77], he publishes an important paper in *Le Radium*, in which he concludes:

> *When a tube is filled with the emanation of radium, at least 23 beams of β rays escape. If one adds the additional beams observed by Hahn, Bayer and Meitner (2 for Ra at its lowest activity; 2 for RaB and RaC, 2 for RaD) the number of beams reaches the value 29 at least [78].*

Indeed, if each line (beam) corresponded to one and the same element, that would really make a lot of new elements! But how can one understand this proliferation of lines, that is, of groups of electrons with different velocities?

Rutherford came to Manchester in 1907. Until that time, he had mostly studied α-particles, unveiling their nature. In 1912, he begins to study β radioactivity. He notices that γ radiation often occurs at the same time as β radioactivity. At first, he accepts the hypotheses of Hahn and Meitner, and he speculates that some electrons, emerging from the atom, might lose some of their energy due to collisions, thereby emitting γ-rays. This could explain the multiplicity of the observed lines [79, 80]. But all these speculations will be countered by a young Englishman, a student of Rutherford, who is visiting Berlin, James Chadwick.

James Chadwick: A Continuous β Spectrum!

On October 20, 1891, a boy is born in Bollington, a small industrial town close to Manchester. His father, John Joseph Chadwick, works in a textile factory, and his mother Anne Mary is a servant. Their son is named James [81]. In spite of difficult conditions, the young James excels in primary school and begins secondary school. He is able to attend university because he obtains a grant and also because the admission to university had become somewhat more democratic. He is admitted to the University of Manchester in 1908 and immediately begins research. Possibly due to his modest origin, the young Chadwick is timid and reserved. To enter university, he needs to pass an oral examination. He decides to seek admission for mathematics, but he makes a mistake and enters the room where the physics examination is being held. He only realizes his mistake when he hears the questions which are addressed to him, but he dares not mention it. He is admitted nonetheless, and he becomes the physicist which we now discover.

He obtains his *Master of Science* in 1912. Thanks to a recommendation of Rutherford, who noticed what an exceptional experimentalist he was, he is awarded the prestigious *1851 Exhibition Scholarship*, which had allowed Rutherford to come to England in 1897. The grant allows him to go abroad, and he chooses to go to the laboratory of Hans Geiger, the *Physikalisch-Technische Reichanstalt*, in Charlottenburg, in a suburb of Berlin. Recall that in 1908 Hans Geiger[1] had constructed with Rutherford the first particle "counter," which made it possible to count particles, one by one, a feat which appeared to be impossible before.[2] Geiger stayed in Manchester until 1912, and he performed there the experiments which led Rutherford to formulate his model of an "atom with a nucleus."

In 1912, Geiger becomes the director of the radioactivity laboratory of the *Physikalisch-Technische Reichsanstalt*. He tries to improve his counter, which consists of a metallic tube with a thin wire running along its axis and connected to a positive voltage. The electrons which are liberated by the collisions of a particle passing through the enclosed gas are accelerated towards the thin wire and, on their way, they liberate further electrons as they collide with the molecules in the gas. This "snowball effect" ends up in precipitating an avalanche of electrons onto the wire. This amplified current becomes observable and measured with an electrometer. The new counter of Geiger is considerably smaller in volume and the central wire is replaced by a very sharp needle: it is the "point counter" which we shall return to later.[3] It detects not only α-particles, but also β-particles that is, electrons.

With the help of Geiger, Chadwick undertakes new measurements of the "spectrum" of β-rays. He uses the same spectrometer as Hahn, Meitner, and von Baeyer, but instead of collecting the traces of the β-particles on a photographic plate, he counts them using Geiger's point counter. In Hahn's spectrometer, the particles with different velocities first follow a semicircle trajectory and then impinge on different points of a photographic plate because the radii of their trajectories depend on their velocities. The black spot formed on the photographic plate makes it possible to measure their velocities, especially when groups of particles with similar velocities form *lines* on the photographic plate. By placing a Geiger counter at different points in the spectrometer, on the plane where the photographic plate laid, Chadwick was able to determine *how many* particles had given velocities. For practical reasons, Chadwick kept the point counter at a fixed position and varied the magnetic field, which amounted to the same.

The result was surprising: he did not observe "lines," that is, groups of β electrons with the same velocity. What he observed was a *continuous spectrum of electrons*, that is, β electrons which had a *continuous distribution of velocities, ranging from zero to a maximum*. For some velocities, those which corresponded to the previously observed lines on the photographic plate, the number of electrons was indeed somewhat larger, but such electrons amounted to barely a few percent of the total number.

Frustrated by this result, he sends a letter to Rutherford on January 14, 1914:

> *I have not made much progress as regards definite results. We wanted to count the β-particles in the various spectrum lines of Ra B+C and then to do the scattering of the strongest swift group. I get*

---

[1] See p. 67.
[2] See p. 67.
[3] See p. 214.

photographs very quickly and easily, but with the counter I can't find even the ghost of a line. there is probably some silly mistake somewhere [82].

However, there was no mistake. After convincing himself that the continuous spectrum was a reality, he sends a communication to the German Physical Society, in which he presents the purpose of the experiment:

The purpose of the present investigation was to determine quantitatively the intensity conditions by directly counting the β-rays in the individual groups. Thereafter it was intended to examine closely the laws determining the passage of β-rays through material by using the individual extremely homogeneous radiation groups. The counting of the individual β-particles was carried out by the method proposed by Geiger. At first sight these results appear in part to contradict the photographic measurements. The difference is however explained by the fact that the photographic plate is extremely sensitive to small changes in the radiation intensity [83].

The word "quantitative" is essential in Chadwick's wording. The photographic plate displayed zones of various darkness, but it did not reveal *how many* particles were incident on any point. In fact, the photographic plate exaggerates the contrast: a small zone on which only slightly more particles fall appears much darker than a neighboring zone, which appears as a grey haze. One can easily imagine that in this case, physicists adjust, with a clear conscience, the chemical development of the photographic plate, in order to increase the contrast and to make the lines more apparent.

Is It Really a Continuous Spectrum?

Was one therefore obliged to admit that the β-particles, emitted during the decay of an atomic nucleus, did not have a definite velocity, and therefore a definite energy, and that they could be emitted with any velocity below a certain maximum limit? This appeared to be most embarrassing. It implied that energy was not conserved during β-decay. If the nucleus was initially (before it decayed) in a quantum state with a definite energy and that the same was true for the residual nucleus after the decay, then the energy of the emitted β-particle should be exactly equal to the difference in energy on the two nuclear states. The emitted electrons, after passing through a magnetic field, should all pass through the same spot, and they should not spread out. *A continuous spectrum was impossible!* And yet, there it was.

Lise Meitner could not believe the continuous spectrum. There had to be some explanation which fitted "her" theory. For example, the observed β-rays could well have all been initially emitted with the same energy, but they may have lost some energy as they passed through the electron cloud surrounding the nucleus.

In Berlin: The War

Chadwick submits his paper on April 2, 1914. On June 28, the archduke Franz Ferdinand, heir to the throne of the Austro-Hungarian Empire, is assassinated by a Serbian terrorist in Sarajevo. Between July 28 and August 4, the great powers declare war. Soldiers are drafted in all the countries, and physicists are sent to the front. In Germany, the onset of the war is almost joyful, everyone believing that the war would be short and, of course, victorious. With the notable exception of Einstein, most physicists, together with most of the population, approve the war and believe that it is justified.

Otto Hahn is sent to the French front at the end of September, and in January 1915, he is assigned to the laboratory of Fritz Haber at the *Kaiser Wilhelm Institut*. He works on warfare gases. Lise Meitner remains in Berlin for some time, as assistant of Planck at the university. At the same time, she gives X-ray lessons to the military doctors. From mid-1915 to the fall of 1917, she becomes a volunteer radiologist in the Austrian army, commuting between the front and the hospitals. In 1917, she is asked to set up a department on radioactivity at the physics *Kaiser Wilhelm Institut*.

Chadwick is trapped in Germany at the onset of the war. In response to a similar measure taken in England, he is interned in a camp, in Ruhleben near Berlin. He remains there during the war, under difficult conditions. Nonetheless, he manages to communicate with the outside world and, with the approval of the camp director, he constructs a small lab, helped by some German physicists he knew well, such as Nernst and Geiger. In the camp, he meets a 19-year-old English officer, Charles Ellis, who was also taken by surprise by the war while he was on vacation. Ellis helps Chadwick to set up a rudimental apparatus. He proves to be very clever and quickly develops a passion for physics.

The war is to be neither short, joyful, nor victorious. When the armistice is signed in November 1918, Germany is drained. The strict conditions dictated by the Versailles Treaty in 1919 will ruin the young Weimar Republic, born in the ashes of the German Empire.

Lise Meitner Returns to β Radioactivity

It was not until 1922 that Lise Meitner returned to the problem of β radioactivity. The French physicist of Polish origin, Jean Kasimierz Danysz, who was active in this research before the war, was killed on the front in 1914 at the age of 34. Chadwick returned to England but no longer studied β radioactivity. However, the young Charles Ellis took a liking to physics and abandoned his military career. As we shall see, he will become the hero of a thriller concerning β radioactivity.

Without doubting the results of Chadwick's experiments, Lise Meitner remains skeptical concerning their interpretation. Contrary to what Chadwick claims, she believes that the continuous spectrum is a *secondary* effect: the β electrons all have the same velocity upon being emitted, but it is reduced when they collide with the electrons surrounding the nucleus. Why should one imagine that electrons can be emitted at any energy in a radioactive process? It seems even less likely in view of the remarkable success of Bohr theory of quanta. His theory should apply equally well to the nucleus, and therefore, electrons, emitted during a decay, must all have velocities determined by the discrete energy levels of the nucleus, and not a continuum of random velocities. She therefore decides to redo the experiments, taking advantage of the improvements achieved by Jean Danysz. She notes also, as Rutherford did a few years earlier, that a considerable number of β spectral lines correspond to electrons which are ejected by γ-rays. She deduces this from the observation that their energies correspond to the differences between the energies of electrons orbiting around the nucleus. A γ-ray is indeed a quantum of electromagnetic radiation which, according to Einstein, is emitted by a nucleus when it makes a transition from one quantum state to another. Lise Meitner then engages in a scientific controversy with Charles Ellis, now working at the Cavendish, directed by Rutherford. Ellis does not hesitate to support his data, as well as those of his colleagues in the Cavendish. However, Lise Meitner is more inclined to remain guided by a theory which leads her to suspect and even to misbelieve data on the grounds that they are incomprehensible. This is precisely the case of the continuous spectrum of β-rays. She states:

> In any case, I believe that the experiments of Chadwick do not allow us to infer the existence of a primary continuous β spectrum [84].

Her position becomes increasingly difficult to defend. However, in 1923, the Compton effect is discovered.[1] The American physicist Arthur Compton observes that when X-rays are scattered by electrons, they lose energy and therefore frequency, just as any particle colliding with an electron would. Lise Meitner is delighted. She believes that she has the explanation of the continuous spectrum. The observed β electrons would be orbiting electrons which are ejected by γ-rays emitted by the nucleus. Such electrons would emerge at different energies according the angle at which they would be deviated, whence the continuous spectrum [85]. In a second paper, published in 1924, she writes:

> [The Compton effect] is manifest in the continuous β spectrum of radioactive substances, as I showed previously [86].

The Decisive Experiment of Charles Ellis

Ellis decides to make a frontal attack on the problem. What experiment could put an end to this controversy? He imagines an experiment which is quite different to

---

[1] See p. 124.

the ones performed so far. If all the radiation, both primary and secondary, can be captured in an enclosure, the latter will heat up. The increase in temperature will correspond to the *total* energy which is liberated by the radioactive decay, that is, to the primary energy emitted during the β-decay. If part of the primary energy is transferred to other particles, the latter will participate in the heating of the enclosure and this energy will also be detected. If Lise Meitner is right, the β electron has a single energy, but the energy which it loses by colliding with other particles is not lost. It is simply transferred. The total energy delivered to the enclosure must therefore correspond to the initial electron energy, which is the largest energy observed in the continuous spectrum. If, on the other hand, the primary electrons have energies ranging from zero to a maximum energy, the energy delivered to the enclosure will lie somewhere between zero and the maximum energy. Together with another physicist from the Cavendish, William Wooster, Ellis sets up a very delicate calorimetric experiment, which consists in measuring the increase in temperature of an enclosure in which a source of radium E is deposited.[1] The advantage of radium-E is that it does not display β lines. It emits what one could call a "pure" continuous ·spectrum with energies ranging from zero to about 1 MeV$^\diamond$, with an average energy of 0.39 MeV. The experiment is very difficult because the temperature of the enclosure rises by only 1/1000 of a degree and it must be measured with a precision sufficient to distinguish the two theories. It is understandable that they required 2 years to finally obtain the result: 0.34 MeV with a maximum 10 % uncertainty. The case was solved:

We may safely generalize this result for radium E to all β-ray bodies and the long controversy about the origin of the continuous spectrum of β-rays appears to be settled [87].

This result came as a shock to Lise Meitner, who immediately repeated the experiment. Two years later, she confirmed the result of Ellis. She was fair play and she wrote to Ellis, in July 1929:

We have verified your results completely. It seems to me now that there can be absolutely no doubt that you were completely correct in assuming that β radiations are primarily inhomogeneous. But I do not understand this result at all [88].

This ended a 15-year-long controversy. The result however remained to be understood.

---

[1]Radium E is the isotope 210 of bismuth,^{210}Bi, which decays by β emission, thus becoming polonium (the isotope 210), which, in turn, decays by α emission to become stable lead (the stable isotope 206).

A Scandal: Energy May Not Be Conserved!

Niels Bohr is beginning to doubt whether the famous law of conservation of energy actually holds for every microscopic event, such as the collision of two particles or the radioactive decay of a nucleus. After all, this is what the continuous β spectra seem to indicate. For each individual event, energy may not be conserved and the energy of the emitted electrons might simply take a value between zero and the maximum energy allowed by the energy conservation law. In fact, this possibility had been considered by physicists as eminent as Nernst and Einstein, but they rejected it.

Together with two young physicists, the Dutchman Hendrik Kramers and the American John Slater, Bohr publishes a paper [89] in which he attempts to formulate a theory which energy would only be preserved statistically, that is, on the average, but not in each individual collision. In this theory, Maxwell's equations were maintained without change. The idea was received with mixed feelings. Einstein and Pauli were completely opposed to the idea, whereas Sommerfeld and Schrödinger were rather favorable. As noted by Abraham Pais [90], the fact that the two most famous physicists, Bohr and Einstein, disagreed on the matter, was a cause for dismay for the others! As always, the answer came from experiment.

Geiger and Bothe: A "Coincidence" Experiment

When the Compton effect was discovered, energy did seem to be conserved during the collision, but the experiment of Compton was unable to check this explicitly. Compton only detected the X-ray after the collision of the photon with the electron. The energy of this X-ray was indeed the energy which was calculated assuming that the X-ray photon was a particle with a well-defined energy and momentum. But Compton did not detect the scattered electron. This is where Hans Geiger and a newcomer, Walther Bothe, come in.

Hans Geiger had been directing the *Physikalisch-Technische Reichsanstalt*, in Berlin-Charlottenburg, since 1912. The laboratory had to deliver attestations and make measurements of radioactive sources, for medicine. It is there that Geiger meets a physicist, 10 years younger than he, Walther Bothe.

Walther Bothe was born on January 8, 1891, in Oranienburg, a suburb of Berlin. He studied in the University of Berlin where he followed the lectures of Max Planck. He defended his PhD thesis just before the onset of the war, and he entered the *Reichsanstalt* to work under the direction of Geiger.

When the war breaks out, Bothe is drafted and sent to the Russian front where he is taken prisoner. He spends his time studying mathematics, the Russian language, and he marries a Russian girl, Barbara Below. The couple returns to Germany in 1920, and Bothe resumes his collaboration with Geiger. Although he began his career as a theoretician, it is as an experimentalist that he becomes known.

Bothe did not always get along easily with his colleagues, but his working power and his scientific integrity made him a respected physicist. He liked to invite friends to his home where, just like his wife, he was friendly and warm. He was an excellent pianist, with a special inclination for Bach and Beethoven [91].

Geiger and Bothe begin an experiment aimed at checking whether energy is conserved in the Compton effect. In order to make sure that the photon loses its energy by colliding with an electron, they decide to detect the X-ray photon and the electron *simultaneously*.

Meanwhile, Geiger had developed his "point counter," mentioned above, and which will be described in detail below.[1] The "point counter" detected electrons and therefore also X-rays which liberate electrons while passing through the gas in the counter. The idea consists in using two counters *in coincidence*. The X-rays pass through a small volume of hydrogen, between two point counters facing each other: the first, called the *e*-counter, detects the electron and the other, called the *hν* counter, detects the X-ray. Occasionally, a Compton event, that is, a collision between an X-ray and an electron, occurs in the hydrogen. Then a recoiling electron is detected in the *e*-counter, and an X-ray, which has lost some energy and changed its direction in the collision, enters the *hν* counter, covered with a thin platinum foil. The X-ray collides with an electron within this foil, and this electron triggers the *hν* counter. When a Compton scattering occurs, one counter fires, and the other one should fire simultaneously. Bothe and Geiger registered simultaneously two electrometers connected to the two counters on a continuously moving photographic film. They succeed in proving thus that, with an accuracy of one thousandth of a second, there is a coincidence between the detected photon and electron and that it is therefore reasonable to assume that they are created in a "Compton collision" [92, 93]. The result of this crucial experiment is clear: the energy is conserved at each collision. This disproves the theory of Bohr, Kramers, and Slater. The experiment was the first one in which two particles were detected simultaneously, in coincidence as we say today. We shall return to this.[2]

The Idea of Wolfgang Pauli

The strangest ideas are put forth to explain the enigmatic continuous spectrum of β-rays, which appears to violate the law of conservation of energy. Do the laws of electrodynamics, together with the theory of relativity, no longer hold at distances as small as the radius of a nucleus? Such an assumption appeared hardly stranger than the fact that quantum mechanics was required to explain phenomena at the scale of the atom. What prevents a further change at the nuclear scale which is between 10 000 and 100 000 times smaller?

---

[1] See p. 214.
[2] See p. 219.

It is under such conditions that radioactivity specialists, among whom Lise Meitner and Hans Geiger, held a meeting in Tübingen, on December 6 and 7, 1930. Pauli was detained in Zurich, and he sent the physicists the following text, in the form of an "Open Letter to the Group of Radioactive Persons at the Conference of the District Society in Tübingen" [94]:

> I have, in connection with the "wrong" statistics of the N and Li 6-nuclei as well as the continuous β-spectrum, hit upon a desperate remedy for rescuing the "alternation law" of statistics and the energy law. This is the possibility that there might exist in the nuclei electrically neutral particles, which I shall call neutrons, which have spin 1/2, obey the exclusion principle and moreover differ from light quanta in not traveling with the velocity of light. The mass of the neutrons would have to be of the same order as the electronic mass and in any case not greater than 0.01 proton masses.— The continuous β-spectrum would then be understandable on the assumption that in β-decay, along with the electron a neutron is emitted as well, in such a way that the sum of the energies of neutron and electron is constant [. . .]
>
> I do not in the meantime trust myself to publish anything about this idea, and in the first place turn confidently to you, dear radioactive folk, with the question—how would things stand with regard to the experimental detection of such a neutron if it possessed an equal or perhaps ten times greater penetrating power than a γ ray?

Pauli realizes perfectly well that his idea is risky to say the least:

> I admit that my remedy may perhaps appear unlikely from the start, since one probably would long ago have seen the neutrons if they existed. But "nothing venture, nothing win" [. . .]

Until then, nobody had dared propose the existence of a new "elementary" particle, without it being observed. Pauli was willing to give up a world built with two particles, the proton and the electron, ideally unique and simple. To test his hypothesis, Pauli uses the roundabout method of an "open letter" addressed to experimentalists, rather than a scientific paper which he dares not publish. During a seminar, held in Zurich at the time he sent that letter, Pauli declared:

> I did today what a theorist should never do in his life. I tried to explain something we cannot understand, by something we cannot observe [95].

Pauli explained that a really serious situation of physics can lead a physicist to a desperate act, such as the one of Bohr when he considered giving up the conservation of energy, or his when he proposed a new particle. He sees no other way out. Either some available energy is lost, or it is carried away by another particle. The new particle, which Pauli calls a "neutron," has little in common with the neutron which Rutherford had imagined in 1920, which had a mass close to that of the proton, and which we shall discuss in the next chapter. Pauli's particle is indeed strange: it is electrically neutral, very light (which explains why a loss of mass is not observed in β-decay), and it has a strong penetrating power (otherwise it would have been detected). One might as well coin it as undetectable. It is an invisible particle which manifests itself only by carrying away part of the energy during β-decay, while the electron carries away an energy which ranges between zero and a maximum energy. It is an *ad hoc* particle, a *deus ex machina*! The Italian physicist Enrico Fermi will later call it a *neutrino,* a name which has stuck since.

But Why Are So Many Spectral Lines Observed?
The Key to the Mystery

If the spectrum of electrons emitted in β-decay is continuous, why are so many precise spectral lines observed? As so many other physicists, Lise Meitner believed that the primary spectrum of electrons consisted of spectral lines, that the electrons were emitted with well-defined energies, and that the observed continuous spectrum was due to a smearing of the primary spectrum. It turned out that the exact opposite was true. The primary spectrum of the electrons varied continuously from zero to a maximum energy, and *the spectral lines are due to a secondary effect.*

It is Charles Ellis who proposes this explanation in 1921. He bases his argument on an experiment performed by Rutherford in 1914:

> This fact receives a simple explanation if it be assumed that the energy of the emitted electron is equal to some energy characteristic only of the γ ray, minus the energy necessary to remove the electron from the atom. The difference in the energies of the electrons ejected from gold and lead by the same γ-rays is then explained by the difference in the work of removal of these electrons from their respective atoms [96].

When the nucleus makes a transition to its ground state, after emitting the β electron, then a continuous spectrum of β-rays is observed. If however the nucleus makes a transition to one of its excited states, it subsequently decays to its ground state by emitting a γ-ray, which is a quantum of light. Ellis explains that when one of these γ-rays (which we call today photons) collides with an electron of the atom, the latter can absorb all the energy of the γ-ray and it is thus ejected with a well-defined energy. This energy is equal to the energy of the γ-ray minus the energy required to extract the electron from the atom. *It is a photoelectric effect occurring within the atom.* This explanation is still considered true today, with the exception that the process is no longer assumed to take place in two steps. It occurs globally and it is called *internal conversion.* It is used to measure the energies of the γ-rays.

One can understand why it took time to unravel the mystery of the observed β electrons. The explanation given by Ellis could not have taken place before 1914 because too little was known concerning γ radiation. It is interesting to observe how a physicist, as accomplished as Lise Meitner, could remain in error for such a long time. This does not diminish our admiration. We know she was wrong because we know what turned out to be the truth. But her ideas were rational and even more likely to be true, in view of what was known at the time. We encountered a similar difference of opinion between Pierre Curie and Ernest Rutherford concerning the origin of radioactivity.

The First Nuclear Reactions

During the First World War, research in physics slowed down in England and in France as well. But as soon as the war ended, the tireless Rutherford succeeded in producing a nuclear reaction, somewhat similar to a chemical reaction. Further nuclear reactions soon followed.

When the war breaks out in 1914, the laboratories devoted to the study of radioactivity are working hard. Thanks to Rutherford, the laboratory in Manchester has become one of the best in the world. Rutherford's model of the atom is becoming accepted, thanks to Bohr. In Paris, Marie Curie finally succeeds in creating a real laboratory, the construction of which is decided in 1909 by the University of Paris and by the *Institut Pasteur*, which share the expense. It consists of a physics and chemistry lab, directed by Marie Curie, and a physiology lab, the *pavillon Pasteur*, directed by a doctor, Claudius Regaud, a pioneer of cancer radiotherapy.

Most of the work is interrupted by the war. As mentioned above, James Chadwick, working with Hans Geiger, is confined near to Berlin.[1] Moseley is drafted into the Royal Navy and killed at the Dardanelles. Francis Aston goes to work in an industrial aeronautical laboratory. In July 1915, Great Britain creates the Admiralty Board of Invention and Research and invites Rutherford to participate in its activities. The Navy seeks desperately to find a way to detect German submarines, which are sinking many ships, about one every day. One shipwreck is memorable: the *Lusitania*, a British ship, is sunk by a torpedo on May 7, 1915, when it is returning from the United States. It sinks off the coast of Ireland with 1,198 people on board, among whom 124 Americans (this was a factor which led the United States to enter the war). With his usual energy, Rutherford works on the detection of submarines using ultrasonics. In the summer of 1917, he joins a French mission to the United States, which had just declared war. Paul Langevin also works on the detection of submarines, and he patents several important detection methods using SONAR ultrasonics. In France, Jean Perrin works on the sound localization of German artillery. Marie Curie postpones the opening of the *Institut de Radium*, and she makes the buildings available to the medical services of the army. With her well-known willful ardor, she sets off to the front with X-ray vehicles in order to help the wounded. She does so although the medical services of the army were not interested. The number of X-ray vehicles (the *"petites Curie"*) will reach 50 by the end of the war. Marie Curie is assisted by her eldest daughter, Irène, who is 17 years old when the war breaks out.

---

[1] See p. 190.

The First Nuclear Reaction

Beginning 1917, Rutherford progressively abandons military research and resumes research in his almost deserted laboratory. Alone, he performs the experiments which Marsden relinquished in 1915. This is the time when it is speculated that transmutations occur in stars and Rutherford attempts to make one in the laboratory. At first, there seem to be huge difficulties. He had been invited in April 1914 to deliver a lecture in Washington to the National Academy of Sciences on "The constitution of matter and the evolution of the elements." This lecture was later published in *The Popular Science Monthly*, in August 1915. In his conclusion, Rutherford said:

> There is no doubt that it will prove a very difficult task to bring about the transmutation of matter under ordinary terrestrial conditions [. . .] The building up of a new atom will require the addition to the atomic nucleus of either the nucleus of hydrogen or of helium, or a combination of these nuclei. On present data, this is only possible if the hydrogen or helium atom is shot into the atom with such great speed that it passes close to the nucleus. In any case, it presumes there are forces close to the nucleus which are equivalent to forces of attraction for positively charged masses. It is possible that the nucleus of an atom may be altered either by direct collision of the nucleus with very swift electrons or atoms of helium such as are ejected from radioactive matter. There is no doubt that under favorable conditions, these particles must pass very close to the nucleus and may either lead to a disruption of the nucleus or to a combination with it [97].

Rutherford hopes to succeed by bombarding various substances with α-rays emitted by a radioactive source. He hopes to observe nuclei of the gas ejected in the forward direction by the impinging α-particle or the remains of a nucleus which might have been split up by the collision. Ernest Marsden had begun experiments in which he made α-particles pass through hydrogen [98, 99]. He placed a radioactive source in a brass container, shaped as a parallelepiped 18 cm long, 6 cm deep, and 2 cm wide. He added to the container a gas the pressure of which he could adjust. At one end of the container, he cut a rectangular opening 1 cm long and 3 mm wide, covered by a very thin aluminum foil, which would allow the α-particles to pass through. At 1 or 2 mm from that opening, he placed a zinc sulfide sheet which had the property of emitting a brief flash of light when it was hit by a particle. This scintillation screen was observed through a weakly magnifying microscope.[1] When the pressure of the hydrogen in the container was weak, he observed numerous scintillations caused by the α-particles emitted by the source. As he gradually increased the pressure of the hydrogen, the α-particles were slowed down at first and finally completely stopped before reaching the screen. However, Marsden continued to observe some weaker scintillations, most likely caused by

---

[1]For the scintillation method, see p. 70.

hydrogen nuclei projected forwards by collisions with the α-particles. The nuclei of hydrogen atoms, which were called H-rays, are less slowed down by the gas and can therefore travel further before being stopped. But in 1915, Marsden left England and became a physics professor at *Victoria College*, in Wellington, New Zealand. He was drafted there.

Thus, in 1917, Rutherford resumes the experiments of Marsden. He is assisted only by William Kay, a most appreciated technician, photographer, and electrician, who is also in charge of setting up the experiments for the university courses [100]. The experiments, performed with hydrogen, confirm the results of Marsden. Some hydrogen nuclei are projected mainly in the direction of the α particles, but their angular distribution differs from that which was predicted by Charles Darwin [101] (Charles Galton Darwin was the grandson of Charles Darwin, the British naturalist who developed a theory of evolution and natural selection of species). The latter assumed, to make things simpler, that the α particles and the nuclei were point particles. For Rutherford, however, this discrepancy means that when an α particle gets close enough to a hydrogen nucleus, the interaction between them is no longer the usual electric repulsion. The distance at which this occurs can be considered, at least roughly, as the radius of the α particle.

Rutherford uses the same experimental setup to study collisions in other gases. With oxygen and carbon, nothing special happens. When the density in the container is low, he observes the scintillations caused by the α-particles. Beyond a certain pressure, he observes nothing because the α-particles are stopped before reaching the scintillation screen. But when he fills the container with air, he is surprised to observe scintillations similar to the ones produced by the hydrogen nuclei. The scintillations persist and become even more frequent when the pressure of the air in the container is increased. This means that they are not produced by α-particles emitted by the source. They also cannot be caused by oxygen, because the phenomenon does not occur when the container is filled with carbon dioxide molecules CO_2, formed by a carbon atom and two oxygen atoms. The scintillations are therefore caused by the nitrogen. In 1919 Rutherford publishes four papers on these experiments [102–105]. In the last paper, modestly entitled "An anomalous effect on nitrogen", Rutherford examines very carefully other possible explanations of the anomalous effect, before rejecting them. He makes sure that the scintillations which he observes are due to hydrogen nuclei. The length of their path in air is roughly the same as he had measured with a hydrogen gas. He tries to deviate them with a magnetic field, and this gives him a rough indication in the right direction. He finally concludes:

From the results so far obtained it is difficult to avoid the conclusion that the long-range atoms arising from collision of α-particles with nitrogen are not nitrogen atoms but probably atoms of hydrogen, or atoms of mass 2. If this is to be the case, we must probably conclude that the nitrogen atom is disintegrated under the intense forces developed in a close collision with a swift α-particle, and that the hydrogen atom which is liberated formed a constituent part of the nitrogen nucleus. [. . .]

The results as a whole suggest that if α-particles—or similar projectiles—of still greater energy were available for experiment, we might expect to break down the nucleus structure of many of the lighter elements.

Rutherford had caused and observed the first *nuclear reaction*. When an α-particle, that is, a nucleus of helium (of mass 4 and charge 2),[1] collides with the nucleus of a nitrogen atom (mass 14 and charge 7), a hydrogen nucleus (mass 1 and charge 1) is expelled. Rutherford caused an *artificial transmutation*, the first one ever recorded in history. But which one remained unknown. The α-particle might have simply been scattered after the collision. In this case, the struck nitrogen nucleus would have become a carbon nucleus, since it had lost a proton. But the α-particle might well have stuck to the nitrogen nucleus. In this case, after the emission of the proton, the remaining nucleus would have a charge 8 ($= 7 + 2 - 1$) and would therefore be an oxygen nucleus. For the time being, Rutherford refrains from making such speculations.

Sir Ernest Rutherford, Cavendish Professor of Physics

On April 2, 1919, Rutherford is elected *Cavendish Professor of Physics*, at the University of Cambridge. It is the crowning event of his career. We should now call him "Sir Ernest" because he was conferred the title of nobility in 1914. On June 2, he bids a solemn farewell to his colleagues and settles in the Cavendish. He succeeds to the 63-year-old Joseph John Thomson, who resigned because he was appointed Master of Trinity College. Rutherford reorganizes the physics department. In an important memorandum, he applies for increased funds and gets to work. He brings with him some of his apparatus from Manchester. He also brings along a most precious assistant, namely, James Chadwick, for whom he obtains a *Wollaston Student* position: indeed, Chadwick is still a student who has not yet obtained a PhD.

New Nuclear Reactions

Rutherford and Chadwick try to produce other decay products of elements, as Rutherford calls them. They perfect their apparatus, and they bombard a broader range of elements with α-particles. Already in 1921, they send a short paper to *Nature*, in which they report that they have caused new transmutations:

In this way we have obtained definite evidence that long-range particles are liberated from boron, fluorine, sodium, aluminum, and phosphorus, in addition of nitrogen [106].

---

[1]See p. 69.

A second more detailed paper [107] was sent to the *Philosophical Magazine* in the fall of 1921. The produced particles were observed after passing through an aluminum foil which would have stopped the α-particles or heavier nuclei. It is therefore very likely that they are H-rays, in other words, protons. The protons have velocities which are greater than those which the impinging α-particles would convey to some residual hydrogen in the container (which could always contain some residual water vapor and therefore hydrogen nuclei). Rutherford and Chadwick observe that the nuclei of boron, nitrogen, fluorine, sodium, aluminum, and phosphorus can be broken by the collision with an α-particle, but not nuclei of carbon, oxygen, or sulfur. In the case of aluminum, they note that the protons are liberated in all directions, in the direction of the incoming α-particles as well as in the opposite direction. There seems to be no relation between the direction of the liberated protons and that of the incoming α-particles. To make sure that they are observing protons, Rutherford and Chadwick make them pass through a magnetic field, as Rutherford had done in his first work with nitrogen. In their paper, published in 1922, they are very cautious because they have no absolute proof that they are indeed observing protons (particles of mass 1 and charge 1), but they are convinced of it. They are entangled in a subtle game, in which what is likely must be treated as true or at least as a working hypothesis, in order to pursue further. The main thing is to remain conscious of the fact and to check that all the details of the picture are coherent and in agreement with experiment.

In a third and important paper, published in 1924, Rutherford and Chadwick examine other elements [108]. The apparatus is the same except for one feature: having ascertained that the particles, which they assume to be protons, are emitted about equally in all directions, they decide to observe the ones which are emitted at right angles relative to the incoming α-particle, originally emitted by the radioactive source. This eliminates a large number of undesirable particles and increases the sensitivity of the detection. They detect this way transmutations which they had not observed before with neon, magnesium, silicon, sulfur, chlorine, argon, and potassium. They observe no effect with heavier elements such as calcium, nickel, copper, zinc, selenium, krypton, molybdenum, palladium, silver, tin, xenon, gold, and uranium.

Rutherford and Chadwick believe that the α-particles do not have enough energy to cause the disintegration of target nuclei heavier than aluminum or phosphorus. When the electric charge of the nucleus is larger than 13 or 14, the α-particle is repelled too strongly and it cannot come close enough to the nucleus. This explanation still holds today.

A Controversy Between Vienna and Cambridge

In the summer of 1923, papers began to be published by physicists from the *Institut für Radiumforschung* in Vienna, which was directed by Stefan Meyer, an old friend of Rutherford and one year younger. The results of the experiments performed in Vienna contradicted those obtained by Rutherford and Chadwick in the *Cavendish*. The Vienna publications were signed by two young physicists, the Swedish Hans Pettersson and the Austrian Gerhard Kirsch, who claimed that they had observed the disintegration of silicon bombarded by α-particles, which Rutherford did not observe [109]. They stated that they would perform similar experiments with other elements. The physicists in Vienna used the same detection apparatus which Rutherford had been using for the past 15 years. It consisted in viewing, through a microscope, the small scintillations occurring on a zinc sulfide screen when particles pass through them.[1] The disagreement soon provoked a controversy because the physicists from Vienna continued publishing results which the Cambridge physicists found most surprising [110]. The somewhat aggressive and arrogant attitude of the young Viennese physicists did not make things easier. Rutherford and Chadwick then repeated their experiments with various elements, using their new method of observing particles emitted at 90°. This enabled them to detect relatively slow H-rays. The latter travelled a distance of about 7 cm in the air before being stopped. This was easily measured by progressively moving away the zinc sulfide screen [111]. But their results were again contradicted by the Viennese who observed the effect with many nuclei, whereas the effect with the same nuclei was not observed in the Cavendish. In particular, the Viennese observed the effect with heavy elements such as titanium, chromium, iron, copper, selenium, bromine, zirconium, etc.

Until then, Rutherford and Chadwick politely quoted the results of Pettersson and Kirsch without expressing doubts. But they adopt a different tone in a new publication:

> *Kirsch and Pettersson investigated the disintegration of some light elements by our previous method, taking special precautions to avoid hydrogen contamination both in the source and in the bombarded materials. They found that beryllium, magnesium and silicon gave very large disintegration effects, three or four times greater even than that of aluminum, while sulphur and chlorine gave little or no effect. The particles from beryllium had a range of about 18 cm., those from magnesium and silicon about 12 cm.*
>
> *These results cannot be reconciled with ours, and the probable explanation, in view or the number of particles and their range, is that the particles they observed were the long range α-particles emitted by the source [108].*

The quarrel continued, and each one maintained his opinion. They began claiming more and more explicitly that the others were in error. However, Rutherford and Stefan Meyer continued to exchange polite letters. In order to resolve the conflict, Meyer invited Rutherford to visit his institute. Rutherford replied by proposing to send his assistant Chadwick to Vienna [112]. Chadwick went in December 1927. From the outset, the atmosphere was friendly with Meyer

---

[1] See p. 70.

but quite tense with Pettersson. He quickly noticed that the way measurements were taken in the two laboratories were different. He had trouble in obtaining satisfactory explanations from Pettersson. On Monday December 12, he wrote to Rutherford:

> Not one of the men does any counting. It is all done by 3 young women. Pettersson says the men get too bored with routine work and finally cannot see anything, while women can go on for ever.

On that day, Pettersson is absent most of the time because his family is arriving from Sweden. Chadwick then begins an experiment which will deliver the key to the mystery. He writes in his letter:

> Today, therefore, I arranged that the girls should count and that I should determine the order of the counts. I made no change whatever in the apparatus, but I ran them up and down the scale like a cat on a piano—but no more drastically than I would in our own experiments if I suspected any bias. The result was that there was no evidence of H particles [from carbon] [. . .] The results do not prove that there is nothing from carbon but I think they make it doubtful that there is much [113].

The cause of the disagreement was simple. The young women *knew* what their bosses hoped they would see, and they therefore saw it! They did so without really cheating, probably unconsciously, in all innocence. The only intervention of Chadwick consisted in not telling them what the expected result was. But it required the caution and pitiless rigor which Chadwick imposed on himself in his experimental work. Chadwick concluded from this unpleasant incident that visual counting of scintillations was too unsure, and that it depended on human factors which were difficult to master. This marked the end of scintillation techniques which were progressively replaced by electrical counters, derived from the original counter of Geiger, to which we shall shortly return.

How Do the Transmutations Occur?

From the outset, Rutherford attempted to understand how a nuclear reaction and a transmutation can occur. Let us write a nuclear reaction as we would a chemical reaction:

$$\alpha + N \rightarrow H \text{ ray} + ?$$

This states that an α-particle hits a nitrogen nucleus N and produces a proton (H ray) and... we do not know what else. The proton was detected, but what happened to the struck nitrogen nucleus? Did it explode? Was the α-particle first absorbed by the nitrogen nucleus before emitting a proton with the resulting production of

an oxygen nucleus?[1] Or did something else happen? In the paper quoted above,
Rutherford and Chadwick state the problem:

> The fate of the alpha-particle is a matter about which we have no
> information. It is unlikely that the field of force remains central at
> very close distances. It is possible that the α-particle is in some way
> attached to the residual nucleus. Certainly it cannot be re-emitted
> with any considerable energy, or we should be able to observe it.

A year later, in the April 4th issue of *Nature* [114], Rutherford recalls that Jean
Perrin had considered such a possibility at the Solvay Council in 1921. In the
proceedings, Jean Perrin wrote:

> The very experiments of Rutherford seem to show that we must
> give up the idea of a simple collision. Due to its high velocity, the
> α projectile can, in spite of a strong electric repulsion, approach
> the immediate vicinity of the nucleus with a low velocity. At that
> instant, a "transmutation" occurs, consisting probably of an internal
> rearrangement of the nuclei, with a possible capture of the incident
> α-particle, the emission of a hydrogen nucleus which gives rise to
> the observed H-ray, and possibly with other less important emissions.
> There is no reason why, in this scenario, the emitted H-projectile,
> should "remember" the direction of the initial impact nor why its
> energy should be lower than that of the incident projectile [115].

In one of his dazzling intuitions, Jean Perrin laid down the foundations of a
theoretical description of the *nuclear reaction* observed by Rutherford. The idea is
that a nuclear reaction is a two-step process. In the first step, the α-particle is incident
with a high velocity, but it reaches the vicinity of the aluminum nucleus with a low
velocity. It merges with the target nucleus which swallows it, so to speak. In the
second step, after a few jolts of this sorcerer's pot, an *H-ray* is ejected. In view of
the likely agitation of this compound nucleus formed by the absorbed α-particle and
the aluminum nucleus, one can well imagine that the direction in which the *H*-ray
is emitted bears little correlation with the direction of the incident α-particle. It is
therefore normal that one should observe *H*-rays emitted in all directions. Although
this is not a proof, it is a least a presumption that this is the way the reaction occurs.
This hypothesis will soon be confirmed by an experiment which uses a completely
different technique, namely, the cloud chamber of Charles Thomson Rees Wilson,
which will be described below.[2]

---

[1] Indeed it would have eight electric charges, the seven initial ones of the nitrogen nucleus, plus the
two of the incoming α-particle, minus the one of the ejected proton.

[2] See p. 222.

The Nucleus in 1920 According to Rutherford

We are in 1920. Isotopes of most of the elements have been discovered. It is generally agreed that nuclei are composed of protons (nuclei of the hydrogen atom) and of electrons. However Rutherford speculates that it might contain a neutral particle, which has a mass similar to the one of the proton and which he calls a "neutron".

In 1920, Rutherford was again invited to deliver the Bakerian lecture[1] at the Royal Society. He decided to speak of the present knowledge of the "nuclear constitution of the atom" [116]. Less than a year later, Rutherford presented a report on the "structure of the atom" at the third Solvay Council [117], held in Brussels, April 1–6, 1921. It was devoted to "atoms and electrons," and the proceedings were published in 1923. Among the physicists who attended were Charles Barkla, William Lawrence Bragg, Marcel Brillouin, Maurice de Broglie, Marie Curie, Paul Ehrenfest, Wander de Haas, Heike Haas, Martin Knudsen, Paul Langevin, Joseph Larmor, Robert Millikan, Jean Perrin, Owen Richardson, Manne Siegbahn, Pierre Weiss, and Pieter Zeeman. There is not a single German or Austrian among the invited participants. The 4 years of war left tenacious hard feelings among the scientists of the victorious camp and they decided to isolate the German scientists and to deprive them of communication with the world. An *International Research Council* was founded whose purpose was to develop scientific exchange, but exclusively among the "allies." The boycott lasted for several years and progressively disappeared after the Locarno agreements in 1925. However, Langevin raised the blockade when he invited Einstein to Paris in 1922, a most symbolic and successful visit. In 1923, Langevin himself went to Berlin [118]. But in 1921, three years after the First World War, the German physicists were still excluded from international meetings.

Rutherford's report presented at the Solvay Council and his Bakerian lecture have much in common, in spite of a few differences. We will follow the thread of the Bakerian lecture, occasionally adding some excerpts from the Solvay council.

Rutherford described what was known about the atomic nucleus. He recalled that the experiments of Geiger and Marsden led him to propose the idea of an "atomic nucleus," which was very small, positively charged, and surrounded by electrons. That allowed him to understand the observed backward "rebound" of α-particles which impinged on a thin foil. The nucleus, as Rutherford expressed it, was "incredibly small" because its radius was smaller than 3×10^{-12} cm, three hundred billion times smaller than a millimeter and 30 000 times smaller than an atom.

---

[1]See p. 35.

He recalls that the charge of the nucleus had been determined by Moseley and that, for each element, it almost miraculously coincided with the number it had in the ordered Mendeleev table. Rutherford then reaches the heart of the matter, namely, the structure of the atomic nucleus.

The Size of the Nucleus

From his observations of collisions of α-particles with nuclei of hydrogen, oxygen, carbon, and nitrogen, Rutherford estimates the size of the nuclei. For this, he first assumes that the nuclei are point particles. This allows him to make an exact calculation of the trajectory of the incident α-particle. For a given scattered angle, he can estimate the distance of closest approach of the α-particle. He notices however that the calculated trajectory becomes wrong whenever the α-particle approaches the nucleus at a distance smaller than 3×10^{-13} cm. At that distance, he concludes that something else happens. The α-particle probably penetrates the nucleus. This gives a crude estimate of the size of the nucleus. More generally, he concludes for light nuclei:

> The diameters of light nuclei, except hydrogen, are probably of the order of 5×10^{-13} cm.

In his Solvay report, Rutherford puts forth another idea for measuring the radius of heavy radioactive nuclei. He assumes that in the radioactive α-decay of uranium, the α-particle is ejected from the nucleus with close to zero velocity. It had so far been withheld by strong internal forces which suddenly cease to act, as if a spring had broken. The α-particle, which carried two positive electric charges is then subject to a strong repulsion caused by the positive charge of the nucleus. It is accelerated, and it reaches the velocity which is measured in the laboratory. The initial repulsion is greater if the point of departure of the α-particle is closer to the nucleus. Rutherford shows that the initial distance from which the α-particle is ejected must be at least 7×10^{-13} cm, somewhat less than one thousandth of the size of the atom, although about ten times larger than the hydrogen nucleus. In a footnote added later, Rutherford makes a correction: in the vicinity of the nucleus, the forces might be different so that the size of the nucleus might be 1.5×10^{-12} cm.

We see that knowledge about the nucleus was scant in 1921. Even its size was uncertain. One thing was sure: it was very small, at least a 1000 or even 10 000 times smaller than the atom inside which it was sitting.

The Constitution of the Nucleus and of Isotopes

On this issue, Rutherford emits an opinion shared by a large number of physicists:

> In considering the possible constitution of the elements, it is natural to suppose that they are built up ultimately of hydrogen nuclei and

electrons. On this view the helium nucleus is composed of four hydrogen nuclei and two negative electrons with a resultant charge of two.

That appears to be settled. The nucleus of helium, *alias* the α-particle, has a mass close to four times that of hydrogen, and it carries two electric charges. It is therefore assumed that it is composed of four hydrogen nuclei and two electrons, negatively charged, which yields a total of two positive charges. Rutherford then discusses the mass of hydrogen and begins to speculate, a thing he rarely does.

Rutherford the Visionary: The Neutron

Rutherford is an experimentalist who only believes well-established facts. However, on the occasion of the prestigious Bakerian lecture, he indulges in a daring speculation:

If we are correct in this assumption it seems likely that one electron can also bind two H nuclei and possibly also one H nucleus. In the one case, this entails the possible existence of an atom of mass nearly 2 carrying one charge, which is to be regarded as an isotope of hydrogen. In the other case, it involves the idea of the possible existence of an atom of mass 1 which has zero nucleus charge. Such an atomic structure seems by no means impossible [. . .]. Such an atom would have very novel properties. Its external field would be practically zero, except very close to the nucleus, and in consequence it should be able to move freely through matter. Its presence would probably be difficult to detect by the spectroscope, and it may be impossible to contain it in a sealed vessel. On the other hand, it should enter the structure of atoms, and may either unite with the nucleus or be disintegrated by its intense field, resulting possibly in the escape of a charged H atom or an electron or both.

This is a concrete description of a strange and highly speculative object: an atom which is neutral as all atoms but without an external electron cloud. Its only electron would be captured inside the nucleus. Rutherford even thought of a way to check whether it exists or not:

If the existence of such atoms be possible, it is to be expected that they may be produced, but probably only in very small numbers, in the electric discharge through hydrogen, where both electrons and H nuclei are present in considerable numbers.

One of the reasons which led Rutherford to assume the existence of a neutral particle, the "neutron," is the formation of heavy elements, *nucleosynthesis*, as we call it today. He also mentions this point in his Bakerian lecture:

The existence of such atoms seems almost necessary to explain the building up of the nuclei of heavy elements; for unless we suppose the production of charged particles of very high velocities it is difficult to see how any positively charged particle can reach the nucleus of a heavy atom against its intense repulsive field.

Immediately after his lecture, Rutherford incites a young physicist at the Cavendish, J. L. Glasson, to hunt for the neutron. He is not successful. His unsuccessful attempts will be published later [119].

Chadwick Hunts for New Forces

Upon his return from captivity, Chadwick sets to work. He is still a student without a PhD. Rutherford obtains a Wollaston grant which allows him to work with him at the Cavendish. Chadwick does not pursue the study of β radioactivity which he had brilliantly conducted in Germany, by showing that the β spectrum was continuous. Instead, he embarks on a series of experiments aimed at determining the charge of the nucleus, in order to confirm, in an independent way, the results of Moseley. He measures the scattering of α-particles on various nuclei, at specific scattering angles [120], and he determines this way their electric charge. For the nuclei of platinum, silver, and copper, he finds that the charges are equal to 77.4, 46.3, and 29.3, respectively, with an error between 1 % and 1.5 %. The atomic numbers of these nuclei are 78, 47 and 29. He concludes:

There can, however, be little doubt that the nuclear charge does really increase by unity as we pass from one element to the next, and that its value is given by the atomic number.

Chadwick then pursues the experiments of Rutherford on provoked transmutations. In his thesis, *The Atomic Nucleus and the Law of Force*, defended on July 21, 1921 (at the age of 30), he studies how α-particles are deviated by hydrogen and he shows that the usual electric repulsive force is altered at short distances. Some of his experiments were performed with Étienne Bieler, a young physicist from the University of McGill who was recommended by an old friend of Rutherford, Arthur Eve. They conclude their paper [121] by stating:

As regards the structure of the α-particle, it will be apparent at once that no system of four H nuclei and two electrons united by the inverse square law force could give a field of force of such intensity over so large an extent. We must conclude either that the α-particle is not made up of four H nuclei and two electrons, or that the law of force is not the inverse square in the immediate neighborhood of an electric charge. It is simpler to choose the latter alternative, particularly as other experimental, as well as theoretical considerations point in this direction. The present experiments do not seem to throw any light on the nature or the law of variation of the forces at the seat of an electric charge, but merely show that the forces are of very great intensity.

Thus, already in 1921, Chadwick believes that a simple electromagnetic force is insufficient to maintain the constituents of a nucleus at such short distances. He later confided:

Any idea one might have about the structure of the nucleus, particles had to be held together somewhere. So that in addition to the repulsive force between the positively charged particles, there had to be an attractive force somewhere. And I played around with various forms of the force with an attraction varying as the inverse fourth power of the distance [122].

Indeed, how should one imagine the internal structure of the nucleus, as well as the forces, most likely extremely strong, which hold the constituents together? What could be the nature of these forces? Étienne Bieler pursues his experiments with aluminum and magnesium. He bombards them with α-particles and he measures their deviation [123]. He also notices that when they come close to the nucleus, the force between the α-particle and the nucleus is no longer the known electric repulsion. Bieler calculates a quantity which he considers to be a rough measure of the radius of an aluminum nucleus. He finds a value equal to 3.44×10^{-13} cm. This deviation from the known electric repulsion is called "anomalous scattering" of α-particles on light nuclei.

The Rapid Expansion of Experimental Means

Visual means of detection of particles is given up in favor of "electric" methods. The Geiger counter becomes the Geiger-Müller counter. A young Scottish physicist, with his head in the clouds, discovers a way to see the tracks left by fast moving particles. The development of radio gives birth to new instruments. Physicists begin to make coincidence measurements.

Laboratories are changing between 1910 and 1930, the years during which the first descriptions of the atomic nucleus arise. The measuring apparatus improves slowly, occasionally also in a spectacular fashion. Physicists develop increasingly finer "microscopes" able to observe these tiny objects. Let us visit several of these laboratories where some of these developments arise. They led to a prodigious progress, in the thirties, due to the Geiger–Müller counter, the electronic amplification of electric signals, coincidence measurements, and Wilson's cloud chamber, a most extraordinary apparatus, which allows us to see, with the naked eye, the tracks left by a passing particle, somewhat like the tracks left by airplanes flying in the sky. They will play a crucial role in our exploration of nuclei.

Scintillation Methods

While he was the director of the Manchester laboratory, Ernest Rutherford made most of the so-called scintillation method, which made it possible to detect visually, through a microscope, the impact of an α-particle on a zinc sulfide screen. This is how Geiger and Marsden observed the deviation of α-particles which led Rutherford to discover the nucleus of the atom. That is also how Rutherford observed for the first time a nuclear reaction[1] in 1919. But this way of observing did not last long after the controversy which lasted several years between the Cavendish and the Institute in Vienna. Although the Cavendish experiments turned out to be the right ones, the controversy showed how fragile this detection method could be. The method proved reliable as long as it was limited to counting α-particles which caused a strong enough flash and as long as they produced the flash exclusively. But when it was applied to the detection of protons, which produced weaker signals, the result depended too much on the judgment of the observer, even on his mood or his willingness to please his boss, as in the case of the women in Vienna.[2] And to

---

[1] See p. 200.
[2] See p. 203.

distinguish an α flash from a proton required almost divine inspiration. The method was thus quickly abandoned in favor of electrical methods, more powerful and more objective.

The Point Counter

In 1908, Rutherford and Geiger[1] constructed a gas detector, which was able to count α-particles one by one. However, the instrument could only count particles slowly because each detection caused an electric perturbation which lasted several seconds, so that one could only count events which did not occur more frequently than about ten every minute. The electric avalanche[2] produced in the counter by a particle passing through was detected by quadrant electrometer. In 1912, Rutherford and Geiger replaced the latter by a wire electrometer, with a much smaller inertia and they were thus able to record about 1000 particles per minute. They further replaced the central wire by a small sphere acting as an anode, itself placed inside a sphere acting as the cathode. The advantage of this configuration was that the α-particles would all travel about the same distance in the gas and they would each produce similar signals. Another innovation consisted in recording the displacement of the electrometer on a photographic film which unwound regularly. The other features were quite similar to their 1908 counter [124].

When he returned to Germany in 1912, Geiger constructed a new counter. It consisted of a short metal tube (4 cm long and a 2 cm diameter). The central wire was replaced by a fine needle the extremity of which (the "point") was at about 0.8 mm from the disc which closed the tube. A circular 2-mm hole, covered by a thin mica sheet, was bored in this disc, allowing the particles to enter the tube. This *point counter* will have several versions. The intense electric field, which prevails in the vicinity of the point of the needle, induced an electric discharge each time a particle passed through the small volume of a few cubic millimeters, shaped as a cone whose summit is at the extremity of the needle and the base a circular hole in the facing wall. This was a delicate and even capricious instrument. According to Geiger:

> *For the effectiveness of a counter the quality of the needle is decisive [...] One cannot give a rule as how to prepare a good needle [125, 126].*

In other words, a good needle requires a know-how, it is an art. The counter is so sensitive, that it fires whatever particle passes through, be it α or β. And since it can detect a β-particle, it can effectively detect a γ-ray, provided the latter collides with an electron and gives it a certain velocity: of course, it is the electron which is detected.

---

[1] See p. 67.
[2] See p. 68.

Normally, the point counter either fires or not: a particle passing through could tear a few electrons off the atoms of the gas, and the latter would multiply thereby forming an electric avalanche detected by the electrometer. The number of electrons which are produced does not depend on the number of electrons initially torn off. Therefore, the counter does not specify the nature of the particle which made it fire and even less its energy. However, Geiger showed that if the needle point is not too fine (between 0.08 and 2 mm in diameter) and if the electric voltage is correctly adjusted, the electric signal becomes proportional, the number of initially created electrons and therefore to the energy lost by the particle as it passed through the counter. This enabled him to distinguish α-particles and electrons. This is called a "proportional counter" [127]. But, as mentioned above, the apparatus was delicate and difficult to use.

The Geiger–Müller Counter

In 1925, Hans Geiger obtains a chair in Kiel. This is where, together with his student Walter Müller, who is both inventive and keen on new techniques, he constructs a new counter. He makes a few modifications to his 1908 counter. They appear minor at first sight, but they change everything. He manages to increase the voltage of the wire (which is thinner) using a special method:

> The thin wire is stretched along the axis of the metal tube; the wire is covered by a thin layer of poorly conducting material of constant thickness. The isolation produced by this layer permits one to increase the potential between the wire and the tube beyond the discharge potential. If a few ions are created somewhere inside the tube, their multiplication by collisions produces an appreciable amount of electricity which flows towards the wire. If the wire were bare, the voltage would create a permanent discharge; with the isolating cover, a charge is created for a short while on the surface of the layer, which interrupts the electric field and breaks the current [128].

A further advantage is that the precise value of the applied potential is not critical (usually 1200 or 1300 V) and this makes the Geiger–Müller much more stable than the earlier Rutherford-Geiger counter and the point counter. Another change is due to the fact that the number of electrons which reach the central wire and which therefore are detected is *grosso modo* proportional to the number of electrons created by the passing particle and therefore to the energy it loses while passing through the tube. By contrast, in the Geiger–Müller counter, there is no connection between the number of electrons created by the passing particle and the number which reach the central wire. In this respect, the Geiger–Müller counter is similar to the point counter. It is a *simple counter*, which either fires or does not . But it is very sensitive (it detects all radiation), stable, and easy to use. This is why it will prevail as a universal counter, which is still used today to detect the presence of radiation.

A Digression: The Birth and Development of Wireless Radio

Let us part from counters for a while and return to the beginning of the twentieth century, even a little earlier, in order to witness the birth and the rapid development of the wireless telegraph, not yet called the *radio*. In 1885, Heinrich Hertz showed the existence of electromagnetic waves which had been predicted by Maxwell. The possibility of causing electric discharges at a distance fascinated physicists as well as engineers and even amateurs, who made numerous and varied experiments, such as those of Oliver Lodge in England and Édouard Branly in France. It seems that it was Aleksandr Popov, physics professor at the school of officers of the Russian Navy in Kronstadt, who was the first to think of using electromagnetic waves to *transmit information*. He used an alphabet which had been patented in 1840 by the American painter and inventor Samuel Morse for the electric telegraph which transmitted messages through an electric wire, as the usual telephone does. In 1896, the young Guglielmo Marconi succeeds in transmitting Morse messages at a distance of 2 400 m. He then embarks for England where he secures the first patent of wireless telegraphy. He transmits a first message across the Channel in 1899 and across the Atlantic in 1901. From then on, wireless telegraphy develops rapidly in Europe and in the United States due to industrialists and the Navy which discovers that it is a heaven-sent means of communicating between ships at sea. On November 16, 1904, the English electrician, John Fleming, working for the Marconi Company, secures a patent for a special vacuum tube which he calls a "valve" and which allows an electric current pass through in only one direction. The valve is a small cathode ray tube: it is a tube, inside which a vacuum is made and in which a heated cathode emits electrons and causes a current to flow towards the anode. Such a current can obviously only flow in that direction. The valve is a good detector of low-frequency signals which modulate the amplitude of Hertzian waves (e.g., audio frequencies in AM radio broadcasting), and it makes it possible to translate Morse signals into deviations of a electrometer, or to listen to them with a headphone.

In 1907, it is an American, Lee De Forest, who secures a patent for a *three-electrode tube*, which he calls an "Audion." His idea consists in connecting the reception antenna to a zigzag-shaped wire, which he places between the cathode and the anode, which has the shape of a sheet, the "plate." Under certain conditions, which were admittedly very unstable in the first tube, the received signal could be amplified in a far more efficient manner than could be achieved with the valve of Fleming. After numerous perfections, this *triode* tube became the basis for the fantastic development of what will be called the *radio* and later *electronics*. The triode tube consists of a tube in which a vacuum is maintained. At first, De Forest believed that it should contain at least some air, and that caused his first tube to be very unstable. Inside the tube, there is a heated cathode which emits electrons and an anode consisting of a metal sheet maintained at a positive potential so as to attract the electrons emitted by the cathode. This allows the current to pass through in only one direction. Between the anode and the cathode there is a metal grid which receives the signal to be amplified. According to the potential of the grating, the electrons emitted by the cathode, will either pass in large numbers or be stopped. A signal sent to the grating can then allow electrons to pass, and the current detected on the sheet will be more or less intense depending on the potential of the grating. This is the basis of electronic amplification.

The Audion tube, that is, the triode, will allow Lee De Forest to make a wireless transmission of the voice of the soprano singer and film actress Geraldine Farrar in 1907 and even of an opera.

The first radio broadcast aimed at unknown listeners was conducted by the Frenchman Raymond Braillard and the Belgian Robert Goldschmitt. For some time, they would broadcast a concert every week. But the first real radio station was KDKA, created on November 1920 on the occasion of the American presidential election. In 1921, the first radio stations with daily broadcasts were created, first in the United States and later worldwide.

During this development of radio broadcasts, an increasingly important production of radio receptors took place. Amateurs developed a passion for radio and began constructing their own radios and amplifiers, as witnessed by the large number of journals aimed at them. For many, the radio and what was not yet called electronics was the most marvelous of all inventions.

Curiously, this capacity of amplifying weak signals remained neglected by physicists until the late twenties.

The Electronically Amplified Ionization Chamber

The ionization chamber had been used from the outset in radioactivity studies by Pierre and Marie Curie and by Rutherford. The one used by Pierre and Marie Curie in 1898 consisted of two horizontal plates connected to the poles of an electric battery. A radioactive substance was placed on the lower plate. The charged particles, which it emitted, ionized the air between the plates, and this produced a weak electric current of the order of 10^{-11} A (100 000 times weaker than a micro-ampere) which could be detected by an electrometer. However, the current produced by a single particle was much too weak to be detected by an electrometer. This led Rutherford and Geiger to the idea to amplify the signal by a proliferation of electrons. However, the Rutherford-Geiger counter was simply a *counter*, which could detect a particle passing by but which gave no information concerning its nature or its energy.

In 1924, the Swiss physicist Heinrich Greinacher had the idea of sending the signal emanating from a point counter onto the grid of a triode tube in order to amplify the signal which was sent to a telephone which would emit an audible "tick" each time a particle was detected [129]. He could also send the amplified signal to a sensitive electrometer in order to keep a photographic record of its motion. He finally noticed that the signals emitted by α-particles were stronger than those produced by electrons (β-particles). Was that a valid reason to abandon the counter and to return to the ionization chamber?

The unamplified signal produced by a particle, without the amplification produced by an avalanche, is admittedly much weaker (about a 1000 times weaker), but this is also its advantage, because most of the spontaneously triggered events, which are usually due to parasites and which frequently occur in ultrasensitive point counters, are eliminated. Greinacher believed that the counting of particles would become more reliable and less subject to parasites if he managed to amplify the signal of the ionization chamber in order to make it detectable. But if the avalanche is to be avoided, it should be by a different amplification of the same order of magnitude. Since the amplification produced by one triode is not sufficient, Greinacher constructs an amplifier possessing three successive stages. This allows him to produce an audible sound in an amplifier or a perceptible signal in an electrometer *for a single α particle*. In the introduction of his paper, he explains:

> No matter how elegant the counting methods are [...] their reliability always raises doubts [...] In a recent work, Geiger and Bothe have tackled this problem. The [fortuitous discharges] [...] make it difficult to interpret the counts. Because of the great importance of these methods, their foundation is the subject of numerous works. The point is to determine under which circumstances a particle induces a discharge and to what extent can discharges occur in the absence of radiation. Clearly, a method which could avoid these questions would be preferable. I attempted therefore to avoid using a self-sustained discharge and to conceive a method which would avoid fortuitous signals. In the following we shall see how it is possible to record α-particles both acoustically and with a galvanometer, using only the amplification produced by electron tubes [130].

In 1927, Greinacher replaces the galvanometer by an oscilloscope [131], an apparatus widely used in radio and which allows one to visualize on a cathode ray tube rapid variations of the electric current. He thus avoids the weak inertia of the wire of the electrometer which did not allow one to observe excessively weak currents or too rapidly varying currents. As on a cathodic screen, the image appearing on the oscilloscope is drawn by an electron beam which is deviated by the current one wishes to measure. It has an incomparably smaller inertia. Greinacher is thus able to photograph the marks left by α-particles and protons and to make a first quantitative estimate of their ionizing power. For the first time, it becomes possible to measure directly the amount of electricity produced by a passing particle. *This amount turns out to be proportional to the energy of the particle*. This is a considerable step forwards. However, Greinacher continues to activate a loudspeaker each time a particle passes through, but he notes that the amplification is so strong that he can hear when they begin even if the loudspeaker in switched off.

In the early 1930s, tube amplifiers are extensively used. There is one in each commercial radio. As a result, Greinacher's purely electronic amplification becomes widespread in Europe; in Vienna [132]; at the Cavendish, under the leadership of Rutherford [133]; and also in Paris, in the laboratory of Maurice de Broglie, thanks to a young 30-year-old physicist, Louis Leprince-Ringuet [134, 135]. Electronics has become part of physicists instrumentation, particularly of nuclear physicists. It will soon become their queen.

Coincidence Measurements

Recall how Bothe and Geiger could detect simultaneously an electron and a photon in two point counters.[1] The method consisted in observing the simultaneous recording of two galvanometers, and they could guarantee the simultaneity to within a thousandth of a second. When Geiger left Berlin, Bothe continued to work with coincidences applied in particular to the study of cosmic rays [136]. To ensure the simultaneity of two actions or at least their precise temporal succession is an age-old problem. For example, two observers would observe scintillations through a microscope and would activate a switch whenever they saw a signal. In such cases, simultaneity could only be guaranteed to within about 1/10 of a second.

But the advent of radio and amplification tubes caused a rapid change. In 1929, Bothe conceives a tube which could detect the simultaneous triggering of two point counters. One of the counters detects a particle and produces a short electric current which Bothe feeds into a grid of *a special tube endowed with two successive grids*, the second grid of which receives the signal from the other counter. Bothe sets up the apparatus in such a way that the double-grid tube only triggers when it receives the signal of the two counters *at the same time*. He obtains thus a signal which he can record, see, or hear from a loudspeaker [137]. He sends his paper in November to *Zeitschrift für Physik*, and it is published in early 1930.

Upon reading this paper, an Italian physicist from the Arcetri Institute of Physics, near Florence, has a brilliant idea. Bruno Rossi was born in 1905. His father was an electrician. He is in the university, working on his PhD, which he expects to obtain in 1927. Upon feeding the signals of two Geiger–Müller counters to the grids of triode tubes (also called *valves*) he can *automatically* detect the presence of two simultaneous signals. Recall that a triode tube is composed of a cathode (heated in order to liberate electrons) usually connected to the earth, of a grid, and of an anode (or plate) set to a positive potential. If the grid is set to a negative potential, it repels the electrons which do not reach the anode, and no current passes through. If the grid is set to a positive potential, it attracts the electrons which pass through the holes of the grid and reach the anode, thereby producing an electric current. But just between these two situations, small variations of the grid potential can produce large variations of the electric current reaching the anode. This enables the tube to amplify the electric current and therefore the signal. Rossi has the idea of using the triode tube as an *automatic switch*: when the grid is set to a positive potential, the current passes through and the switch is off; when it is set to a negative potential, the current

---

[1] See p. 193.

is stopped and the switch is on. Rossi sends the signals coming from the counters (negative electric impulses since they are produced by electrons) to the grids of two triode tubes, and *he connects the plates* (the conventional name of the anodes). He maintains the grids of the two tubes at a continuous positive potential, and the current passes. When a signal is sent by one of the counters, the tube behaves as a switch in the off position and no longer lets the current flow although it continues to flow in the other tube. However, if both tubes receive a signal at the same time, the current is stopped. A "coincidence" is detected [138]. Rossi's coincidence counter will soon be adopted all over the world.

The original method of Bothe is in principle equivalent, but it is not so simple to use because it requires a special tube and mainly because it cannot be extended to more than two counters, whereas Rossi's method can, easily. Because Walther Bothe preceded Rossi, he received the Nobel Prize in 1954 for "the coincidence method and for the discoveries made therewith." In his Nobel lecture (which he could not deliver personally because of an illness), Bothe recalls the first coincidence measurements he made with his friend Geiger, and he then speaks of his electronic device:

I used a circuit employing a multiple-grid coincidence valve as early as 1929. Rossi was the first to describe another system working with valves in parallel; it has the advantage that it can easily be extended to coincidences between more than two events, and is therefore predominantly used to-day.

He concludes with these words:

Many applications of the coincidence method will therefore be found in the large field of nuclear physics, and we can say without exaggeration that the method is one of the essential tools of the modern nuclear physicist [139].

Walther Bothe died less than 3 years later, in 1957.

The later development of instrumentation will show how right Bothe was. The coincidence method became the basic method used in nuclear physics. Indeed, physicists are always hunting for rare events buried amidst numerous well-known ones. It is like searching for a smaller and smaller needle in a growing hay stack. Without the coincidence measurements it would be impossible to detect the events searched for.

The Measurement of the Energy of γ Radiation

γ Radiation is an electromagnetic radiation, just like light. It *is* light and so are X-rays except that the γ radiation has a shorter wavelength. A quantum of γ light has therefore a much higher energy than a quantum of visible light. But how can one measure its wavelength?

Absorption Measurements

Charles Barkla was able to discover the essential properties of X-rays by simply observing what made them more or less absorbed in a given thickness of matter.[1] He could establish a rough but useful law: the higher was its penetration power, the higher was the energy of the ray.

Diffraction on Crystals

It was William Henry Bragg and his son William Lawrence Bragg who, after Max von Laue,[2] developed a precise manner to measure the wavelength of X rays. They observed their diffraction on a crystal.[3] The X-rays are reflected more intensely at certain angles because the photons which are reflected by different atoms of the crystal arrive "in phase." In principle, this method could also be applied to γ rays, however the reflection angles would then have to be smaller than 1°, which would make it difficult to do a precise measurement. Worse, photographic plates are less sensitive to γ-rays than to X-rays: whereas a 10-minute exposure suffices for X-rays, a 24-h exposure is required for γ-rays. The first to try nonetheless is no other than Rutherford who performed the experiment with Edward Neville da Costa Andrade, a 27-year-old English physicist who obtained his PhD in 1911 in Heidelberg, Germany. They perform their first measurements with "soft" γ-rays which have a low energy, similar to that of X-rays, and then they pursue with higher energy γ rays [140]. The angles at which they must make their observations in order to determine the wavelength shrink from 10° to less than 1°. Even so the energy of the γ-rays is only 180 000 eV (0.18 MeV). In order to save time, they slowly vary the deflection angle by turning the crystal in order to cover the widest possible energy range. This "revolving crystal" method had been used for the first time by Maurice de Broglie [141] to measure the wavelength of X-rays. It was perfected by his students, Jean Thibaud [142] and Marcel Frilley [143]. The latter succeeded in measuring the energy of 770 000 eV γ-rays which required a grazing reflection since the angle was only 10 s of an arc.

The Photoelectric Effect

Another method to measure the energy of γ-rays used the photoelectric effect. When a photon incident on a metal collides with an electron, it can either be deviated or scattered (with a corresponding change of its wavelength and therefore

---

[1] See p. 60.
[2] See p. 78.
[3] See p. 78.

of its energy). This is the Compton effect.[1] Alternatively, it can be absorbed by the electron in which case it transfers *all its energy* to the electron which is ejected from the metal. The energy of the ejected electron is reduced by a quantity which depends only on the metal and which is the energy required to extract an electron from the metal. That is the photoelectric effect which was known since the nineteenth century and which Einstein explained[2] in 1905. It suffices then to measure the energy of the electron in order to determine the energy of the γ-ray. However, this method was not convenient to determine the energies of γ-rays emitted by radioactive substances. The effect was weak and required very long exposure times on photographic plates. It also displayed an important background and not very narrow lines.

Conversion Electrons

Whenever it was possible, it was more convenient to use *internal conversion* electrons. The latter are ejected by an internal photoelectric effect. The γ-ray emitted by the nucleus transfers its energy to an electron of the surrounding electron cloud, and the latter is ejected with the energy of the γ-ray minus the energy required to extract the electron from the atom. This was the method most often used by physicists who studied radioactivity. But in order to study γ-rays emitted during artificially induced nuclear reactions, the method was not particularly useful and new ones had to be invented.

A Unique Detector: Wilson's Cloud Chamber

Clouds and Charles Thomas Rees Wilson

In September 1894, a young student on vacation climbs to the top of Ben Navis, 1,343 m, the highest mountain in Scotland. He is fascinated by the luminous effects of the setting sun on the clouds, glorious iridescent arcs. He is a physicist and cannot not help wondering how such colorful clouds are formed, and he attempts to reproduce the effect in his laboratory. Charles Thomson Rees Wilson was born on February 14, 1869, in Glencorse, near Edinburg. His parents were farmers, but his father died when he was 4 years old. His mother then moved to Manchester where the young Charles attended university in view of becoming a doctor. But he finally decided to do physics instead. In 1892, he obtained a position at the Cavendish to work under the direction of J. J. Thomson, who was then doing most productive research on cathode rays.

---

[1] See p. 124.
[2] See p. 96.

To understand how clouds form, Wilson studies how water vapor condenses into the tiny droplets which fog and clouds consist of. He knows the physical law which governs the condensation of water vapor in air: if the pressure (or the temperature) is suddenly reduced, part of the vapor in the air condenses in the form of water droplets.

But this condensation takes place only in the neighborhood of so-called *condensation nuclei* such as fine dust particles present in the air. This had been discovered in 1888 by the Scottish physicist John Aitken [144]. A similar phenomenon is witnessed by amateurs of Champagne wine. As long as the wine remains in the bottle, no bubbles form. But as soon as the bottle is opened, the pressure of the liquid inside is reduces and bubbles arise. But not anywhere: when poured into a glass, the bubbles originate from one or several points on the inside surface of the glass. This happens because the glass surface has irregular scratches which play a rôle similar to condensation points. Some even go so far as to file down ever so slightly the bottom of the glass so as to make all the bubbles originate there!

Wilson constructs a chamber the volume of which he is able to suddenly expand. He thus suddenly reduces the pressure of the enclosed gas, which is air plus water vapor. The first "expansions," as he calls them, display fog, which is a condensation around the "Aitken nuclei" which are fine dust particles. Wilson then lets the water condense and fall to the bottom of the chamber, dragging along the dust particles which served as condensation points. He repeats the process several times to make sure that all dust particles have been thus eliminated. As expected, he then observes that fog is no longer formed during an expansion. However, when he expands the volume by more than 25 %, fog again forms when he thought he had got rid of all the dust particles! It is as if there remained some condensation nuclei which could not be dragged out by the condensation. And if he expands the volume by more than 38 %, it is not a cloud, but a real rain which is formed! It almost looked as if every molecule in the gas could play the role of a condensation point [145]. What puzzles Wilson is his inability to get rid of the condensation points. They appear to be continuously *created*. But how? This was just the time, at the beginning of 1896, when news of X-rays had reached laboratories and J. J. Thomson began an active study of their properties, especially their capacity of favoring the flow of an electric current through gases. For Thomson, the X-rays must be *ionizing* the gas by tearing out electrons from some of the atoms, the latter being pushed by the electric field. When Wilson exposes his chamber to X rays, he obtains a dense cloud:

Instead of a shower settling in 1 or 2 s, a fog lasting for more than a minute was produced [145].

In further experiments, Wilson compares the effect produced in his chamber by X-rays, ultraviolet light, and "uranic" rays. He finds that the various radiations produce condensation nuclei, and he proves that the condensation nuclei are electrically charged:

When air exposed to X-rays is enclosed by two parallel plates, between which a sufficient difference of potential is maintained, the fogs obtained on expansion are very much less dense than in the absence of the electric field, and if the rays be turned off before expansion all nuclei are found to have been removed, whereas without any electric field a fog is obtained even if the expansion be not made till some seconds after the rays have been cut off. This behavior of the nuclei proves them to be charged particles or "ions" [146].

Wilson notes that the rays, which were still called "uranic" at the time, produce the same effect as X-rays. He notes, in passing, that:

Expansion experiments probably furnish one of the most delicate methods of detecting these rays.

An Outstanding Detector

For a few years, Wilson is mainly concerned with meteorology. It is only in 1910 that he begins to investigate how to use his cloud chamber to detect and photograph particle trajectories. His first attempt, in 1911, involves a simple chamber which he irradiates with X-rays and later with α- and β-rays emitted by a radioactive substance:

I was delighted to see the cloud chamber filled with little wisps and threads of clouds—the tracks of the electrons ejected by the rays. The radium-tipped metal tongue of a spinthariscope was placed inside the cloud chamber and the very beautiful sight of the clouds condensed along the tracks of the α-particles was seen for the first time. The long thread of the β-particles were also seen when a suitable source was brought near the cloud chamber [147].

He quickly has an improved chamber constructed. His first results impress Rutherford who qualifies them as "the most wonderful experiment in the world." Wilson obtains the expansion by quickly opening a communication between the chamber and an evacuated reservoir. The procedure had to be repeated several times in order to get rid of all dust and to eliminate with an electric field the ions which were produced permanently [148].

This incredible apparatus displays directly the particle trajectories in the form of condensation lines, similar to the white tracks which airplanes leave in the sky. They can be either seen with the naked eye or be photographed. What is wonderful, almost incredible, is to see *a single particle*, more precisely the track it leaves behind. In spite of this spectacular feat, Wilson's cloud chamber will not be used before the 1920s. It is its capacity of detecting *all the charged particles emitted after a collision* which will make it famous.

The Compton Effect Seen in the Cloud Chamber

Bothe and Geiger had shown that in the Compton effect which involves a collision of an X-ray photon and an electron, the photon and the electron come out in different directions, with energies which can easily be calculated in terms of the angle by which the photon is deviated. In order to check this calculation and to show that energy is conserved in each collision, Bothe and Geiger set up an experiment in which the photon and the electron are detected simultaneously, in a coincidence experiment. In principle, the Wilson cloud chamber could do the same since it would display not only the path of the electron but also that of the photon, because of the electrons it would collide with (if it did collide, which was not always the case). This was the task tackled by Arthur Compton and Alfred Simon in 1925. The experiment was difficult and the events rare, but the result was unambiguous [149]: the energy and the momentum were conserved at each collision.

A Nuclear Reaction, Actually Seen

A young physicist tackled a similar problem at the Cavendish. Patrick Blackett [150, 151] was born on November 18, 1897. He began a career as a Navy officer. He fought at the battles of the Falkland Islands and of Jutland. When the war was over, he resigned with the rank of lieutenant and began to study physics in Cambridge under the direction of Rutherford. He began research in 1921 when Rutherford and Chadwick were observing nuclear reactions using scintillation counters. Blackett tried to observe the nuclear reaction in a Wilson cloud chamber. If it were possible to observe *all* the particles involved in the reaction, one could understand what actually happens to the α-particle, as Rutherford used to say. But in addition to the usual difficulties of regulating the cloud chamber, the main problem was the rarity of the events. Remember how small nuclei are. When α-particles pass through a gas, the probability for them to hit a nucleus is very small. Blackett took about 23,000 photographs of the Wilson's cloud chamber [152]. Each photograph contained about 18 tracks, altogether 400,000 tracks. Most of the tracks were made by α-particles. Among these tracks, one could occasionally observe a sudden deviation of the α-particle but with no change of the total length of the path. Such events were interpreted as "elastic" collisions, in which the α-particles bounced off the target with no energy transfer (more precisely, with a small energy transfer). One could also observe, at the point where the α particle changed its path, a short track produced by the recoiling nucleus. Finally, among all these tracks, Blackett discovered eight events which he could identify to transmutations. The pattern of the paths were quite similar to the ones attributed to "elastic" collisions of α-particles with nitrogen nuclei with one difference: the track originating from the collision point was both longer and much finer than that of an α-particle. *It was the track of a proton.*[1] No other track originated at the collision point [152]:

> *The study of the photographs has led to the conclusion that an alpha-particle that ejects a proton from a nitrogen nucleus is itself bound to that nucleus [...] Of the nature of the integrated nucleus little can be said without further data. It must however have a mass 17, and provided no other nuclear electrons are gained or lost in the process, an atomic number 8.*

That was the interesting point: it appeared that the α-particle had been absorbed by the nitrogen nucleus, in what seemed to be a fusion process, similar to the merging of two water droplets when they come into contact. After this fusion, a proton emerged leaving an oxygen nucleus. One could also imagine that the nucleus formed by the absorption of the α-particle by the nitrogen nucleus would first emit a proton and break up later. Blackett mentioned that, if this were the case, the event would give rise to further tracks which were not observed. This confirmed Rutherford's views and the intuition of Jean Perrin.[2] The theory of nuclear reactions had begun.

---

[1]Protons have an electric charge which is only half the charge of the α-particle. They therefore ionize the gas less, thus producing a finer track in the cloud chamber. They are also less slowed down, thereby leaving a longer track.

[2]See p. 206.

The Atomic Nucleus in 1930

On February 25, 1927, Rutherford is invited to deliver a talk on "Atomic nuclei and their transformations" to the Physical Society. He first notes what appears to be obvious:

> *Since both helium nuclei (α-particles) and swift electrons are hurled from the nucleus, it could safely be deduced that the nucleus of these heavy atoms must contain electrons and helium nuclei as part of their structure, unless it be supposed that the helium nucleus is in some way formed of simpler constituents at the moment for its expulsion from the main nucleus [153].*

But he goes further and sketches a real model of the nucleus:

> *While it is impossible that positively charged particles like the proton or the α-particle can remain in equilibrium under a coulomb law of repulsive force, the case is quite different if the particles are electrically neutral [. . .] We thus arrive at a general conception of nuclear structure in which the central charged nucleus is surrounded by a number of uncharged particles. In a paper before the Franklin Institute in 1924 I put forward a suggestion that the central nucleus was a closely ordered arrangement of α-particles and electrons in a semi-crystalline formation, and showed that certain simple arrangements were in fair accord with the charge and mass of some of the atoms.*

In the following months, Rutherford developed further the model and he described it in detail in a paper which he sent to the *Philosophical Magazine* on August 10, 1927. The paper begins by describing the situation:

> *The electrical forces from the nucleus on a positively charged body are repulsive and the experimental evidence on the scattering of α-particles by aluminum and magnesium indicates that the attractive forces which come into play due to the distortion or polarization of the main nuclear structure by the charged α-particle only become of importance at distances of about 1×10^{-12} cm [154].*

In fact, the model proposed by Rutherford is a kind of miniature atom, with the size of the nucleus, in which the α-particles orbit around a "center" of the nucleus:

> *On the views put forward in this paper, the nucleus of a heavy atom has certain well defined regions in its structure. At the centre is the controlling charged nucleus of very small dimensions surrounded at a distance by a number of neutral satellites describing quantum orbits controlled by thè electric field from the central nucleus [. . .] It is fairly certain that the central nucleus of a heavy element is a very compact structure, occupying a volume of radius not greater that 1×10^{12} cm. The region around the central nucleus extending to about $r = 1.5 \times 10^{-12}$ cm is probably occupied by electrons and possibly*

also charged nuclei of small mass which are held in equilibrium by the attractive forces which arise from the distortion of polarization of the central nucleus. The electrons in this region, circulating round the nucleus, must have velocities close to that of light.

Rutherford does not specify the nature of the "neutral satellites." They might be α-particles together with two electrons. The attractive forces which hold the structure together are so intense that they overcome the Coulomb repulsion. Rutherford sees them as electromagnetic forces modified at very short distances. He attempts to explain radioactive decay in terms of his model. He recalls a disturbing observation made by Hans Geiger in Manchester in 1911, while he was working with John Mitchell Nuttall [155]. They measured the radioactive half-life of about 20 elements together with the penetration length of the corresponding α-particles, from which they deduced their initial velocities. They notice a surprising correlation between the *radioactive half-life* and the *penetration length* of the emitted particles: the shorter the radioactive half-life, the greater is the penetration length![1] All attempts to explain this had failed, and Rutherford's new model did no better. Rutherford himself noted:

The fact that the energy of the issuing particle must be in part due to the gain of energy in escaping through the repulsive field, makes it doubtful whether the empirical relation found by Geiger has any exact fundamental significance [154].

One is never too careful, and that holds also for Rutherford. All that is needed is to find a "precise fundamental meaning" to the so-called Geiger-Nuttall relation, as we shall see shortly.

This attempt of Rutherford is typical of his way of understanding physics, which so often proved to be successful. But here, it had reached its limits. In his paper, he mentions quantum orbits, quantum numbers which specify the orbits, but his "quantum" description of the nucleus is very similar to Bohr's 1913 description of the atom, namely, a classical description with quantum rules added. In the following years, a new generation of physicists will make an impressive succession of discoveries. The description of the structure of the nucleus will change radically.

Some Certainties and One Enigma

A Certainty...

In 1930 the problem of the composition of nuclei seemed to be settled. Let us repeat the main arguments which appeared to confirm it:

---

[1] There is an approximately linear relation between the logarithm of the penetration length and that of the radioactive half-life.

- The masses of nuclei are always close to a multiple of the proton mass, with a missing mass which could be explained in terms of the binding energy of the nucleus. Therefore, *nuclei are composed of protons*, the number of which is given by their mass.
- But this would make the electric charge of nuclei too large, by a factor of about two in the case of light nuclei (with mass less than 40).
- *The nucleus contains therefore also negatively charged electrons, as shown by the fact that radioactive nuclei emit electrons* (in β-decay). The number of electrons in the nucleus is equal to the difference between the number of protons and the electric charge of the nucleus. This yields both the correct mass and electric charge of the nucleus. The mass of the electron is 1836 times smaller than that of the proton, so that the electrons in the nucleus do not modify significantly its mass.

The arguments appeared to be irrefutable. But were they really?

... And An Enigma: The Nitrogen 14 Nucleus

One problem, which might have appeared to be a detail, did worry the physicists: was the nitrogen 14 nucleus a *boson* of a *fermion*? To understand what was at stake, let us recall this fundamental distinction of microscopic particles, concerning their identity. Bose and Einstein and later Fermi and Dirac had shown that particles such as protons, electrons, and photons could be divided into two families, *bosons* and *fermions*.[1] Two electrons, in an atom or a molecule, for example, cannot have all their quantum numbers equal: they are fermions. This is a statement of the Pauli exclusion principle.[2] However, there can be any number of photons bearing the same quantum numbers: they are bosons. But what about composite particles, such as the α-particle or more generally atoms and atomic nuclei? The question arises when one considers an atom of a nucleus as a single object, without considering its internal structure. This is the case of the hydrogen nuclei (protons) in the hydrogen molecule or of the nitrogen nuclei in the nitrogen molecule.

A reasonable rule had been adopted and later proved by Paul Ehrenfest and Robert Oppenheimer [156]: when the number of protons plus the number of electrons in a nucleus is *even*, the nucleus behaves as a *boson*. When this number is *odd*, it behaves as a *fermion*. The nitrogen 14 nucleus was assumed to contain 14 protons and seven electrons, and *it therefore had to be a fermion*.

However, experiment showed that it was a boson! Indeed, Franco Rasetti, the friend of Fermi, had observed the rotation of the nitrogen molecule, which consists of two nitrogen 14 nuclei, bound by their 14 peripheral electrons [157, 158]. The molecule is similar to a kind of dumbbell which can rotate and also vibrate around

---

[1] See p. 147.
[2] See p. 126.

its equilibrium position. The rotations and vibrations are quantized. One can observe the photons which are emitted when the molecule makes a transition between different rotational states. This allowed Rasetti to measure the spectrum of the rotational states [157, 159], when he was at the *California Institute of Technology*, in Pasadena, with Robert Millikan.

Two young German physicists in Göttingen, Walter Heitler and Gerhard Herzberg, found Rasetti's findings most surprising because *they implied that the nitrogen 14 nucleus behaved as a boson.* In a paper, dated July 28, 1929, they wrote:

> *This is an extremely surprising result. Indeed, the nitrogen nucleus contains a total of 14 protons and 7 electrons [. . .] But quantum mechanics implies that systems which consist of an [odd] number of protons obey [Fermi] statistics since protons and electrons themselves obey the statistics of Fermi. If the observations of Rasetti are correct, this rule loses all its validity inside the nucleus [160].*

When he returned to Rome, Rasetti repeated his measurement and he confirmed, without any doubt, that nitrogen 14 obeys Bose-Einstein and not Fermi-Dirac statistics [161, 162]! Two Dutch physicists, Leonard Ornstein and William Wijk [163], had shown previously that it had a spin equal to 1. This raises a tough question [164]: how can one make an integer spin by adding or subtracting 1/2 an odd number of times?

Should One Consider a Radical Solution?

In 1930, a Russian physicist, Jakov Dorfman, suggested another possibility. He pointed out that very light nuclei, from helium to nitrogen, had no magnetic moment (or a very small one):

> *That means that the electron angular momentum does not manifest itself in the nucleus[. . .]*
> *This experimental fact may be explained in three different ways:*
>
> 1° *The angular momenta of spinning electrons are compensated in some way in the nucleus [. . .]*
> 2° *The electrons in the nucleus lose their magnetic moment, as Heitler and Herzberg have assumed.*
> 3° *There are no electrons in atomic nuclei. [D. Ivanenko and V. Ambartsumian have developed such a theory, based on the latest ideas of Dirac [165]].*

This is a radical solution which makes reference to the work of two other Russians, Victor Ambartsumian and Dmitri Ivanenko [166], who attempted to construct a theory of β decay in which electrons did not exist in the nucleus, but were created at the onset of their radioactive decay. So far, it was assumed that since a nucleus emits an electron during β-decay, electrons were initially inside the nucleus. But Ambartsumian and Ivanenko point out that γ photons do not preexist in the nucleus, but, instead, *they are created at the onset of the radiation*:

There exists an analogy between the emission of β rays by radioactive substances and the emission of light quanta by atoms. Recent data [. . .] seem to support the idea, suggested a few months ago by Heitler and Herzberg, that electrons lose their identity inside the nucleus [. . .] These considerations have led us to attempt to construct a theory of β rays which is analogous to the theory of light quanta, proposed by Dirac.

According to the two Russian physicists, the emission of electrons by nuclei is not an irrefutable proof of their existence inside the nucleus: they might be created at the onset of their emission, as light quanta are. They did not pursue this idea all the way, but it was an interesting idea, a grain thrown into the wind.

At the Beginning of 1932, the Enigma Remains

On April 28, 1932, at the Royal Society in London, a discussion took place, chaired by Rutherford. Ralph Fowler reviewed the experimental evidence and concluded:

We find, too, in this way the most unambiguous evidence for the type of statistics satisfied by the nuclei, in particular that N_{14} has the Einstein-Bose statistics which forces us (along with other evidence) to the deep and disturbing conclusion that the electrons in the nucleus no longer contribute to the spin of the statistical type [167].

References

1. Bohr, N., "On the Constitution of Atoms and Molecules. Part II : Systems containing only a single nucleus", *Philosophical Magazine* **26**, 476–502, 1913.
2. *La structure de la matière : rapports et discussions du Conseil de physique Solvay tenu à Bruxelles du 27 au 31 octobre 1913*, Gauthier-Villars, Paris, 1921, p. 55.
3. Rutherford, E., "The Structure of the Atom", *Nature* **92**, 423, December 11, 1913.
4. Boltwood, B. B., "The origin of radium", *Nature* **76**, 544–545, September 26, and 589–589, October 10, 1907.
5. Merricks, L., *The world made new : Frederick Soddy, science, politics, and environment*, Oxford University Press, Oxford; New York, 1996.
6. Trenn, T. J. (editor), *Radioactivity and Atomic Theory. Annual Progress Reports on Radioactivity 1904–1920 to the Chemical Society by Frederic Soddy F.R.S.*, Taylor & Francis, London, 1975.
7. Soddy, F., "Annual Report for the Chemical Society for 1910, vol. 7", in *Radioactivity and Atomic Theory*, edited by Trenn, T. J., pp. 256–286.
8. Soddy, F., "The Chemistry of Mesothorium", *Journal of the Chemical Society. Transactions* **99**, 72–83, 1911.
9. Mendeleev, D. I., "On the Correlations between the Properties of the Elements and their Atomic Weights", *Zhurnal Russkoe Fiziko-Khimicheskoe Obshchestvo (Journal of the Russian Chemical Society)* **1**, 60–77, 1869. English translation in William B Jensen, ed., *Mendeleev on the Periodic Table, selected writings, 1869–1905*, New York, Dover, 2005.
10. Mendeleev, D. I., "Ueber die Beziehungen der Eigenschaften zuden Atomgewichten der Elemente", *Zeitschrift für Chemie und Pharmacie* **12**, 405–406, 1869.
11. Meyer, J. L., "Die Natur der chemischen Elemente als Function ihrer Atomgewichte", *Annalen der Chemie und Pharmacie. Supplementband* **7**, 354–364, 1870.
12. Mendeleev, D. I., "A natural system of the elements and its use in predicting the properties of undiscovered elements", *Journal of the Russian Chemical Society* **3**, 25–56, 1871.
13. Crookes, W., "Opening address to the Chemical Section of the British Association at Birmingham", *Nature* **34**, 423–432, September 2, 1886.
14. Soddy, F., "Intra-atomic Charge", *Nature* **92**, 399–400, December 4, 1913.
15. Fajans, K., "Über eine Beziehung zwischen der Art einer radioaktiven Umwandlung und dem elektrochemischen Verhalten der betreffenden Radioelemente", *Physikalische Zeitschrift* **14**, 131–136, February 15, 1913.
16. Fajans, K., "Die Stellung der Radioelemente im periodischen System", *Physikalische Zeitschrift* **14**, 136–142, February 15, 1913.
17. Fajans, K., "Position des éléments radioactifs dans le système périodique", *Le Radium* **10**, 57–65, 1913.
18. Fajans, K., "Sur une relation entre la nature d'une transformation radioactive et le rôle électrochimique de l'élément radioactif correspondant", *Le Radium* **10**, 57–60, 1913.
19. Fajans, K., "La place des éléments radioactifs dans le système périodique", *Le Radium* **10**, 61–65, 1913.
20. Fajans, K., "Remarques sur le travail 'Position des éléments radioactifs dans le système périodique' ", *Le Radium* **10**, 171–174, 1913.
21. Soddy, F., "The Radio-elements and the Periodic Law", *Chemical News* **107**, 97–99, February 28, 1913.
22. Thomson, J. J., "On Rays of positive Electricity", *Philosophical Magazine* **13**, 561–575, May 1907.
23. Thomson, J. J., "Rays of Positive Electricity", *Philosophical Magazine* **14**, 359–364, September 1907.
24. Thomson, J. J., "Rays of Positive Electricity", *Philosophical Magazine* **20**, 752–767, October 1910.

234 A Timid Infancy

25. Watson, H. E., "The Densities and Molecular Weights of Neon and Helium", *Journal of the Chemical Society, Transactions* **97**, 810–833, 1910.
26. Aston, F. W., *Mass Spectra and Isotopes*, E. Arnold & Co, London, 1933, p. 30.
27. Thomson, J. J., *Rays of Positive Electricity and their Application to Chemical Analyses*, Longmans, Green and Co., London, 1913.
28. Aston, F. W., "The distribution of intensity along the positive ray parabolas of atoms and molecules of hydrogen and its possible explanation", *Proceedings of the Cambridge Philosophical Society* **19**, 317–323, session of May 19, 1919.
29. Aston, F. W. and Fowler, R. H., "Some Problems of the Mass-Spectrograph", *Philosophical Magazine* **43**, 514–528, March 1922.
30. Aston, F. W., "The Constitution of Atmospheric Neon", *Philosophical Magazine* **39**, 449–455, April 1920.
31. Aston, F. W., "The Mass-Spectra of Chemical Elements", *Philosophical Magazine* **39**, 611–625, May 1920.
32. Aston, F. W., *Isotopes*, Edward Arnold & Co, London, 1922.
33. *Ibid.*, p. 90.
34. "The British association, Cardiff Meeting, September 7–14,1920", *Engineering* **110**, 382–383, September 17, 1920.
35. Jammer, M., *Concepts of Mass in classical and modern physics*, Harvard University Press, Cambridge, 1961, p 136.
36. Abraham, M., "Prinzipien der Dynamik des Elektrons", *Physikalische Zeitschrift* **1**, 57–63, October 10, 1902.
37. Aston, F. W., *Isotopes*, p. 101.
38. Einstein, A., "Zur Elektrodynamik bewegeter Körper", *Annalen der Physik, Leipzig* **17**, 891–921, 1905.
39. Einstein, A., "Ist die Trägheit eines Körpers von seinem Energieinhalt abhängig ?", *Annalen der Physik, Leipzig* **18**, 639–641, 1905.
40. Langevin, P., "L'inertie de l'énergie et ses conséquences", *Journal de Physique (Paris)* **3**, 553–591, 1913.
41. Kaufmann, W., "Über die Konstitution des Elektrons", *Annalen der Physik, Leipzig* **19**, 487–553, 1906.
42. Dempster, A. J., "A new method of positive ray analysis", *Physical Review* **11**, 316–325, April 1918.
43. Dempster, A. J., "Positive-Ray Analysis of Lithium and Magnesium", *Physical Review* **18**, 415–422, December 1921.
44. Dempster, A. J., "Positive-Ray Analysis of Potassium, Calcium and Zinc", *Physical Review* **20**, 631–638, December 1922.
45. Costa, J.-L., "Spectres de masse de quelques éléments légers", *Annales de Physique (Paris)* **4**, 425–456, 1923.
46. Costa, J.-L., "Détermination précise de la masse du lithium 6 (méthode d'Aston)", *Comptes Rendus de l'Académie des Sciences* **180**, 1661–1662, session of June 2, 1925.
47. Aston, F. W., "A New Mass-Spectrograph and the Whole Number Rule", *Proceedings of the Royal Society, London* **A114**, 487–514, 1927.
48. Aston, F. W., *Mass Spectra and Isotopes*, E. Arnold & Co, London, 1933.
49. *Ibid.*, p. 170.
50. Jensen, C., *Controversy and consensus : nuclear beta decay 1911-1934*, Basel, Birkhäuser Verlag, 2000.
51. Rutherford, E., "Uranium Radiation and the Electrical Conduction Produced by It", *Philosophical Magazine* **47**, 109–163, January 1899.
52. Bouguer, P., *Traité d'optique sur la gradation de la lumière*, H.L. Guérin et L.F. Delatour, Paris, 1760, p 232.
53. Soddy, F., "Annual Report for the Chemical Society for 1908-09, vol. 6", in *Radioactivity and Atomic Theory*, edited by Trenn, T. J., pp. 184–219.
54. Hahn, O., *A scientific autobiography*, Charles Scribner's Sons, New York, 1966.

55. Hahn, O., *My life*, Macdonald, London, 1970. Translated from *Mein Leben*, F. Bruckmann, München, 1968.
56. Gerlach, W. and Hahn, D., *Otto Hahn. Ein Forscherleben unserer Zeit*, Wissenschaftliche Verlagsgesellschaft MbH, Stuttgart, 1984.
57. Hahn, O., "A new Radio-Active element, which evolves thorium emanation. Preliminary communication", *Proceedings of the Royal Society, London* **A76**, 115–117, 1905.
58. Boltwood, B. B., letter to E. Rutherford, dated May 4, 1905, in *Rutherford and Boltwood: letters on Radioactivity*, edited by Badash, L., p. 72.
59. Boltwood, B. B., letter to E. Rutherford, dated September 22, 1905, *Ibid.*, p. 81.
60. Rutherford, E., letter to B. Boltwood, dated October 10, 1905, *Ibid.*, p 90.
61. Hahn, O., "Ein neues Zwischenprodukt in Thorium", *Physikalische Zeitschrift* **8**, 277–281, May 1, 1907.
62. Hahn, O., "Ein kurzlebigen Zwischenprodukt zwischen Mesothor und Radiothor", *Physikalische Zeitschrift* **9**, 246–248, April 15, 1908.
63. Sime, R. L., *Lise Meitner, a life in Physics*, University of California Press, Berkeley, 1996.
64. Rife, P., *Lise Meitner and the Dawn of Nuclear Age*, Birkhäuser, Boston, 1999.
65. Meitner, L., "Über die Absorption der α- und β-Strahlen", *Physikalische Zeitschrift* **7**, 588–590, 1906.
66. Meitner, L., "Looking back", *Bulletin of the Atomic Scientists* **20**, 2–7, November 1964.
67. Hahn, O. and Meitner, L., "Über die Absorption der β-Strahlen einiger Radioelemente", *Physikalische Zeitschrift* **9**, 321–333, May 15, 1908.
68. Hahn, O. and Meitner, L., "Über die β-Strahlen des Aktiniums", *Physikalische Zeitschrift* **9**, 697–702, October 25, 1908.
69. Rutherford, E. and Hahn, O., "Mass of the α Particles from Thorium", *Philosophical Magazine* **12**, 371–378, October 1906.
70. Hahn, O., *A scientific autobiography*, p. 55.
71. von Baeyer, O. and Hahn, O., "Magnetische Linienspektren von Beta-Strahlen", *Physikalische Zeitschrift* **11**, 488–493, June 1, 1910.
72. von Baeyer, O., Hahn, O. and Meitner, L., "Über die β-Strahlen des aktiven Niederschlags des Thoriums", *Physikalische Zeitschrift* **12**, 273–79, April 15, 1911.
73. Soddy, F., "Annual Report for the Chemical Society for 1908-09, vol. 6", in *Radioactivity and Atomic Theory*, edited by Trenn, T. J., p. 230.
74. Franklin, A., "William Wilson and the Absorption of Beta Rays", *Physics in Perspective* **4**, 40–77, 2002.
75. Wilson, W., "On the Absorption of Homogeneous Beta Rays by Matter and on the Variation of the Absorption of the Rays with velocity", *Proceedings of the Royal Society, London* **A82**, 612–628, session of January 24, 1909.
76. Danysz, J., "Sur les rayons beta de la famille du radium", *Comptes Rendus de l'Académie des Sciences* **153**, 339–341, session of July 31, 1911.
77. Danysz, J., "Sur les rayons β de la famille du radium", *Comptes Rendus de l'Académie des Sciences* **153**, 1066–1068, session of November 27, 1911.
78. Danysz, J., "Sur les rayons β de la famille du radium", *Le Radium* **9**, 1–5, 1912.
79. Rutherford, E., "The Origin of β and γ Rays from Radioactive Substances", *Philosophical Magazine* **24**, 453–462, October 1912.
80. Rutherford, E. and Robinson, H., "The Analysis of the β Rays from Radium B and Radium C", *Philosophical Magazine* **[6] 26**, 717–29, October 1913.
81. Brown, A., *The Neutron and the Bomb. A biography of Sir James Chadwick*, Oxford University Press, Oxford, 1997.
82. Chadwick, J., letter to Rutherford, dated January 14, 1914, in Brown, A., *The Neutron and the Bomb*, p. 24.
83. Chadwick, J., "Intensitätsverteilung im magnetischen Spektrum der β-Strahlen von Radium B+C", *Verhandlungen der deutschen physikalischen Gesellschaft* **16**, 383–391, 1914.
84. Meitner, L., "Über die β-Strahl-Spektra und ihren Zusammenhang mit der γ-Strahlung", *Zeitschrift für Physik* **11**, 35–54, 1922.

85. Meitner, L., "Über eine mögliche Deutung des kontinuierlichen β-Strahlenspektrums", *Zeitschrift für Physik* **19**, 307–312, 1923.

86. Meitner, L., "Über eine notwendige Folgerung aus dem Comptoneffekt und ihre Bestätigung", *Zeitschrift für Physik* **22**, 334–342, 1924.

87. Ellis, C. D. and Wooster, W. A., "The Average Energy of Disintegration of Radium E", *Proceedings of the Royal Society, London* **A117**, 109–123, 1927.

88. Sime, R. L., *Lise Meitner*, p. 105.

89. Bohr, N., Kramers, H. A. and Slater, J. C., "The quantum theory of radiation", *Philosophical Magazine* **47**, 785–802, May 1924.

90. Pais, A., *Subtle is the Lord...*, p 420.

91. Segrè, E., *From X-rays to quarks, modern physicists and their discoveries*, W. H. Freeman, San Francisco, 1980.

92. Bothe, W. and Geiger, H., "Experimentelles zur Theorie von Bohr, Kramers und Slater", *Naturwissenschaften* **13**, 440–441, May 15, 1925.

93. Bothe, W. and Geiger, H., "Über das Wesen des Comptoneffekts; ein experimenteller Beitrag zur Theorie der Strahlung", *Zeitschrift für Physik* **32**, 639–663, 1925.

94. Pauli, W., *Writings on Physics and Philosophy*, edited by Charles P. Enz and Karl von Meyenn, Springer Verlag, Berlin, 1994, p. 198.

95. Pauli, W., *Wissenschaftlicher Briefwechsel mit Bohr, Einstein, Heisenberg, u.a. Wolfgang Pauli, Scientific correspondence with Bohr, Einstein, Heisenberg a.o.*, Springer-Verlag, New York, 1979, p. 39 footnote 2.

96. Ellis, C. D., "The Magnetic Spectrum of the β-Rays Excited by γ-Rays", *Proceedings of the Royal Society, London* **A99**, 261–271, 1921.

97. Rutherford, E., "The constitution of matter and the evolution of the elements", *Popular Science Monthly* **87**, 105–142, August 10, 1915. First course of lectures on the William Ellery Hale Foundation, National Academy of Sciences, delivered at the Washington Meeting, April, 1914.

98. Marsden, E., "The Passage of α-particles through Hydrogen", *Philosophical Magazine* **27**, 824–830, May 1914.

99. Marsden, E. and Lantsberry, W. C., "The Passage of α particles through Hydrogen. II", *Philosophical Magazine* **30**, 240–243, August 1915.

100. da Costa Andrade, E. N., "Rutherford at Manchester, 1913-14", in *Rutherford at Manchester*, edited by Birks, J. B., pp. 27–42.

101. Darwin, C., "Collisions of α Particles with Light Atoms", *Philosophical Magazine* **27**, 499–507, 1914.

102. Rutherford, E., "Collision of α-Particles with Light Atoms. I. Hydrogen", *Philosophical Magazine* **37**, 537–561, June 1919.

103. Rutherford, E., "Collision of α-Particles with Light Atoms. II. Velocity of the Hydrogen Ions", *Philosophical Magazine* **37**, 562–571, June 1919.

104. Rutherford, E., "Collision of α particles with light atoms. III. Nitrogen and Oxygen Atoms", *Philosophical Magazine* **37**, 571–580, June 1919.

105. Rutherford, E., "Collision of α particles with light atoms. IV. An anomalous effect in nitrogen", *Philosophical Magazine* **37**, 581–587, June 1919.

106. Rutherford, E. and Chadwick, J., "The Disintegration of Elements by α-Particles", *Nature* **107**, 41, March 10, 1921.

107. Rutherford, E. and Chadwick, J., "The Artificial Disintegration of Light Elements", *Philosophical Magazine* **42**, 809–825, November 1921.

108. Rutherford, E. and Chadwick, J., "Further Experiments on the Artificial Disintegration of Elements", *Proceedings of the Physical Society* **36**, 417–422, August 1924.

109. Brown, A., *The Neutron and the Bomb*, pp. 77–88.

110. Kirsch, G. and Pettersson, H., "Experiments on the Artificial Disintegration of Atoms", *Philosophical Magazine* **47**, 500–12, March 1924.

111. Rutherford, E. and Chadwick, J., "The Bombardment of Elements by α-Particles", *Nature* **113**, 457, March 29, 1924.

112. Brown, A., *The Neutron and the Bomb*, p. 84.
113. Chadwick, J., letter to Rutherford, dated December 12, 1927, in Brown A., *The Neutron and the Bomb*, p. 87.
114. Rutherford, E., "Disintegration of Atomic Nuclei", *Nature* **115**, 493–494, April 4, 1925.
115. Perrin, J., in *Atomes et électrons, rapports et discussions du Conseil de Physique Solvay tenu à Bruxelles du 1er au 6 avril 1921*, p. 68.
116. Rutherford, E., "Bakerian Lecture : Nuclear Constitution of Atoms", *Proceedings of the Royal Society, London* **A97**, 374–400, 1920.
117. Rutherford, E., "La Structure de l'Atome", in *Atomes et électrons, Rapports et discussions du Conseil de Physique tenu à Bruxelles du 1er au 6 avril 1921 (Troisième Conseil Solvay)*, pp. 36–79, Gauthier-Villars, Paris, 1923.
118. Bensaude-Vincent, B., *Langevin*, Belin, Paris, 1987, p. 98.
119. Glasson, G. L., "Attempts to detect the presence of neutrons in a discharge tube", *Philosophical Magazine* **42**, 596–600, July 1921.
120. Chadwick, J., "The Charge of the Atomic Nucleus and the Law of Force", *Philosophical Magazine* **40**, 734–746, December 1920.
121. Chadwick, J. and Bieler, É. S., "The Collisions of α Particles with Hydrogen Nuclei", *Philosophical Magazine* **42**, 923–40, December 1921.
122. Charles Weiner, *Sir James Chadwick, oral history*, American Institute of Physics, 1969; cited by Andrew Brown, *The neutron and the bomb*, p. 51.
123. Bieler, É., "The Large-Angle Scattering of α-Particles by Light Nuclei", *Proceedings of the Royal Society, London* **A105**, 434–450, April 1, 1924.
124. Geiger, H. and Rutherford, E., "Photographic Registration of α particles", *Philosophical Magazine* **24**, 618–623, October 1912.
125. Geiger, H., "Über eine einfache Methode zur Zählung von α und β-Strahlen", *Verhandlungen der deutschen physikalischen Gesellschaft* **15**, 534–539, session of July 27, 1913.
126. Geiger, H., "Demonstration einer einfachen Methode zur Zählung von α- and β-Strahlen", *Physikalische Zeitschrift* **14**, 1129, November 15, 1913.
127. Geiger, H. and Klemperer, O., "Beitrag zur Wirkungsweise des Spitzenzählers", *Zeitschrift für Physik* **49**, 753–760, 1928.
128. Geiger, H. and Müller, W., "Das Elektronenezählrohr", *Physikalische Zeitschrift* **29**, 839–841, November 15, 1928.
129. Greinacher, H., "Über die akustische Beobachtung und galvanometrische Registrierung von Elementarstrahlen und Einzelionen", *Zeitschrift für Physik* **23**, 361–378, 1924.
130. Greinacher, H., "Eine neue Methode zur Messung der Elementarstrahlen", *Zeitschrift für Physik* **36**, 364–373, 1926.
131. Greinacher, H., "Über die Registrierung von α- und H-Strahlen nach der neuen elektrischen Zählmethode", *Zeitschrift für Physik* **44**, 319–325, 1927.
132. Ortner, G. and Stetter, G., "Über den elektrischen Nachweis einselner Korpuscularstrahlen", *Zeitschrift für Physik* **54**, 449–476, 1929.
133. Ward, F. A. B. and Wynn-Williams, C. E., "The Rate of Emission of Alpha Particles from Radium", *Proceedings of the Royal Society, London* **A125**, 713–730, 1929.
134. Leprince-Ringuet, L., "L'amplificateur à lampes et la détection des rayonnements corpusculaires isolés", *Annales des Postes Télégraphes et Téléphones* pp. 480–492, June 1931.
135. Leprince-Ringuet, L., "Relation entre le parcours d'un proton rapide dans l'air et l'ionisation qu'il produit. Application à l'étude de la désintégration artificielle des éléments", *Comptes Rendus de l'Académie des Sciences* **192**, 1543–1545, session of June 15, 1931.
136. Bothe, W. and Kolhörster, W., "Das Wesen der Höhenstrahlung", *Zeitschrift für Physik* **56**, 751–777, 1929.
137. Bothe, W., "Zur Vereinfachung von Koinzidenzzählungen", *Zeitschrift für Physik* **59**, 1–5, 1930.
138. Rossi, B., "Method of Registering Multiple Simultaneous Impulses of Several Geiger's Counters", *Nature* **125**, 636, April 26, 1930.

139. Bothe, W., "The coincidence method", in *Nobel Lectures 1942-62*, Elsevier, Amsterdam, 1954.
140. Rutherford, E. and da Costa Andrade, E. N., "The Spectrum of the Penetrating γ Rays from Radium B and Radium C", *Philosophical Magazine* **28**, 263–273, August 1914.
141. de Broglie, M., "La spectroscopie des rayons de Röntgen", *Journal de Physique* **4**, 101–116, 1914.
142. Thibaud, J., "La spectrographie des rayons gamma. Spectres β secondaires et diffraction cristalline", *Annales de Physique (Paris)* **5**, 73–152, 1926.
143. Frilley, M., "Spectrographie par diffraction cristalline des rayons γ de la famille du radium", *Annales de Physique (Paris)* **11**, 483–567, 1929.
144. Aitken, J., "On the number of dust particles in the atmosphere", *Transactions of the Royal Society of Edinburgh* **35**, 1–20, session of February 6, 1888.
145. Wilson, C. T. R., "Condensation of Water Vapour in the Presence of Dust-free Air and other Gases", *Transactions of the Royal Society* **A189**, 265–307, session of April 8, 1897.
146. Wilson, C. T. R., "On the Condensation Nuclei produced in Gases by the Action of Röntgen Rays, Ultra-violet Light and other Agents", *Proceedings of the Royal Society, London* **64**, 127–129, session of November 24, 1898.
147. Wilson, C. T. R., "On the cloud method of making visible ions and the tracks of ionizing particles, Nobel Lecture, December 12, 1927", in *Nobel Lectures, Physics 1922–1941*, Elsevier, Amsterdam, 1965.
148. Wilson, C. T. R., "On an Expansion Apparatus for making Visible the Tracks of Ionising Particles and some results obtained by its Use", *Proceedings of the Royal Society, London* **A87**, 277–292, september 19, 1912.
149. Compton, A. H. and Simon, A. W., "Measurements of β-Rays associated with scattered X-Rays", *Physical Review* **25**, 306–313, March 1925.
150. Lovell, B., *P.M.S. Blackett: a biographical memoir*, The Royal Society, London, 1976. Reprinted from the *Biographical Memoirs of the Royal Society* **21**, 1-115, 1975.
151. Nye, M. J., *Blackett: Physics, War and Politics in the Twentieth Century*, Harvard University Press, Cambridge, Mass. and London, 2004.
152. Blackett, P. M. S., "The Ejection of Protons from Nitrogen Nuclei, Photographed by the Wilson Method", *Proceedings of the Royal Society, London* **107**, 349–360, 1925.
153. Rutherford, E., "Atomic Nuclei and their Transformations", *Proceedings of the Physical Society* **39**, 359–372, 1927.
154. Rutherford, E., "Structure of the Radioactive Atom and Origin of the α-Rays", *Philosophical Magazine* **4**, 580–605, September 1927.
155. Geiger, H. and Nuttall, J. M., "The Ranges of the α particles from Various Radioactive Substances and a Relation between Range and Period of Transformation", *Philosophical Magazine* **22**, 613–621, October 1911.
156. Ehrenfest, P. and Oppenheimer, J. R., "Note on the statistics of nuclei", *Physical Review* **37**, 333–338, February 1931.
157. Rasetti, F., "Selection Rules in the Raman Effect", *Nature* **123**, 757–759, May 18, 1929.
158. Rasetti, F., "On the Raman effect in diatomic gases. II", *Proceedings of the National Academy of Sciences (USA)* **15**, 515–519, June 15, 1929.
159. Rasetti, F., "On the Raman effect in diatomic gases", *Proceedings of the National Academy of Sciences (USA)* **15**, 234–237, March 15, 1929.
160. Heitler, W. and Herzberg, G., "Gehorchen die Stickstoffkerne der Boseschen Statistik?", *Naturwissenschaften* **17**, 673–674, August 23, 1929.
161. Rasetti, F., "Alternating Intensities in the Spectrum of Nitrogen", *Nature* **124**, 792–93, November 23, 1929.
162. Rasetti, F., "Über die Rotations-Ramanspektren von Stickstoff und Sauerstoff", *Zeitschrift für Physik* **61**, 598–601, 1930.
163. Ornstein, L. S. and van Wijk, W. R., "Untersuchungen über das negative Stickstoffbandenspektrum", *Zeitschrift für Physik* **49**, 315–322, 1928.

164. Kronig, R., "Der Drehimpuls des Stickstoffkerns", *Naturwissenschaften* **16**, 335, May 11, 1928.

165. Dorfman, J. G., "Le moment magnétique du noyau de l'atome", *Comptes Rendus de l'Académie des Sciences* **190**, 924–925, session of April 14, 1930.

166. Ambarzumjan, V. A. and Iwanenko, D., "Les électrons observables et les rayons β", *Comptes Rendus de l'Académie des Sciences* **190**, 582–584, session of March 3, 1930.

167. Rutherford, E., Chadwick, J., Ellis, C. D., Fowler, R. H., McLennan, J. C., Lindemann, F. A. and Mott, N. F., "Discussion on the Structure of Atomic Nuclei", *Proceedings of the Royal Society of London* **A136**, 735–762, June 1932.

1930–1940: A Dazzling Development

Kennst du den Berg und seinen Wolkensteg?
Das Maultier sucht im Nebel seinen Weg,
In Höhlen wohnt der Drachen alte Brut
Es stüürzt der Fels und über ihn die Flut.

Goethe, *Mignon*

Know'st thou the mountain bridge that hangs
on cloud?
The mules in mist grope o'er the torrent loud,
In caves lie coil'd the dragon's ancient brood,
The crag leaps down and over it the flood.

Translation of William Allan Neilson

The Nucleus: A New Boundary

The physics of the atomic nucleus becomes a primary concern of physicists. One Russian and two Americans show how the recent quantum mechanics can explain an enigma in α-decay: the Geige-Nuttall law. A Frenchman discovers that α-particles emitted by radioactive substances do not all have the same velocity.

B. Fernandez and G. Ripka, *Unravelling the Mystery of the Atomic Nucleus:*
A Sixty Year Journey 1896 — 1956, DOI 10.1007/978-1-4614-4181-6_5,
© Springer Science+Business Media New York 2013

Quantum Mechanics Acting in the Nucleus

On August 2, 1928, the prestigious German journal *Zeitschrift für Physik* receives a paper written by a Russian physicist, George Gamow, who is working in the theoretical physics laboratory of Max Born in Göttingen. Gamow uses quantum mechanics to explain the famous Geiger–Nuttall law. Recall that Hans Geiger and John Mitchell Nuttall had observed[1] that the velocity of the emitted α-particles is greater when the radioactive substance decays faster [1]. By coincidence, on September 22, the review *Nature* publishes a paper sent on July 30 by two American physicists, Ronald Gurney and Edward Condon, working in the Palmer Physical Laboratory in Princeton [2]. They made essentially the same calculation as Gamow, independently.

George Gamow

George Gamow, a young Russian physicist, was still a student. Born in Odessa in 1904, he studied in the University of Leningrad and obtained a scholarship which allowed him to spend the 1927–1928 academic year in Göttingen, in the famous laboratory of Max Born, which saw the birth of quantum mechanics. This is where he published his calculation of α-decay. He then travelled to Copenhagen (1928–1929), to Cambridge (at the Cavendish in 1929–1930), a second time to Copenhagen (1930–1931), to Leningrad (1931–1933), and to Paris (*Institut du Radium*, 1933–1934). In 1934, he was invited to the United States and remained there thereafter. In the years 1934–1956, he was professor in the George Washington University, and in 1956–1968, professor in the University of Colorado. He became a US citizen in 1940 and died in 1975. Gamow was a colorful personality, a giant 1.96 m high, weighing 220 pounds. He spoke six languages with a strong accent and loved laughing and joking. He wrote several very successful popularized science books, among which the *Mr. Tomkins* series and *The 30 years which shook physics, A history of quantum mechanics*.

Gamow begins by stating some general problems concerning the existence of nuclei:

> It is often stated that attractive non-Coulombic forces play an important role in the nucleus. We can make several hypotheses regarding their nature[. . .] In any case, they diminish strongly outside the nucleus but inside they are stronger than the Coulomb repulsion [1].

The particles which form the nucleus must somehow hold together, a thought present in the minds of all physicists at the time. As stressed by Gamow, the forces which bind the nucleus act only at very short distances.

---

[1] See p. 228.

Consider the inverse α radioactivity: an α-particle travels close to the nucleus. To which forces is it exposed? As long as it remains well outside the nucleus, it only feels the electric Coulomb repulsion. The repulsion increases as the α-particle gets closer to the nucleus, somewhat as if a golf ball needed to climb a little hill before reaching the hole. But as soon as the α-particle reaches a critical distance, roughly the radius of the nucleus, it becomes trapped by the particles in the nucleus and falls inside (as the golf ball would do when it reaches the hole). The α-particle "sees" the nucleus as a deep well surrounded by a "potential hill." Physicists call it a *potential well* surrounded by a *Coulomb barrier*$^\diamond$. Once it is trapped in the potential well, the α-particle requires some energy to overcome the Coulomb barrier in order to escape.

But how does the α-particle actually escape during an α-decay? According to classical mechanics, it can only escape if it is given sufficient energy to overcome the Coulomb barrier. Failing this, it cannot escape by passing through the barrier. However, *quantum mechanics does allow the α-particle to pass through the barrier.* In regions of space, which classical mechanics forbids the particle to enter (the region of the potential barrier), the wavefunction of the particle does not vanish completely, but it diminishes at a rate which depends on the height of the barrier. Thus, if the barrier is not too wide, some nonvanishing wavefunction will remain outside of the barrier, thereby providing the α-particle with a small but nonvanishing probability to be outside the barrier, from where it can escape. Thus, according to quantum mechanics, the particle can pass through the barrier. The process is called *tunneling* through the barrier. Only a very small part of the wavefunction tunnels through the barrier thereby providing a very small probability for the α-particle to escape. For example, the nucleus of a uranium atom has a 50% probability of emitting an α-particle in 4.5 billion years! However, since there are $2,5 \times 10^{21}$ uranium nuclei in 1 g of uranium, about 10 000 nuclei will emit an α-particle every second. This is an admittedly small fraction, but it is nonetheless an observable effect!

This is precisely what Gurney and Condon state in their paper sent to *Nature*:

In classical mechanics, the orbit of a moving particle is entirely confined to those parts of space for which its potential energy is less than its total energy. If a ball is moving in a valley of potential energy and does not have enough energy to get over a mountain on one side of the valley, it must certainly stay in the valley for all time, unless it acquires the deficiency in energy somehow. But this is not so in the quantum mechanics. It will always have a small but finite chance of tipping through the mountain and escaping from the valley [2].

We can add that if the ball can pass through the mountain, the speed it will have on the other side will be larger if it starts off inside with a higher energy. Gurney and Condon show that this description of the emission of α-particles yields a qualitative explanation of the Geiger–Nuttall law. If the escaping α-particle is close to the top of the barrier, it starts off with a higher energy and it will therefore escape with a higher energy. Furthermore, when it starts off close to the top of the barrier, it passes through a thinner barrier, thereby increasing its probability to escape. That is precisely what the Geiger–Nuttall law states.

Gurney and Condon give this qualitative argument, but Gamow takes one further step. He calculates the relation between the speed of the emitted α-particle and the radioactive half-life of the emitting nucleus. He thus explains why short-lived α emitters yield more energetic α-particles. Gurney and Condon end their paper in *Nature* by an amusing remark, most likely aimed at Rutherford:

> Much has been written of the explosive violence with which the α-particle is hurled from its place in the nucleus. But from the process pictured above, one would rather say that the α-particle slips away almost unnoticed.

In other words, it is not a violent nuclear explosion which emits the impressively fast α-particle, it is rather the fact that there exists a small but finite probability that the α-particle should lie outside the range of nuclear forces and then it is repelled by the strong Coulomb force.

Salomon Rosenblum and the Fine Structure of α Radioactivity

Ever since the memorable work of William Henry Bragg at the turn of the century,[1] it was assumed that α-particles were emitted by radioactive substances at a constant velocity which was a characteristic of the emitting substance. But in 1929, this certainty is challenged in a paper sent to the *Comptes Rendus de l'Académie des Sciences* by a young physicist working in the laboratory of Marie Curie at the *Institut du Radium* [3].

Salomon Rosenblum was born in Ciechanoviec, near Minsk, on June 14, 1896, into a well-to-do family. His university studies are interrupted by the war, and he emigrates to Denmark. He studies philosophy in the University of Copenhagen, and then Hebrew, Armenian, and Arabic in the University of Lund. One day, while he is working on a thesis on oriental languages, he meets an assistant of Niels Bohr in a café. The latter tells him about the early success of quantum mechanics and explains to him how Bohr succeeded in explaining the spectrum of hydrogen. Without hesitating, Rosenblum stops working on his thesis and devotes himself to physics. He works in Copenhagen, Berlin, and finally in Paris where Marie Curie admits him into the *Institut du Radium* in 1923, following a recommendation by Niels Bohr. On July 3, 1928, he defends his thesis on the penetration of α-particles through matter. He then turns his attention to a problem which had puzzled him a few years earlier, while he was observing the magnetic spectra of α-particles emitted by ThC.[2] He notices that the spectral line is rather wide as if it consisted of two close

---

[1]See p. 63.

[2]ThC is the isotope 212 of bismuth, $^{212}_{83}$Bi.

spectral lines, that is, as if the α-*particles did not all have exactly the same velocity.* But the small magnet which he uses at the *Institut du Radium* does not allow him to go further.

However, a large magnet was constructed and installed in Bellevue, at the *Office des Inventions.* Its construction had been proposed in 1912 by Aimé Cotton, a specialist in magnetization and optics. Because of the war, the magnet was not built until 1928. It was the largest magnet built in France and one the first of the kind in the world [4, 5]. It was huge and weighed over a hundred tons. It produced very intense magnetic fields, a forerunner of future large equipments such as accelerators, spectrometers, etc. Rosenblum uses it to make precision measurements of the velocities of α-particles, as Lise Meitner and Jean Danysz had done for electrons emitted in β radioactivity. Instead of making the α-particles deviate by some 10°, he makes them deviate by 180° so that their trajectory is a semicircle. He obtains thus a focalization, meaning that particles arriving in the region between the poles of the magnet (where the magnetic field is present) at slightly different angles, but with a given velocity, all impinge on the same point of a photographic plate after describing the semicircle.[1] Rosenblum observes that what everyone so far had taken to be a single spectral line, corresponding to α-particles propagating with the same velocity, was in fact a set of four distinct lines, corresponding to α-particles propagating with four different velocities:

> In the region of the spectrum corresponding to α-rays of thorium C, we observed not one but four spectral lines, two lying very close to each other and two other weak ones further away [3, 6].

Rosenblum discovered what he called a *fine structure of α-rays.* Further experiments showed that this multiplicity of the spectral lines was a general feature.

What does it mean? When Bragg performed his experiments,[2] the existence of a nucleus in the center of the atom was unknown. But when the nucleus was discovered, as well as quantum mechanics, the ideas of Niels Bohr were applied to the nucleus. It was assumed that, like atoms, the nuclei existed only in specific and relatively few "quantum states." The simplest example is the hydrogen atom in which the electron describes a few quantized trajectories at various distances from the nucleus. If the electron describes a trajectory far from the nucleus, the atom is in an "excited state" and it can "fall" into a lower energy orbit, thereby "de-exciting" the atom while emitting *a quantum of light,* a *photon.* This is a *cooling effect,* similar to the process during which a hot body cools down by radiating visible light if it is hot enough, otherwise infrared radiation. Once the electron reaches the lowest energy orbit, it can no longer radiate and the atom is said to be in its *ground state.*

In 1928, almost nothing was known about the inner structure of the nucleus. It was however assumed that quantum mechanics also applied to the nucleus and that the latter existed in certain "quantum states." The "ground state" was the

---

[1] See p. 170.
[2] See p. 63.

quantum state with lowest energy. This provided a simple explanation of the fact that radioactive nuclei emit α-particles with a definite energy or velocity: since the energies before and after the α-decay had to be the same for all the emitting nuclei of a given species, one and the same energy was available for the emitted α-particle.

The observations of Rosenblum shed doubt upon this. Indeed, after the emission of the α-particle, a *residual* nucleus can be formed in various "quantum states" with different energies and the α-particle is emitted with correspondingly different energies.

Rutherford made the following comment:

This new method of attack on the problem of the homogeneity of the α-rays is of much interest and importance and promises to give us valuable data on this question [7].

He sets out immediately to study this fine structure with the electronically amplified ionization chamber which has just been set up in his laboratory.[1] With this ionization chamber, he can measure the velocities of the α-particles more precisely than by observing the particles propagating through air, as had been done so far. Several papers [8–11] are published by the Cavendish concerning the energies of α-particles emitted by radium C, thorium C, and actinium C.[2] The spectra indeed turn out to be more complex than had been expected before the experiments of Rosenblum. Furthermore, Rutherford embarks on the construction of a magnetic spectrometer which yields the first results in 1932 and confirms the observations of Rosenblum [12].

The fine structure of the α spectra also explained why the γ-rays had such high energies. They were most likely emitted after the emission of the α-particle by a nucleus which was not in its "ground state" and which subsequently de-excited to its ground state by emitting a γ-ray. This idea, first stated by Gamow, was confirmed by Rutherford and Ellis [13].

1931: The First International Congress of Nuclear Physics

A congress, organized by the *Fondazione Alessandro Volta* and funded by a gift of the Italian Edison Society to the Royal Academy of Italy, was held on October 11–18, 1931, in Rome. The honorary president was Guglielmo Marconi, the president was senator Orso Corbino and the secretary general Enrico Fermi. It was the first international meeting devoted to nuclear physics. The proceedings were quickly published by the Italian Royal Academy, under the direction of Fermi [14]. The participants included physicists who had made important contributions to quantum mechanics, radioactivity, and the new field of nuclear physics: Francis

---

[1] See p. 217.
[2] Namely, three isotopes of bismuth, $^{214}_{83}$Bi, $^{212}_{83}$Bi, and $^{211}_{83}$Bi.

Aston, Patrick Blackett, Walther Bothe, Niels Bohr, Léon Brillouin, Marie Curie, Arthur Compton, Charles Ellis, Paul Ehrenfest, Ralph Fowler, Hans Geiger, Samuel Goudsmit, Werner Heisenberg, Ettore Majorana, Lise Meitner, Robert Millikan, Wolfgang Pauli, Jean Perrin, Enrico Persico, and Franco Rasetti. The talks were given in the morning, and the rest of the day was left free for discussions. Some talks assessed present problems, and other talks announced new results. Three talks were precursors of imminent discoveries, namely, those of Goudsmit, Bothe, and Gamow.

Goudsmit and the Magnetic Moments of Nuclei

Samuel Goudsmit talked about the so-called hyperfine structure of atomic spectra, which is due to the *magnetic moment* of the nucleus in the atom [15]. Progress in atomic spectroscopy had revealed already 10 years earlier that the optical lines of certain heavy elements were split into multiplets which could not be accounted for. In 1924, Pauli proposed a simple explanation [16]: if the nucleus has a magnetic moment, it will behave as a magnet which will produce an internal Zeeman effect, similar to the normal Zeeman effect.[1] By measuring the splitting of the spectral lines, one can deduce the value of the magnetic moment of the nucleus. This explains the observed splitting of several elements, but in some cases, the results are most puzzling. For example, the isotope 6 of lithium was thought to consist of six protons and three internal electrons. Its spin, that is, its intrinsic angular momentum, could then be only $\frac{1}{2}$ or $\frac{3}{2}$ at best (in units of $h/2\pi$) but certainly not zero! One cannot obtain a zero spin with an odd number of spin $\frac{1}{2}$ particles. However, the absence of a hyperfine structure in its spectrum indicates that the magnetic moment of the nucleus is in fact zero. There were other troubling examples. How could they be accounted for? Goudsmit concluded:

> *In spite of the uncertainty in part of the experimental results, it seems that the above discussed disagreements between theory and data are well founded and real. They are very likely due to an insufficient knowledge of the structure of the nucleus. It is my belief that the mechanics applicable to the nucleus must differ considerably from the quantum mechanics now used for the atom, in the same way as the latter differs from the classical mechanics for large masses. Classical mechanics has been partially successful in explaining atomic properties and similarly it is possible to describe at present some properties of the nucleus with the language of atomic mechanics, but one should not be surprised at finding great difficulties.*

---

[1] The Zeeman effect is a splitting of atomic spectral lines when the atom is subject to an external magnetic field. The reason is that the magnetic field modifies the motion of the electrons in the atom. See p. 110.

One can well imagine how reluctant physicists were to give up quantum mechanics, which Goudsmit rightly called the mechanics of atoms and which had been conceived a few years earlier in order to explain the atom. Goudsmit however was ready to consider a new mechanics applicable in the nucleus which is much smaller than the atom.

Walther Bothe: The Mystery of the Penetrating Radiation

Walther Bothe summarized the experiments in which α-particles, emitted by radioactive substances, collided with nuclei [17]. He ended by describing the results which he obtained with his assistant Herbert Becker a year and a half earlier and which appeared to be difficult to interpret [18,19]. They exposed several substances to the radiation of an intense α source of polonium, and they studied the produced γ-rays. It was a difficult experiment: the point counter which they used could not detect all the γ-rays, but only those which had interacted with an electron (because the detector detects only charged particles). Furthermore, polonium itself emits γ-rays which interfere with the measurements. Finally, the intense polonium source produces a parasite radiation which has to be stopped by a thick layer of lead. However, Bothe and Becker detected a very strange radiation whenever beryllium was subject to an intense α radiation. This radiation was not composed of charged particles. It was electrically neutral, which suggested that it consisted of γ-rays. But when they measured its penetrating power in matter, they found it unusually large. A 7 cm thick iron plate was transparent to 61% of this radiation, whereas the same plate was only transparent to 5% of the γ radiation from radium. These experiments of Bothe and Becker were the precursors of one of the greatest discoveries of the century.

Georges Gamow: The Nucleus Behaves as a Liquid Drop

Georges Gamow is then 27 years old and a professor in the University of Leningrad. He begins by a general talk on the structure of the atomic nucleus. He puts forth, for the first time, the idea that the nucleus can be conceived as a liquid drop. This *liquid drop model* will soon become a basis to understand the nucleus. To arrive at this description, Gamow considers first that there must be attractive forces acting between the nuclear constituents, be they protons or α-particles (Gamow does not commit himself to one or the other), because the constituents are electrically positively charged and would repel each other if only electromagnetic forces were at play. The attractive force must be more intense than the Coulomb repulsion between the positively charged particles. But it must disappear fast when the articles separate by even modest distances, that is, when they are practically no longer in contact with one another and when only the strong Coulomb repulsion remains. Furthermore, the force must prevent the constituent particles from penetrating each other. This is similar to the interaction between molecules in a liquid which slide along each other without interpenetration: a so-called "van der Waals" force keeps the molecules apart in the liquid.

We should therefore observe a "surface tension" in nuclei. Well inside the nucleus, the constituents are subject to forces acting in all directions so that they cancel each other. One can say that no force acts on constituents which are well inside the nucleus. But at the surface of the nuclear liquid, this is no longer the case. There, a constituent is attracted to the constituents inside the nucleus, and the latter generate a force which prevents the constituent from escaping and which therefore compresses the liquid. This is what causes liquid drops to be spherical. It is energetically favorable to make the surface as small as possible. Gamow concludes that the *potential energy* of a particle inside the nucleus should be approximately constant throughout the nuclear volume, within which the particle can move without loosing energy. But as soon as the particle reaches the nuclear surface, the potential rises steeply and a considerable energy is required to extract the particle from the nucleus. In this model, the nucleus acts as a potential which is roughly constant throughout the nuclear volume. This description of the nucleus differs considerably from the one which had been proposed by Rutherford.[1]

The Discovery of an Exceptional Isotope: Deuterium

In 1931, the American physicist Raymond Birge noticed that the chemical measurements of the mass of the hydrogen atom yielded a somewhat larger mass than the physical measurement of Aston.[2] However, chemical measurements involve natural mixtures of the isotopes of a given element. Birge concluded that natural hydrogen should contain a small amount of a heavier isotope of hydrogen with mass 2: about one among 4,500 atoms [20].

On December 5, 1931, Harold Urey, Ferdinand Brickwedde, and George Murphy, American physicists working in the University of Columbia in New York and in the National Bureau of Standards in Washington, write a letter to the editor of the *Physical Review* in which they announce their discovery of this rare isotope of hydrogen [21]. A more detailed paper is published a month later [22]. Urey and his collaborators could not simply use a mass spectrometer, as Aston did, because the amount of mass 2 isotope was really too small, especially compared to the

---

[1] See p. 227.

[2] Recall that the chemical measurements yield relative values of the atomic masses: 8 g of oxygen and 1 g of hydrogen are required to form 9 g of water. Since one knows that each molecule of water contains one oxygen atom and two hydrogen atoms, one deduces that the oxygen atom is 16 times heavier than the hydrogen atom. With similar methods, the relative masses of other atoms can be determined.

number of ionized hydrogen molecules,[1] which also have a mass 2 and were far more abundant. They had to devise another method. They used the fact that since the mass 2 isotope has a mass double of the usual isotope, it should liquefy at a slightly lower temperature than the lighter one. They therefore liquefied natural hydrogen and proceeded to perform a fractionalized distillation: since the lighter isotope evaporates first, the liquid acquires a larger proportion of the heavier isotope.[2] They then examined the X-ray spectrum of the sample, and they observed weak but unquestionable spectral lines next to the known hydrogen lines. The difference in mass causes a small shift in the spectral lines of the heavier isotope. They found an abundance of 1/4000 of the hydrogen isotope, which agreed with the estimate made by Birge.

Their paper appeared in the *Physical Review* dated April 1, 1932. On February 27 of that same year, Chadwick published a paper which shattered nuclear physics: he announced the discovery of the neutron, the subject of the following chapter. For the time being, the heavier isotope of hydrogen was believed to consist of two protons and one internal electron.

A Fight for a Name

A fight began between the two sides of the Atlantic ocean concerning the name which should be given to this new isotope. It is described by Roger Stuewer [23]. The Berkeley physicists proposed to call it a *deuton*, but this did not please Urey, Brickwedde, and Murphy, who had discovered the isotope and who considered it their right to choose a name for it. Now, Rutherford did not like the word *deuton* which he considered too close to *neutron*, and he proposed *diplon* (from the Greek *diplôn*, meaning double). Further names, such as *dygen, deutum, and diplogen*, were occasionally used, causing further dispute. The affair was settled in 1934: the isotope 2 of the hydrogen nucleus was branded *deuteron* (from the Greek *deuteros*, second), and the atom formed by a deuteron and an electron was called *deuterium*.

The Spin of the Deuteron

Two years later, George Murphy and Helen Johnston succeeded in measuring the spin of the deuteron by studying the spectral lines of the molecule formed by two heavy hydrogens, that is, by two deuteron nuclei [24]. The same method had been

---

[1]This is a molecule consisting of two hydrogen nuclei bound by a single electron. It is produced when an ordinary hydrogen molecule loses one of its electrons.

[2]The same principle is applied when alcohol is obtained from the distillation of wine: the alcohol is lighter, and it evaporates first.

used by Rasetti to show that the nucleus of nitrogen 14 had a spin 1, a feature which caused quite a stir among physicists.[1] The spin 1 of the deuteron was even more surprising: how could a system consisting of two protons, each one with spin 1/2, and one electron, which also has spin 1/2, produce a total spin equal to 1? The answer was soon to come.

---

[1] See p. 229.

The Discovery of the Neutron

> *Rutherford hunts for the neutron but in vain. Frédéric and Irène Joliot-Curie observe a surprising property of the penetrating radiation of Bothe and Becker. Chadwick triumphantly exclaims: it's the neutron! A question is raised: is the neutron composed of a proton and an electron?*

Rutherford enjoyed company and always liked to talk with his collaborators and to discuss his ideas and current problems. In the 1920s at the Cavendish, James Chadwick was the one he enjoyed talking to most. In 1969, Chadwick confided to Charles Weiner:

> Before the experiments, before we began to observe in these experiments, we had to accustom ourselves to the dark, to get our eyes adjusted, and we had a big box in the room in which we took refuge while Crowe, Rutherford's personal assistant and technician, prepared the apparatus. That is to say, he brought the radioactive source down from the radium room, put it in the apparatus, evacuated it, or filled it with whatever, put the various sources in and made the arrangements that we'd agreed upon. And we sat in this dark room, dark box, for perhaps half an hour or so, and naturally talked [...] it was these conversations that convinced me that the neutron must exist [25].

The "neutron" conceived by Rutherford was a primary concern in his laboratory, ever since he had put forth the idea in his famous Bakerian lecture.[1] In September 1924, while vacationing in Scotland, Chadwick writes to Rutherford, who was visiting in Canada. He tells him about some recent results, and he describes the controversy which opposed the Cavendish and the *Institut für Radiumforschung* in Vienna.[2] He adds:

> I think we shall have to make a real search for the neutron. I believe I have a scheme which may just work but I must consult Aston first [26].

We do not know what the scheme was, but the fact that he wanted to consult Aston suggests that he was probably worried about the masses of light nuclei. We also don't know whether he did consult Aston, but we do know that Aston's measured masses were not accurate enough at the time to be useful.

In 1925, the reserved, timid, and distant Chadwick falls in love with a self-confident and lively young girl, Aileen Stewart-Brown. They get married in August 1925. She will gradually help him acquire the self-confidence he lacked.

---

[1] See p. 209.
[2] See p. 203.

Frédéric and Irène Joliot-Curie

It is at this point that two new nuclear physicists join the band.

Frédéric Joliot was born on March 19, 1900, into a well-to-do family [27–30]. His father was a successful wholesale calico dealer who married in 1878 Émilie Roederer, a young girl from Alsace. Frédéric Joliot had an uneventful youth, obtained good results in school, and excelled in physics and football. He attended the *École Municipale de Physique et de Chimie Industrielle*[1] in 1920 and followed the lectures of Paul Langevin, the great physicist whose lectures fascinated all those who had the luck of hearing them. Joliot soon becomes the best student in his class, and physics becomes his passion. He displays an exceptional talent as an experimentalist. After completing his military service, he tells Paul Langevin that he wants to do research. Without hesitating, Paul Langevin sends him to Marie Curie with a recommendation. They meet in 1925:

> I still see her there, at her desk, short with very vivid eyes. I sat in front of her in my officer's uniform [. . .] quite intimidated. She heard me out and suddenly asked: "Can you begin tomorrow?"
>
> My military service was to last three more weeks. "I will write to your colonel," she decided. The following day I was her assistant [31].

A dream had come true. When he was younger, Joliot had pinned a photograph of Pierre and Marie Curie, the mythical scientists, on the wall of his room! But he still had several exams to pass before obtaining the Bachelor of Science required to prepare a PhD thesis.

At the same time, he embarks on laboratory work with great drive. He is bursting with energy and imagination, has excellent relations with others, and displays an elegance in manner, and countless women are seduced. But Frédéric knows nothing about radioactivity. To ensure he does, Marie Curie entrusts him to a confirmed physicist, her daughter Irène.

Irène Curie [29, 32] was born on September 12, 1897. She was an uncommon child whom Marie Curie called "her wild one." She was taken care of by her grandfather while Pierre and Marie worked in their laboratory. During the First World War, Irène first assists her mother with her traveling X-ray facility for the soldiers in the army. Then she acts on her own successively as a radiologist and a teacher of radiologists. She also completes her B.Sc.

Upon her return to Paris after the war, Irène Curie is awarded the *Medaille Militaire*. She begins to work on her PhD thesis at the *Institut du Radium* which was built before the war but which only began to function when peace was restored. Calm and reserved in manner, paying little attention to her dress, Irène spoke little

---

[1]The *École Municipale de Physique et de Chimie Industrielle* was founded in 1882. Its name was changed to *École Supérieure de Physique et de Chimie Industrielle (ESPCI) in 1948.*

but directly, often abruptly, giving the impression of being haughty. In a way, she was the exact opposite of Frédéric Joliot. But this could be a misleading impression: she liked to dance, occasionally to flirt, and was an accomplished sportswoman.

When Frédéric Joliot arrived at the *Institut du Radium*, Irène was about to complete her thesis on the α-rays of polonium and she obtained her PhD in March 1925. Frédéric detects a great sensibility and an ardent spirit, hiding behind her somewhat distant manner. They fall in love and get married on October 9, 1926. Frédéric obtains his B.Sc. diploma in 1927 and defends his Ph.D. thesis in 1930 on "Electrochemical studies of radioactive elements" [33]. From 1928 onwards, Frédéric Joliot and Irène Curie begin a most fruitful collaboration.

In 1931, they publish a paper on the manufacture of an intense source of polonium [34]. Intense polonium sources were a tradition at the *Institut du Radium*. When they read the paper of Bothe and Becker on the bombardment of beryllium by α-particles stemming from a polonium source, they have one advantage over the German team: their source is ten times more intense. They do the experiment with a somewhat different experimental setup. Instead of using a point counter, they use an ionization chamber the current of which is detected by a very sensitive Hoffmann electrometer, but not quite sensitive enough to detect a single particle. Electronic amplification had not yet been developed at the *Institut du Radium*. On December 28, 1931, Frédéric Joliot and Irène Curie send each a paper to the Académie des Sciences [35]. They measured the penetration power of what they thought, as Bothe did, were γ-rays. The paper of Irène concerns beryllium (which the French called *glucinium* at the time) and lithium. She finds that the radiation stemming from beryllium had a particularly strong penetrating power:

> *Besides the absorption of the γ radiation of polonium, the effect of which becomes negligible beyond 15 mm of lead, we observe no filtering effect; the radiation appears to be homogeneous; half of it is absorbed by $4^{cm}.7^1$ of lead [36].*

She makes a rough estimate of the energy of this radiation, and she finds that it lies between 15 and 20 MeV (million electron volts), a very high energy. The paper of Frédéric concerns boron. The absorption coefficient would correspond to an energy of about 11 MeV. According to the explanation given by Bothe, a boron nucleus, or its decay product, would be formed in an excited state and would make a transition to its ground state by emitting a γ-ray. If the nucleus is governed by quantum mechanics, it can only cool down by emitting photons of definite energy, as it makes transitions from one state to a lower energy one. Now, 11 MeV photons really appear to be too energetic. For this reason, Joliot suggests that the α-particle is first absorbed by boron and that the nucleus which is thus formed subsequently decays by emitting a γ photon.

---

[1] We respect the original notation. Today, we would write 4.7 cm.

Protons Are Ejected

The most important was yet to come. The note presented to the *Académie des Sciences* on January 18, 1932 contains a surprising observation, which will turn out to be crucial:

> *The rays enter the chamber after passing through an aluminum sheet. We observed that the ionization current produced by these rays, filtered by a 1.5 cm lead sheet, remains essentially the same when very different substances (C, Al, Cu, Ag, Pb) are placed at the entrance of the chamber. On the contrary, the current increases substantially when it is shielded by substances containing hydrogen, such as paraffin, water, cellophane; the current almost doubles in this case[...] Since this additional current was only observed with hydrogenated substances, we assumed that they were H rays [37].*

This penetrating radiation, which they still believe to be γ-rays (photons) is able therefore to eject particles, which are detected in the ionization chamber. Which particles? They do not prove that they are protons (which they still call "H-rays"), but they eliminate other possibilities: they are neither photons (because silver absorbs them less than aluminum) nor electrons (because they are not eliminated by an electromagnetic field[1]). But how can one interpret the presence of protons which they believe to be ejected by γ-rays?

> *If one assumes that photons can transfer part of their energy to protons by a process similar to the one in which electrons are ejected in the Compton effect, one finds that the Be and B rays should have energies respectively of the order of 50×10^6 and 35×10^6 eV.*

These energies are much too high. Something is wrong. Irène and Frédéric continue working very hard, and they perform another experiment, this time using a Wilson cloud chamber which Frédéric had constructed, one of his favorite instruments. Their results are presented at a session of the *Académie des Sciences* on February 22, 1932. They do indeed observe proton trajectories but also ejections of nuclei such as helium. This leads them to the conclusion:

> *All these experiments show that the ejection of atomic nuclei by high energy γ-rays is most likely a very general effect.*

This mysterious radiation is able to set into motion particles as heavy as helium atoms, four times heavier than hydrogen!

---

[1] An electromagnetic field does not really eliminate electrons. Instead, it deviates them by making them follow circular trajectories which have very small radii in view of the small electron mass. The electrons continue going round and round and therefore remain close to their starting point. They do not reach detectors placed further away.

The Neutron Is Revealed

When Chadwick reads the January 18 paper of Frédéric Joliot and Irène Curie, he jumps up:

> *Then one morning I read the communication of the Curie-Joliot in the* Comptes Rendus, *in which they reported a still more surprising property of the radiation from beryllium, a most startling property. Not many minutes afterwards Feather came to my room to tell me about this report, as astonished as I was. A little later that morning I told Rutherford. It was a custom of long standing that I should visit him about 11 am to tell him any news of interest and to discuss the work in progress in the laboratory. As I told him about the Curie-Joliot observation and their views on it, I saw his growing amazement; and finally he burst out "I don't believe it". Such an impatient remark was utterly out of character, and in all my long association with him I recall no similar occasion. I mention it to emphasize the electrifying effect of the Curie-Joliot report. Of course, Rutherford agreed that one must believe the observations; the explanation was quite another matter [38].*

For Chadwick, this penetrating radiation could only consists of the "neutrons" imagined by Rutherford, who had searched for them in vain for 10 years [39]. Because he believed that the neutron was composed of a proton and an electron, intimately bound, he had assumed that the neutron had a certain ionizing power, meaning that it could eject some electrons from atoms and thus be detected by a Geiger counter. Unfortunately, all his attempts to do so had failed. But he had not thought of the possibility that his hypothetical neutron might be able to eject hydrogen atoms (protons) with a given velocity and which could be detected. Chadwick immediately sets to work. He now wants to show that this so-called γ radiation consists of neutrons. He has at his disposal an ionization chamber coupled to an electronic amplifier which can detect a single particle. He repeats the experiments of the Joliot-Curies. He works day and night (mostly at night because his amplifier is extremely sensitive to surrounding noise). On February 17, he sends a letter to *Nature* entitled "Possible existence of a neutron." He describes his results which confirm and extend those obtained by the Joliot-Curies:

> *These experiments have shown that the radiation ejects particles from hydrogen, helium, lithium, beryllium, carbon, air, and argon. The particles ejected from hydrogen behave, as regards range and ionizing power, like protons with speeds up to about 3.2×10^9 cm. per sec. The particles from the other elements have a large ionizing power, and appear to be in each case recoil atoms of the elements [40].*

Chadwick thus confirms the results of the Joliot-Curies: when it passes through hydrogen, the penetrating radiation ejects protons. What he calls "recoil atoms" are the other ejected nuclei. For Chadwick, it is difficult, even impossible to understand these observations if this radiation consists of γ-rays because that would imply, as Joliot already noticed, that the γ-rays should have energies of the order of 50 MeV which is huge. But even such energies could not explain the velocities of the nitrogen nuclei which are ejected when the radiation passes through air. The nitrogen nuclei travel 3 mm whereas one can estimate that they would only travel 1.5 mm if they were produced by 50 MeV γ-rays. Chadwick then comes to his cherished conclusion:

> The difficulties disappear, however, if it is assumed that the radiation consists of particles of mass 1 and charge 0, or neutrons.

Chadwick pursues his argument by showing that all his observations are compatible with the following scenario: when an α-particle (of mass 4 and electric charge 2) collides with a beryllium nucleus (of mass 9 and electric charge 4), it is absorbed by the latter thereby forming a carbon nucleus (of mass 13 and electric charge 6). Then a neutron (of mass 1 and electric charge zero) is ejected. He calculates the velocity of protons ejected after being struck by the neutron, and he finds about 30 000 km/s. This is close to the observed velocity. One can also explain why the protons, which are ejected in the same direction as the α-particles, have higher velocities than those ejected backwards. He comes to a finely formulated conclusion:

> It is to be expected that many of the effects of a neutron in passing through matter should resemble those of a quantum of high energy, and it is not easy to reach the final decision between the two hypotheses. Up to the present, all the evidence is in favor of the neutron, while the quantum hypothesis can only be upheld if the conservation of energy and momentum be relinquished at some point.

What Chadwick expresses as an understatement is for him almost a certainty, as witnessed by a letter he sends to Niels Bohr, on February 24:

> Dear Bohr,
>
> I enclose the proof of a letter I have written to "Nature" and which will appear either this week or next. I thought you might like to know about it beforehand. The suggestion is that α-particles eject from beryllium (and also from boron) particles which have no net charge, and which probably have a mass about equal to that of the proton. As you will see, I put this forward rather cautiously, but I think the evidence is really rather strong. Whatever the radiation from beryllium may be, it has most remarkable properties. I have made many experiments which I do not mention in the letter to "Nature" and they can all be interpreted readily in the assumption that the particles are neutrons [41].

Is the Neutron Lighter or Heavier than the Proton?

A few months later, Chadwick sent a 17-page paper to the Royal Society, in which he repeated, but in greater detail, what he wrote in his letter to *Nature* [42]. He added an important new result, namely, an estimate of the neutron mass. He considered the reaction in which a neutron is produced by a collision of an α-particle with a boron nucleus. He wrote it as if it were a chemical reaction:

$$B^{11} + He^4 \rightarrow N^{14} + n^1$$

It was known that boron was a mixture of two isotopes, one of mass 10 (about 20% of the nuclei) and the other of mass 11 (about 80%). Why did Chadwick decide that it was the isotope 11 which produced the neutron? Because it had been observed that the isotope 10 emitted protons and, therefore, he thought, not neutrons. Knowing the masses of boron, helium, and nitrogen, he was able to calculate the mass of the neutron, taking into account the energies of the α-particle and of the neutron, which he knew approximately. According to the theory of relativity of Einstein, the total energy, both before and after the collision, equals the mass energy plus the kinetic energy. Therefore, the energy before the collision is equal to:

the mass of boron + the mass of the α particle + the kinetic energy
of the α particle

$$= 11.00825 + 4.00106 + 0.00565$$

(It is understood that the masses are multiplied by the square c^2 of the speed of light, according to Einstein's formula $E = mc^2$). The energy after the collision is equal to:

the mass of nitrogen + the kinetic energy of nitrogen

+the mass of the neutron + the kinetic energy of the neutron

In order to make the energies before and after the collision equal, the neutron had to be assigned a mass equal to 1.0067 units,[1] in any case in the range 1.005–1.008. *The mass of the neutron was therefore smaller than the mass of a proton plus an electron.* This was compatible with the idea that a neutron is a bound state of a proton and of an electron. The difference between the mass of the neutron and the masses of its two constituents would then be their binding energy, that is, the energy required to separate them:

We find therefore that the mass of the neutron is 1.0067. The errors quoted for the mass measurements are those given by Aston. They are the maximum errors which can be allowed in his measurements, and the probable error may be taken as about one-quarter of these. Allowing for the errors in the mass measurements it appears that the mass of the neutron cannot be less than 1.003, and that it probably lies between 1.005 and 1.008. Such a value for the mass to the neutron is to be expected if the neutron consists of a proton and an electron,

---

[1] In 1932 the mass unit for atoms was 1/16 of the mass of oxygen, which was taken as reference.

and it lends strong support to this view. Since the sum of the masses
of the proton and electron is 1.0078, the binding energy, or mass
defect, of the neutron is about 1 to 2 million electron volts. This is a
quite reasonable value.

But in July 1933, Irène Curie and Frédéric Joliot send a note to the *Académie des Sciences* which contradicts the conclusion of Chadwick [43]. They give a convincing argument: if one assumes that the neutron has the mass proposed by Chadwick, then the isotope 9 of beryllium, the mass of which had been determined in February 1933 by Kenneth Bainbridge [44] in the United States, should not exist. Indeed, this nucleus can be considered as an assembly of two α-particles and of one neutron. However, the observed mass of beryllium 9 is *greater* than the sum of the masses of its constituents, and beryllium would therefore decay while emitting energy. No nucleus can exist under such conditions! Irène Curie and Frédéric Joliot propose another solution: Chadwick had assumed that the neutron was emitted after a collision of an α-particle (of mass 4) with a boron nucleus of mass 11. They suggest that the neutron is emitted by the boron isotope 10 and not 11. They suggest that boron 10 can decay *by emitting either a proton or a neutron*. If this is the case, one can calculate the difference between the proton and neutron masses, without knowing the masses of the nuclei (which is a source of uncertainty), simply in terms of the energies of the emitted particles. In this case, the mass of the neutron turns out to be 1.011: *the neutron would have a mass larger than the proton*, contrary to what Chadwick had found. With this mass of the neutron, the mass of the beryllium 9 nucleus is smaller than the mass of two helium nuclei and one neutron, so that the beryllium 9 nucleus is bound and can exist. The Joliot-Curies suggest that possibly it is the proton which is made out of a neutron and a positively charged electron:

Finally we think that one should consider the proton to consist of
a neutron and a positive electron; the binding energy is then about
5×10^6 eV and it is therefore very stable.

In 1934, the Joliot-Curies confirm their findings in a paper sent to *Nature* and written in English, which shows how important they considered their result to be [45]. They observed similar reactions with two other nuclei: aluminum and magnesium. From three independent measurements, they deduce the neutron mass to be, respectively, 1.0098, 1.0092, and 1.0089. These masses differ little, and they are all larger than the mass of the proton (1.0081 in the same units, namely, 1/16 of the mass of oxygen).

A few months later, Chadwick makes a new and completely independent measurement of the mass of the neutron [46, 47]. His experiment follows a suggestion of Maurice Goldhaber, a Jewish physicist born in Austria in 1911 and living in exile in Great Britain. Chadwick and Goldhaber cause the deuteron to split up into its components (a proton and a neutron) by subjecting it to γ-rays emitted by a thorium C" radioactive source.[1] When the deuteron splits up, the proton and the neutron are

---

[1] Thorium C" is the isotope 208 of thallium, which today is denoted as ^{208}Tl.

emitted in opposite directions with approximately equal velocities. They measure approximately the energy of the proton and, knowing the energy of the γ-ray, they can deduce the mass of the neutron:

> *This leads to a value for the mass of the neutron of 1.0084 if we take the value for the deuteron mass given by Oliphant, Kempton, and Rutherford, or Bethe, and 1.0090 if we take the new value of Aston. It seems that the neutron is definitely heavier than the hydrogen atom (1.0081).*

This very precise measurement was performed in 1934, and it was the subject of the PhD thesis of Goldhaber. It establishes the fact that *the neutron is heavier than the proton*, confirming the results of Frédéric and Irène Joliot-Curie. Numerous other measurements followed in which the masses of nuclei became known with ever-increasing accuracy. The fact remained that the neutron is heavier than the proton.[1]

The inescapable conclusion is that the *neutron is not composed of a proton and an electron*. Should it be considered as a *new elementary particle*? In 1932, this remained an open question.

---

[1] The value accepted today is 1.0089856 in units of that time equal to 1/16 of the mass of oxygen.

Nuclear Theory After the Discovery of the Neutron

Electrons are no longer believed to be present inside the atomic nucleus which is understood to be composed of protons and neutrons. Heisenberg formulates the first theory of the nuclear forces which bind them to form the nucleus. A Sicilian genius appears on the scene: Ettore Majorana. Speculations arise about the configurations of protons and neutrons in the nucleus.

The discovery of the neutron sets the imagination of physicists aflame. On April 21, 1932, Dmitri Ivanenko (transliterated as "Iwanenko" in German journals), a Russian physicist from the Physico-Technical Institute in Leningrad, sends a letter to *Nature* which is published on May 28 and in which he states:

> Dr. Chadwick's explanation of the mysterious beryllium radiation is very attractive to theoretical physicists. Is it not possible to admit that neutrons play also an important role in the building of the nuclei, the nuclear electrons being all packed in α-particles or neutrons? [. . .] The chief point of interest is how far the neutrons can be considered as elementary particles (something like protons or electrons). It is easy to calculate the number of α-particles, protons, and neutrons for a given nucleus, and form in this way an idea about the angular momentum of nucleus (assuming for the neutrons an angular momentum 1/2) [48].

As early as 1930, Ivanenko suggested that the electrons may be created at the onset of their radioactive decay, and Jakov Dorfman considered the possibility that there would be no electrons in the nucleus.[1] Now, Ivanenko is ready to give up the idea that there should be electrons in the nucleus and that the neutron is composed of a proton and an electron. To support the idea, he appeals to the strange behavior which electrons seem to have in losing all their properties as soon as they enter the nucleus. He suggests that neutrons could be as "elementary" as protons and electrons.

Werner Heisenberg

In the beginning of June 1932, Werner Heisenberg sends to *Zeitschrift für Physik* the first of three papers dealing with the structure of nuclei [49]. He adopts the idea of Ivanenko that nuclei are made up of protons and neutrons, and he examines

---

[1] See p. 230.

the consequences of this assumption on the masses of nuclei, their binding energy, the stability of the helium nucleus, and the instability of some of them leading to radioactivity. It is the beginning of a new era in nuclear physics in which a real theory of nuclear structure arises. For Heisenberg, the problem is the following:

The experiments of Curie and Joliot and their interpretation by Chadwick have shown that in the structure of nuclei a new, fundamental component, the neutron, plays an important part. This suggests that atomic nuclei are composed of protons and neutrons, without the assistance of electrons. If this assumption is correct, it means a very considerable simplification of nuclear theory. The fundamental difficulties of the theory of β-decay and the statistics of the nitrogen-nucleus can then be reduced to the question: in what way can a neutron decay into a proton and an electron and what statistics does it satisfy? The structure of nuclei, however, can be described, according to the laws of quantum mechanics, in terms of the interaction between protons and neutrons.

The proton and neutron have masses close to 1 unit. Oxygen, for example, has a mass 16, and its nucleus has an electric charge equal to 8. It would then be composed of eight protons and eight neutrons.[1] The nucleus of the isotope 58 of iron would be composed of 26 protons (because it has an electric charge equal to 26) and of 32 neutrons.

The nitrogen problem[2] *was solved*: according to the new scheme, its nucleus contains simply seven protons and seven neutrons so that it must have integer spin (instead of half-integer). Furthermore, it contains an even number of fermions (either protons or neutrons), and it must therefore obey "Bose-Einstein statistics".[3]

The "Exchange" Interaction of Heisenberg

Heisenberg then addresses the fundamental question: what is the force which binds the protons and neutrons so as to form a nucleus? What experimental evidence could he be guided by in 1932 to determine this force?

The first data were the binding energies which had been carefully measured during the past 10 years by Aston and a few others. They showed that the binding energy of a proton or a neutron, that is, the energy required to extract it from the nucleus, did not vary appreciably from one nucleus to the other. There were however a few notable exceptions: the binding energy of the deuteron (composed of one proton and one neutron) was smaller than the average, while the binding energy of the α-particle was considerably larger.

---

[1] At that time, the atomic mass unit was 1/16 of the mass of oxygen.
[2] See p. 229.
[3] See p. 147.

The second set of data concerned the sizes of nuclei, which in fact were not well known. Some values of the size of aluminum were obtained by Étienne Bieler[1] who studied the anomalous scattering of α-particles. The sizes of some other nuclei were estimated by Gamow, who studied heavy radioactive nuclei and found sizes of about 7×10^{-13} cm deduced from the radiation of actinium, thorium, and radium. The estimate of Gamow was based on his quantum theory of α radioactivity with which he was able to explain the relation between the energy of the emitted particles and the radioactive half-life of the emitting element. Thus, the *volume* of nuclei appeared to be roughly proportional to the number of its constituents (protons and neutrons). This pleaded in favor of the *liquid drop* model proposed by Gamow: protons and neutrons dwell quite closely packed in the nucleus, separated by a fixed distance, and they do not penetrate one another. Each constituent interacts only with its closest neighbors, and this leads to a constant binding energy, a phenomenon called "saturation."

Heisenberg imagined a completely new kind of interaction, inspired by the way in which a single electron manages to bind two hydrogen nuclei so as to form an ionized H_2^+ molecule. The intensity of this type of force had been calculated a few years earlier by Walter Heitler and Fritz London, two young collaborators of Max Born in Heidelberg [50]. They showed that the attraction was caused by a strange process, which involves the spin of the proton and the exclusion principle of Pauli, two effects which occur in quantum mechanics and which have no analogue in classical mechanics.[2] It leads to a so-called *exchange interaction* which rests on the fact that the wavefunction only changes sign when the two protons are exchanged.

They initiated a real theory of the binding of atoms which form chemical compounds, a quantum theory of chemical binding. In 1933, Heitler was invited to give a series of talks at the *Institut Henri Poincaré* in Paris. He began by making a general statement:

> The problem could not be solved with classical mechanics for two reasons. First, it was not possible to find a force which could attract two neutral atoms, such as two hydrogen atoms. In classical mechanics we know of only gravitational, electric and magnetic forces. The former, as well as the magnetic forces are much too weak to explain the attraction revealed by chemistry [51].

Heitler recalled the fact that electric forces could produce some attraction but that it was much too weak. He then showed how the spin of the protons and Pauli's restrictive exclusion principle entailed a special phenomenon proper to quantum mechanics, namely, an attraction or a repulsion between the protons depending on whether their two spins point in the same or opposite directions. This is due to the fact that two protons are indistinguishable particles which can therefore be

---

[1] See p. 211.
[2] See p. 126.

exchanged and that they obey the Pauli principle. This is why the induced interaction was called an "exchange interaction."

Heisenberg does not attempt to make a complete analogy with the H_2^+ ion because he considers the nucleus to be composed of protons and neutrons without electrons. But he postulates an exchange binding energy between a proton and a neutron. By exchanging the electric charge, a proton-neutron pair would become a neutron-proton pair:

> *Here also one can intuitively conceive the exchange of positions in terms of an electron without spin and obeying Bose statistics. However it is preferable to consider this permutation integral J (r) as a fundamental property of a neutron-proton pair, without having recourse to the motion of an electron.*

Heisenberg thus introduces a new interaction of which he knows practically nothing except for the fact that it is an "exchange" interaction. He restricts this interaction to proton-neutron pairs. He assumes that two protons have only a Coulomb repulsion. He assumes a weak attraction between two neutrons. The exchange interaction has an interesting property: it acts only at short distances.

Heisenberg goes on to show that this force explains roughly the stability of various nuclei. Light nuclei have almost the same number of protons and neutrons. This is true up to calcium 40 which has 20 protons and 20 neutrons. Heavier nuclei have more neutrons than protons. For example, the stable lead isotope 208 has 82 protons and 126 neutrons. Heisenberg believes that the binding energy of a nucleus increases with the number of neutron-proton pairs it contains. This holds as long as the Coulomb repulsion between protons is weak which is the case of light nuclei. But as the number of protons increases, the Coulomb repulsion becomes more important and tends to break up the nucleus. It actually causes nuclei heavier than bismuth 209 (83 protons and 126 neutrons) to become unstable. Uranium is almost stable: its half-life is about 4.5 billion years. Nuclei between calcium and lead can be stable, thanks to the attractive force which acts between neutrons and protons.

A Further Argument in Favor of Considering the Neutron to Be an "Elementary" Particle

At the end of June 1932, Heisenberg sends his second paper [52] in which he pursues the discussion of the stability of nuclei and in which he considers the nature of the neutron: should it be considered as an elementary particle, such as the proton and the electron, or should it be considered as a combination of a proton and an electron? Heisenberg shows that the second hypothesis contradicts quantum mechanics which would predict a binding energy a hundred times smaller than the one measured by Chadwick. If the neutron was a composite particle, one would have to assume that quantum mechanics does not apply to the neutron!

Do Neutrons and Protons Repel at Short Distances?

In his third paper [53], received by *Zeitschrift für Physik* on February 16, 1933, Heisenberg tries to formulate the equations of motion of protons and neutrons in the nucleus. However, he is not satisfied with the result because the exchange interaction, which he assumed between neutrons and protons, leads to a binding energy which increases fast with the mass of the nucleus. This contradicts the experimental data of Aston, which showed that the binding energy of a proton or a neutron in the nucleus remains almost constant. Heisenberg shows that any attractive force will yield a similar result. In order to obtain something similar to the liquid drop model of Gamow, he postulates a strong repulsion between particles which come closer than a certain critical distance.

The description of the nucleus given by Heisenberg in his three articles may appear somewhat confusing because he occasionally changes his mind and because he postulates a strange exchange interaction without any apparent justification. Recall however that Heisenberg was groping his way forwards in unknown territory, guided only by his intuition. We shall see how his as yet uncertain ideas will become a basis for an understanding of the nucleus.

Ettore Majorana

This is where a most atypical actor enters, Ettore Majorana, a young Italian physicist who is working in the team of Fermi in Rome. Majorana was born on August 5, 1906, in Catania, Sicily. His father was an engineer, and he had an uncle who was a physics professor in the University of Bologna [54]. He began to study engineering at the *Bienno di Studi di Ingegneria* of the University of Rome. There, he meets Emilio Segrè [55], who is following the same courses but who, after meeting Franco Rasetti and Enrico Fermi, decided to switch to fundamental physics. Segrè convinces Majorana to see Fermi. Edoardo Amaldi recalls the encounter:

> He came to the Physics Institute in via Panisperna and was taken by Segrè to Fermi's office, where Rasetti was also present. This was the first time I saw him. From a distance he looked slender with a timid, almost hesitant bearing; close to, one noticed his very black hair, dark skin, slightly hollow cheeks and extremely lively and sparkling eyes. Altogether he looked like a Saracen. Fermi was then working on the statistical model later known as the Thomas-Fermi model [54].

The Thomas-Fermi model is an efficient approximation to the theory of atoms with several electrons, that is, to all atoms excepting hydrogen. It is not possible to solve this problem exactly because one should, in principle, take into account the Coulomb repulsion between each pair of electrons in addition to the dominant force, which is their attraction to the central nucleus. Fermi had devised a solution to this problem by taking into account, for each electron, the average Coulomb repulsion which was created by the other electrons and which reduced somewhat the attraction to the central nucleus. Fermi always sought simple and practical approximations. He showed that the potential resulting from his approximation was universal and could be calculated once and for all. Today, we refer to it as the universal Fermi potential. Fermi's calculation was spelled out in a preprint which he showed to Majorana [56, 57]. At the same time, the English physicist Llewellyn Thomas had made a similar calculation [58] so that today the method is called the Thomas-Fermi method. Amaldi continues the story:

> *Majorana listened with interest and, after having asked for some explanations, left without giving any indication of his thoughts or intentions. The next day, towards the end of the morning, he again came into Fermi's office and asked him without more ado to show him the table which he had seen for a few moments the day before. Holding this table in his hand, he took from his pocket a piece of paper on which he had worked out a similar table at home in the last twenty-four hours, transforming, as far as Segrè remembers, the second-order Thomas-Fermi non-linear differential equation into a Ricatti equation, which he had the integrated numerically. He compared the two tables and, having noted that they agreed, said that Fermi's table was correct: he then went out of the office and left the Institute. A few days later he switched over to physics and began to attend the Institute regularly.*

Majorana soon proved to be an exceptional physicist and mathematician, even in the proximity of Fermi. He was timid and did not relate easily with others. Nonetheless, he made good friends in the Institute, such as Edoardo Amaldi, Giovanni Gentile, and Emilio Segrè. He was extremely critical, of his own work to begin with. At the same time, he was very kind and generous with his friends. At the end of June 1932, the paper of Irène Curie and Frédéric Joliot was published, showing that the penetrating radiation, discovered by Bothe,[1] could project protons at high velocities. After reading the paper, Majorana declared [54]:

> *They haven't understood a thing. They are probably recoil protons produced by a heavy neutral particle.*

Soon after, the paper of Chadwick was published announcing the existence of the neutron[2] as well as the paper of Ivanenko[3] who suggested that nuclei consist of protons and neutrons and no electrons and that the neutron was perhaps an elementary particle.

At Easter, Majorana developed a theory according to which the interaction between protons and neutrons was an exchange interaction, similar to the one soon to be proposed by Heisenberg. Fermi urged him to publish his result, but Majorana

---

[1] See p. 256.
[2] See p. 258.
[3] See p. 263.

refused on the grounds that his work was still incomplete. He even did not allow Fermi to mention it in the talk he gave on July 7th at the Paris Congress on "the present state of the physics of the atomic nucleus" [59]. Nonetheless, Majorana was convinced by Fermi to go to Germany for some time. In January 1933, he went first to Leipzig where he met Heisenberg, and then to Copenhagen. He arrived in Leipzig just at the time when Heisenberg sent the third of his above-mentioned papers to *Zeitschrift für Physik*.

Majorana is not satisfied with Heisenberg's neutron-proton interaction. He does not like the fact that Heisenberg needs to give this interaction a complicated form in order to obtain the right result. Majorana then publishes a paper in *Zeitschrift für Physik*, in which he gives his version of the neutron-proton interaction. He begins by recalling that Heisenberg started by making an analogy between the neutron-proton interaction and the interaction between a hydrogen atom and a H^+ ion (which is a proton). He claims that this implies, in spite of what Heisenberg stated, that the neutron is composed of a proton and an electron. He adds:

> If we assume that nuclei consist of protons and neutrons we have to formulate the simplest law of interaction between them which will lead, if the electrostatic repulsion is negligible, to a constant density for nuclear matter. We have to find three laws of interaction: one between protons, one between protons and neutrons, and one between neutrons. We shall assume, however, that only Coulomb's force acts between protons [...] We assume that there is no noticeable interaction between the neutrons for there is no proof of the contrary [60].

Majorana then considers the liquid drop model of the nucleus. He discusses the possibility of an interaction similar to the one between the molecules in a liquid: a "long-range" attraction and a very short-range repulsion which prevents the molecules from penetrating each other. But he is not satisfied with the idea:

> Such a solution would be aesthetically unsatisfactory, however, since we would have not only attractive forces of unknown origin between the particles, but also, for short distances, repulsive forces of enormous magnitude corresponding to a potential of several million volts. We shall, therefore, try to find another solution and introduce as few arbitrary elements as possible. The main problem is this: how can we obtain a density independent of the nuclear mass without obstructing the free movement of the particles by an artificial impenetrability?

Like Heisenberg, Majorana assumes that there is no interaction between neutrons and only a Coulomb repulsion between protons. His proton-neutron interaction differs from the one proposed by Heisenberg as much for *aesthetic* as for rational reasons. He seeks the *simplest solution* because he believes, as most physicists often do, that physical laws should be simple. Majorana also chooses, like Heisenberg, an exchange interaction which is attractive in states which do not change when the positions of the proton and neutron are exchanged but not their spin. Thus, he does not need to refer to an electron which would oscillate between the two particles thereby exchanging their electric charge. Thus, a neutron and a proton with spins pointing in opposite directions attract each other, and they repel if their spins are parallel. This is how Majorana explains the strong binding of the α-particle (which has spin zero because the proton–neutron pairs have opposite spins), whereas the deuteron which has a spin equal to one is only weakly bound. This exchange interaction is still called today the "Majorana exchange" interaction.

Majorana returns to Italy in the fall of 1933. In 1937, he becomes professor of theoretical physics in the University of Naples where he maintains, as he did in Rome, a cloistered life. On March 23, 1938, he embarks on a boat heading to Palermo where he stays for 2 days. On March 25, he takes a boat to return, and disappears before reaching Naples. He had sent a letter to his friend Antonio Carelli in which he stated his intention of putting an end to his life. In spite of intense searches, no trace of Majorana's body was ever found. He was 32 years old.

Eugene P. Wigner

Jenö Pal Wigner was born on November 17, 1902, in Pest, the eastern part of Budapest. He will not be the only Hungarian endowed with an important role in our story. His family is prosperous, Jewish but not religious, and will later convert to Lutheran Protestantism [61, 62]. He obtains his PhD in chemical engineering at the *Technische Hochschule* of the University of Berlin. His thesis advisor is another Hungarian, Michael Polanyi. At the age of 22, he returns home to work in the family tannery, but soon after, he is offered a job as crystallographer at the *Kaiser Wilhelm Institut*. He immediately accepts and embarks on the career of a physicist, which he had always dreamt of. From the outset, he is interested in *group theory*$^\diamond$, a branch of mathematics which was created by the French mathematician Évariste Galois in 1832 and developed by the German mathematician Ferdinand Georg Frobenius and the Norwegian mathematician Marius Sophus Lie. Wigner is impressed by the work of Heisenberg, who had brilliantly solved the problem of the spectrum of the helium atom [63] taking into account Pauli's exclusion principle. Wigner succeeds in solving the three electron problem [64], but the calculation becomes too complicated for a larger number of electrons. Following a suggestion of his friend, the mathematician János von Neumann, he applies group theory methods to the problem of an atom with any number of electrons, as well as to the problem of the vibrations of molecules and crystals [64–66]. His work promotes group theory to one of the major tools of theoretical physics.

In 1930, Wigner is offered a tempting position in the University of Princeton. He accepts the offer and settles in Princeton where he finds his friend von Neumann, who had anglicized his Christian name to John. So Wigner also decides to change his name to Eugene Paul Wigner. In 1932, after the discovery of the neutron and the initial papers of Heisenberg, Wigner studies the forces acting between neutrons and protons and makes major contributions.

He first studies the binding energy of the deuteron and of the helium nucleus, which was then understood as consisting of two protons and two neutrons [67]. He tries to understand why the binding energy of the deuteron is so much smaller than that of the helium nucleus, about 17 times smaller according to the data available at the time.[1] To understand this, Wigner begins by considering the structure of the deuteron, composed of one neutron and one proton. He postulates an attractive potential which depends on the distance between the two particles and which decreases fast beyond a certain distance. He adjusts the parameters of the potential, that is, its depth and its range so as to fit the observed binding energy of the deuteron. He notices that the binding energy of the helium nucleus can become much larger if the range of the potential is made very small.

Wigner sends his paper to the publisher in December 1932, before he could read Heisenberg's third paper [53], sent at about the same time and published on February 16, 1933. Wigner obtains a surprising result, which is a purely quantum mechanical feature with no analogue in classical mechanics: in the deuteron, *the neutron and the proton spend most of their time separated by distances which are greater than the range of the very potential which binds them together*. This is due to the small binding energy of the two particles. The weaker the binding energy, the more their wavefunction spreads out, and this implies that the probability of finding them far apart increases. This is an example which demonstrates that the notions of a precise position and orbit no longer apply in quantum mechanics because, indeed, the deuteron does not decay, which it would in classical mechanics if, at any instant, the two particles were at a distance beyond the range of the potential. Wigner then studies a problem related to the deuteron, namely, the collision of a neutron and a proton [68]. If his calculation of the structure of the deuteron is correct, the same potential should allow him to calculate the *elastic scattering* of the two particles. He can calculate the distribution of angles at which they scatter. Wigner finds an interesting phenomenon: his results do not depend on the precise shape of the potential, as long as it has the right depth and range. He found one could even compensate a larger (or smaller) depth by a smaller (or larger) range.

Do the Protons and Neutrons form Shells as Electrons Do in the Atom?

William Harkins, Before the Discovery of the Neutron

Ever since 1915, the American physicist William Harkins had been studying the increasing list of known nuclei [69,70]. In 1931, he published a paper, summarizing his observations [71]. In order to understand how the nuclear constituents (which were then assumed to be protons and electrons) build up a nucleus, Harkins was

---

[1] Today's data make it "only" 12.7 times smaller.

inspired by the work of Mendeleev: he listed the nuclei in the order of increasing mass and deduced what he boastfully called *laws* and which at least appeared to be regularities. He felt that it was the manifestation of a "periodic system of atomic nuclei" but could go no further because he only took into account the parity and the number of constituents.

James Bartlett

On July 3, 1932, the American physicist James Bartlett sent a letter to the editor of the *Physical Review*, in which he clearly suggested the possibility of a shell structure of nuclei:

> For some time, there has been speculation as to whether or not the atomic nucleus can be regarded as consisting of shells of protons, just as the external structure [of the atom] is known to consist of shells of electrons [72].

He suggested that the protons, to which he later added neutrons, form successive shells, as the electrons do around the nucleus. Each shell was labelled by an "azimuthal" quantum number l which is an angular momentum. Shells with angular momentum $0, 1, 2, \ldots$ would contain, respectively, $2, 6, 10, \ldots$ protons and as many neutrons. He pursued this model in another letter [73], dated August 30, in which he set up a kind of domino game, consisting in building up successions of nuclei by adding protons and neutrons and even occasionally some electrons, in order to understand why certain nuclei had more isotopes than others. However, he went no further than count the number of protons and neutrons.

Walter Elsasser and Kurt Guggenheimer

A year later, a paper is published in the *Journal de Physique* signed by Walter Elsasser, a German physicist who fled to Paris [74]. Elsasser was born in 1904 in Mannheim, into a Jewish family [75]. He worked with James Franck in Munich and with Ehrenfest in Leyden. As many Jewish physicists, he left Germany in 1933, passed through Switzerland, where Pauli told him that Frédéric Joliot was looking for a theorist. He went to Paris and, in the fall of 1934, he obtained a job at the newly founded *Caisse Nationale des Sciences*, the ancestor of today's CNRS.[1] He finally emigrated to the United States, partly to avoid the 2-year military service in France. Once there, he changed fields and became a well-known expert of terrestrial magnetism.

---

[1] The CNRS stands for *Centre National de Recherche Scientifique*. It is one of the major research organizations in France.

In his 1933 paper, he raises the question of the structure of the nucleus. Does it contain α-particles in addition to neutrons and protons? True, the α-particle is robust, and it has a large binding energy. However, Elsasser notes that a neutron is hardly more bound in the α-particle (the helium nucleus) than it is in other nuclei. (The binding of a neutron is a measure of the energy required to extract it from the nucleus.) In his paper, he makes the following conclusion:

Obviously we cannot under these circumstances consider it reasonable to assume that the nucleus is a system composed of α-particles and a few free neutrons [. . .] We will therefore treat all the neutrons and all the protons on the same footing, with no other clustering of particles than that which follows from Pauli's exclusion principle [74].

So how is one to picture the structure of the nucleus? Elsasser introduces a new concept: he knows that the nucleus does not possess an attractive center which could play the role of the nucleus in the atom. Indeed, all the particles in the nucleus, namely, the protons and the neutrons, play equivalent roles. But Elsasser says that a particle in the nucleus feels an *average attractive force* which is due to its interaction with all the others. If this average force is the same in all directions, then each neutron and each proton in the nucleus can be labelled by quantum numbers. Elsasser even claims that *each particle, neutron or proton, describes an orbit* so that the particles can thus be grouped into *successive shells*, which he calls *envelopes*. The title of his paper is "The Pauli principle in nuclei." Indeed, Elsasser, like Bartlett, considers that it is Pauli's exclusion principle (which states that two particles cannot bear the same quantum numbers) which is responsible for the formation of shells in nuclei, as it is for the shells formed by electrons in atoms. Each shell can only contain a given number of particles. For the electrons in an atom, the "principal quantum number" determines the spatial extent of the orbit, whereas the "azimuthal quantum number" fixes the *angular momentum*.

In 1933, a physicochemist, Kurt Guggenheimer, also fled Nazi Germany and found a temporary refuge at the *Collège de France*, in the laboratory of Paul Langevin. Guggenheimer had a thorough knowledge of the structure of molecules and of the manner in which the atoms were bound in the molecule. He meets Elsasser, and they both raise the burning question: what keeps the neutrons and protons bound to the nucleus and how are they configured in the nucleus? They consider collaborating, but, as they never manage to agree, they publish their results separately.

In his first paper, dated May 9, 1934, Guggenheimer examines stable nuclei and makes a few simple remarks [76]. In the lightest nuclei, the numbers of protons and neutrons increase in parallel fashion: one goes to the next nucleus by adding successively a neutron and a proton, then two neutrons and two protons, and so forth until one reaches neon. It was known that, as the nucleus becomes heavier, the number of neutrons increases faster than the number of protons. Thus, in lead, the nucleus ends up containing 82 protons and 126 neutrons. However, while Heisenberg had considered the average increase of the number of protons and neutrons,[1] Guggenheimer notices that *the neutron excess is established in a discontinuous manner*. The numbers of neutrons and protons remain equal until they reach 20 each, thereby forming calcium 40, but there a sudden change occurs: calcium has isotopes which range from 40 to 48, which means that the 20 protons in calcium can bind up to 28 neutrons. A similar discontinuity is observed for nuclei with 50 neutrons: stable nuclei can be formed with very different numbers of protons, ranging from 36 (in krypton) to 42 (in molybdenum). The same phenomenon occurs yet again when the nucleus contains 82 neutrons. It can then bind any number of protons ranging from 54 (in the 136 isotope of xenon) to 62 (in the 144 isotope of samarium). Guggenheimer interprets these discontinuities in the number of protons and neutrons as discontinuities in their binding energies, similar to the ones observed for electrons in atoms: once a shell is filled by electrons, no further electron can be added to the shell because of Pauli's exclusion principle. Therefore, the extra electron must go into a higher energy shell, and it is therefore less bound in the atom:

> The discontinuities in the binding energies display evidence of quantum effects. In particular, the shape of the curves suggest that large jumps in the binding energies occur when a shell is closed and a new one is born [76].

Guggenheimer notices discontinuities for:

- 20, 36, 54, 84 protons
- 50, 82 neutrons

In a paper dated July 9, 1934, Guggenheimer attempts to define a kind of "affinity" of the neutron for the nucleus, based on the number of isotopes and the abundance of various isotopes. Certain elements, such as calcium or tin, have a large number of isotopes: 20 protons (in the case of calcium) and 50 protons (in the case of tin) can form stable nuclei with a number of neutrons ranging, respectively, from 20 to 28 in the case of calcium and from 62 to 74 in the case of tin. But he can go no further with these ideas. However, he makes an interesting remark which seems to contradict the generally accepted idea of Heisenberg, who had postulated that the attractive force between neutrons and protons was much larger than the force between two protons or two neutrons. Why then, asks Guggenheimer, are there so few nuclei which have odd numbers of both protons and neutrons? Why are nuclei with even numbers of protons and/or even numbers of neutrons so numerous? It is as if some force attempted to pair up protons as well as neutrons. Guggenheimer then left Paris for a position in the University of Glasgow.

---

[1] See p. 263.

In a new paper, Elsasser attempts to make a theoretical calculation of various "envelopes" or "shells" [77]. He describes the nucleus as a spherical *potential well*, similar to the one used by Gamow to bind an α-particle to a nucleus in α radioactivity.[1]

This allows him to solve the Schrödinger equation for each proton and neutron in the potential well and to calculate its "wavefunction." Elsasser can thus determine the number of particles which have a principal quantum number equal to 1,2,3,.... To each one of these quantum numbers corresponds a shell, which Elsasser calls an envelope. He finds that filled shells occur for nuclei with numbers of protons (or neutrons) equal to 2, 8, 18, 32, 50, 60, and 82. Recall that Guggenheimer had suggested 20, 36, and 54 for protons and 50 and 82 for neutrons. It is not conclusive in spite of some common predictions of the two approaches.

Heisenberg and the Hartree Method

A year later, Heisenberg publishes a paper in which he attempts to calculate the masses, therefore the binding energies of light nuclei [78]. This will be his last paper devoted to nuclear structure. He still assumes that the interaction between neutrons and protons is similar to the one proposed by Wigner, and he applies the method of Hartree which works so well for atoms.

Douglas Hartree is an English physicist, born in Cambridge in 1897. In 1928, he becomes assistant at the Christ College after obtaining his PhD in 1926. He proposes a method to calculate the structure of atoms with many electrons which proves to be most successful. What makes the problem difficult is that the electrons not only feel the attractive Coulomb potential produced by the central nucleus but also the Coulomb repulsion exerted by the other electrons. An approximate solution to this problem was proposed in 1927 by Enrico Fermi and by Llewellyn Thomas who used an average potential of the electrons.[2] Hartree suggests to first calculate the wavefunctions of the electrons while neglecting the Coulomb repulsive interactions between them and then to treat the effect of the latter as a perturbation. This allows him to take into account approximately the average repulsion exerted on an electron by the others. This repulsion modifies slightly the Coulomb potential produced by the nucleus. He then recalculates the electron wavefunctions with this new potential which he hopes to be closer to the truth. He reiterates the process until the potential no longer changes. This ensures the internal consistency of his calculation. The method became called the "self-consistent field" method [79–81]. In 1929, he obtains the chair of applied mathematics in the University of Manchester, and during the summer of 1933, he visits the Massachusetts Institute of Technology (MIT), in Cambridge, USA. There, he uses an analogue computer which was conceived by Vannevar Bush [82], a physicist and engineer at MIT who will play an important

---

[1] See p. 243.
[2] See p. 267.

role by organizing technological developments during the Second World War. When Hartree returns to Manchester, he also constructs an analogue computer, and after the war, he becomes a pioneer in electronic numerical computation.

Heisenberg therefore applies the method of Hartree to neutrons and protons in the nucleus, while still assuming that the force acting between a neutron and a proton is considerably larger than the one acting between two neutrons or two protons. He chooses a Majorana-type force and performs the Hartree calculation approximately because he performs the calculation by hand. He concludes that he can explain the anomalies in the binding energies, that is, the shell closure effects, in light nuclei, but that the method fails in heavy nuclei. Heisenberg explains that each proton can only feel the attraction of the neighboring neutrons because of the short range of the force and that this makes the nucleus different from the atom.

Wigner and Feenberg Use the Hartree-Fock Method

Wigner makes a calculation with a young promising American physicist [83], Eugene Feenberg [84], who was born in 1906 in Ford Smith, Arkansas. His parents were Polish Jews who emigrated to the United States in 1883. He is a brilliant pupil in school, and he is able to study in Harvard during the Great Depression thanks to a part-time job. In 1931, he obtains a scholarship enabling him to travel in Europe, where he works with Sommerfeld, Pauli, Fermi, and Heisenberg. When Hitler comes to power, he returns to Harvard where he defends his PhD thesis. He then becomes interested in the interaction between neutrons and protons. In 1936, he works with Gregory Breit, and he meets Wigner in 1937. Feenberg was a man of great integrity, modest and devoted to his students.

His work with Wigner concerned light nuclei with masses ranging between helium (two protons and two neutrons) and oxygen (eight protons and eight neutrons). Indeed, if the neutrons and protons form shells (as electrons do in atoms), there can be two neutrons and two protons in the lowest shell (which spectroscopists call the s shell) and up to six neutrons and six protons on the next shell (which spectroscopists call the p shell). Feenberg and Wigner use the Hartree method, as Heisenberg did: each particle describes an "orbit" in a potential, and the orbit is designated by well-defined quantum numbers. This, at first sight, appears to be a crude approximation.

Wigner and Feenberg improve the Hartree method using a method conceived by a Russian physicist, Vladimir Fock [85]. Born in 1898 in St. Petersburg, Vladimir Alexandrovich Fock graduated from St. Petersburg State University in 1922 and remained associated all his life with the university, where he taught for more than 40 years. He made fundamental contributions to quantum mechanics, quantum field theory, theory of gravity, and mathematical physics, including applied mathematics. He died in 1974. The Fock method respects Pauli's exclusion principle, whereas the Hartree does so only approximately. The method of Fock consists in *antisymmetrizing* the wavefunction, consisting of independent neutron and proton orbits, thereby ensuring that the Pauli principle is correctly enforced. This leads to what is called the *Hartree-Fock* method, and it is used today to make calculations of the structure of all nuclei. Wigner and Feenberg differed further from Heisenberg, in that they assumed that the nuclear interaction is *charge independent*, meaning that the force acting between a neutron and a proton is the same as that which acts between two neutrons or two protons. In this sense, the only difference between a neutron and a proton is the electric charge of the latter. This causes an additional Coulomb repulsive force between the protons which is much weaker than the nuclear interaction. The binding energies calculated this way by Wigner and Feenberg are quite close to the experimental values. Wigner and Feenberg find further evidence for their assumption that the nuclear interaction is charge independent: if the interaction is really the same between all the nucleons,[1] then the binding energies of the two nuclei, boron 11 (five protons and six neutrons) and carbon 11 (six protons and five neutrons), should be almost the same. The only difference would be due to the weak Coulomb repulsion between the protons which makes boron 11 slightly lighter than carbon 11. They calculate the energy difference due to the Coulomb force, and they find it equal to 4.06 MeV (million electron volts). The measured value is 4.90 MeV, a most encouraging agreement for what is only an approximate calculation.

Friedrich Hund

Independently of Wigner and Feenberg and practically at the same time, Friedrich Hund publishes a paper making similar assumptions and he makes similar conclusions, in particular concerning the comparison of nuclei with the same mass [86]. Hund searches for a simple mass formula which would yield the binding energies of various nuclei. Born in 1896 in Karlsruhe, Friedrich Hund obtained his PhD with Max Born in 1922. From then on, he worked mainly on solid state and molecular physics. In 1927, he took up a professorship at Rostock, and he spent the 1930s at the University of Leipzig, where he remained until the end of the Second World War. He left East Germany in 1951, and he obtained a position in Frankfurt. In 1957, he returned to Göttingen. He died in 1997, at the age of 101.

The Nuclear Shell Model: An Idea for the Future?

The idea that the neutrons and protons in a nucleus form successive "shells," as electrons do in an atom, appears to be promising, at least for light nuclei, in spite of the big difference between nuclei and atoms. Indeed, in an atom, there is a central strong electric charge, carried by the nucleus, and it appears natural to expect electrons to describe orbits around this central charge. In nuclei however, there is no central charge, and the interaction between the particles is different and not well known. In addition, the shell structure of nuclei becomes increasingly problematic for nuclei heavier than calcium 40. We shall see however that the idea of a shell structure in nuclei will be vigorously attacked by a fearful foe, Niels Bohr in person!

---

[1] In 1941, Christian Møller coined the word "nucleon" to designate both protons and neutrons, see p. 425.

A New Particle: The Positron

The study of cosmic rays reveals the existence of a positively charged electron, thus confirming a brilliant prediction of Dirac. It is soon discovered that the presence of this particle is quite a common occurrence.

Cosmic Rays

We often mentioned "cosmic rays" without stating what they are. It is time to do so because they are about to reveal the existence of several new particles.

In 1897, Charles Thomson Rees Wilson[1] was working on the condensation of water vapor in the air. He noted that, no matter how hard he tried to get rid of the dust in the air container, some condensation always seemed to form around centers, the number of which he estimated to be less than 100 per cubic centimeter. He then noticed that X-rays produced numerous centers of condensation [87]. In 1900, Hans Geitel observed that a charged electroscope would slowly discharge when placed in air. Air is therefore slightly conducting, and it must therefore contain charged particles such as ions and electrons [88]. Charles Coulomb had already observed a similar phenomenon in the eighteenth century: a charged sphere hanging on a silk thread would progressively lose its charge [89]. Wilson noted that the loss of electric charge was the same when the apparatus was placed in the laboratory (close to radioactive substances), or in the countryside, or when it was exposed to light, kept in the dark, or subject to 120 or 210 V. He estimated that about 20 ions were formed every second in a cubic centimeter of air which had previously been cleared of dust particles [90]. He therefore tried to test experimentally the hypothesis, according to which:

> The continuous production of ions in dust-free air could be explained as being due to radiation from sources outside our atmosphere, possibly radiation like Röntgen rays or like cathode rays, but of enormously greater penetrating power [91].

Wilson confirmed that about twenty ions per second were continuously formed in a cubic centimeter and that this number appeared to be the same everywhere, even underground. He concluded that this radiation most likely did not originate outside the earth but that it could be a property of air itself.

---

[1] See p. 222.

Rutherford confirmed these results in 1902. He found that about 15 ions were formed every second in 1 cm^3 of air, a number which is smaller but not far from that of Wilson [92].

What is the origin of this radiation? Numerous measurements existed indicating that the earth was weakly radioactive.[1] This made it difficult to determine whether the observed ionization was due to the weak radioactivity of the earth and of the atmosphere.

A decisive step was taken in 1911 by the Austrian physicist Victor Hess, who was twice able to reach an altitude of 1,000 m in a balloon, thanks to the help of the Austrian Aeronautical Club. During this ascension, he measured the number of "spontaneously" produced ions per second and per cubic centimeter and he failed to observe a decrease in the radiation at this altitude. A year later, he made seven further balloon flights reaching an altitude of 5,350 m. This time, he made a startling observation: between 4,000 and 5,000 m, the intensity of the radiation doubles [93]!

The German physicist Werner Kolhörster reached an altitude of 6,200 m in 1913 and 9,300 m in 1914, also in a balloon. He observed that the radiation almost tripled in intensity between these two altitudes [94, 95]. Thus, the radiation was most probably not emitted by substances in the earth or in the atmosphere.

What struck the physicists at that time was that the radiation persisted everywhere on the earth, on the sea, and even in tunnels. This proved that, if it originated in outer space, it was an extremely penetrating radiation, far more penetrating than the radiations which had been observed so far. It was called the "very penetrating radiation" (*durchdringenden Strahlung*) or the "altitude radiation" (*Höhenstrahlung*) until the American physicist Robert Millikan, who was famous for having measured the electric charge of the electron, called it "cosmic rays" [96] in 1926, a name that stuck.

At first, it was believed that this penetrating radiation consisted of γ-rays. The same had been assumed about the penetrating radiation which Bothe and Becker had discovered and which turned out to consist of neutrons. But in 1929, Walther Bothe and Werner Kolhörster showed that the cosmic rays consisted of "material" particles and not of γ-rays [97]. From then on, physicists began to study intensively cosmic rays.

Blackett and Occhialini

Patrick Blackett and Giuseppe Occhialini, two physicists working in the Cavendish, take great strides in the study of cosmic rays [98, 99]. Blackett was the first to

---

[1]Indeed, the earth contains some uranium and therefore radium and therefore also radon, the residue of the disintegration of radium. Since radon is a gas, it is continuously escaping from the earth. There is more of it in granitic regions which are richer in uranium. The radioactive half-life of radon is 3.8 days.

observe a nuclear reaction in a Wilson cloud chamber.[1] In 1924–1925, he spent a sabbatical year in the laboratory of James Franck in Göttingen. In 1931, he admitted into his laboratory, at the Cavendish in Cambridge, a young Italian physicist who had been awarded an Italian scholarship. Giuseppe Occhialini was born on September 5, 1907, in Fossombrone, in the Marches. He was artistic (he wanted to be a painter) and sporty (he was a speleologist). However, he went into physics and became a friend of Bruno Rossi, who began to study cosmic rays. When he arrived in Cambridge, he knew all about the coincidences noted by Rossi.[2]

So far, cosmic rays were studied by observing their track in a Wilson cloud chamber. The chamber was expanded and a photograph taken at regular intervals. With some luck, a particle would have passed through just at that time. Only one out of 10 to 100 photographs would reveal tracks, depending on the size and the disposition of the chamber. Blackett and Occhialini then have an idea: since the Rossi circuit could be set up so as to detect the *simultaneous* triggering of two Geiger counters, they use it to *expand the cloud chamber only when a particle passes through*. Their cloud chamber has the shape of a pie, about 3 cm thick and 13 cm in diameter. They place vertically above and below two Geiger–Müller counters set up in coincidence. They produce an electric signal whenever a particle passes through the two Geiger–Müller counters and therefore also through the cloud chamber. This becomes the signal to expand the cloud chamber and to take the photograph. The time taken by the particle to pass through the first counter, through the cloud chamber, and finally through the second counter is about one billionth of a second. It therefore passes through almost simultaneously. That gives ample time to expand the cloud chamber because the small liquid drops, formed by the ions upon the expansion of the cloud chamber, last several tenths of a second. With this method, 80% of their photographs display the track of a particle which had passed through the cloud chamber. Blackett and Occhialini are thus able to make a resounding confirmation of the observations made a few years earlier in Leningrad by the Russian physicist Dmitri Skobelzyn.

Skobelzyn was studying β radioactivity by observing the photographs of trajectories in a Wilson cloud chamber [100]. Some of these trajectories appeared to be caused by electrons with energies far greater than those produced by β radioactivity. Two years later, after having analyzed over 600 photographs, Skobelzyn confirmed the existence of very high-energy electrons: their trajectories were almost straight lines, barely deviated by a magnetic field of 1500 Gauss. This meant that they had very high velocities, almost reaching the speed of light and therefore very high energies, which Skobelzyn estimated to be more than 15 MeV (15 million electron volts):

The 613 photographs obtained so far confirm the regularity of the phenomenon and they clearly establish the existence of an "ultra β radiation". Since this radiation cannot be attributed to local sources,

---

[1] See p. 225.
[2] See p. 219.

*no other possibility remains but to relate the observed effect to the
"very penetrating radiation" or "altitude radiation" [101].*

Skobelzyn was thus the first to observe a *particle* in the cosmic radiation: it was
an ordinary electron. But he makes a further observation:

*A track of the type of a non-deviated β ray occurs on the average in
one case out of 20. There are however some photographs displaying
a group of two or even three trajectories of this type. Among the
27 photographs displaying "ultra β radiation" there are three with
double trajectories and one with triple trajectories. The trajectories
of such a group lie within an rather small angle [. . .] If one considers
the angle and the average frequency of such trajectories, one can
evaluate the probability of a simultaneous occurrence of independent
trajectories. The probability is extraordinarily small, even evanescent.
One cannot doubt that the components of such group originate from
a common source, which is confirmed by a stereoscopic observation
of the photographs.*

Skobelzyn observes that groups of tracks (trajectories) clearly occur more
frequently than by pure chance. Such an event is however rarely observed when
the Wilson cloud chamber is expanded at random times. But when Blackett and
Occhialini manage to expand their cloud chamber only when particles pass through,
they quickly discover several instances of groups of tracks, some of which have
more than ten tracks [98]. They call such groups "showers."

There is legend [102] stating that when Occhialini observed the first shower on a
photograph which he had just developed, he ran to show it to Rutherford, and in his
enthusiasm, he kissed the servant who opened the door for him. The cloud chamber,
triggered by the two Geiger–Müller counters set in coincidence, led Blackett to
discover a new particle in 1945. He was awarded the Nobel Prize in 1948 for "for
his development of the cloud chamber method, and his discoveries therewith in the
fields of nuclear physics and cosmic radiation."

Carl Anderson Discovers a Positive Electron

The study of cosmic rays led a young American physicist, Carl Anderson, to an
important discovery. Born in New York in 1905, Anderson studied at the California
Institute of Technology, known as *Caltech*, and he defended his PhD thesis in
1930. Working with Robert Millikan in Caltech, he constructed a vertical Wilson
cloud chamber[1] with which he could take photographs of tracks caused by cosmic
rays. The cloud chamber was placed between the poles of a magnet which was
strong enough to curb the trajectories of the particles. Indeed, the trajectory of a
charged particle in a magnetic field is circular, and its radius determines its velocity,
provided its mass is known. On August 1932, Anderson notices that, among all the

---

[1] See page 222.

trajectories formed by small droplets of condensed water in the Wilson chamber, one trajectory differed from the others in that it was deviated to the right instead of the left. This could be due to a trivial cause: the electron might be traveling in the opposite direction, a feature the instrument could not specify. However, Anderson placed a lead sheet in the middle of the chamber, which had the effect of considerably slowing down the electrons. The photograph clearly displayed a fast particle (only slightly deviated by the magnetic field) impinging on the lead plate and a considerably slower particle (more deviated by the magnetic field) emerging from the other side. The direction in which the particle was traveling was thus clearly revealed. Indeed, it is hard to imagine how the lead plate could have *accelerated* the particle! The trajectory could only be caused by a *positively charged* particle, traveling like the other cosmic ray particles, from above to below. However, the only positively charged particle known at the time was the proton. But the track could not have been caused by a proton which would have been slowed down considerably more by the lead plate. Except for its positive electric charge, the particle behaved exactly as an electron: *it was indeed a new particle*, which Anderson called a *positive electron*. Anderson then checked his observation by examining photographs taken by other physicists and by taking further photographs himself. Among 1,500 photographs, he discovered 15 tracks of positive electrons. He considered this to be sufficient proof, and he sent a short paper to *Science* [103] on September 1, followed by another paper also sent to *Physical Review* [104], which contains his famous photograph, which was taken on August 2, 1932, and which became one of the most famous pictures in physics. A glance at the picture reveals the presence of this positive electron, which we call today a *positron*.

The Positive Electron of Anderson and that of Dirac

Anderson was far from realizing that Paul Dirac, the English theoretician who created relativistic quantum mechanics, had predicted the existence of this positive electron!

Dirac had deduced the existence of this particle from his famous "Dirac equation" which was a relativistic quantum mechanics equation.[1] His equation has "normal" positive energy solutions which give a correct description of the electron, including its half-integer spin. But in addition to the normal solutions, there exist solutions in which the electron has negative energies. In his first paper, Dirac thought that this was a defect of the theory. Recall that in nonrelativistic quantum mechanics, the energy of a free particle is its positive *kinetic* energy, which is equal to the square of its velocity multiplied by half its mass: $\frac{1}{2}mv^2$. It is difficult to imagine how this kinetic energy could become negative, since the mass is a positive quantity.

---

[1] See p. 146.

In relativistic mechanics, the energy of a free particle of mass m is given by the formula $\sqrt{m^2c^4 + p^2c^2}$ where p is the momentum of the particle and c the velocity of light. It is also a positive quantity. So Dirac soon raises the question: what do the negative energy solutions of his equation mean? And what prevents an electron from falling into one of the negative energy states, which is also a solution of the Dirac equation? Once there, what prevents it from cascading down to states with even lower energy? In a paper sent to the *Royal Society* on December 6, 1929, he offers an interpretation [105]. It is really a fantastic idea, which sounds more like science fiction: let us suppose, he says, that *all* the negative energy states *are already occupied by electrons*. Then Pauli's exclusion principle will forbid any electron from jumping into one of the negative energy states because there is already an electron in that state. The state in which all the negative energy states are filled with particles is called today the *Dirac sea*. One cannot observe it directly. The observed particles in our universe are those which are in positive energy states. However, a particle might somehow be ejected from the Dirac sea, and it would then create what Dirac calls a "hole":

> These holes will be things of positive energy and will therefore be in this respect like ordinary particles. Further, the motion of one of these holes in an external electromagnetic field will be the same as that of the negative-energy electron that would fill it, and will thus correspond to its possessing a charge +e.

Thus, Dirac shows that a "hole," that is, an electron which is missing from a negative energy state, *will appear as a positive energy particle which will propagate in an electromagnetic field as if it had a positive charge*. But what could this positively charged particle be? The only positively charged particle known at the time was the proton. At first, Dirac thought that the "hole" would indeed be a proton:

> We are therefore led to the assumption that the holes in the distribution of negative-energy electrons are the protons.

In a paper published in an October issue of *Nature*, Dirac discusses the matter further. He begins with a general consideration:

> Matter is made up of atoms, each consisting of a number of electrons moving round a central nucleus. It is likely that the nuclei are not simple particles, but are themselves made up of electrons, together with hydrogen nuclei, or protons as they are called, bound very strongly together. There would thus be only two kinds of simple particles out of which all matter is built, the electrons, each carrying a charge −e, and the protons, each carrying a charge +e [106].

After mentioning the problem of the nitrogen atom,[1] he makes a philosophical remark:

---

[1] See p. 229.

It has always been the dream of philosophers to have all matter built up from one fundamental kind of particle, so that it is not altogether satisfactory to have two in our theory, the electron and the proton."

He goes on to explains that the solutions of his equation may explain the existence of both the electron and the proton:

There are, however, reasons for believing that the electron and proton are really not independent, but are just two manifestations for one elementary kind of particle. This connection between the electron and proton is, in fact, rather forced upon us by general considerations about the symmetry between positive and negative electric charge, which symmetry prevents us from building up a theory of the negatively charged electrons without bringing in also the positively charged protons.

However, this view faces serious difficulties. If the electron and the proton are kind of mirror images, why should the mass of the proton be 1,836 times greater than the mass of the electron? Worse, nothing seems to prevent an electron from jumping into the "proton" hole, the liberated energy being emitted in the form of electromagnetic radiation. In other words, an electron and a proton could annihilate, and yet no such thing is observed.

Dirac mentions a solution which was proposed by a young American physicist, Robert Oppenheimer [107]. Born in New York in 1904 into a well-to-do family, Oppenheimer studied first in Harvard and then in Europe, in Göttingen [108]. Thus, he was trained by Heisenberg, Rutherford, and Dirac [108]. In 1929, he became professor at the California Institute of Technology. Oppenheimer proposes not to relate the electron and the proton and that each has its own Dirac sea. The negative energy states of both electrons and protons are filled by electrons and protons. They then have no reason to have the same mass, nor can an electron annihilate with a proton. Dirac admits that the solution of Oppenheimer solves the difficulties mentioned above, but he feels that it presents the disadvantage of giving up what he calls "a unitary theory of electrons and protons."

A year and a half later, Dirac adheres to the solution proposed by Oppenheimer, and in the September 1, 1931, issue of the *Proceedings of the Royal Society*, he takes one further step: he believes that the "hole" states should indeed exist, but that they are not protons:

It thus appears that we must abandon the identification of the holes with protons and must find some other interpretation for them [. . .] A hole, if there were one, would be a new kind of particle, unknown to experimental physics, having the same mass and opposite charge to an electron. We may call such a particle an anti-electron[. . .] The protons on the above view are quite unconnected with electrons. Presumably the protons will have their own negative-energy states, all of which normally are occupied, an unoccupied one appearing as an anti-proton [109].

Dirac thus predicts the existence of two particles: the antielectron and the antiproton. In the Solvay Council of 1933, Dirac identifies his antielectron to the positive electron discovered by Anderson:

Let us admit that in the universe as we know it the negative energy states are almost all occupied by electrons and that this distribution of electrons is not accessible to our observation because of its uniformity which extends over all space. Under these conditions, any non-occupied negative-energy state will be observed as a kind of gap. It is possible to admit that these gaps are positrons [110].

The electromagnetic equations apply to this "gap," that is, to the absence of an electron in a negative energy state, and this state has all the properties of an electron except that it has a positive electric charge:

Thus the gap assumes exactly the form of an ordinary particle with a positive electric charge and its identification with the positron is quite plausible. An ordinary positive energy electron cannot jump into one of the occupied negative energy states in virtue of the Pauli principle; it can, on the contrary, jump into a gap in order to fill it. Thus an electron and a positron can destroy each other. Their energy must reappear in the form of photons.

This is an example of the gradual evolution of quantum physics. Dirac attempts to understand and interpret a mathematical equation, and he comes up with the extraordinary idea that in our universe, the infinite set of negative energy states is occupied by electrons. They form a Dirac sea the presence of which we cannot detect. We can however imagine that one of the electrons in a negative energy state of the Dirac sea "jumps" into a positive energy state. In order for this to happen, the electron must be supplied with a sufficient energy, which must exceed twice its mass, or about a million electron volts. The electron then leaves a "hole" in the Dirac sea, and this hole behaves as the positive electron discovered by Anderson. Inversely, an electron can jump into a "hole" in the Dirac sea and thus occupy the corresponding empty negative energy state. This process is known as the *annihilation* of an electron with a positron which is its antiparticle. The process liberates an energy equal to at least twice the mass of the electron, and the energy appears in the form of photons or γ-rays. The equations thus describe several observed phenomena which go beyond our common experience. The mathematical formalism becomes an indispensable tool to express a physical reality which far exceeds the one perceived by our common-day senses.

Irène and Frédéric Joliot-Curie

Frédéric and Irène Joliot-Curie had made the crucial experiments which led Chadwick to show that the famous penetrating radiation, emitted by beryllium bombarded by α-particles, consisted of neutrons. Although they had not understood their results this way, they begin studying neutrons assiduously. They use a Wilson cloud chamber to study the "projection" of light nuclei by neutrons. They observe numerous tracks of electrons which can be identified by the curvature of their trajectory in a magnetic field.[1] On April 11, 1932, they note:

> Several trajectories, similar to electron trajectories, display a curvature pointing in the opposite direction: they are probably electrons which are emitted in a direction opposite to that of the incident beam and their energy is sometimes very high [111].

When Anderson's discovery is announced and confirmed by Blackett and Occhialini, Irène Currie and Frédéric Joliot reexamine their previous photographs. It then appears likely that the tracks are due to positrons which pass through the cloud chamber at the same time as the penetrating radiation emitted by beryllium bombarded by α-particles. This is soon confirmed by experiments performed by Chadwick, by Blackett and Occhialini [112] in the Cavendish, and by Lise Meitner in Berlin [113]. The Joliot-Curies then repeat their experiments and show that the positrons are produced by the γ-rays and not by the neutrons [114, 115]. This was almost simultaneously observed by Lise Meitner and Kurt Philipp in Berlin [116], as well as by Carl Anderson. The Joliot-Curies describe the phenomenon thus:

> We can picture the phenomenon as follows: a high energy γ photon impinging on a heavy nucleus is transformed into two electrons of opposite charge [117].

They go further. In addition to this production mode, which they call "materialization," they show that positrons can also be formed directly during a collision between an α-particle and an aluminum or beryllium nucleus and that this process produces more positrons than normal electrons [117]. Therefore, it is not an electron-positron pair which is formed but a real emission of positrons related to the collision between the α-particle and the nucleus, which suggests an explanation of the emission of a neutron:

> One knows that aluminum [. . .] subjected to α-rays [emits] transmutation protons. Occasionally, the transmutation might occur with an emission of a neutron and a positive electron instead of a proton.

They suggest to call these positive electrons "transmutation electrons." This interpretation, if it is confirmed, allows them to make a precise evaluation of the mass of the neutron, far more precise than that of Chadwick, which confirmed that, contrary to his first evaluation, the neutron had a *higher* mass than the proton.[2]

---

[1] The protons are 1,836 times heavier than electrons, and the curvature of their trajectory is therefore 1,836 times smaller.

[2] See p. 258.

The Birth of Particle Accelerators

A new instrument is born, ever increasing in size: the particle
accelerator, which will completely change nuclear physics.

Until 1932, all the discoveries concerning the atomic nucleus were made using α-particles emitted by radioactive substances. There was a simple reason for this: in order to interact, particles must come sufficiently close to the nucleus and they need to overcome the strong electric repulsion. They need sufficient energy to do this. Although this does not apply to neutrons which are electrically neutral, neutron sources were very weak because they were produced by collisions of α-particles, typically with beryllium. The energy of α-particles emitted by radioactive nuclei is quite high, about 8 MeV (8 million electron volts) and they travel with velocities equal to about 7% of the speed of light. There existed no other way to produce particles with such velocities nor a fortiori with higher ones.

The energies involved in the nucleus are of the order of millions of electron volts, or MeV, the units used by nuclear physicists. For example, an energy of about 8 MeV is required to extract a proton from a nucleus. Therefore, physicists needed an apparatus able to produce protons, α-particles, or neutrons with energies of several MeV, preferably tens of MeV. In 1919, Rutherford concluded his famous paper which described the first nuclear reaction with the words[1]:

> The results as a whole suggest that if α-particles—or similar projectiles—of still greater energy were available for experiment, we might expect to break down the nucleus structure of many of the lighter elements.

About 8 years later, on November 30, 1927, Rutherford declared, in the annual speech he delivered as president of the Royal Society:

> It would be of great scientific interest if it were possible in laboratory experiments to have a supply of electrons and atoms of matter in general, of which the individual energy of motion is greater even than that of the α-particle. This would open up an extraordinary interesting field of investigation which could not fail to give us information of great value, not only on the constitution and stability of atomic nuclei but in many other directions [118].

How could this be achieved? One knew at the time how to produce charged particles such as protons or helium nuclei, as well as many other nuclei. Indeed, Aston needed to do this in order to measure nuclear masses with his mass

---

[1] See p. 200.

spectrometer.[1] In order to endow the charged particles with a given velocity, it was sufficient to place the source of the charged particles at a sufficiently high voltage. But nobody at the time had produced voltages reaching millions of volts.

With both imagination and tenacity, physicists will eventually succeed in building very special instruments, unknown until 1932 and able to transmit high energies to particles such as protons, deuterons, and α-particles. These instruments are *particle accelerators*. This chapter is devoted to the conception and construction of the first of these instruments which will play a key role in the development of nuclear physics, mainly after the Second World War [6, 120, 121].

Direct Acceleration: A High-Voltage Race

The most straightforward solution of the problem was to produce high voltages, thousands or even millions of volts. But such high voltages raise new problems. How can one prevent such an apparatus from spontaneously discharging by emitting long and destructive sparks? A well-known example is the lightning emitted by clouds. Towards the end of the 1920s, it seemed inconceivable to produce voltages higher than a million volts and such voltages would accelerate particles to energies still well below those emitted by radioactive substances. Nor was it quite sure that such particles would do a better job at disintegrating nuclei.

The Use of Lightning

Three young physicists in Berlin, Arno Brasch, Fritz Lange [122], and Kurt Urban, thought of using lightning as a high-voltage source. During the summer of 1927, they began an experiment at the Monte Generoso (1704 m) in the Swiss Alps, near Lugano. A long cable joined the mountain and the deep valley [123]. Unfortunately, Kurt Urban died after being struck by lightning in the summer of 1928, and the experiment was stopped. Brasch and Lange then attempted to produce an artificial lightning in the laboratory. In 1924, a German engineer, Erwin Marx, working in the great Siemens firm, had invented a method to produce very high-voltage discharges [124], which were used to test isolating materials. His method consisted in charging condensers connected in parallel and to discharge them connected in series. This way, he was able to reach voltages up to 3 million volts (Brasch and Lange had observed 16-million-volt discharges produced by natural lightning). Upon the discharge of the condensers, a more or less long-lasting electric discharge (surge) was obtained. Brasch and Lange tried to construct such a "Marx generator,"

---

[1]See p. 168.

also called a "surge generator" (*Stoßspannungsgenerator* in German). But the duration of the discharge was too short to use this instrument in order to accelerate protons, and the idea was given up.

The Tesla Coil

The American physicist Merle Tuve had another idea. He used a transformer. Tuve, born in 1901, was the grandson of Norwegian immigrants. He studied in the University of Minnesota, and in 1926, he defended his PhD thesis in the John Hopkins University in Baltimore. He then obtained a job at the Carnegie Institution, in Washington, in the department of Terrestrial Magnetism, directed at the time by a physicist he had known in Minnesota, Gregory Breit.

Together with Gregory Breit, he attempts to produce high voltages using an apparatus somewhat similar to the old Rühmkorff coil. Their apparatus consists of a primary coil with just a few turns into which the current of a sudden discharge of a condenser is sent. This produces a high voltage in a secondary coil consisting of an isolated wire wound around a closed glass tube, which had been evacuated, a dozen centimeters in diameter and a meter long. This glass tube was one of the very first accelerating tubes. The secondary circuit is connected to two spheres with a 25 cm diameter each. The whole thing is immersed in oil which is a better insulator than air. Tuve succeeds this way in obtaining voltages of about 1.2 million volts. However, his apparatus proved difficult to stabilize, and the highest voltage could only be applied for 10^{-6} s (a millionth of a second). Worse, it was an alternating voltage. He managed to accelerate electrons with this machine but not even the lightest ions. The method was finally abandoned in favor of simpler and more efficient ones. Nonetheless, Tuve did succeed in mastering the technique of constructing a so-called accelerating tube, through which the accelerated particles pass. This technique is essential for the construction of high-voltage accelerators.

John Cockcroft and Ernest Walton: The First Nuclear Reaction Produced in an Accelerator

The physicists in the Cavendish also tackle the problem. In order to obtain high voltages, the young physicist John Cockcroft thinks of using a *voltage multiplier*, inspired by the electrical circuit which had been invented by a German engineer, Moritz Schenkel [125], in 1919 and, independently, by the Swiss physicist Heinrich Greinacher [126] in 1921. The circuit consists of condensers and valves set up in a way such that an alternating current charges the condensers to a voltage which is twice that of the alternating current. Starting with a modest alternating voltage, Cockcroft hopes to obtain *high direct current* voltages by setting up a cascade of such circuits.

John Cockcroft was born on May 27, 1897, in Todmorden, England. His family was in the cotton industry. He studied mathematics in the University of Manchester in 1914–1915, but his studies were interrupted by the First World War when he served as an artilleryman. After the war, he continued to study electrical engineering and he worked in the Metropolitan Vickers Electrical Company, a fact which had consequences later. He then decided to return to the university, and in 1924, he won the famous English competition called the "mathematical Tripos." He then obtained a position at the Cavendish which was directed by Rutherford. He first worked on the production of high magnetic fields at low temperature. He got interested in the production of high voltages in 1928, while working with a young Irish physicist, Ernest Walton who had just joined the Cavendish.

Walton was born on October 6, 1903, in Dungarvan, a small town in Southern Ireland. He obtained his Master of Science in 1927 in the University of Belfast. He then continued his studies at the Cavendish, thanks to a research grant obtained from the Royal Commission of the 1851 Exposition, which had previously made it possible for Rutherford to move to England. The research project of Cockcroft and Walton seemed most uncertain: to accelerate α-particles to energies comparable to those observed in radioactive decay, voltages reaching several million volts needed to be obtained.

In 1928, George Gamow visits the Cavendish. He presents his work on α-decay which he explains in terms of a quantum mechanics effect: the tunneling of the α-particle through a Coulomb potential barrier.[1] Cockcroft discusses with him the possibility that a particle might penetrate into a nucleus by passing through the Coulomb potential barrier. He asks: what is the probability that a 300 000 eV proton might penetrate into a nucleus, which is surrounded by a Coulomb potential barrier a million electron volts high? Gamow makes the calculation and finds that if a thousand protons are incident on a nucleus, a few should indeed tunnel through the barrier and penetrate the nucleus. In order to observe this, a sufficiently intense proton beam is required.

Frequently, in the forefront of research, risks need to be taken. Cockcroft sends a *memorandum* to Rutherford suggesting the construction of a three hundred thousand volt generator. Rutherford accepts, and Cockcroft sets to work together with Walton. During the summer of 1930, after overcoming endless difficulties, they finally succeed in constructing a working 300 000 V generator. They immediately use it to send 280-keV (280 000 eV) protons onto a lithium target. They expect to observe γ-rays emitted by the disintegration of the target nuclei, but they don't. No γ-rays are observed. They conclude that the proton energy is too low, and they are determined to raise it.

The laboratory then moves into a larger building. At the end of 1931, they succeed in creating 800 000 V and they repeat the experiment. Fearing a repetition of their previous failure, they decide to detect emitted α-particles using the simple and well-tested method of using a microscope to observe scintillations caused by α-particles passing through a zinc sulfide screen. This time, the experiment works and they send a letter to *Nature,* dated April 16:

---

[1] See p. 242.

On applying an accelerating potential of the order of 125 kilovolts, a number of bright scintillations were at once observed, the number increasing rapidly with the voltage up to the highest voltages used, namely 400 kilovolts. At this point many hundreds of scintillations per minute were observed using a proton current of a few micro-amperes [. . .] The brightness of the scintillations and the density of the tracks observed in the expansion chamber suggest that the particles are normal α-particles. If this point of view turns out to be correct, it seems not unlikely that the lithium isotope of mass 7 occasionally captures a proton and the resulting nucleus of mass 8 breaks into two α-particles, each of mass four [127].

Gamow's estimate actually was not so bad when he predicted that some nuclear reactions would occur at about 300 kV (300 000 V). Indeed, Cockcroft and Walton observed some signals between 125 and 400 kV without even using the 800 kV which their apparatus allowed them to reach, in principle.

This was an important discovery which Cockcroft and Walton carefully checked in further experiments. They observed the tracks in a Wilson cloud chamber and recognized the tracks of α-particles. They then attempt to check whether the two α-particles are emitted at the same time. To do this, they place on both sides of the lithium target sheet a zinc sulfide screen. Each physicist watches one of the screens with a microscope and pushes a button each time he sees a scintillation marking the arrival of an α-particle. And indeed, most of the time, they report seeing a scintillation at the same time. This was a first confirmation that lithium first absorbs the proton and then decays into two α-particles. Finally, they repeat the experiment in an ionization chamber with an electronic amplifier, and they confirm that the observed particles are α-particles. Their article is a historic landmark. Cockcroft and Walton had observed the first nuclear reaction caused by artificially accelerated particles.

Robert Van de Graaff

Shortly before, in the United States, Robert Van de Graaff proposed another method of attaining a really constant high voltage. Van de Graaff was born in 1901 in Tuscaloosa, Alabama. After studying engineering in the University of Alabama, Van de Graaff goes to Europe. During the academic year 1924–1925 he follows the lectures of Marie Curie in the *Sorbonne* and becomes interested in nuclear physics. He spends some time in Oxford where he realizes how important the experiments of Rutherford are and how useful it would be to produce high-energy particles. When he returns to the United States, he works in the Palmer Physics Laboratory of the University of Princeton, and in the fall of 1929, he constructs the first prototype electrostatic accelerator which reaches 80 000 V. He then perfects his apparatus, and in 1931, he presents it at the inaugural dinner of the American Institute of Physics. His accelerator then reaches a voltage of one million volts.

Van de Graaff makes use of a very simple idea inspired by the "electric machines" of Ramsden, Holtz, Carré, or Wimshurst. These electric machines also inspired Richard Vollrath, a physicist in Los Angeles. His machine consisted in charging up a metallic body by blowing onto it air containing small charged silica particles, in order to obtain high voltages, but it never worked properly [128].

Van de Graaff uses an isolating belt which is electrified by passing close to an electric conductor raised to a low voltage. The belt connects two pulleys, one of which is placed *inside a hollow, conducting, but isolated sphere*. There, sharp needles transfer the electric charge of the belt to the sphere. By transporting in this manner electric charges on an isolating belt, one creates an "inverted" electric current: the positive charges are forced to flow *in the opposite direction*, and they progressively increase the positive charge of the positive sphere, thereby increasing its voltage. One cannot increase the voltage of the sphere indefinitely because it eventually discharges by ionizing the surrounding air and by emitting sparks which can be devastating. That is why the voltage of the sphere is voluntarily limited by conducting needles which collect the excess charge when it reaches a certain threshold.

Van de Graaff then accepts a research position at the MIT offered to him by the new director Karl Compton, the brother of Arthur Compton who had discovered the Compton effect. This is where, together with Compton and Lester Van Atta, he undertakes the construction of an impressive machine placed in an airplane shed in Round Hill, near South Darmouth (Massachusetts) [129]. Two towers, built with an isolating material, are placed on carts which move on rails. Each one supports an aluminum sphere 4.60 m in diameter. The construction is 13 m high. There are belts which transport electric charges into each aluminum sphere, positive charges in one and negative charges in the other. The spheres are connected by a horizontal tube which allows particles to acquire high speeds when they pass from one sphere to the other. Unfortunately, the tube caused many difficulties. Furthermore, the humid atmosphere caused by the neighboring ocean made it difficult to maintain high voltages. As a result, the constructed machine never served as a particle accelerator. It is described in a paper written in 1936 which reports several methods of attaining high voltages but which does not mention accelerated particles [130]. The machine was transferred to MIT in 1937 and installed in an adequate place after undergoing substantial modifications. The spheres were placed side by side. One was used to collect the electric charge brought by the belt, and the other contained the source of ions and an extremity of the accelerating tube. The machine began to function in 1940. It accelerated particles to 2.75 million volts (MeV) [131].

Meanwhile, the group working with Merle Tuve gave up using a Tesla coil. Together with Van de Graaff, they began the construction of an electrostatic accelerator. The first accelerator had a sphere 1 m in diameter mounted on an isolating tripod with a vertical accelerating tube [132]. The first physics experiments began in 1933 with 600 000 V protons [133]. In the spring of 1934, the machine was succeeded by a more powerful one, with a sphere 2 m in diameter and reaching 1.3 million volts.

This is how Van de Graaff, who was the first one to attain a voltage of one million volts, ended up by being overtaken not only by Cockcroft and Walton, but also by the Merle Tuve team. Van de Graaff was more a constructor of accelerators than a nuclear physicist. He made the mistake of seeking too much too fast.

Acceleration in Steps

Gustaf Ising

Because it was difficult to produce voltages reaching several hundred thousand volts, another method was proposed in 1924 by the Swedish physicist Gustav Ising, who was working in the *Tekniske Högskola* in Stockholm [134]. The idea was to transmit the required energy to accelerated particles not in one go but progressively in small steps. Ising thought of sending the particles through conducting tubes and setting a modest voltage between successive tubes. While the particles propagate inside a conducting tube, they are shielded from all electromagnetic fields. But they are given a boost in energy each time they pass from one tube to the next.

Ising simply suggested the idea, but he did not construct the apparatus.

Rolf Wideröe

A Norwegian student was the first man who attempted to construct it in 1928. Rolf Wideröe, born on November 7, 1902, studied in the Technical University of Karlsruhe in Germany. He then began to work on his PhD thesis in Aachen. He attempted to construct an accelerator which he called a *radiation transformer* and which accelerated electrons. Today, it is called a *betatron*, and it is mainly used for radiotherapy. The betatron of Wideröe can be compared to a voltage transformer the outer coil of which is replaced by a glass tube in the form of torus. A magnetic field makes the electrons circulate in this tube, and it accelerates them as it varies. At almost the same time, the same method was attempted by Ernest Walton, who followed a suggestion of Rutherford, but without success [135]. Wideröe also fails to make his apparatus work because he does not have the required high-frequency alternator. He then returns to the idea of Ising. He uses a high-frequency electromagnetic field, the period of which is equal to the time taken for the particle to pass through the tube. The particle is thus accelerated each time it passes from one tube to the next. Since the particle travels faster and faster, the successive tubes have to be longer and longer. Wideröe constructs a machine which has a classical ion source similar to the one used by Aston for mass measurements. In his experiments, Wideröe sends potassium and sodium ions through the tubes because they are heavier and therefore slower than protons or α-particles. The voltage between the first and second tube is 25 000 V, and the same voltage is set between the second and third tube. Thus, with an alternating voltage of 25 000 V, he is able to transmit 50 000 V to the ions. Wideröe publishes his thesis [136] in 1928. The apparatus constructed by Wideröe is the ancestor of what today we call *linear* accelerators.

An Idea of Ernest O. Lawrence

Far away, on the West Coast of the United States, a young, dynamic, and ambitious physicist in the University of Berkeley is about to attempt to improve the apparatus of Wideröe, so as to make it a useful accelerator for physics experiments. Ernest Orlando Lawrence was born on August 8, 1901, in Canton, South Dakota. His mother and father were both children of Norwegian immigrants and teachers.[1] Ernest Lawrence studied first in Canton and later in the universities of South Dakota, Chicago, and Yale, where he obtained his PhD in 1925. He became known for his work on ionization potentials, and in 1927, he was considered as one of the most brilliant experimentalists of his generation. The universities of Yale and Berkeley both offer him an associate professor position, without requiring him to become an instructor first, which is the usual procedure. In 1928, Lawrence chooses Berkeley which offers him a good research budget and light teaching duties. Lawrence is tall, dynamic, bursting with energy, and plays tennis.

One day, while browsing through the journal *Archiv für Elektrotechnik* in the lab library, he finds the thesis of Wideröe. Although he does not read German, Lawrence grabs the essentials by looking at the mathematical formulas and the figures. In his thesis, Wideröe describes first his unsuccessful attempt to construct a "radiation transformer" using circular orbits of electrons subject to a magnetic field and next his successful "linear" accelerator. Then Lawrence has a brilliant idea: why not use repeatedly the same space to accelerate the particles? If the conducting tubes of Wideröe were circular instead of being linear, the particles could circulate indefinitely from one tube to the next while being accelerated each time they pass from one to the other. This requires a strong magnetic field. Wideröe insisted on this difficulty and on the impossibility of obtaining stable orbits, the reason why he gave up the idea. Because he did not understand German, Lawrence did not read the reservations of Wideröe. Anyway, the task was difficult: even if the particles can be made to follow circular trajectories, how can one ensure that they will be accelerated after completing their trajectory? Consider a particle which propagates in a container which has the form of a camembert cheese box, inside of which a vacuum is maintained in order to prevent the particle from being deviated by collisions with air molecules. The container is placed between the opposite poles of a magnet so that the particle is exposed to a magnetic field, and it describes a circular orbit. Now, imagine that the container is split into two halves, which are electrically isolated one from the other,[i] and that each half is connected to an

---

[1] Wideröe was Norwegian, Tuve and Lawrence had Norwegian parents or grandparents, and Ising was Swedish. Scandinavia played a key role in the construction of the first accelerators!

alternating voltage. Each time the particle passes from one half of the container to the other, it will undergo an acceleration provided that the electric field points in the right direction precisely at the moment when it passes from one half to the other. After describing a circular orbit in one half, it enters the other half. If the alternating current is correctly adjusted, the particle can again be accelerated at the precise moment when it passes from that half to the other. The alternating voltage must change sign just in time for the particle to be accelerated each time it passes from one half to the other. Lawrence makes a simple calculation and realizes a crucial fact: *the time taken for a particle to describe a semicircle does not depend on its velocity.* If its velocity increases, the radius of its trajectory also increases so that it takes the same time to complete a semicircle. Thus, the frequency of the alternating current can be adjusted so as to accelerate the particle each time it passes from one half of the camembert to the other. The particle progressively increases its velocity and energy while increasing the radius of its circular trajectory. It does so until the radius becomes so large that the particle hits the wall of the container. The two semicircular containers had the shape of a "D" and became known as "dees." Physicists called this machine a *whirling device*. In a paper published a few years later in 1935, Lawrence still calls this machine "an apparatus of the type developed by Lawrence and Livingston" or "the apparatus of Sloan and Lawrence." He adds the following footnote:

> Since we shall have many occasions in the future to refer to this apparatus, we feel that it should have a name. The term "magnetic resonance accelerator" is suggested. In this, the last two words imply the essential principle of operation, while the first word is added to distinguish it from the apparatus of Sloan and Lawrence (Phys. Rev. 38, 2021 (1931), which can be called a "linear resonance accelerator". The word "cyclotron", of obvious derivation, has come the be used as a sort of laboratory slang for the magnetic device [137].

It is the name "cyclotron" which was adopted throughout the world.

The idea of Lawrence is interesting, but it needs to be materialized. Lawrence convinces Niels Edlefsen, a student who had just completed his PhD and was about to leave Berkeley, to try to construct such a machine. Edlefsen sets up a glass vacuum container which he places between the two poles of a magnet, 10 cm in diameter. However, he does not obtain convincing results, but Lawrence finds them sufficiently interesting to be presented [138] at a meeting of the National Academy of Sciences held in Berkeley on September 19, 1930.

David Sloan: A Linear Accelerator for Heavy Ions

Lawrence had not forgotten Wideröe's linear accelerator, and he asks another student, David Sloan, to try to construct one. Sloan succeeds in 1931: the accelerator is composed of 30 successive tubes, and it transmits an energy of 1.25 million volts to mercury ions, using an alternating voltage of only 42 000 V [139]. The accelerator is 1.14 m long. This technique however is not well suited to accelerate protons or α-particles which travel much faster than mercury ions of similar energy. A 16-m accelerator would be required for this purpose, and this seems quite prohibitive at the time. Sloan continues to improve his accelerator, and in 1932, he produces 2.85-million-volt ions. Working with the young physicist Wesley Coates, he added 6 tubes, reduced their diameter, and increased the frequency of his alternator. His accelerator became 1.85 m long. However, such energies were still much too low to produce nuclear reactions with the accelerated mercury ions [140, 141].

Stanley Livingston: The Cyclotron

A third student of Lawrence, Stanley Livingston, seeking a subject for his PhD thesis, is assigned the task to make the cyclotron work. He sets up an apparatus similar to the one of Edlefsen, improves it, and succeeds in making it work. Indeed, he places a proton detector on the trajectory which has the largest possible diameter (10 cm), and he observes a clear increase in the number of detected particles when the magnetic field corresponds exactly to the frequency of the alternating voltage, that is, when the magnetic field acquires an intensity such that the particles complete a semicircle just in time for the alternating voltage to reverse its sign. This is what we call today a *cyclotron resonance*. Livingston defends his PhD thesis on April 14, 1931. He is able to accelerate protons to 80 000 eV using a 1800-V alternator. The protons undergo 80 successive accelerations.

His results are disclosed in a short publication[1] sent to the *Physical Review* on July 20. He insists on the superiority of such an accelerator:

These experiments make it evident that with quite ordinary laboratory facilities proton beams having great enough energies for nuclear studies can be readily produced with intensities far exceeding the intensities of beams of alpha-particles from radioactive sources [142].

and he announces:

Possibly the most interesting consequence of these experiments is that it appears now that the production of 10 000 000-volt protons can be readily accomplished when a suitable larger magnet and high frequency oscillator are available. The importance of the production of protons of such speeds can hardly be overestimated and it is our hope that the necessary equipment for doing this will be made available to us.

---

[1]It was a *letter to the editor*, signed by E. Lawrence and S. Livingston. Urgent communications could be sent to the journal provided they were short. They were published fast, within about 1 month, whereas 6 months were often required for the publication of a full-fledged article.

Similar fast publications were accepted by *Nature* in England, by *Physikalische Zeitschrift* and *Naturwissenschaften* in Germany, and by the *Comptes Rendus de l'Académie des Sciences* in France.

The high-frequency oscillator is indeed one of the major obstacles. The required frequency is of the order of 10 MHz (ten million oscillations, that is, double reversals of the electric current, per second). That was reaching the limit of available alternators at the time. The final energy of the particles is directly linked to the number of turns they make in the accelerator, and their spiral-shaped trajectory has to fit entirely within the space between the poles of the magnet. Lawrence obtains funds to construct a magnet with poles 25 cm in diameter. He immediately embarks on its construction with Livingston during the summer of 1931. The new accelerator becomes functional in the spring of 1932: it accelerates protons to an energy equal to 1.2 million electron volts (1.2 MeV). *It was the only machine in the world able to do that at the time.* It was the first cyclotron which could be used for nuclear physics experiments. It was described in a detailed paper which became a classic and which marked the birth of a great family of cyclotrons [143].

In May 1932, a sensational event became known: Cockcroft and Walton had succeeded in observing the first nuclear reaction induced by accelerated protons. In Berkeley, the cyclotron of Lawrence was not only working, but it could accelerate protons to an energy of 1.2 MeV which was considerably higher than that produced by the accelerator available to Cockcroft. However, Lawrence was obsessed by the construction of more and more powerful cyclotrons, and he somewhat neglected physics experiments. For example, he lacked proper detectors. The Berkeley team then quickly set to work, and they also observed the disintegration of lithium by protons [144]. A large number of other nuclear reactions were subsequently observed [137, 145, 146].

Acting as a industrial manager with a gift for public relations, Lawrence, a forerunner of directors of the large postwar laboratories, has great ambitions. He is able to collect funds, not an easy task during the economic crisis, especially in the United States. Now that he is able to construct a cyclotron, he can increase the energy of the accelerated particles, apparently only limited by the diameter of the poles of the magnet. Indeed, the radii of the trajectories of particles between the poles of the magnet increase with their energy. The radii can in principle be reduced by increasing the magnetic field, but that was not an easy task because it required more intense electric currents which heated the coils of the magnet. Furthermore, a stronger magnetic field would increase the velocity of the particles, for a given radius of their trajectory, and this in turn would require a higher frequency of the alternator, which was another major technological difficulty. The only available way to increase the energy of the accelerated particles appeared to be the construction of huge magnets, but they could not produce magnetic fields exceeding about 1.5 Tesla. Even before his 25-cm cyclotron began functioning, Lawrence undertook the construction of a cyclotron with magnetic poles 68 cm in diameter: it will become his "27 $\frac{1}{2}$ inch" accelerator which will accelerate protons to an energy of 3 MeV and deuterons to 5 MeV. The magnet, which was huge for that time, was built thanks to a gift of the Federal Telegraph Company. This new accelerator begins to function in the summer of 1932. It is described in detail in a paper sent to the *Physical Review* [147] on March 12, 1934:

Meanwhile we have constructed a larger model which has proved to be capable of accelerating hydrogen ions to voltages as high as five million [electron-volts]. It has been used almost continuously during the past six months in certain preliminary investigations of nuclear phenomena in the range up to three million volts.

Lawrence and his team chiseled the magnetic field by adding thin wedges to the poles of the magnet. They called them *shims*, a term used by technicians to designate a small thickness of iron. When it is placed between the poles of the magnet, the wedge decreases the distance separating the poles and increases this way the magnetic field; this allows a local fine-tuning of the field. The technique was called *shimming*.

In his paper, Lawrence announces an even bigger accelerator[1]:

However, to go to higher voltages it seems more desirable to build a larger apparatus, in which the full available diameter of the magnet pole faces, i.e. 45 inches, is used. Such a larger apparatus, the construction of which we are now commencing, should produce hydrogen molecule ions and deuterons with energies of about ten million volts or more.

The construction is however delayed. Until 1936, the Berkeley teams work on improving the existing "$27\frac{1}{2}$ inch" accelerator. In the paper which he publishes in 1936, together with a newcomer, Donald Cooksey [148], Lawrence describes the improvements which are notable. The major improvement concerns the "extraction" of the beam. Until then, the particles accelerated in the cyclotron remained in the camembert-cheese-shape chamber and ended up hitting the wall of the chamber. To perform physical experiments, one had to place the substance to be bombarded inside the chamber. The new development made in 1936 consisted in allowing the particles, which had attained their maximum velocity, to escape from the chamber and to impinge on a the target placed at a certain distance from the chamber. The accelerated particles could travel to the target either in air or in a vacuum tube. Several other modifications improved considerably the performance of the accelerator.

But the project for a large cyclotron was not given up. In fact, an even bigger one will be built. In 1939, the "60-inch" cyclotron, 152 cm in diameter, is completed in Berkeley [149]. It accelerates protons to an energy of 10 MeV and α-particles to 40 MeV. The paper which describes it [149] is signed by eight people. The team is enlarged, as all teams will be thereafter.

The cyclotron was an immediate success and became famous in the whole world. The first to be launched outside Berkeley was the one in Cornell, under

---

[1] We did not wish to modify the text of Lawrence who mentions millions of volts instead of electron volts. It is an abuse which physicists often indulge in (see the word in the Glossary).

the supervision of Livingston; the following year saw the birth of eight new accelerators in Bartol, Columbia, Illinois, Michigan, Princeton, Purdue, Rochester, and Washington [150].

The race to reach higher energies, and to construct gigantic accelerators, has begun [6, 120, 151]. Barely 7 years earlier, in his first publication [142], Lawrence spoke of "ordinary laboratory facilities." In the span of 10 years. a major change has taken place. Before, a laboratory was equipped with a variety of instruments enabling physicists to make all sorts of experiments, namely. radioactive sources, Geiger–Müller detectors, amplified ionization chambers, electrometers. and so on. Now. the laboratory of Lawrence is centered on a single instrument, the accelerator, which is a huge machine which requires a team to run it as well as a much larger budget.

"Charge Independence" of the Nuclear Force

*The accelerators reveal an important fact: the interaction be-
tween two protons or two neutrons is the same as the interaction
between a neutron and a proton: the nuclear interaction is
charge independent.*

In his first paper on the structure of nuclei,[1] Heisenberg had assumed that
an attractive force existed between a neutron and a proton, but not between two
neutrons nor between two protons, with the exception of the weaker Coulomb
repulsion between the positively charged protons. Heisenberg argued that his kind
of force would have the tendency to form nuclei with equal numbers of protons
and neutrons. In 1935, two American physicists, Eugene Feenberg and Julian
Knipp, study the binding energies$^\diamond$ of deuterium (composed of one proton and
one neutron), tritium (composed of one proton and two neutrons), and helium
(composed of two neutrons and two protons). They calculate the wavefunctions
of these nuclei assuming that *the interaction between protons and neutrons is the
same as the interaction between two protons or two neutrons* [152]. Their results
actually agreed with those obtained by Wigner.[2] Does that imply that the interaction
between protons and neutrons is the same as the interaction between two protons
or two neutrons? The question was settled by a series of experiments in which
the scattering of protons by protons was observed; more precisely, the number of
protons scattered at different angles after colliding was measured. This experiment
could only be done with sufficiently energetic protons such as those produced by
the first accelerators.

The idea is quite simple. If the only interaction acting between protons is
the Coulomb repulsion, one can calculate exactly the scattering process. In the
department of Terrestrial Magnetism of the Carnegie Laboratory in Washington,
William Wells attempts to measure the scattering by taking photographs of the
trajectories of protons ejected by α-particles in a Wilson cloud chamber. He
observes the ejected protons which collide with other protons [153]. However, the
protons are ejected with various velocities and the photographs of proton–proton
collisions are rare, so that he cannot reach a conclusion. A physicist in Berkeley,
Milton White, uses protons accelerated in a small cyclotron. His data is scarce, and
he also fails to conclude [154].

In the Carnegie Laboratory in Washington, an important attempt had been
made to construct an accelerator using a Tesla coil.[3] The physicists involved were

---

[1] See p. 264.
[2] See p. 270.
[3] See p. 291.

Merle Tuve, Norman Heydenburg, Lawrence Hafstad, and Gregory Breit. After recognizing their failure, they succeed in constructing and running an electrostatic accelerator similar to the one of Van de Graaff, which they use to study the scattering of protons by protons. The result is published in two papers in the November 1, 1936, issue of the *Physical Review*. The first describes the experiment, and it is signed by Tuve, Heydenburg, and Hafstad [155]. They accelerated protons with a well-defined energy and injected them into a container filled with hydrogen. They used an electronically amplified ionization chamber to measure the number of protons scattered at various angles. The data thus obtained were analyzed in a theoretical paper signed by Gregory Breit, Edward Condon, and Richard Present [156], who compared the interaction between the protons to the interaction between a neutron and a proton. The latter had been determined by Fermi and Amaldi who had measured the mean free path of neutrons in paraffin, that is, the average distance travelled between two collisions, either with carbon nuclei or, more frequently, with hydrogen nuclei which are protons [157].

Breit, Condon, and Present reach an important conclusion: *the interaction between two protons or two neutrons is the same as the interaction between a neutron and a proton*. The Coulomb repulsion, which acts only between protons, is much weaker, but it acts at larger distances beyond the range of the nuclear force. This property is called the *charge independence* of the nuclear force. It is a fundamental property of the nuclear force as we know it today. It could not have been observed without an accelerator.

The Discovery of Artificial Radioactivity

By bombarding aluminum with α-particles, Frédéric and Irène Joliot-Curie produce the first artificially radioactive element. It soon appears that hundreds of radioactive isotopes of known elements exist.

The seventh physics Solvay Council [158] is held in Brussels on October 22–29, 1933. Since 1911, the Council was presided by Hendrik Anton Lorentz, who was universally respected but who died on February 4, 1928. Paul Langevin was designated as the new president.

Paul Langevin was born in 1872 into a modest family [159]. He was a brilliant pupil, and he topped the list of students admitted to the *École Municipale de Physique et de Chimie Industrielle*, where Pierre Curie[1] was teaching. He is admitted to the *École Normale Supérieure*, where he gets to know Jean Perrin, and he passes the *agrégation*[2] in 1897. He then obtains a scholarship from the city of Paris which allows him to spend a year at the Cavendish, the famous laboratory of J. J. Thomson. There, he meets Rutherford (who is 1 year older), Townsend, and Charles Wilson. He returns to Paris, obtains his PhD in Physics in 1902, and embarks on a brilliant career, working first on the ionization of gases and on magnetism. He is elected member of the *Collège de France* in 1910. During the First World War in 1914–1918, he studies the detection of submarines and he invents the SONAR method. After the war, Langevin becomes increasingly militant for peace and opposed to the rise of fascism. On October 30, 1940, when the Germans begin the occupation of Paris, he is arrested, put into jail, dismissed from his position in the *Collège de France*, and then assigned to a forced residence in Troyes. Thanks to the Resistance, of which Frédéric Joliot became one of the leaders, Langevin is able to escape and to reach Switzerland in 1943. He dies shortly after the war, on December 19, 1946.

The Solvay Council discusses *the structure and the properties of atomic nuclei*, a hot subject at the time. Since the Rome Congress, held in 1931, the discovery of the neutron had changed the way nuclei were conceived. Nuclei were thought of consisting of protons and neutrons, without electrons. The first theories of nuclear structure appeared in the works of Heisenberg, Majorana, and Wigner.[3] The number of participants to the Council had increased. In addition to Niels Bohr and Marie Curie, a new generation arose: Enrico Fermi, Irène and Frédéric Joliot-Curie, Paul

---

[1]See p. 10.

[2]The highest competitive examination for teachers in France.

[3]See p. 263.

Dirac, Charles Ellis, George Gamow, Wolfgang Pauli, Francis Perrin, Salomon Rosenblum, Werner Heisenberg, Erwin Schrödinger, Ernest Lawrence, and John Cockcroft.

Among the six reports on the current understanding and ongoing research, the one presented by Frédéric Joliot and Irène Curie [160] was met with skepticism, especially when they presented their interpretation of results obtained by bombarding aluminum with α-particles: they observed a penetrating neutral radiation which they believed to be caused by neutrons [161].

This interpretation faced an important difficulty. It had been known for a long time that α-particles incident on aluminum produced protons. This was understood as the following process: the α-particle collides with the aluminum nucleus thereby producing a proton which is emitted and a residual nucleus which must be the isotope 30 of silicon in order to account for the total number of protons and neutrons.[1] This isotope of silicon was known to exist. However, if this reaction emitted a neutron, the resulting nucleus would be the isotope 30 of phosphorus and this isotope had never been observed. It did not exist in nature! That was indeed embarrassing, and it shed doubt on the claim that neutrons were emitted.

Another observation suggested an interpretation which they submitted to the Solvay Council. After the discovery of the positron by Carl Anderson, the Joliot-Curies realized that several tracks observed in their Wilson chamber were due to positrons. And they noticed that positrons were emitted when aluminum was bombarded by α-particles. They gave the following interpretation of the emission of neutrons:

> One knows that aluminum or the isotope 10 of boron emit trans-mutation protons under the action of α-rays. Occasionally the transmutation might occur with the emission of a neutron and a positive electron, instead of a proton [162].

They suggest therefore that the neutron and the positron are emitted together and at the same time. Together, they have a positive charge and a mass similar to that of a proton.

Lise Meitner then intervenes to confirm that positrons are produced, but she adds:

> It is interesting to compare the number of emitted positive electrons and the number of recoiling H rays[2] produced at the same time as the neutrons. The comparison of results obtained for aluminum and fluorine proves that for Al [aluminum], in spite of the fact that the number of positive electrons is four times greater than for F [fluorine], no neutron could be detected. This is not in favor of the idea that, in this case, the emission of the neutron takes place at the same time as that of the positive electron [163].

---

[1]Respectively, 13 and 14 for aluminum, 2 and 2 for the α -particle, and 14 and 16 for the isotope 30 of silicon.

[2]Ionized hydrogen atoms, namely, protons, were still occasionally called H-rays at the time.

This increases the skepticism. The beautiful hypothesis of the Joliot-Curies appears to be on shaky ground. If Lise Meitner is right in claiming that no neutrons are emitted, all has to be reconsidered. But when she returns to Berlin, Lise Meitner realizes that her data do not allow her to be so affirmative. In December, she sends a note which is added to the proceedings:

> A careful examination of our photographs[1] obtained with Al [aluminum] and Fe [iron] incited me to ask if our statistics were not too scant to make the above conclusion. That is why I used a more intense polonium source to make a series of photographs on Al and Fe and I found that among 230 photographs with Al there were 11 recoil H rays, and 4 out of 200 for F. The objection I raised against the views of Mr and Mrs Joliot, who claimed that positive electrons are emitted by the aluminum nucleus, therefore no longer holds.

But nobody took notice of her added remark.

The Joliot-Curies After the Solvay Council

Irène and Frédéric Joliot-Curie were quite shaken when they returned from the Solvay Council. In an article published later in 1951, they wrote:

> Finally the great majority of the physicists present did not believe that our experiments were accurate. After the session we were rather distressed, but just at that moment, professor Niels Bohr told us privately that he considered our results very important. Shortly after, Pauli gave us a similar encouragement [164].

Back in Paris, they set to work. How can one check that a neutron and a positron are emitted at the same time? A direct measurement of the coincidence of two particles was beyond the reach of technology, even using a Bruno Rossi circuit. They find another way out. The nuclear reaction takes place only if the α-particles have a sufficient energy, above a given threshold. They decide to measure the threshold, that is, the minimum velocity of the α-particles, for the emission of both positrons and neutrons. If the threshold turns out to be the same, it will at least prove that the two processes are related. In December 1933, they repeat the measurement of the threshold for neutrons. The radioactive source is placed in a container of gas, the pressure of which can be regulated. At the other end of the container, they place an aluminum foil. If a vacuum is maintained in the container, the α-particles hit the aluminum nuclei with their full energy. But if the container contains a carbonic gas the pressure of which is gradually increased, the α-particles are slowed down and they therefore impinge on the aluminum foil with a slower velocity. Above a certain pressure, the nuclear reactions no longer take place because the α-particles

---

[1] These are Wilson cloud chamber photographs.

are too slow. The neutrons are detected because they project forwards the hydrogen nuclei, that is, protons. The projected protons are in turn detected in an electronically amplified ionization chamber, which became available to them at the *Institut du Radium*.

"A New Kind of Radioactivity"

Early in 1934, the Joliot-Curies begin to measure the threshold for the emission of positrons. In the afternoon of January 11, Joliot is working in his laboratory in the *Institut du Radium* [165]. He begins with low-velocity α-particles, and then he progressively decreases the pressure of the gas container in order to increase their velocity. The Geiger–Müller counter, set up to detect the positrons, begins then to crackle: indeed, the α-particles had just exceeded the threshold energy. In order to determine the threshold more precisely, he then increases the pressure of the gas container in order to slow down the α-particles. To his great surprise, *the counter continues to crackle*! Joliot immediately realizes that this is a new phenomenon and sets up a simple apparatus. He simply makes radioactive source adhere for some time to the aluminum foil and then takes it away. The foil continues to emit positrons at a decreasing rate. He recognizes the well-known phenomenon observed with heavy radioactive substances: the α-particle impinging in the aluminum nucleus has produced a short-lived radioactive substance.

He has to make sure that the Geiger–Müller, built by the visiting young German physicist Wolfgang Gentner, is working properly. He is invited to dinner that evening with Irène, and he asks Gentner to repeat the experiment in order to check the counter. The next day, Friday January 12, Frédéric and Irène find a note from Gentner saying that the counter works perfectly well. They are aware of having made an important discovery. Working day and night, Frédéric and Irène repeat the experiment with aluminum and several other elements. They induce radioactivity in boron and magnesium. However, no effect is observed on most of the exposed elements, which range from hydrogen to silver.

On Monday, January 15, they present a note to the *Académie des Sciences*:

We have discovered the following phenomenon: the emission of positive electrons by some light elements, irradiated by the α-rays of polonium, persists for a more or less long time, which can exceed one half hour in the case of boron, after having removed the source of α-rays [166].

In fact, they have discovered two new phenomena:

- An artificially induced radioactivity, which will soon be called *artificial radioactivity*
- Radioactivity with the emission of positive electrons or positrons which will soon be called β^+ *radioactivity*, whereas the well-known radioactivity with the emission of electrons will be called β^- *radioactivity*

They specify the radioactive half-lives: 3 min and 15 s in the case of aluminum, 14 min in the case of boron, and 2 min and 30 s in the case of magnesium.

They present what seems to be an obvious interpretation:

We think that the emission process should be the following for aluminum:

$$^{27}_{13}\text{Al} + ^4_2\text{He} = ^{30}_{15}\text{P} + ^1_0 n.$$

The isotope $^{30}_{15}P$ of phosphorus would be radioactive with a half-life of $3^m 15^s$ and it would emit positrons following the reaction

$$^{30}_{15}\text{P} = ^{30}_{14}\text{Si} + \varepsilon^+.$$

Their notation of nuclei is still used today: the upper number on the left of the atomic symbol (Al, He, P, ...) is the total number of protons and neutrons, also called the *mass number*, the lower number on the left is the number of protons in the nucleus. Thus, aluminum 27 (which has 13 protons and 14 neutrons) is denoted $^{27}_{13}\text{Al}$, helium 4 (which has two protons and two neutrons) is denoted ^4_2He, phosphorus 30 (which has 15 protons and 15 neutrons) is denoted $^{30}_{15}\text{P}$, and the neutron (consisting obviously of 0 protons and 1 neutron) is denoted ^1_0n.

The initial intuition of the Joliot-Curies was in fact quite close to the truth. When an α-particle collides with a nucleus of aluminum, a neutron is emitted and the remaining nucleus is phosphorus. *But it is not any phosphorus*: it contains 15 neutrons, and so *it is not a known nucleus*. The measurements of Aston attributed only one isotope to phosphorus, namely, phosphorus-31. Until then, everyone had implicitly assumed that nuclear reactions could only produce nuclei which exist in nature, which explains the embarrassment, skepticism, and incredulity caused by the results of the Joliot-Curies. They had produced the isotope 30 of phosphorus which does not exist in nature because its half-life is 3 min and 15 s, so that *any such nucleus which might have been formed has disappeared a long time ago*. After a half-life, only half of the nuclei survive, about one out of 3 million survives for an hour, and none survive more than 1 day. The phosphorus 30 nucleus decays by emitting a positron, a process in which a proton is transformed into a neutron so that the nucleus of phosphorus-30 (containing 15 protons and 15 neutrons) has become a silicon nucleus (containing 14 protons and 16 neutrons).

The Chemical Proof

The effect is spectacular and the interpretation convincing. The Joliot-Curies want to go further to obtain a direct proof that the reaction proceeds as they thought. To do this, they want to show that the radioactive substance which emits the positrons is indeed an isotope of phosphorus, that it is *chemically* phosphorus. In the nuclear reactions previously observed, one cannot identify chemically the emitted substances because they are produced in minute quantities. But if a radioactive substance is produced, one can use nuclear chemistry which had been invented 36 years earlier by Pierre and Marie Curie and improved by many nuclear chemists. They therefore dissolve the irradiated aluminum foil in hydrochloric acid, and then they use an appropriate chemical substance to separate the phosphorus which becomes an insoluble precipitate. And indeed, they find that the precipitate is radioactive, whereas the acid bath is not. The radioactive substance therefore reacts chemically as phosphorus: it can only *be* phosphorus. The experiment had to be completed quickly because the radioactive phosphorus disintegrates rapidly (89% of it disappears in 10 min). The results obtained are presented as the weekly meeting of the *Académie des Sciences*, on January 29, 1934, and to show the importance which they confer to this discovery, the Joliot-Curies immediately send a paper to *Nature* in which they repeat the essentials of the two publications. The paper is published on February 10, 1934, and the concluding remarks are:

> *These experiments give the first chemical proof of artificial transmu-tation, and also the proof of the capture of the α-particle in these reactions [. . .] These elements and similar ones may possibly be formed in different nuclear reactions with other bombarding particles: protons, deuterons, neutrons. For example, $_7N^{13}$ could perhaps be formed by the capture of a deuteron in $_6C^{12}$, followed by the emission of a neutron [167].*

It Spreads like Wildfire

Physicists immediately react. In Berkeley, Lawrence runs into his laboratory with the proceedings of the *Académie des Sciences* which he had just received and in which the Joliot-Curies write that other radioactive elements are likely to be produced by bombarding substances with deuterons. Lawrence feels that they had written this *with him in mind* because he was the only one who had a deuteron beam. Less than an hour later, he observes artificial radioactivity when deuterons bombard carbon: he sends a paper [146] to the *Physical Review* already on February 27th. That same day, three physicists from the California Institute of Technology send a paper to *Science* [168]. They also bombarded carbon with 0.9 MeV deuterons which were produced by a small electrostatic accelerator[1] which they had constructed in their lab in order to produce an intense neutron source [169, 170]. They observed the radioactivity of nitrogen 13. In fact, the physicists in Berkeley had produced it without realizing it, each time that their proton or deuteron beam hit the wall or a metallic object. But in order not to record what they believed were parasite signals, or "noise," they systematically shut off the counters whenever the beam was stopped. They therefore observed nothing.

---

[1] See p. 291.

The Joliot-Curies receive letters from physicists all over the world: from Pauli who thinks that the β^+ spectra must be continuous as the known β^- spectra and from Rasetti who wants to repeat the experiments in Rome. But the first and heartiest letter is written by Rutherford:

> I am delighted to see an account of your experiment in producing a radioactive body by exposure to α-rays. I congratulate you both on a fine piece of work which I am sure will ultimately prove of much importance.
>
> I am personally very much interested in your results as I have long thought that some such an effect should be observed under the right conditions. In the past I have tried a number of experiments using a sensitive electroscope to detect such effects but without any success. We also tried the effect of protons last year on the heavy elements but with negative results.
>
> [...]
>
> We shall try and see whether similar effects appear with proton or diplon bombardment [171].

It is amusing to note that it was very easy to produce artificially radioactive substances. All that was needed was a sufficiently intense radioactive source. Irène and Frédéric Joliot-Curie made their discovery by pursuing a well-defined goal. They wanted to show that positrons and neutrons were emitted simultaneously when aluminum was bombarded with α-particles. As Becquerel had done 38 years earlier, they discovered an unexpected phenomenon, a feature of any discovery, by pursuing a well-defined experiment. They were almost bound to make this discovery, which owed nothing to chance.

The Importance of the Discovery

It is not easy to realize the impact of the discovery. For all physicists, it was a revelation. So far, they had believed that nuclear reactions could only produce nuclei which already exist in nature and that only heavy nuclei were radioactive. Thus, for example, 15 protons and 16 neutrons could form a phosphorus nucleus, but 15 protons and 15 neutrons could not, even for a short time, form the isotope 30 of phosphorus which had not been observed in nature and which therefore did not exist. There was no compelling reason to believe this. It was simply an implicit mental barrier. It was as in a Scrabble game in which a general agreement existed not to admit certain words. The Joliot-Curies had discovered that among nuclei which did not exist in nature, certain could exist for short periods of time before decaying by radioactivity. Rutherford had thought of this possibility, but he failed to demonstrate it experimentally as he confessed in his letter with the frankness and humility of a great mind.

Soon, physicists all around the world began producing radioactive nuclei [172–174]. Irène and Frédéric Joliot-Curie were invited to give a talk at the international physics meeting held in London. After reviewing the known nuclear reactions, they describe their experiments. In their conclusion, they consider possible applications, in particular in biology and medicine:

These radioelements may be applied in medicine and perhaps in other practical fields. When introduced into the living body, these substances may behave very differently from ordinary radio-elements on account of their different chemical properties and because of their disintegration without leaving any radioactive residue [. . .] Finally, we must anticipate a considerable development in the use of these radioactive nuclei as indicators in the study of the behavior of their inactive isotopes in certain chemical reactions or in biological phenomena [175].

New Perspectives for Radioactive Indicators

The Joliot-Curies refer to the use of radioactive elements as *indicators*, a method which had already been used for 20 years with naturally radioactive elements.[1] It had been conceived in 1913 by the Hungarian physicist George de Hevesy, born in Budapest on August 1, 1885 (1 month before Niels Bohr). After studying first in Budapest and then in Berlin, he obtained his PhD in Freiburg in Brisgau in 1908. He worked as an assistant in Switzerland and then went to Manchester for 2 years to work with Rutherford. He tried in vain to separate lead from what was called radium *D* (*i.e.*, the isotope 210 of lead) which is a decay product of radium. He was finally forced to admit that there was no *chemical* difference which could distinguish them. As described above,[2] this and other data led Frederick Soddy to suggest that all atoms of a given element did not necessarily have the same mass and he called "isotopes" substances which were chemically equivalent but which had different masses.[3] It was impossible to separate radium *D* from lead *because it was lead*.

From Manchester, Hevesy goes to the Institute of Radium in Vienna where he meets Friedrich Paneth, 2 years younger than he. Paneth had also tried to separate lead from radium *D*. After proving that it was really impossible and that, once they are mixed, two compounds of lead and of radium *D* maintain a constant proportion whatever chemical reaction they participate in [176], Hevesy and Paneth showed how this property could be used to measure the amount of lead salts in a solution:

---

[1]This method, which is widely used in chemistry and in biology, is called today the "labelled molecules" or "tracer" method.

[2]See p. 165.

[3]Later it was found that "radium *D*" was the isotope 210 of lead, which today is denoted as $^{210}_{82}$Pb.

If one mixes a given amount of Ra D with a given amount of lead salts, then, once a perfect mixture is obtained, the ratio of the concentrations remains the same, no matter how small the quantity of lead which is extracted from the solution. Because of its radioactivity, one can measure incomparably smaller amounts of Ra D than of lead. Ra D can thus serve as a quantitative and qualitative measure of the added lead: Ra D becomes an "indicator" of lead [177].

Hevesy then becomes professor in Budapest. After the war, he goes to Copenhagen where he discovers, with the Dutch chemist Dirk Coster, the element with atomic number 72, which had not been observed before [178] and which is given the name of *hafnium* in honor of Copenhagen (the Latin name for Copenhagen is *Hafnia*). Hevesy then becomes interested in botany, and he has the idea of following the circulation of lead in a plant by *tagging* the lead. He adds to the lead, contained in the water, a tiny but known amount of radium *D*, and he is able to observe, thanks the radioactivity of the radium *D*, how the lead spreads into the living organism [179]. He then applies the same method using a radioactive isotope of bismuth to follow the trace of bismuth which has been eaten by a rat [180].

So far, this method had been limited to natural radioactive elements such as lead and bismuth. The newly discovered artificially induced radioactivity opens the possibility of making radioactive isotopes, *indicators*, possibly of all the elements.

The Death of Marie Curie

When artificial radioactivity was discovered, Marie Curie was still, at the age of 67, director of the *Institut du Radium*. She was very happy that the discovery of artificial radioactivity was made in the laboratory which she had created with so much effort. As Joliot later recalled:

Marie Curie had followed our research and I will never forget her joyful expression when Irène and I showed her the small glass tube containing the first artificial radio-element. I can still see her holding this tube of radio-element with her fingers which had already been burnt by radium. To check what we told her, she brought it close to a Geiger–Müller counter and she could hear the numerous clicks of the ray counter. It was probably the last deep satisfaction of her life. A few months later, she died of leukemia [181].

Marie Curie died on July 4, 1934, in the *Sancellemoz* sanatorium, situated in the Savoie mountains where she had been transported. A major figure of science left the scene, a woman who had imposed her authority in the whole world.

The 1935 Nobel Prizes Are Attributed to Chadwick and to the Joliot-Curies

The 1935 Nobel Prize in physics was attributed to James Chadwick for "his discovery of the neutron" and the Nobel Prize in chemistry to Frédéric Joliot and Irène Joliot-Curie "for the synthesis of now radioactive elements." After the discovery of the neutron, some had suggested to share the Nobel Prize between Chadwick and the Joliot-Curies. It is possible that the Swedish Academy of Sciences chose the opportunity of a great discovery made by the Joliot-Curies to reward Chadwick at the same time. In the official ceremony, Chadwick made a brief sketch of the history of the neutron and of the importance of the neutron in the theory of the nucleus. Frédéric and Irène Joliot-Curie both made a speech. In order to cloud the issue, Irène, who was more a chemist than Frédéric, spoke of the physics of their discovery, whereas Frédéric insisted on the chemical aspects. In his conclusion, he evoked the future:

> If, turning towards the past, we cast a glance at the progress achieved by science at an ever-increasing pace, we are entitled to think that scientists, building up or shattering elements at will, will be able to bring about transmutations of an explosive type, true chemical chain reactions. If such transmutations do succeed in spreading in matter, an enormous liberation of usable energy can be imagined [182].

In France, the press insisted on the fact that the Nobel Prize had been attributed to the daughter of Marie Curie, who had received two Nobel prizes! Several newspapers announced that the Joliot-Curies will be able to produce "artificial radium," an extraordinary achievement because radium was used to treat cancer and it was a very expensive substance. To have at one's disposal intense and cheap radioactive sources led some to dream of the disappearance of cancers.

The School of Rome

Enrico Fermi becomes professor in the University of Rome. He builds up a group of young and enthusiastic physicists which becomes one the most brilliant in Europe and which discovers strange properties of slow neutrons. The question of transuranic elements is raised for the first time.

We left Fermi in Florence when he discovered what we now call the *Fermi-Dirac statistics*: two particles, which today we call fermions and which, for example, electrons in an atom or a metal, cannot bear the same quantum numbers.[1] Shortly later, senator Corbino obtained the creation of a theoretical physics chair in the University of Rome, the first of its kind in Italy. It was subject to a competitive examination which Fermi won with flying colors in November 1926 at the age of 26. He quits his job in Florence, where he is replaced by his old friend Enrico Persico, and begins to work in the *Istituto di fisica* of the University of Rome, located in the old buildings at *via Panisperna, 89*. In the biographical introduction to the *Scientific Works* of Fermi, Emilio Segrè makes the following description of the Institute:

> The old physics building in via Panisperna, although built around 1880, was still perfectly adequate for scientific work at that time and compared favorably with other major European laboratories. The equipment was fair and mainly included instruments for optical spectroscopy with good modern Hilger spectrographs and adequate subsidiary apparatus. The shop was old fashioned with rather poor machines; the library, on the other hand, was excellent. The location of the Institute, surrounded as it was by a small park on a hill in a rather central part of Rome, was convenient and beautiful at the same time. The gardens landscaped with several palms and bamboo thickets, the silence prevailing, except at dusk when many sparrows populated the greenery, made it a most peaceful and attractive center of study [183].

One of Fermi's first tasks was to bring Franco Rasetti from Florence by obtaining for him a position as assistant to Corbino. This allowed him to begin experimental work without delay. And, thanks again to Corbino, he recruits Emilio Segrè, Edoardo Amaldi, and Ettore Majorana. Realizing that they needed to learn modern experimental techniques, he sent them to various laboratories abroad. Rasetti spent a year in Pasadena in the lab of Millikan where he made the famous experiment on the spin of nitrogen.[2] He then spent the academic year 1931–1932 in the lab of

---

[1] See p. 143.

[2] See p. 229.

Otto Hahn and Lise Meitner, where he learned the techniques of radioactivity and familiarized himself with the construction of Geiger–Müller counters and Wilson cloud chambers. Segrè went to work in the lab of Pieter Zeeman in Amsterdam and after to Hamburg in the lab of Otto Stern. Amaldi went to Leipzig to work in the lab of Debye.

At the end of the 1920s, quantum mechanics has attained maturity and it is able to explain the structure of the atom, at least in principle. The research in *via Panisperna* is concentrated mainly on optical spectrometry, that is, on the structure of the atom. However, Fermi believes that the field to tackle is the structure of the atomic nucleus, that is, *nuclear physics*. He raises the question: should one orient the research towards radioactivity and nuclear physics? After rather vivid discussions, the decision is taken in 1929. Fermi is in charge of the 1931 Nuclear Physics Congress in Rome, and this allows his team to become familiar with the modern problems.[1]

The Theory of β Decay

We saw how the continuous spectrum of electrons emitted in β radioactivity had puzzled physicists.[2] Pauli had pulled out of his hat a very light hypothetical and undetectable particle endowed with the missing energy. At first, he called it a "neutron." However, at the Rome congress, in 1931, Fermi discussed it privately with him and proposed to call it a *neutrino* (from the Italian word *neutrone)* in order to distinguish it from the neutron, the particle Rutherford had imagined [184]. The presence of the neutrino, emitted at the same time as the electron, made it possible to conserve energy in the process, provided the neutrino shared the energy with the electron in a random fashion. The hypothesis of the neutrino was a bold one. And it was hazardous because it was an *ad hoc* assumption which explained only what it was introduced for.

When he returned from the Solvay Council in 1933, Fermi tackled the problem and came out with a real theory of β radioactivity (which today we call β decay) which was quite different from what had been conceived so far. He discusses it with some of his friends in the team during a Christmas vacation in the Alps in 1933. He submits a paper to *Nature*, and it is refused on the grounds that it concerned "speculations too far remote from reality" [185]. The paper is translated into German and sent to *Zeitschrift für Physik* [186]. The paper became a classic, and it contains the essentials of the theory of β decay. After reviewing the problems faced in β radioactivity, Fermi sets the basis of a new theory:

> It appears therefore appropriate to assume, as Heisenberg did, that nuclei are composed exclusively of heavy particles—protons and

---

[1] See p. 246.
[2] See p. 188 to 196.

> neutrons. In order to explain the possibility of light particles being
> emitted by the nucleus, we shall attempt to construct a theory of
> the emission of the light particles which is analogous to the theory
> of emission of light quanta by an excited atom, according to the
> usual radiation process. In this theory of radiation, the total number
> of quanta of light is not constant: light quanta appear when they
> are emitted by an atom and they disappear when they are absorbed.
> By analogy, we base the theory of β radioactivity on the following
> hypotheses:
>   a.) The total number of electrons as well as neutrinos is not
> necessarily constant. The electrons (or the neutrinos) can appear and
> disappear.

The idea that electrons may not preexist in the nucleus before its decay but
that they may be created just at that moment had already been proposed by the
Russian physicists Victor Ambartsumian and Dmitri Ivanenko.[1] But they did not
set up a theory of the process. Fermi constructs a formal theory in which the
process is described as a transformation of a neutron into a proton together with
the *creation* of an electron and a neutrino. He is inspired by the process in which
light is emitted by an atom of a substance at high temperature and which depends
on a universal constant, *the elementary electric charge e*. He therefore introduces a
universal constant, which he calls g and which governs the simultaneous emission
of the electron and the neutrino. According to the value of this constant, the decay
is more or less probable and this in turn determines whether the radioactive half-life
is shorter or longer. Fermi shows that certain decays are, in a first approximation,
impossible, meaning that in practice, they are possibly a hundred times less
probable. In general, after it has decayed, a nucleus can be in several possible
quantum states. One says that there are several possible *transitions* from the initial
to the final state of the nucleus. Fermi distinguishes two groups of transitions: the
allowed transitions and the *forbidden transitions*. He continues by showing that in
both cases, a simple relation exists between *the maximum energy of the electrons*
and the radioactive *half-life*, and he checks that the relation holds in the case of
the known β decays. His new theory allows him to calculate the *shape of the β
spectra*, that is, the relative number of electrons emitted at a given energy. This was
in good agreement with the results obtained a few months earlier by the Canadian
physicist Bernice Sargent, who observed the shape of the electron spectra [187] and
discovered empirically that the radioactive half-life was longer when the maximum
energy of the emitted electrons was smaller [188].

But what is the meaning of his constant g? It is the analogue of the elementary
electric charge of charged particles, which determines the intensity of electromagnetic interactions (the emission of radiation, the interaction between charged
particles, magnetism, etc.). Fermi's constant g is *the charge of the new interaction*
which governs β decay. Fermi evaluates its value. This interaction is much weaker

---

[1] See p. 231.

than electromagnetic interactions. The interaction between particles carrying the charge g has little in common with the interactions which were known at the time, namely, gravitational, electromagnetic, and nuclear interactions. Although Fermi's interaction is considerably weaker than electromagnetic interactions, it is much larger than gravitational ones. The neutrino is endowed with this *weak charge* and no other. It is thus an electrically neutral particle which interacts only very weakly. The neutrino is like a ghost which can pass through matter without interacting with it. Its interaction with protons and neutrons is so weak that, according to a rough estimate, it would have a 50% probability of passing through a solid lead target *about one-light-year thick*.

Fermi's theory was recognized to be a considerable step forward. It provided for a *quantitative* explanation of the observed β radioactivity. It explained not only the observed shape of the β spectra but it also provided for a relation between the maximum energy of the electrons and the half-life of the β radioactive element. It also made the existence of the neutrino more likely. The neutrino was detected only in 1956 by Frederick Reines and Clyde Cowan in the United States [189]. Although the theory of β decay will undergo several modifications, its foundation is still the one formulated by Fermi, who however never published any further paper on the subject. True, he became involved with another phenomenon.

Neutron Physics in Rome

Dozens of Radioelements

As soon as artificial radioactivity became known in Rome, Fermi understood that this was a most promising field and that he had been well advised to orient his research team in the new field of nuclear physics. Fermi did not have an α-particle source as intense as the one of the Joliot-Curies, but the latter pointed out in their publication in *Nature* that one should be able to produce radioelements by bombarding substances with protons, deuterons, or neutrons. Fermi thought that neutrons had a great advantage: they have no electric charge so that they do not feel the Coulomb repulsion of the positively charged nuclei. They can therefore come close enough to the nuclei in order to be captured by the nuclear forces and provoke nuclear reactions. New isotopes would thus be formed and those which were unstable would decay by β radioactivity.

By a fortunate coincidence Giulio Trabacchi, a professor in the *Laboratorio delle Sanità Pubblica*, actually in the same building as the Physics Institute, had in his possession 1 g of radium as well as the necessary equipment to extract the emanation of radium, namely, the isotope 222 of radon. Radon 222 decays by emitting an α-particle with a half-life of 3.8 days. If one inserts radon gas into a glass which contains a powder of beryllium, the beryllium nuclei are bombarded by the α-particles and neutrons are emitted. Fermi constructs a Geiger–Müller counter,[1] the external tube of which is a simple medical metal tube, and he begins the measurements. He decides to use these neutrons in order to bombard all the elements he can get hold of: hydrogen, lithium, beryllium, carbon, nitrogen, and oxygen. It does not work. But on March 25, 1934, he bombards fluorine and the counter finally begins to crackle! That same day, he sends a paper to *La Ricerca Scientifica* [190] in which he interprets the observation by the nuclear reaction:

$$F^{19} + n^1 \rightarrow N^{16} + He^4$$

A fluorine nucleus of mass 19 (nine protons and ten neutrons) absorbs a neutron, and the nucleus which is thus formed immediately decays into a nitrogen 16 nucleus (the usual isotope of nitrogen has a mass 14) and a helium nucleus (an α-particle). The nitrogen 16 nucleus is unstable and decays by emitting a β particle (a negative electron). In this process, one of the neutrons of nitrogen 16 is transformed into a proton so that the resulting nucleus is the stable oxygen 16 nucleus.

In a second paper, Fermi announces the creation of 13 new radioactive isotopes which he obtained bombarding iron, silicon, phosphorus, chlorine, vanadium, aluminum, copper, arsenic, silver, tellurium, iodine, chromium, and barium [191]. All these results are also reported in a short paper sent to *Nature* [192], a preprint of which he sent to Rutherford, who reacted quickly:

> Dear Fermi,
> I have to thank you for your kindness in sending me an account of your recent experiments in causing temporary radioactivity in a number of elements by means of neutrons. Your results are of great interest, and no doubt later we shall be able to obtain more information as to the actual mechanism of such transformations. It is by no means clear that in all cases the process is as simple as appears to be the case in the observations of the Joliots.
> I congratulate you on your successful escape from the sphere of theoretical physics! You seem to have struck a good line to start with. You may be interested to hear that Professor Dirac also is doing some experiments. This seems to be a good augury for the future of theoretical physics!
> Congratulations and best wishes!
> Yours sincerely
> Rutherford [193].

In order to pursue the experiments and to identify with more certainty the produced radioactive elements, Fermi had to make a chemical analysis, as the Joliot-Curies had done. It so happened that a young chemist, Oscar D'Agostino of the *Laboratorio delle Sanità Pubblica*, had obtained a scholarship and was working at that time at the *Institut du Radium* with Marie Curie in order to learn the techniques of nuclear chemistry. When he returns to Italy for an Easter vacation,

---

[1] See p. 215.

he is immediately offered a job by Fermi and he does not return to Paris. The whole team then sets to work with a sense of urgency. They want to be the first to discover all that can be discovered in the field. Their publications succeed each other at a fast rate until the summer [194–197] 1934, when Fermi makes a series of lectures in Argentina and Brasil.

He is given a hearty and royal welcome there in spite of lecturing in Italian. While he is traveling in the Americas, Segrè and Amaldi spend some time at the Cavendish where they discuss in detail theory and experiments with Rutherford. They also show him a draft of a paper which they wish to send to the Royal Society and which Rutherford immediately presents there [198]. In this paper, Fermi and his collaborators make a general and detailed description of their experiments. They note that all the radioelements which are produced decay by β^- radioactivity, which was known since the onset of the century, and in which an ordinary negatively charged electron is emitted. The reason is that the radioelements are produced by adding a neutron to the nucleus of the target so that the latter has an excess of neutrons. It is therefore natural for it to decay by transforming one of its neutrons into a proton while emitting an electron (and a *neutrino*, this evasive particle which is always emitted in β decay). They display an impressive table containing over forty radioelements the half-lives of which span from less than 1 min to 2 days! When he returns from South America, Fermi stops in London in order to present his latest results in an international physics conference [199].

Transuranic Elements?

The case of uranium is special. When Fermi and his team bombard it with neutrons, they detect first a radioactive substance with a half-life of about 1 h and a half. But they are unable to identify it. A thorough chemical analysis allows them merely to exclude certain possibilities. What can this substance be? An idea begins to emerge: the most abundant isotope of uranium, which has a mass of 238 (92 protons and 146 neutrons), absorbs a neutron, thereby becoming the isotope 239 of uranium. A neutron of uranium 239 then transforms into a proton by β decay, and a hitherto unknown element is formed: the element 93 which has 93 protons. The element is heavier than uranium; it is a *transuranic element*. This possibility is presented with great caution in a letter sent on June 6, 1934, to *La Ricerca Scientifica* [200]. It is followed by another letter sent to *Nature* [201] and which also proposes the idea with caution. The team in Rome detected several radioelements with half-lives ranging between 10 s and 2 days. They also could not be identified. Chemical analysis now excludes several elements close to uranium, namely, uranium, protactinium, thorium, actinium, radium, bismuth, and lead. However, the paper does not reach a definite conclusion.

Two days before, on June 4, 1934, during the traditional meeting of the *Accademia dei Lincei* which is held at the end of the Italian academic year, professor Corbino gives a lecture entitled "Results and perspectives in modern physics." He praises the results obtained in his institute:

The case of uranium with atomic number 92 is of particular interest. It appears that, after absorbing a neutron, it is quickly transformed, by emitting an electron, into an element which has a position above uranium in the periodic table of elements, that is, into a new element with atomic number 93 [...] It is clear that further tests are required; several have been performed, all with positive results. Such tests are however very delicate and this justifies the reserved tone of Fermi and the pursuit of experiments before the discovery is announced. As far as my opinion may have some value, I have followed these works daily and I believe that the production of this new element is a certainty [184, p. 76].

The speech produced a terrific effect much to the dismay of Fermi, who had suggested the possible existence of this element with great caution because he never claimed anything without being absolutely certain. And now, the discovery is mentioned in the press! One newspaper even claimed that Fermi had offered the queen a bottle filled with the famous element 93. The *New York Times* also got hold of the news with a two-column headline "An Italian produces the element 93 by bombarding uranium." Fermi was upset and could not sleep. He went to see Corbino in order to set things right. They published a press release stating that the words of senator Corbino had been badly interpreted [202].

However, the experiment was not proven wrong. It was soon confirmed by Lise Meitner and Otto Hahn [203]. But did the experiment involve nuclei heavier than uranium?

"Slow" Neutrons

During the summer 1934, Bruno Pontecorvo, a student in the University of Rome, obtains his PhD and applies to work in the laboratory of Fermi. Franco Rasetti knew his family well in Pisa, when Bruno was still a child. He finds it difficult to recognize the man, which Laura Fermi describes thus:

Bruno was unusually beautiful. It is most likely his proportions which made him so seductive. Nobody would have wanted to swell his chest or his shoulders nor lengthen his arms or legs [204].

Bruno Pontecorvo proves to be brilliant, and he joins the team of Fermi in the summer. Edoardo Amaldi and he are entrusted with the task of identifying more precisely and quantitatively the radioactivity induced in different elements. In the paper published in the *Proceedings of the Royal Society,* the radioactivity was simply qualified as weak, average, or strong.

They set up a simple apparatus: the neutron source (a glass bulb filled with the emanation of radium and beryllium powder) is placed inside a hollow cylinder, made of the substance under study. The cylinder is put into a lead box. After irradiating the cylinder for some time, the neutron source is replaced by a Geiger–Müller counter which detects the artificial radioactivity induced in the cylinder by the neutrons.

One day in November, while they are studying silver, Pontecorvo notices some anomalies: the induced radioactivity cannot be reproduced. It appears to depend on the position of the cylinder in the lead box. They then discover that the radioactivity depends on the environment: it is stronger when the apparatus is placed on a wooden table than on a marble table! At first, this causes some incredulity and even sarcastic remarks in the lab. They finally tell Fermi, and they decide to place a lead sheet between the neutron source and the silver, to see what effect it could have. This was in the morning of October 20, 1934. Fermi then suggests to use paraffin. To everyone's surprise, the radioactivity becomes a hundred times stronger!

Soon after 1 p.m., they all go home for lunch. When they return at about 3 p.m., Fermi, as usual, has found an explanation: when a neutron passes through a heavy substance such as lead, it occasionally collides with a the nucleus of the lead atom. In general, it will bounce off as in an elastic collision. Since the lead nucleus is 208 times heavier than the neutron, the neutron loses very little (kinetic) energy in the collision. If, however, a neutron passes through a lighter material, such as wood, or better still, paraffin which contains many light nuclei such as hydrogen which have the same mass as the neutron, the situation is quite different. When a neutron collides with a hydrogen nucleus (a proton), it transfers a much greater fraction of its (kinetic) energy to the proton so that it progressively slows down and loses its (kinetic) energy until it reaches the thermal energy of the protons which is due to the temperature of the substance. Fermi explains that these *"slow" neutrons must have a much greater probability to be absorbed by nuclei than fast neutrons.*

This is an important discovery which contradicts the way in which the interaction of a neutron with a nucleus had been understood until then. It was thought that when a neutron came sufficiently close to a nucleus, it became subject to global attraction caused by the protons in the nucleus. This attraction could either deviate the neutron or absorb it. According to classical mechanics, the nucleus, seen from a certain distance, appears to be a disc. If the trajectory of the neutron passes through the disc, an interaction occurs. The apparent surface of the disc is called the *cross section*$^\diamond$. The word is in fact used in a more general sense and, in general, the radius of the disc is not the same as the radius of the nucleus. The neutrons which impinge on the disc are absorbed and not the others. The radius, and therefore the surface of the disc, that is, the cross section, grows when the probability of absorbing a neutron increases. When the process is treated in quantum mechanics, it is found that the cross section varies with the energy of the neutron. But the experimental results suggest a far greater growth, as if the cross section was a hundred times greater than the size of the nucleus. This seemed incomprehensible!

Two days later, Fermi sends a letter to *La Ricerca Scientifica* in which he writes:

A possible explanation of these facts appears to be the following: because of the numerous collisions with the hydrogen atoms, the neutrons quickly loose their energy [205].

Two weeks later, he sends another letter to *La Ricerca Scientifica* in which he describes a new experiment which shows that much more radioactivity is caused by neutrons which pass through water before bombarding a substance. He writes:

The fact that slow neutrons are efficient in activating these substances leads one to think that they must also be strongly absorbed [206].

Progressing slowly, Fermi now suggests that the probabilities of being absorbed and of causing radioactivity are related. Finally, he writes a complete paper in 1935 and sends it the *Proceedings of the Royal Society*. He formulates the basis of the physics of slow neutrons although it is but a beginning. He shows that the reason why slow neutrons produce so easily radioelements is indeed because they are strongly absorbed. The *cross sections* (or the probabilities) for absorption were measured for numerous elements. For some, namely, boron, yttrium, and especially cadmium, they are huge.

Fermi gives a theoretical explanation of the increase in absorption in terms of quantum mechanics: he assumes that the neutron "sees" the nucleus as a strongly attractive zone which has the same size as the nucleus. As the neutron is slowed down, its de Broglie wavelength becomes progressively larger. The probability that it gets caught by a nucleus increases, as if the neutron became larger:

Let us admit, as it has been generally assumed, that the forces acting between the neutron and a nucleus extend about as far as the nuclear radius itself. If it is so, the de Broglie wave-length is, for fast neutrons, of the order of the radius of action, and consequently for slow neutrons is much larger [207].

All this is true, but the observed effect is still much larger and a further problem remains: the slow neutrons should also have a greater probability of making an elastic collision, that is, of bouncing off a nucleus. But this does not happen! Two years will pass before this is explained. Be patient! Fermi's paper ends with a table displaying the measurements made by the team in Rome: *all the elements* have been systematically bombarded by slow neutrons (by neutrons which have passed through a hydrogenated substance, in general, paraffin). About forty radioelements have been observed even if all have not been reliably identified.

A New Field in Nuclear Physics

Slow neutrons become a fully fledged field of nuclear physics to which many physicists devote their research. The uncontested experts are Fermi and his team, including Rasetti, Segrè, and Amaldi. At the *Institut du Radium* in Paris, Frédéric Joliot and Irène Curie also produce radioelements by exposing substances to slow neutrons [208]. A large number of radioelements can be produced with the neutrons. Each element has not only several stable isotopes (as had been thought previously) but a considerably larger number of radioactive isotopes. Many experiments are therefore performed in order to produce them. Biologists and chemists, who use *radioactive indicators*,[1] think of many new applications.

The spectacular absorption of neutrons by certain nuclei was a real challenge to physicists. If it could be understood, it would most likely tell a lot about the as yet little known internal structure of the nucleus.

A by-product of this branch of physics is the possibility of detecting slow neutrons by simply using, for example, a piece of rhodium. When rhodium absorbs a neutron, the substance which consists of the sole isotope 103 suddenly contains the isotope 104 which is radioactive with a half-life of 42 s. If a sheet of rhodium is exposed to the slow neutrons for several minutes, the latter can be detected by measuring the induced radioactivity. Twenty minutes later, the radioactivity disappears and the rhodium sheet is ready for another measurement.

Resonances

Slow neutrons had not yet said their last word. Further surprises lay ahead. Several experimental teams measure both the absorption and the elastic scattering (which occurs when a neutron simply bounces off a nucleus) in various substances. They obtain a wide variety of results. In several cases, Fermi's law, stating that the rate of absorption of neutrons increases as the inverse of their velocity, does not seem to apply [209, 210]. More strangely, the measured absorption is different when the detector consists of a sheet of the element under study or another element [211,212].

In the fall of 1935, Fermi and Amaldi are alone in Rome. Rasetti, who felt very hostile to the Mussolini fascist regime, went to the United States for at least a year. Pontecorvo joined the Joliot-Curies in Paris, thanks to a scholarship granted by the *Ministère de l'Éducation Nationale* of France. D'Agostino obtained a position at the *Istituto di Chimica del Consiglio Nazionale delle Ricerche*. Finally, Segrè spent 3 months in the United States after which he became professor in Palermo. So Fermi and Amaldi try to sort out the incoherences observed in the absorption of neutrons.

In early November 1935, they make a detailed study of neutron absorption by eleven different elements, combined in every possible way and using seven detectors. They confirm that the absorption of neutrons does not follow the simple law stated by Fermi [213]. They publish a succession of papers [157, 214–217]. Meanwhile, other papers are published by physicists in other laboratories. In Oxford, Leo Szilard makes a completely independent observation: neutrons which pass through a cadmium sheet 1.6 mm thick are not, or at least barely, absorbed by a second sheet of cadmium, although they are strongly absorbed by a second sheet of indium. It seems as if the first cadmium sheet has absorbed all the neutrons which are absorbable by cadmium and remains transparent to the other neutrons. For Szilard, the reason appears obvious: cadmium absorbs only those neutrons

---

[1] See p. 312.

the velocities of which lie within a very narrow range, and it allows the others to pass through [218]. While they were visiting the University of Columbia in New York, Rasetti and Segrè made an experiment in collaboration with three American physicists, George Pegram, John Dunning, and George Fink. They measured directly how the absorption of neutrons varies with their velocity.[1] Fermi's law appeared to apply to silver, but not to cadmium [219].

Amaldi and Fermi then discuss their results and their conclusions in a paper [220] sent to *La Ricerca scientifica* on May 29, 1936. They use the notation which they had used in their second paper [214], dated December 14, 1935. They notice that the slow neutrons emitted by their source can be divided into several groups; the neutrons belonging to a particular group are absorbed in a very selective manner by certain substances. The groups are labelled by letters: C stands for neutrons absorbed by cadmium, A for neutrons absorbed by silver (*argento* in Italian), and D for neutrons absorbed by rhodium.

The idea which comes to mind is that each group corresponds to neutrons with a given velocity and therefore energy. But at the time, no method existed to measure the energy of neutrons. However, painstakingly, they succeed in showing that the C neutrons are *thermal* neutrons the velocity of which is determined by the motion of molecules at the ambient temperature.[2] They then show that the neutrons belonging to the other groups have higher energies, and they succeed in estimating their energy distributions and even the *width* of each distribution, which is the narrow range of energies which the neutrons of a given group may have in order to be selectively absorbed. During the summer, both Fermi and Amaldi are invited to the University of Columbia. Amaldi translates their paper into English and sends it to the *Physical Review* [221]. The paper soon becomes a classic.

Within a few months, the way in which neutrons were understood to interact with nuclei had completely changed. The interaction was a *resonance* process: a nucleus bombarded by neutrons has a large probability of absorbing a neutron only if the latter has precisely the energy corresponding to a given state of the newly formed nucleus. It follows that the probability of absorbing a neutron varies *very quickly* with the neutron velocity. The opposite was believed a few months earlier. We shall shortly encounter further upheavals in nuclear theory.

---

[1]They did this by fixing the substance to a fast rotating disc. The latter was then bombarded tangentially by neutrons. This enabled them to vary the *relative* velocity of the neutrons and the target nuclei within a certain range.

[2]At a temperature of 20°C, thermal neutrons have velocities ranging from about 1000 to 4000 m/s, the average velocity being 2200 m/s. However, neutrons emitted by neutron sources (such as the beryllium + radon source used by Fermi) have energies of several million electron volts (MeV). For example, 5-MeV neutrons have velocities of 31 000 km/s. They travel more than a thousand times faster than thermal neutrons.

Fermi Is Awarded the Nobel Prize. The End of the Rome Team

The 1938 Nobel Prize in physics is attributed to Fermi "for his demonstration of the existence of new radioactive elements obtained by neutron irradiation, and for his concomitant discovery of nuclear reactions induced by slow neutrons." He was discretely informed of this by Niels Bohr before the official announcement. Because his wife Laura was Jewish, he decided to emigrate. He set off to Stockholm with his family with only a few suitcases as if he was going to stay for a few days only. After receiving the Nobel Prize, he went to New York where he had been offered a chair at the University of Columbia. Italy had lost one of the greatest physicists of the century. At the same time, Emilio Segrè, who was Jewish, was expelled from the University of Palermo. He set off to Berkeley where he obtained a job as assistant. He was awarded the Nobel Prize in physics in 1959.

Within a few years, the brilliant team in Rome had simply evaporated.

The Great Exodus of Jewish Scientists Under Nazism

At the end of the First World War, Germany is blood-drained. The young Weimar Republic experiences a difficult start and faces both a violent communist resistance and a young Nazi party. It goes through, however, a calmer period in the years 1924–1929. The 1920s give rise to particularly brilliant intellectual and artistic activities in philosophy, literature, motion pictures, and music. But the financial crisis in 1929 gives rise to terrible unemployment and, helped by the short-sighted and arrogant governing class, Hitler comes to power in January 1933.

Jews and other "non-Aryans" are expelled from the universities and from research in several steps [222, 223]. The first expelled are the lower-rank university employees, then the professors, and finally the researchers. About 15,000 scientists leave Germany, most in 1933. Even the old Fritz Haber is not spared. He was considered to be a hero by his fellow citizens and as a war criminal by the allies because he had perfected the warfare gas used in the 1914–1918 war. In 1933, he resigns, goes to England, and dies in Basel a year later.

These departures are disastrous, and their effect will last well after the war. Among the scientists who left Germany or the neighboring countries, let us simply list those who are mentioned in this book, in the order of their date of birth:

- Lise Meitner (1878–1968), belonging to an Austrian Jewish family. She was working with Otto Hahn in Berlin since 1907, and she remained in Germany until 1938, protected by her Austrian passport. But after the *Anschluss*, she flees to Sweden, passing illegally through the Dutch border.
- Albert Einstein (1879–1955), who refused to go back to Germany when Hitler took power in 1933; he emigrated to the United States.
- James Franck (1882–1964, Nobel Prize in physics in 1925), a German Jew, gives up ostensibly his chair as professor in the University of Göttingen and emigrates to the United States.
- Max Born (1882–1970, Nobel Prize in physics in 1954), a German Jew, one of the founders of quantum mechanics, emigrates to England.
- Victor Hess (1883–1964, Nobel Prize in physics in 1936), an Austrian with a Jewish wife, is expelled from the university after the *Anschluss* and emigrates then to the United States.
- Niels Bohr (1885–1962, Nobel Prize in physics in 1922), whose mother was Jewish, escaped from Denmark to Sweden in 1943 and then travelled to London. Subsequently, he joined the Manhattan Project in the United States.
- George de Hevesy (1885–1966, Nobel Prize in chemistry in 1943), Hungarian chemist of Jewish descent, emigrated to Denmark and fled to Sweden.
- Friedrich Paneth (1887–1958), an Austrian chemist of Jewish descent, fled to Britain in 1933 and became a British citizen in 1939. He returned to Germany as director of the Max Planck Institute for Chemistry in 1953.
- Kazimierz (Kasimir) Fajans (1887–1975), a Polish Jew working in Germany, emigrates to the United States in 1935.

- Otto Stern (1888–1969, Nobel Prize in physics in 1943), a German Jew, the author of the famous Stern and Gerlach experiment, emigrates to the United States in 1933.
- Marietta Blau (1894–1970), an Austrian Jew, emigrates in 1938 first to Sweden, then to Mexico, and finally to the United States.
- Leo Szilard (1898–1964), a Hungarian Jew, defends his PhD thesis in Berlin in 1922 and flees Germany in 1933; he emigrates first to England and then to the United States where he will play an important role in the Manhattan Project.
- Lothar Nordheim (1899–1985), a German Jew, emigrates in 1934 first to Holland and then to the United States.
- Fritz London (1900–1954), a German Jew, flees Germany in 1933 and emigrates first to France, then to England, and finally to the United States.
- Eugene Wigner (1902–1995, Nobel Prize in physics in 1963), a Hungarian Jew, goes to the United States in 1930 and remains there after 1933.
- Samuel Goudsmit (1902–1978), a Dutch Jew, goes to the United States in 1927 and remains there after 1933.
- John von Neumann (1903–1957), a Hungarian mathematician of Jewish origin goes to the United States in 1930 and remains there after 1933.
- Walter Heitler (1904–1981), a German Jew, emigrates in 1933 to England and then to Ireland.
- Gerhard Herzberg (1904–1999, Nobel Prize in chemistry in 1971), emigrates to Canada with his Jewish wife in 1933.
- Walter Elsasser (1904–1991), a German Jew, emigrates to Paris in 1933 and then to the United States.
- Kurt Guggenheimer (1902–1975), a German Jew, emigrates to Paris in 1933 and then to Scotland.
- George Placzek (1905–1955), a Czech Jew, emigrates to Denmark in 1932; he then becomes professor at the University of Jerusalem and at the University of Karcoc (USSR) and then emigrates to the United States.
- Felix Bloch, (1905–1983, Nobel Prize in physics in 1952), one of the founders of nuclear magnetic resonance (NMR), a Jew of Swiss origin, leaves Leipzig in 1933 and settles in Stanford in the United States.
- Otto Robert Frisch (1904–1979), an Austrian Jew, the nephew of Lise Meitner, emigrates to England and joins the Manhattan Project as part of the British delegation.
- Hans Bethe (1906–2005, Nobel Prize in physics in 1937), a German Jew, expelled from the university of Tübingen, emigrates first to England and then to the United States where he takes an active part in the Manhattan Project.
- Rudolf Peierls (1907–1995), a German Jew, a student of Heisenberg, emigrates to England and joins the Manhattan Project as part of the British delegation.
- Edward Teller (1908–2003), a Hungarian Jew, leaves Germany in 1933 and emigrates to the United States, where he joins the Manhattan Project.
- Victor Weisskopf (1908–2002), an Austrian Jew, leaves Austria in 1937 and emigrates to the United States where he participates in the Manhattan Project.

- Joseph Rotblat (1908–2005), a Polish Jew, emigrates to Great Britain in 1939 and later participates in the Manhattan Project. In 1944, he leaves the Manhattan Project when it becomes known that Hitler do not possess an atomic bomb. He is the only physicist to do so. He becomes a British subject in 1946. In 1995, the Nobel Peace Prize is awarded jointly to Joseph Rotblat and Pugwash Conferences on Science and World Affairs "for their efforts to diminish the part played by nuclear arms in international politics and, in the longer run, to eliminate such arms."
- Arno Brasch (1904–1963), a German Jew, emigrates to the United States in 1934.
- Maurice Goldhaber (born in 1911), an Austrian Jew, emigrates to England and then to the United States in 1939.
- Gertrude Scharff-Goldhaber (1911–1998), a German Jew, emigrates to England, where she meets Maurice Goldhaber. They marry in 1939 and emigrate to the United States.

We should add the Italians who fled their country after the adoption of the anti-Jewish laws in 1938:

- Enrico Fermi (1901–1954, Nobel Prize in physics in 1938), leaves Italy because his wife is Jewish; he will play a leading role in the Manhattan Project.
- Emilio Segrè (1905–1989, Nobel Prize in physics in 1959), an Italian Jew, emigrates in 1938 to the United States.
- Bruno Pontecorvo (1913–1993), an Italian Jew, emigrates to Paris in 1936. In 1940 he flees to Spain, then to the United States, and to Canada. He emigrates to England in 1949, and goes to Russia in 1950.

Finally, some non-Jewish scientists voluntarily leave Germany or countries under German influence. Among these we have mentioned:

- Erwin Schrödinger (1887–1961, Nobel Prize in physics in 1933), Austrian, professor in Berlin, decides to leave Germany in 1933 in spite of not being Jewish and returns to Austria. But after the *Anschluss*, he flees from Austria and goes to Rome, where Fermi helps him to go to Dublin. He returns to Austria in 1955.
- Franco Rasetti (1901–2001), leaves Italy and heads for the United States and Canada. He refuses Fermi's offer to work on the atomic bomb.
- Fritz Lange (1899–1987), a German communist, emigrates to the USSR in 1934. He works there on the problem of separating uranium 235 with the purpose of making an atomic bomb. He returns to East Germany in 1959.

A Proliferation of Theories: Yukawa, Breit and Wigner, Bohr

A new generation of Japanese theorists has matured: Hideki Yukawa proposes a revolutionary form of the nuclear interaction which predicts a new particle. Gregory Breit and Eugene Wigner formulate a theory of nuclear reactions, and Niels Bohr describes the nucleus as a liquid drop.

We have already encountered several attempts made by Japanese physicists to model the atom. In 1904, Hantaro Nagaoka published his saturnal model of the atom[1] and he continued to be known for his work on spectroscopy, on the Zeeman effect in particular. In 1926–1927, Takeo Hori, also a spectroscopist, went to Denmark to perfect his knowledge.[2] In 1928, Yoshio Nishina, together with Oskar Klein, formulated the theory of the collision between a photon and an electron, and they derived a famous formula which allows one to calculate the relative number of photons scattered at a given angle [224]. These physicists came to Europe to perfect their knowledge, mostly in Germany. Upon their return to Japan, they began teaching the "modern" physics, meaning quantum mechanics, to a new generation of physicists.

Hideki Yukawa

One of them was soon to become famous. Hideki Yukawa was born in Tokyo on January 23, 1907. He proved to be brilliant in physics at the University of Kyoto where his father, Takuji Ogawa, taught geography [225]. 1932 was a busy year for the young Hideki: he got married (and adopted the name of his wife Sumi Yukawa, a rather frequent custom in Japan). He then obtained a position as assistant in the University of Kyoto. This was just the time when the neutron was discovered, when nuclear physics consequently made much progress, and when Heisenberg formulated a model for the interaction between neutrons and protons. On November 1934, Yukawa gives a talk on "the interaction between elementary particles" at the monthly meeting of the Japanese Society of Mathematical Physics. He proposes a new theory of the forces which act between particles such as protons and neutrons. He predicts the existence of a new particle, and he predicts that its mass should

---

[1] See p. 58.
[2] See p. 148.

be about 200 times the mass of the electron. His work is published [226] in 1935 in the *Proceedings of the Japanese Society of Mathematical Physics*. His theory is endowed with an exceptional destiny.

The Theory of Yukawa

In his autobiography written 20 years later [225], Yukawa explains how the theory was born. He was inspired by the electromagnetic interaction which is mediated by the electromagnetic field carried by the photons. Would it be possible for the electron to mediate the interaction between a proton and a neutron? Possibly, except for the fact that the electron, emitted by a neutron which decays into a proton, does not have a well-defined energy, a fact that led Pauli to postulate the existence of the neutrino, a very light particle which interacts very weakly with matter but which carries away part of the energy.[1] The existence of the neutrino was not easily accepted neither by Bohr nor by Heisenberg, at least not until 1934 when Fermi formulated his theory of β radioactivity.[2] This led Yukawa to speculate that the proton and the neutron might exchange *a pair of particles*, but the Russian physicists Igor Tamm and Dmitri Ivanenko soon showed that this led to an interaction which was much too weak [227]. In his autobiography, Yukawa tells the rest of the story:

The crucial point came to me one night in October. The nuclear force is effective at extremely small distances, on the order of 0.02 trillionth of a centimeter. That much I knew already. My new insight was the realization that this distance and the mass of the new particle that I was seeking are inversely related to each other. Why had I not noticed that before? The next morning, I tackled the problem of the mass of the new particle and found it to be about two hundred times that of the electron. It also had to have the charge of plus or minus that of the electron. Such a particle had not, of course, been found, so I asked myself, "Why not?" The answer was simple: an energy of 100 MeV would be needed to create such a particle, and there was no accelerator, at that time, with that much energy available [228].

In fact, Yukawa's idea is not so different from that of Heisenberg who suggested that an electron is exchanged between a proton and a neutron: the neutron emits an electron which is absorbed by a proton. In this process, the neutron becomes a proton and vice versa. It is this oscillation which is responsible for the "exchange" force of Heisenberg, who simply considered it as a convenient image.

The new idea of Yukawa consisted in replacing the electron by a new quantum which was electrically charged in order to transport the electric charge from the proton to the neutron. *Yukawa's innovation was to propose a mass for the exchanged particle.*

---

[1] See p. 194.
[2] See p. 316.

Why this mass? Because the force which the particle mediates is short ranged and decreases fast beyond a certain distance. As the Italian physicist Gian Carlo Wick showed shortly after [229], the relation between the range R of the force and the mass of the exchanged particle can be estimated from Heisenberg's uncertainty principle.[1] Yukawa assumed a range corresponding to a particle with a rest mass Mc^2 equal to 200 MeV. This corresponds to a range of about one femtometer (10^{-15} m).

In the paper he published, Yukawa proposes a new mathematical formula for the force acting between the nucleons. It is somewhat similar in form to the electromagnetic Coulomb interaction at small distances, but it decreases exponentially at larger distances. It is much stronger, and its dependence on the distance is determined by the mass of the exchanged particle, which he calls a "hypothetical quantum." He summarizes the paper thus:

> The interaction of elementary particles are described by considering a hypothetical quantum which has the elementary charge and the proper mass and which obeys Bose's statistics. The interaction of such a quantum with the heavy particle should be far greater than that with the light particle in order to account for the large interaction of the neutron and the proton as well as the small probability of β-disintegration [226].

Can the "Hypothetical Quantum" Be Observed?

How could one confirm the existence of the massive quantum, which Yukawa in 1935 dares not yet call a "particle"? One could think of producing it in collisions between nuclei. But this would require to collide nuclei at energies of at least 200 MeV which was far above the energy which could be provided by the available accelerators at the time. In his paper, Yukawa suggests another possibility:

> The massive quanta may also have some bearing on the shower produced by cosmic rays.

It was not easy to accept the idea of yet another new particle, after the neutron, the positron, and even the "hypothetical neutrino." It would certainly help if the new proposed particle could be observed.

---

[1] Consider a proton interacting with a neutron. The proton begins by emitting a quantum (today, we would say "a particle"). The system then consists of the proton and the neutron with, in addition, the quantum (the particle) of mass M, which increases the energy of the system by the amount $\Delta E = Mc^2$ where c is the speed of light. According to Heisenberg's uncertainty principle, this change of energy cannot last longer than the time interval $\Delta t = \hbar/\Delta E = \hbar/Mc^2$ where $\hbar$ is Planck's constant h divided by 2π. This means that the quantum must be absorbed by the neutron at a time no later than Δt. During this time, the exchanged quantum cannot travel a distance greater than $R = c\Delta t = \hbar/Mc$ because it cannot travel faster than light. The interaction between the proton and the neutron therefore has a range of the order of $\hbar/Mc$ which decreases as the mass of the exchanged quantum (particle) increases.

But just at that time, Carl Anderson and Seth Neddermeyer announced that certain tracks caused by cosmic rays in their Wilson cloud chamber were caused by particles which differed from those of electrons in that they had a far greater penetrating power, as well as from protons because they lacked sufficient ionization power [230]. Their observation was confirmed by two physicists in Harvard, J. Curry Street and Edward Stevenson [231]. Thus, a particle with a mass somewhere in between the mass of the electron and of the proton seemed to exist. Was it the particle proposed by Yukawa?

Four Fundamental Interactions in Nature

The theory of Yukawa marked the beginning of the modern theory of interaction between protons and neutrons, which we call a *strong interaction* to distinguish it from the other interactions which are weaker. Until 1934, only two interactions were known to occur in nature: *gravitation* which is the attraction between two massive objects and the *electromagnetic interaction* which acts between electrically charged particles. It does not act on neutral particles such as the neutron or the neutrino.[1]

In his theory of β radioactivity, Fermi added to the list a further interaction which acts between particles endowed with a different kind of "charge" which today we call the *weak* charge. The corresponding *weak interaction* acts only between particles carrying a weak charge. Neutrinos have only weak interactions and because such interactions are so weak, matter is practically transparent to neutrinos. Yukawa added to the list a nuclear interaction which is much stronger than the electromagnetic interaction. He endowed the protons and neutrons with a "strong charge." Electrons, positrons, and neutrinos do not feel the strong interactions.

The Name of the New Particle

The names suggested for Yukawa's "massive quantum" proliferated: dynatron, penetron, barytron, heavy electron, yukon, etc. Because the particle was expected to have a mass between that of the electron and the proton, Carl Anderson and Seth Neddermeyer [232] discard these names and propose to call it a *mesotron*, meaning an *intermediate particle* (from the Greek word *mesos, medius* in Latin, which means middle or intermediate). The suffix *tron* was an imitation of the electron and the neutron. But the Indian physicist Homi Jehangir Bhabha pointed out that the group of letters *tr* in the words "neutron" and "electron" were already present in the roots *neutr-* and *electr-* of these words, and he proposed to call the new particle a *meson* [233]. It is the latter which finally prevailed.

---

[1]However, today, we know that the neutron has a magnetic moment in spite of having zero electric charge. The magnetic moment of the neutron allows it to interact with a magnetic field.

The First Theories of Nuclear Reactions

We saw above[1] that Fermi and his collaborators in Rome had discovered that slow neutrons could be absorbed by nuclei with a very high probability, that is, with a large *cross section*, which far exceeds that which would arise from the average attraction of the protons and neutrons in the nucleus. How could one explain the fact that the neutrons are strongly absorbed only in certain cases? An explanation which comes naturally to mind is that the absorption is related to a *resonance*, that is, that when the neutron may have exactly the energy required for it to be captured in a quantized orbit, the other nucleons remaining passive. Such a process had been observed in the capture of electrons by atoms, which is much more probable at certain electron energies.[2] Several physicists published papers based on this idea [234]. However, the theory had a defect: it predicted that a strong absorption of neutrons was inevitably accompanied by strong elastic scattering, in which the neutron bounces of the nucleus as a rubber ball does. But this was not observed. In 1936, two papers appeared which changed this.

Breit and Wigner

The *Physical Review* dated April 1, 1936 contains a paper signed by Gregory Breit[3] working in the University of Princeton and by the Hungarian physicist Eugene Paul Wigner, who was also in Princeton and who had already made significant contributions to nuclear physics.[4] Until that time, the strong absorption of thermal neutrons[5] was understood as a process in which the neutron is first captured by the average potential produced by the target nucleus and remains there for a certain time, bound as a satellite in its orbit, to use a classical analogy. The neutron then either escapes (thereby being elastically scattered) or it remains in the nucleus and makes quantum jumps to lower energy orbits while emitting γ photons [234, 235]. The trouble was that such a theory predicted that if the probability of being absorbed was strong, so was the probability of being elastically scattered. This contradicted experimental observations.

In the process proposed by Breit and Wigner, the neutron is indeed captured by the nucleus with which it forms a well-determined quantum state. But this state is not composed of a neutron on one hand and the nucleus in another: *the neutron and the nucleus merge to form a new system*, which is a new nucleus containing one extra neutron. The newly formed nucleus is in a "stationary state," and it will lose

---

[1]See p. 324.

[2]See p. 110 for Franck and Hertz experiment.

[3]See p. 291.

[4]See p. 270.

[5]See p. 325.

its energy either by emitting γ-rays (photons) or by ejecting the neutron. However, the latter process is but one of many possibilities, and it may well have a small probability:

It will be supposed that there exist quasi-stationary (virtual) energy levels of the system nucleus+neutron which happen to fall in the region of thermal energies as well as somewhat above that region. The incident neutron will be supposed to pass from its incident state into the quasi-stationary level. The excited system formed by the nucleus and neutron will then jump into a lower level through the emission of γ-radiation or perhaps at times in some other fashion [236].

The fact that the state does not have a precise energy but a certain spread is precisely due to the fact that it does not exist for a long time and that it "decays" by making transitions to lower energy states while emitting γ-rays which carry the excess energy away, or by emitting a neutron. Again, we find that it is Heisenberg's uncertainty principle which explains this: the longer the lifetime of the excited state, the smaller is the uncertainty in the energy which is usually measured by an energy "width" called Γ which is related to the lifetime τ (or half-life) of the state by the expression $\Gamma \approx \hbar/\tau$. (In this expression, $\hbar$ is Planck's constant h divided by 2π).

A resonance occurs when the energy of the incident neutron is close to the energy of the system nucleus+neutron. Breit and Wigner derive a mathematical formula which gives the cross section, that is, the probability for the process to occur, as a function of the neutron energy. All physicists associate the names "Breit and Wigner" to this famous formula.[1]

Niels Bohr and the Theory of Nuclear Reactions

On February 29, 1936, shortly before the publication of the paper of Breit and Wigner in the *Physical Review, Nature* publishes a paper containing a general theory of the capture of neutrons by Niels Bohr who had presented it to the Danish Royal Academy on January 27.

In the introduction, Bohr states the problem:

Among the properties of atomic nuclei disclosed by the fundamental researches of Lord Rutherford and his followers on artificial nuclear transmutations, one of the most striking is the extraordinary tendency

---

[1]For those who are curious to see the formula, here it is:

$$\sigma = \Lambda^2 \pi S \frac{\Gamma_s \Gamma}{(v - v_0)^2 + \Gamma^2}$$

In this formula, σ is the cross section which is proportional to the probability of the process, Λ and S are constants, v is the energy of the neutron, v_0 is the energy at which the resonance occurs, Γ is the width of the resonance which is the range of energies at which the process can occur with a high probability, and Γ_s is proportional to the probability for the system to decay in certain fashion after the neutron capture.

of such nuclei to react with each other as soon as direct contact is established. In collisions between charged particles and nuclei, contact is, of course, often prevented or made less probable by the mutual electric repulsion; and the typical features of nuclear reactions are therefore perhaps most clearly shown by neutron impacts [237].

Bohr then analyzes the resonances which are very narrow, meaning that the reaction occurs only within a narrow range of neutron incident energies. He explains that this is an indication that the system composed of the initial nucleus and the neutron remains in a quantum state for a "long time," that is, for a time which is far greater than the time required for the neutron to pass through the nucleus. This again is a consequence of Heisenberg's uncertainty principle: the longer a system lives, the more precisely its energy is determined. Bohr then unveils his idea:

The phenomena of neutron capture thus force us to assume that a collision between a high-speed neutron and a heavy nucleus will in the first place result in the formation of a compound system of remarkable stability. The possible later breaking up of this intermediate system by the ejection of a material particle, or its passing with emission of radiation to a final stable state, must in fact be considered as separate competing processes which have no immediate connection with the first stage of the encounter.

This is the idea which dominated and still dominates the physics of nuclear reactions at least at these energies. According to Bohr, a nuclear reaction takes place in two distinct steps: when the neutron[1] comes into contact with the nucleus, it is swallowed so to speak. The new nucleus which is thus formed (and which is an isotope of the initial nucleus because it has an extra neutron but the same number of protons) is in an excited state which has a "long" lifetime, possibly of the order of 10^{-15} s, which is admittedly a short time for us but which is between a thousand and ten thousand times longer than the time for a thermal neutron to pass through a nucleus. In his paper, Bohr adds a remark concerning the structure of nuclei:

It is, at any rate, clear that the nuclear models hitherto treated in detail are unsuited to account for the typical properties of nuclei for which, as we have seen, energy exchanges between the individual nuclear particles is a decisive factor. In fact, in these models it is, for the sake of simplicity, assumed that the state of motion of each particle in the nucleus can, in the first approximation, be treated as taking place in a conservative field of force, and can therefore be characterized by quantum numbers in a similar way to the motion of an electron in an ordinary atom. In the atom and in the nucleus we have indeed to do with two extreme cases of mechanical many-body problems for which a procedure of approximation resting on a

---

[1]But Bohr suggests that the same occurs when other particles, such as α-particles, come into contact with the nucleus.

combination of one-body problems, so effective in the former case, loses any validity in the latter.

Bohr thus denies any validity to the ideas underlying so far the models of nuclear structure. The models assumed, as he rightly points out, that each particle moved in the attractive potential created by the other particles and that, in a first approximation, one could forget the other particles and replace them by an average potential.[1] The problem was thus reduced to the study of the motion of a single particle in a potential. This is called a "one-body problem." For example, the rotation of a planet around the sun reduces to a one-body problem because the motions of the two bodies is perfectly symmetrical and can be mathematically reduced to the motion of a single planet around a virtual point which is the center of gravity of the system and which usually lies within the interior of the sun. However, at the beginning of the century, the works of Henri Poincaré showed that, as soon as a third particle is present, the problem becomes more complicated and no stable solution exists. Today, we qualify such a system as "chaotic" which means that it is not possible to predict its evolution for a long span of time. But in the case of the solar system, the mass of the sun is so much greater than the mass of the planets that one can in a first approximation calculate the orbit of one planet while ignoring the presence of the others. The effect of the latter may be taken in account by calculating a small correction which does not modify significantly the initially calculated orbit. The atom is a somewhat similar system because of the presence of the central positively charged and heavy nucleus. But Bohr states that the nucleus is a completely different system in which *all the particles have the same mass and interact in a similar fashion.* There is no distinct center of the nucleus. And to make things worse, the neutrons and the protons are quite close to one another in the nucleus so that it becomes difficult to understand how a nucleon could describe an "orbit" in the nucleus without immediately colliding with another nucleon. The picture described by Bohr makes the nucleus quite similar to a liquid drop.

The Structure of the Nucleus According to Bohr in 1937

A year after the publication of his famous paper in *Nature*,[2] Bohr goes further in a detailed paper published by the Royal Society of Sciences of Denmark and written together with the young and brilliant Danish physicist Fritz Kalckar, who unfortunately died a year later of cerebral hemorrhage at the age of 28. The paper states the basic idea of the compound nucleus:

The extreme facility of energy exchanges between the densely packed particles in atomic nuclei plays a decisive role in determining the

---

[1] Bohr uses the word "conservative," which simply means that energy is conserved.
[2] See p. 336.

course of nuclear transmutations initiated by the impact of material particles. In fact, the assumption underlying the usual treatment of such collisions, that the transmutation consists essentially in a direct transfer of energy from the incident particle to some other particle in the original nucleus leading to its expulsion, cannot be maintained [238].

Bohr's fundamental idea is reiterated:

On the contrary, we must realize that every nuclear transmutation will involve an intermediate stage in which the energy is temporarily stored in some closely coupled motion of all the particles of the compound system formed by the nucleus and the incident particle. On account of the strong forces which come into play between any two material particles at the small distances in question, the coupling between the particles of this compound system is so intimate that its eventual disintegration—whether it consists in the release of an "elementary" particle like a proton or a neutron, or of a "complex" nuclear particle like a deuteron or an α-ray—must be considered as a separate event, independent of the first stage of the collision process.

Thus, according to Bohr, every nuclear reaction proceeds in two steps. At first, the particle which collides with the nucleus penetrates and forms a *compound nucleus* which exists for a "long" time. This "long time" is in fact a very short time, but not at the scale of the nuclear reaction which involves extremely small distances and high particle velocities. The compound nucleus lives for a time which is much greater than the time required for the incident nucleon to travel through the nucleus. Under these conditions, it is reasonable to assume that the decay of the compound nucleus, which is the second step, is a completely independent process which bears no "memory" of the formation of the compound nucleus in the first step.

Bohr and Kalckar then consider the spin of the nucleus, that is, its intrinsic angular momentum:

In particular any attempt of accounting for the spin values by attributing orbital momenta to the individual nuclear particles seems quite unjustifiable. We must in fact assume that any orbital momentum is shared by all the constituent particles of the nucleus in a way which resembles that of the rotation of a solid body.

One can compare Bohr's model of the nucleus to a track of bumper cars. He refutes the idea that the spin of the system, that is, its rotation, can be the sum of the rotations of each individual nucleon which would rotate without colliding with the others. On the contrary, the motion of each particle is chaotic. The spin of the nucleus must be due to the rotation of the system as a whole as if, in our analogy, the car track itself would rotate thereby making all the cars rotate together.

Once again, Bohr refutes completely a "shell model" of the nucleus which would be similar to the shell model with which he was able to explain the configuration of electrons in an atom. True, the nucleus is quite different from the atom. But what about the attempts of Wigner and Feenberg[1] to describe the nucleus in terms of a shell model and which was successful at least for light nuclei? And what about the regularities which were observed by Guggenheimer and Elsasser?[2] Were they illusions?

Be it as it may, Bohr's theory impressed physicists because it was an intuitive model and it seemed to be obvious as soon as it was formulated. His arguments appeared to be irrefutable. Physicists adopted it, and one consequence was to discourage attempts to describe the motion of protons and neutrons in the nucleus in terms of a shell model similar to the one of the atom with orbits bearing well-defined quantum numbers. Later, Feenberg wrote:

> In the field of nuclear reactions the overwhelming success of the resonance formalism and the idea of a compound nucleus had one unfortunate consequence in that the intellectual climate became unfavorable to further work on shell structure and remained so until the rapid increase in experimental information made the possibility of a useful shell model more and more obvious [239].

The idea that protons and neutrons might form successive "shells" in a nucleus seemed to be given up at the time.

---

[1] See p. 276.
[2] See p. 272.

The Death of a Giant: Ernest Rutherford

In the fall of 1937, Rutherford celebrated his 66th birthday. At the summit of his career and celebrity, he continued to be active in research. Since 1933, he was president of the Society for the Protection of Science and Learning, which helped refugees, mostly Jews who had to flee Nazi Germany or other countries in Central Europe. On October 7, 1937, he wrote a letter to one of the refugees:

I have just returned from a good holiday in the country, and I am leaving for India at the end of November to preside over a joint conference of the British Association and the Indian Science Congress at Calcutta. I have not been in India before, and I shall be glad of the opportunity of seeing something of that country [240].

He had decided to retire in 1941 when he would reach the age of 70. Several of the physicists which he had formed in the Cavendish had left: Chadwick went to Liverpool, Blackett to Birkbech College, Ellis to King's College in London, and Oliphant[1] to Birmingham. This illustrates the quality of the physicists who had been working under his supervision at the Cavendish, but the laboratory became progressively depopulated. Furthermore, Rutherford began the construction of a cyclotron, and he would have liked to leave when all was working well. He also wished to find a successor [241].

On Thursday, October 14, 1937, he did not feel well. The diagnosis was a strangulated hernia, and he was operated on Friday evening. Four days later, he died of septicemia [242]. In its October 30 issue, *Nature* published the tributes of several physicists, such as Arthur Eve, James Chadwick, Niels Bohr, and Frederick Soddy [243]. Chadwick ended with the following words:

Even the casual reader of Rutherford's papers must be deeply impressed by his power in experiments. One experiment after the other is so directly conceived, so clean and so convincing as to produce a feeling almost of awe, and they came in such profusion that one marvels that one man could do so much. He had, of course, a volcanic energy and an intense enthusiasm—his most obvious characteristic—and an immense capacity for work. A "clever" man with these advantages can produce notable work, but he would not be a Rutherford. Rutherford had no cleverness—just greatness. He had the most astonishing insight into physical processes, and in a few remarks he would illuminate a whole subject.

---

[1]Mark Oliphant was born in 1901 in Adelaide, Australia. In 1927, he joined Rutherford at the Cavendish Laboratory, where he received a PhD in 1929 and began nuclear research with Rutherford. In 1937, he became professor of Physics at Birmingham, and in 1939, he began to build a 60-inch cyclotron with the help of Ernest Lawrence. During the Second World War, he worked on radar and joined the Manhattan Project. After the war, he became a founding member of the Pugwash Movement. He returned to Australia in 1950 and became the research director of the Australian National University. He died in 2000.

Chadwick then describes the pleasure and marvel which he experienced when he worked with Rutherford. He considers Rutherford to be the greatest experimental physicist since Faraday. He adds:

I cannot end this tribute to Rutherford without some word about his personal qualities. He knew his worth but he was and he remained, amidst his many honors, innately modest. Pomposity and humbug he disliked, and he himself never presumed on his reputation or position. He treated his students, even the most junior, as brother workers in the same field—and when necessary spoke to them "like a father." These virtues, with his large, generous nature and his robust common sense, endeared him to all his students. All over the world workers in radioactivity, nuclear physics and allied subjects regarded Rutherford as the great authority and paid him tribute of high admiration; but we, his students, bore him also a very deep affection. The world mourns the death of a great scientist, but we have lost our friend, our counsellor, our staff and our leader.

Fermi wrote in Nature:

If we consider most of his experiments, we are impressed by the fact that they are conceived so simply as to be easily understood and appreciated by a layman: their performance does not require a complicated piece of machinery, nor even often exceptional experimental skill. But it is not an exaggeration to state that such simple experiments, as for example the discovery of the positive nucleus inside its cloud of electrons, or the method for producing artificial disintegrations by α-particle bombardment, are milestones in our knowledge of Nature [244].

Rutherford was buried on October 25, 1937 in Westminster Abbey, next to the graves of Newton and of Kelvin.

Hans Bethe Sums Up the Situation in 1936–1937

An impressive amount of experimental data has accumulated, and theory is making rapid progress. A young and talented German physicist, exiled in the United States, sums up the situation.

Barely four years after the discovery of the neutron, the way physicists understood the nucleus had considerably changed. This was due to the increasing amount of experimental data and to the rapidly progressing theory. Three papers appeared, in April 1936 and April and July 1937, in *Reviews of Modern Physics,* a trimestrial journal created in 1929. They were written by the young German physicist Hans Bethe, who had recently emigrated to the United States.

The first paper [245], written with the American physicist Robert Bacher, is devoted to the structure of the nucleus. It reviews the experimental results and summarizes the present theories. The second paper [246] is devoted to a theoretical description of nuclear reactions produced by collisions of nuclei with protons, neutrons, and α-particles. The third paper [247] is written with Stanley Livingston, and it summarizes the experimental results obtained with nuclear reactions. The papers remained references for at least 20 years. They not only review and summarize results but they propose a synthesis to which Bethe often adds a grain of salt without admitting it explicitly.

Hans Albrecht Bethe

Hans Albrecht Bethe was born on July 2, 1906, in Strasbourg, which was then part of Germany. His father was protestant, and his mother was a Jew, converted to Protestantism. However, religion did not play an important role in the family [248]. Hans studied in Frankfurt and Munich. He defended his PhD thesis in 1928. His thesis advisor, Arnold Sommerfeld, considered him as his best student after Heisenberg and Pauli. He then travelled successively to Frankfurt, Stuttgart, and Munich, where he became *Privatdozent* in May 1930; to Cambridge where he met Patrick Blackett; and to Rome in 1931 and 1932 where he worked with Enrico Fermi, a collaboration which influenced him deeply. He was nominated assistant in the University of Tübingen but was expelled for being a Jew when Hitler came to power. He then emigrated first to England where he obtained temporary jobs in Bristol and Manchester. In 1934, he was offered a position of assistant professor at the University of Cornell in New York state. He accepted and became professor in 1937. In Cornell, Bethe created a prestigious school of theoretical physics.

Many first class physicists graduated there, among whom Emil Konopinski, Richard Feynman, Freeman Dyson, Richard Dalitz, Jeffrey Goldstone, and David Thouless. He became an American citizen in 1941.

Ever since he began working with Sommerfeld in Munich, Bethe displayed an incomparable mastery of all physics. He could perform powerful calculations, and he had a prodigious memory. Throughout all his life, he remained both honest and courageous, when, for example, he defended Oppenheimer during the trying MacCarthy period. In 1967, Hans Bethe was awarded the Nobel Prize in physics for his work on nuclear reactions in particular for his discovery of energy production in stars.

Let us describe some of the salient points of his three papers.

The Structure of Nuclei

The first paper, also the most famous, concerns the structure of nuclei. The progress achieved in a few years is impressive.

The Size of Nuclei

In Bethe's own words:

The radii of nuclei range from about 2 or 3×10^{-13} cm to about 9×10^{-13} cm for the uranium nucleus. It seems that the volume of a nucleus is approximately proportional to its mass number, so that the volume per elementary particle is about the same in every nucleus.

The *mass number* mentioned by Bethe is the total number of protons and neutrons in the nucleus. Since the volume of the nucleus is proportional to the mass number, each proton and neutron fills about the same space in all nuclei. This agrees with the liquid drop model of the nucleus in which the protons and neutrons play the role of the molecules of the liquid. The description of a nucleus as a liquid drop, initially proposed by Gamow,[1] gains ground especially since it is defended by another great physicist, Niels Bohr.

The values of the nuclear radii are deduced from the estimates made by Gamow who studied the energies of α-particles emitted by radioactive nuclei,[2] and from the scattering of α-particles by light nuclei observed by Étienne Bieler[3] and compiled by the English physicist Ernest Pollard [249] in 1935. The radii quoted by Bethe

---

[1] See p. 248.
[2] See p. 265.
[3] See p. 211.

are smaller than those estimated by Rutherford in 1920[1] and they will not change thereafter. Thus, in 1936, the fact that nuclei are droplets of matter with a roughly constant density seems well established. Light nuclei are the only exceptions.

The Mass and Binding Energy: The Weizsäcker Formula

Bethe insists on the importance of the mass measurements because the mass of a nucleus is a measure of its binding energy and therefore of its stability, one might even say its solidity. Recall that the mass of a nucleus is smaller than the mass of its constituent nucleons: the missing mass is equal to the energy required to tear it apart into its constituents: it is called the binding energy of the nucleus. At that time, a new generation of more precise mass measurements was initiated by Josef Mattauch in the University of Vienna and in the *Kaiser Wilhelm Institut für Chemie* in Berlin; by Alfred O. Nier, in the University of Minnesota; and by Kenneth Bainbridge, a young American physicist who had constructed his first spectrograph in Princeton in 1929 and later more and more perfected ones in Harvard. After the Second World War, they will become recognized experts in the field.

Bethe insists on the fact that the energy required to extract a proton or a neutron from a nucleus does not vary much from nucleus to nucleus, except for the lightest nuclei. If each neutron or each proton was subject to the attraction of *all the other* protons and neutrons, the separation energy would increase with the nuclear size because, as the nucleus became larger, the number of broken bonds would also increase. The fact that the separation energy does not increase with the nuclear size or mass is a so-called saturation property.

Finally, Bethe quotes the semiempirical formula of Carl Friedrich von Weizsäcker [250]. The formula yields the masses of all nuclei with good average accuracy. Weizsäcker published his formula in 1935, and it became famous in the form in which Bethe published it. The formula allows one to calculate the mass of a nucleus by adding the masses of its constituent nucleons and then making corrections following steps almost as if one were filling a tax form:

- Subtract a quantity proportional to the number of nucleons (protons plus neutrons) in the nucleus. This is a first approximation to the binding energy which was known to be roughly proportional to the number of nucleons.
- Add a quantity which tends to favor energetically nuclei with equal numbers of protons and neutrons.
- Add a quantity which is proportional to the surface of the nucleus and which reduces the binding energy. This term takes into account the fact that nucleons at the surface of the nucleus are only attracted by nucleons on one side, namely, those which are in the nucleus. The same phenomenon gives rise to *surface tension* in liquids.

---

[1] See p. 208.

- Finally add a quantity which reduces the binding energy of a nucleus when its electric charge increases, because like charges repel.

The first three terms are multiplied by parameters the values of which are adjusted so as to reproduce the observed nuclear masses as accurately as possible. The last term due to the Coulomb repulsion can be calculated exactly. Although the three parameters are empirical, meaning that they are not deduced from fundamental physical constants, Weizsäcker's formula yields remarkable fits to nuclear masses and binding energies. This is a further argument in favor of the liquid drop model of the nucleus.

Nuclear Forces

This paper of Bethe and Bacher was written before the revolutionary theory of Yukawa became known.[1] Bethe lists the main features of nuclear forces, namely, their short range and their saturation property. The "exchange" forces of Heisenberg and Majorana fulfill these requirements. He also quotes the work of Wigner who used an ordinary, but short-range force. However, Wigner's force does not give rise to saturation and therefore contradicts experimental observations.[2]

The Structure of the Nucleus

Bethe is quite aware of the arguments put forward by Bohr. Nonetheless, he mentions the efforts of the defenders of the shell model of the nucleus, stressing the fact that the shell model lacks justification but that no other means are known to make calculations. Furthermore, there are disturbing experimental facts stressed by Elsasser and Guggenheimer. Bethe writes:

> This assumption can certainly not claim more than moderate success as regards the calculation of nuclear binding energies. However, it is the basis for a prediction of certain periodicities in nuclear structure for which there is considerable experimental evidence.

The Intrinsic Angular Momenta of Nuclei: The *Spin* of Nuclei

Is this internal "rotational motion," also called spin, due to the rotation of its constituents, caused mainly by a few protons and neutrons, or is it, as Bohr believes,

---

[1] See p. 331.

[2] The book *Nuclear forces*, by David Brink (Pergamon Press, 1965), gives a very lucid account of the history and development of nuclear forces in the period 1932–1952. It contains a collection of 14 original papers by Bohr, Heisenberg, Wigner, Majorana, Yukawa, etc., and an introduction in which the ideas underlying the papers are discussed.

due to the rotation of the nucleus as a whole, as if it were a solid body? Be it as it may, numerous measurements had been made by 1936. Most relied on the perturbation of the nuclear spin on the motion of the surrounding electrons. Bethe and Bacher quote values for the spin of about 40 nuclei.

The Magnetic Moments of Nuclei

Bethe explains that the magnetic moments of numerous nuclei are known, thanks to a remark made by Pauli, who showed that if the nucleus behaves as a small magnet, it can somewhat perturb the observed atomic spectral lines caused by electrons making transitions from one quantum state to another.[1] This was called the "hyperfine structure" of the spectral lines, and the magnetic moment of the nucleus could be deduced from it. The magnetic moments of about thirty nuclei were known in 1936.

Are Some Nuclei Deformed? The Nuclear "Quadrupole Moments"

The *quadrupole moment* of a nucleus is a quantity which gives a measure of how much the electric charge in the nucleus differs from that of a spherical system. A spherical nucleus has a zero quadrupole moment, whereas an ellipsoidal nucleus, with a shape similar to the football used in rugby, has a positive quadrupole moment which increases as its shape is elongated. If, on the contrary, the nucleus has a flattened shape, such as a pumpkin, its quadrupole moment is negative and its size measures how flattened it is.[2] The quadrupole moment only tells us what the shape of the nucleus is on the average. It does not reveal the full shape, but only how much it differs from the shape of a sphere. Two German physicists, Hermann Schüler and Theodor Schmidt, had determined the quadrupole moment of two nuclei, the 151 and 153 isotopes of europium. They achieved this by measuring the hyperfine structure of the corresponding atomic spectral lines.[3] They noted irregularities in the atomic spectra which could only be accounted for by assuming that the nuclei had a nonvanishing quadrupole moment and that they were therefore not spherical [251]. They suggested that a similar phenomenon might explain the observed anomalies in the hyperfine structure of the isotope 115 of lutetium, the isotope 115 of indium, the isotopes 121 and 123 of antimony, and the isotope 101 of mercury. Numerous deformed nuclei may exist. We shall return to this later.

---

[1] See p. 247.

[2] For those who are curious, the exact formula for the quadrupole moment of a nucleus shaped as an ellipsoid, the principal axes of which are a and b, is $Q = \frac{Ze}{10}(a^2 - b^2)$, where Z is the number of electric charges (therefore, the number of protons) and e the elementary electric charge, that is, the electric charge of the proton.

[3] See p. 247.

Nuclear Reactions

The following two papers deal with what Bethe calls "nuclear dynamics," meaning phenomena generated by collisions of protons, deuterons, neutrons, and α-particles with nuclei. A detailed theory is presented in the second paper. It is essentially based on the theory of the compound nucleus of Bohr and on the related theory of resonances, introduced by Breit and Wigner.[1] Shortly before, Bethe had published a paper [252] together with the exiled Czech Jewish physicist, George Placzek. In this paper, they generalized the theory of Breit and Wigner to the case where a particle which penetrates into a nucleus does not form a single state of the compound nucleus, but several such states. Born in Brno, Czechoslovakia, in 1905, George Placzek played an important role in physics in the 1930s and 1940s. He studied in Prague and in Vienna where he obtained his PhD in 1928, after which he travelled a lot: to Utrecht (1928–1931), Leipzig (1931), Rome (1931–1932), Copenhagen (1932–1933), Jerusalem (1933–1934), Kharkov (1935–1936), Paris (1938), and finally to New York in 1939. He spoke many languages and was very cultured. Edoardo Amaldi called him a real European [253]. He was very lively, able to grasp quickly the various aspects of a problem, as well as the weak points. He had a keen critical sense which was appreciated by his colleagues who also appreciated his advice. His family was wiped out by the Nazis during the war. He died in 1955 during a trip to Italy.

---

[1] See p. 335.

The Fission of Uranium

> *The number of "transuranic" nuclei is increasing in Berlin.*
> *Paris voices some disagreement. Chemists in Berlin observe an*
> *incredible phenomenon. The key to the mystery is revealed in the*
> *cold Swedish snow by two Austrian refugees. Fission ignites keen*
> *interest and gives rise to a speculative energy-producing chain*
> *reaction. The shadow of a terrifying weapon begins to spread*
> *across the world.*

The Italian team of Enrico Fermi had brilliantly initiated a new field of nuclear physics by exposing a series of elements to neutrons. Dozens of new radioactive elements were produced this way.[1] Some of the elements produced by bombarding uranium were mysterious because they could not be identified chemically. What were they exactly?

A Fragile Discovery: The Transuranic Elements

In the other cases, it was possible to identify the radioactive nucleus, produced by the incident neutrons, with a nucleus close to the target nucleus. This had a simple interpretation: the target nucleus absorbed the neutron, and the resulting nucleus was unstable. It therefore decayed by emitting β ray thereby transforming one of its neutrons into a proton. The resulting nucleus had one proton more than the target nucleus, and it was therefore an isotope of the following element in the Mendeleev table. It was usually also unstable, and it emitted a further β electron. The resulting nucleus had to be chemically identified using radiochemistry.

Radiochemistry was invented by Pierre and Marie Curie, and it was further perfected by radiochemists such as Frederick Soddy and Otto Hahn. The method used in the 1930s consisted in dissolving the unknown radioactive substance usually into an acid solution and in adding to this solution a known element called the *driver*. Several chemical reactions were then performed with the solution thus obtained, and one sought if the unknown radioelement behaved or not like the driver, if it was driven with it into the chemical reactions. If that was the case, the chemical properties of the unknown substance and of the driver were similar, and it was inferred that they belonged to the same chemical family, that is, to the same column of the Mendeleev table. If it really had identical chemical properties, it was then identified as the same element, possibly an isotope of the driver.

---

[1] See p. 320.

However, in the case of uranium, nobody could identify the chemical nature of the β ray emitter. Fermi and his team had shown that it could not be uranium, nor any of the lighter nuclei between uranium and lead, such as protactinium, thorium, actinium, radium, and bismuth. Could it be a lighter element? That seemed unlikely. The lead nucleus has ten protons and 20 neutrons less than uranium. How could a low-energy neutron manage to extract so many nucleons from the uranium target?

It seemed natural to conclude that a new element had been formed, an element with an atomic number greater than that of uranium (92), which at the time had the highest known atomic number. It could be the element 93 obtained by an β decay of the isotope 239 of uranium, created when a neutron was absorbed by uranium 238. But how could one confirm this? The first step consisted in checking that no experimental error had been made. Indeed, a paper of Aristid von Grosse suggested that it was in fact protactinium [254]. Now, Lise Meitner and Otto Hahn were quite familiar with the chemical properties of protactinium, which they had discovered [255, 256] in 1917. Furthermore, Hahn had an unpleasant disagreement with Grosse, one of his former students, concerning exactly who had discovered protactinium. On the other hand, Lise Meitner had been interested in the discovery of artificial radioactivity by Irène Curie and Frédéric Joliot, and she confirmed their observations in an experiment performed with a Wilson cloud chamber [257]. She also followed the experiments of the Fermi team in Rome. She was quite impressed by their first results:

> *I found these experiments so fascinating that, upon their appearance in* Nuovo Cimento *and* Nature, *I immediately persuaded Otto Hahn to again resume our direct collaborative work—which had been interrupted for several years—in order to dedicate ourselves to these problems [258].*

We thus find them bombarding uranium with neutrons and attempting to identify the admittedly numerous radioactive substances thus produced [259]. This was the outset of a new collaboration which Lise Meitner described in 1954 as follows:

> *This collective work was performed with good spirits and in a cheerful mood, which reflect the personality of Hahn. This proved most favorable for work but it was also present during the Christmas festivities or during anniversaries. On one such occasion our two departments were described, in verse, as a "henhouse" and every one of us was targeted, however with kindness according to the following motto:*
> *Only make fun, do not hurt*
> *That's what we wish, following the rule:*
> *Tact is everybody's law.*
> *Good wit shuns malice [260].*

Lise Meitner and Otto Hahn make a quick judgment of the suggestion of Grosse:

> We thus see by direct observation that the two substances are
> not isotopes of the element 91 as has been suggested by A. v.
> Grosse [203].

Using their expertise in radiochemistry, they extend the measurements of Fermi and they exclude the following elements: francium, radon, astatine,[1] bismuth, lead, and mercury. For Otto Hahn and Lise Meitner, there remained only one possibility which they state with caution, as Fermi did:

> But it is easy for a chemist to see that the reactions which we have
> described exclude not only the elements 90, 91 and 92, but also all the
> elements up to mercury (except possibly eka-iodine), so that it appears
> very likely that the elements with half-lives of 13 and 90 minutes are
> beyond 92 [...] So far the research seem to show that the 13-minute
> substance could be the element 93 and the 90-minute one the isotope
> 94, since the two substances are not isotopes of one another [203].

But are these substances, which are only known for their 13 and 90 min half-lives, *transuranic* elements? Only negative information concerning their chemical properties was available at the time. But to conclude from this that it was the hypothetical element 93 was a step which the German chemist Ida Noddack, known to have discovered rhenium, the element 75, was not ready to take:

> This method of proof is not valid. Fermi compared his new β
> emitter not only with the immediate neighbor of uranium, namely
> protactinium, but he also considered several other elements down to
> lead. That indicates that he thought that a series of consecutive
> decays was possible (with emission of electrons, protons, and helium
> nuclei), which eventually formed the radioelement with the 13 minute
> half life. It is not clear why he did not investigate the element polonium
> (84) which is also between uranium (92) and lead (82), and why
> he chose lead. The old view that the radioactive elements form a
> continuous series which ends at lead or thallium (81) is just what the
> previously mentioned experiments of Curie and Joliot had disproved.
> Fermi therefore ought to have compared his new radioelement with
> all known elements [261].

Compare the known element to all the known elements! Ida Noddack was asking for quite a lot, and indeed, to most physicists, it seemed most unlikely that by adding

---

[1]This is what Hahn and Meitner call *eka-iodine* because it is a hole in the periodic table, which corresponds to a halogen element which is close to iodine for its chemical properties and which will not be observed before 1940. The prefix *eka* means 1 in Sanskrit, and it was used for the first time by Mendeleev in 1872 when he predicted elements which were not known at his time and for which he left blank positions in the periodic table. He gave them provisional names: eka-aluminum, eka-boron, and eka-silicon. These elements were discovered later: gallium in 1875, scandium in 1879, and germanium in 1886. By calling *eka-aluminum* the predicted element, Mendeleev wished to stress that an element, with properties similar to those of aluminum, should find a position in the same column of his table.

a single neutron to a uranium nucleus you would produce much lighter nuclei. Even mercury (with atomic number 80) seemed quite distant. However, Ida Noddack, who was quite aware of such an objection, replied to it in the same paper:

> One could assume equally well that when neutrons are used to produce nuclear disintegrations, some distinctly new nuclear reactions take place which have not been observed previously with proton or alpha-particle bombardment of atomic nuclei [. . .] it is conceivable that the nucleus breaks up into several large fragments, which would of course be isotopes of known elements but would not be neighbors of the irradiated element.

That was indeed a logical possibility, but it looked so unlikely! To invoke these new reactions seemed to be an unfounded speculation. Even Ida Noddack made no attempt to explore this idea experimentally. Her conjecture was forgotten.

Loads of "Transuranic" Elements

Lise Meitner and Otto Hahn pursue their experiments in order to study further these new elements, the transuranic nature of which is likely but not proved. Two months later, they announce that the famous element with a 90 min half-life is probably a mixture of two substances, one with a half-life between 50 and 70 min and the other with a much longer half-life lasting probably for days [262]. After 5 months of work, they publish a paper which states the chemical properties of the observed radioactive substances: the substance with a 13-min half-life has chemical properties similar to rhenium (the element 75), and it could be the first transuranic element, the element 93, which they call *eka-rhenium*; the substance with a 90 min half-life would finally be a single element with chemical properties similar to osmium, and it could be the element 94 which they call *eka-osmium* [263]. It is only in March 1936 that they publish a new paper in which the different radioactive substances are listed (Table 1).

In order to try to put some order into this apparently inextricable maze, they ask the young chemist Fritz Strassmann to help them. Born in 1902, Strassmann obtained a doctoral degree in engineering in Hanover in 1929. He then joined the *Kaiser Wilhelm Institut für Chemie*, hoping that working with Otto Hahn would help him find a job in industry. He enjoyed so much working in the laboratory that he remained there until 1933, in spite of earning nothing since 1932. However, to obtain a job in industry, he would have needed to join the Nazi party. In 1933, he refused to do so, and he was thus also prevented from getting a job in a university.[1]

---

[1] In 1943, in the midst of World War II, Strassmann and his wife Maria Heckter Strassmann saved the life of a Jewish woman, a 46-year-old pianist, Andrea Wolffenstein, by concealing her in their house in spite of the great risk involved [265]. This earned him recognition as a "Righteous Gentile" by Yad Vashem in Israel.

Table 1 The radioactive decays as published in 1936 by Lise Meitner and Otto Hahn [264]

β radiation produced by irradiating uranium with neutrons	
Substance	Radioactive half-life
$^{235}_{90}\text{Th}$	4 min
$^{235}_{91}\text{Pa}$	Very short?
$^{235}_{92}\text{U}$	24 ± 2 min
$^{237}_{92}\text{U}$	40 s
$^{239}_{92}\text{U}$	10 s
$^{237}_{93}\text{EkaRe}$	16 ± 1 min
$^{239}_{93}\text{EkaRe}$	2.2 min
$^{237}_{94}\text{EkaOs?}$	12 h
$^{239}_{94}\text{EkaRe}$	59 min
$^{239}_{95}\text{EkaIr}$	3 days

The first five lines correspond to know nuclei: uranium (symbol U), thorium (Th), and praseodymium (Pa) Otto Hahn and Lise Meitner distinguish three different isotopes of uranium (the upper number, on the left of the symbol, is the total number of neutrons and protons and therefore is different for different isotopes, while the lower number is the number of protons, which characterizes the element uranium). They argue that the last five lines correspond to "transuranic" elements 93, 94, and 95, chemical analogues of rhenium, osmium, and iridium, hence their names: EkaRe, EkaOs, and EkaIr.

Upon the request of Lise Meitner, Otto Hahn offered him a meager salary (50 marks per month!). By 1936, Fritz Strassmann had become an experienced radiochemist. A year later, in May 1937, Otto Hahn, Lise Meitner, and Fritz Strassmann publish in *Zeitschrift für Physik* a 22-page paper in which they present what they believed to be three *radioactive series* which are produced when uranium is bombarded neutrons [266]. There was one seemingly strange thing in the way the various radioactivities were distinguished: the nucleus which is formed when a uranium nucleus (assumed to be the isotope 238, by far the most abundant) absorbs a neutron is the isotope 239 of uranium. Why then, starting with this single nucleus, should one distinguish three radioactive families? Otto Hahn and Lise Meitner believe that they are dealing with a phenomenon which they had discovered a few years earlier. In the succession of substances produced by the decay of uranium, they observed a "protactinium" stage; more specifically, the isotope 234 existed in two forms with different radioactive half-lives although it corresponded to exactly the same nucleus [267, 268]. It consisted of two configurations of protons and neutrons which were sufficiently different to survive long enough without one decaying into the other. These two forms were called *isomers*, and the general phenomenon was called *isometry*. It was explained in 1936 by Carl Friedrich von Weizsäcker on the basis of quantum mechanics [269]. Otto Hahn and Lise Meitner imagined that when a neutron was absorbed by a uranium 238 nucleus, the resulting uranium 239 nucleus could be formed in three different isomeric states, each with its own decay mode, so that each *isomer* would give rise to a *family of isomers*, independent of the other. Another oddity: none of the nuclei assumed to be heavier than uranium seemed to decay by emitting an α-particle. One knew that this was a very frequent decay mode of the natural radioactivity of nuclei with similar masses, such as uranium and thorium. An attempt to detect such α-particles was made in Berlin, with no success [270].

At the Institut du Radium

Radiochemistry had a great tradition at the *Institut du Radium*. It was a heritage left by Marie Curie. When she died in 1934, André Debierne, a long-time collaborator of Pierre and Marie Curie, became the director of the *Institut du Radium*, but it was Frédéric Joliot and Irène Curie who gave most impetus to the scientific work performed there. In addition, a young chemist from the *École Municipale de Physique et Chimie Industrielle*, Bertrand Goldschmidt, joined the *Institut du Radium* as assistant to Marie Curie in 1933, at the age of 21, and he remained there to do his Ph.D. under the direction of Debierne. In a book written in 1987, Goldschmidt describes the laboratory at the time:

> The life in the laboratory was dominated by the Joliots, Fred and Irène, as we usually called them. They inspired and often directed most of the current research. The celebrity of the Institut du Radium attracted young researchers from all over the world. Among the most brilliant arrivals at the end of 1935, were Bruno Pontecorvo, an Italian of my age, who was very charming and who immediately became very popular, and Hans Halban, an ambitious and self-confident Austrian, a few years older. They were soon joined by Lew Kowarski, a russian emigrant who will also play a rôle in this story[. . .]
>
> There were about fifteen different nationalities among these researchers and an exceptional number of young women, who were attracted by the prestige of Mme Curie, a symbol of a successful emancipation of a woman [. . .]
>
> There was no distinction between the French and the foreigners. Among the latter, some were supported financially by their government, others, either political refugees or decided not to return to their home country, obtained grants from French institutions [271].

At that time, Irène Curie is studying the new radioelements obtained by bombarding various substances with neutrons. She begins by forming a research team with two young physicists, Hans von Halban and Peter Preiswerk. Born on July 7, 1908, in Leipzig, Hans von Halban [272] was the son of an Austrian physicochemist who settled first in Frankfurt and then in Zurich, where the young Hans obtained his

PhD in 1935. He decides then to devote himself to nuclear physics and chooses the *Institut du Radium*, which had become famous for the discovery of artificial radioactivity.

The newly formed team begins a study of thorium: Irène Curie, Halban, and Preiswerk observe two unknown radioelements which they try to understand following the lines set by Otto Hahn and Lise Meitner: the neutron is absorbed by the thorium nucleus. An isotope of thorium is formed which decays by emitting either an α-particle or by β radiation. In the latter case, this results in an element heavier than thorium, bearing the number 91 or protactinium [273]. Then Halban and Preiswerk turn to slow neutron physics, which had been initiated by Fermi. They study how neutrons are slowed down by substances containing protons (hydrogen) and how the neutrons are diffracted and absorbed at precise energies [274, 275].

Irène Curie then tackles the hot subject, namely, radioactive substances produced by bombarding uranium with neutrons. Her new collaborator is the 27-year-old Yugoslav chemist Pavle Savić (who used the francized name Paul Savitch in his papers).

Although Hahn and Lise Meitner had attempted to sort out the various radioactive elements produced, Irène finds the situation still most confusing. The experiments are difficult because the β radiation of substances produced by the neutron bombardment needs to be detected and distinguished from the substances which uranium and its decay products are constantly emitting with a thousand times greater intensity. A particularly troublesome substance is *uranium X*.[1] To perform his experiments, Hahn would eliminate the undesired substances by purifying uranium chemically, shortly before the experiment, because the radioactivity of uranium was constantly remaking them. Irène Curie decides to place a screen between the Geiger counter and the uranium sample which has just been exposed to neutrons. The screen is a copper plate sufficiently thick (about 0.5 mm) to stop most of the β electrons naturally emitted by uranium, but allows the faster electrons to pass through. Irène Curie and Pavle Savić then discover several β radiations which are due to radioactivities which Hahn and Lise Meitner had not reported. One substance captures their attention: it has a 3 1/2 h half-life, and it is sufficiently abundant to be easily detected. In a paper sent at the end of July to the *Journal de Physique et Le Radium*, Irène states:

> The substance $R_{3.5h}$ has not been observed by Hahn, Meitner and Strassmann, nor by us, before we used the screens [...] However, it is not a rare mode of transformation [276].

So what can this new substance be? What are its chemical properties? Together with Pavle Savić, Irène Curie attempts to determine the behavior of the substance which she calls $R_{3.5h}$ for the time being, because all she knows about it is its 3.5 h half-life. They first think it is thorium:

---

[1] Which is the isotope 234 of thorium, $^{234}_{90}$Th.

Numerous experiments have shown that R$_{3.5h}$ separates from uranium and transuranic elements and that it accompanies uranium X [...] There is no doubt that R$_{3.5h}$ is an isotope of thorium, probably formed by the capture of a neutrons and the expulsion of an α-particle.

When Otto Hahn and Lise Meitner read the paper, they are surprised and find it hard to believe. In order to have a clear conscience, they make some further experiments to check: they find that it cannot be thorium. They believe that Irène Curie and Savitch are mistaken and that the substance to which they assigned a radioactive half-life of 3 1/2 h is nothing but a mixture of two known radioactive substances, one with a 2 h half-life and the other with a longer one. They explain this in a letter to Irène Curie:

Berlin-Dahlem, 20.1.1938
Dear Mrs Curie,
The purpose of our letter is the work, which you have published with P. Savitch concerning the artificial transmutation products of uranium, and in which you mention an isotope of thorium, which aroused a particular interest among us. Because we believe to have very convincing proof that these substances do not exist, we would like to make some of our arguments known to you. Indeed, we believe that that an exchange of letters can be far more fruitful for these questions than a discussion in journals [277].

They then propose that she should withdraw her statements:

We would like to know your opinion about the arguments developed above. If you accept our opinion, the simplest thing would be that you publish a note in your journal. In this case, we shall publish nothing on the matter.
With our best regards, also to M. Joliot,
Yours Lise Meitner, Otto Hahn

The letter is polite although not very warm. The authors are sure of what they claim, allowing no doubt whatever. Irène repeats the experiments with Pavle Savić, and she discovers that Hahn and Lise Meitner are right in their first point: the R$_{3.5h}$ is not thorium. But what she observes is even more disconcerting: by its chemical properties, the mysterious substance is similar to a rare earth, which makes it even more remote from the transuranic elements of Hahn and Lise Meitner. According to the sacrosanct Mendeleev table, as it was conceived at the time (see Table 2, p. 357), a substance produced by the absorption of a neutron and a β emission could be in the column of rare earths, the heavy counterpart of which is the element 89, namely, actinium. The following columns, in the order of increasing atomic number, corresponded to thorium, praseodymium, and uranium. The element 93 therefore had to be placed in a column to the right of these since it had four electric charges more than actinium. It had to be the counterpart of rhenium. But the chemical facts are stubborn. A new publication is sent to the *Comptes rendus de l'Académie des Sciences:*

Table 2 The beginning of the last two rows of the Mendeleev table as they were conceived in 1938 (the present version is displayed at the end of this book). The third column contains lanthanum and all the lanthanides which are homologous to actinium. As far as heavier elements are concerned, it was admitted that thorium (the element 90) was homologous to hafnium (the element 72), praseodymium to tantalum, uranium to tungsten, and so forth. In this view, the transuranic elements were supposed to be homologous to rhenium, osmium, and iridium, whence the denominations "EkaOs," "EkaRe," and "EkaIr." However, the transmutation of uranium into radium, produced by an incident neutron, was very difficult to imagine because it would entail the loss of four electric charges: such a process would require the emission of two α-particles.

55	56	57	72	73	74	75	76	77	78	79	80
Cs	Ba	La	Hf	Ta	W	Re	Os	Ir	Pt	Au	Hg
Cesium	Barium	Lanthanum	Hafnium	Tantalum	Tungsten	Rhenium	Osmium	Iridium	Platinum	Gold	Mercury
87	88	89	90	91	92	93	94	95	96	(97)	(98)
?	Radium	Actinium	Thorium	Praseo-dymium	Uranium	Eka-Re	Eka-Os	Eka-Ir	Eka-Pt	?	?
						Eka-Rhenium	Eka-Osmium	Eka-Platinum			

It therefore has properties similar to those of rare earths [. . .]
According to its chemical properties, it appears that this substance
could only be an isotope of actinium, or a new transuranic element
with chemical properties which are quite different to the corresponding
higher homologous elements rhenium and platinum. From the physics
point of view these two hypotheses face considerable difficulties. It is
necessary to obtain some further experimental data concerning this
radioelement in order to specify the mode of transmutation which
produces it [278].

It is March 21, 1938. Two months later, Irène Curie and Savić send another note:

We therefore wanted to know if it was or not an isotope of actinium
[. . .] $R_{3.5h}$ is easily separated from actinium [. . .] It appears therefore
that this substance can only be a transuranic element with consider-
ably different properties of the other known transuranic elements, a
hypothesis which raises problems for its interpretation [279].

There lies the dilemma: the substance has to be "transuranic," but its chemical properties assign it no place in the already numerous families constructed by Hahn and Lise Meitner. As it is stated, the problem has no solution.

Otto Hahn and Lise Meitner cannot believe that they could have missed a substance such as this $R_{3.5h}$, which has no place in the structure of the transuranic elements which they had constructed. Does it really exist? They believe that Irène Curie and Pavle Savić are mistaken. They are not only skeptical but also furious because this dispute raises doubts about their work in the worst possible time, when Nazi violence is spreading all over Germany. They hoped to rely, to some extent, on their international reputation to be protected. Of Jewish origin, Lise Meitner is particularly threatened. But until then, she resolutely wished to remain in the institute where she had her friends, where she could continue her research, and where she had reasons to live.

Lise Meitner Flees Nazi Germany

On March 1938, Hitler invades Austria where he is triumphantly received and where he declares the *Anschluß*, the annexation of Austria to Germany. From then on, Lise Meitner was no longer protected by her foreign Austrian passport, and the *Kaiser Wilhelm Institute* itself came under attack. She was dismissed from the institute. Worse still, she was suddenly trapped: without a passport, she was not permitted to leave Germany. None of the efforts made in her favor by the authorities of the *Kaiser Wilhelm Institut* succeeded in obtaining this permission. Physicists abroad then sought a solution which was found finally in Sweden where Manne Siegbahn accepted to offer her a position at the Nobel Institute. One of the most active was the Dutch chemist Dirk Coster who did all he could to welcome her to Groningen. He went in person to Berlin, officially invited to give a talk. When he left Berlin, both he and Lise Meitner boarded independently the same train. Dirk Coster and Lise Meitner crossed the Dutch border without incident on July 13, 1938, thanks to the Dutch guards that Coster had personally met beforehand. He took her to Groningen and gave her some money which he had collected from physicist colleagues. This allowed Lise Meitner to reach Sweden which admitted her without a passport and without a visa. As a provision for her journey, Otto Hahn had given her a ring which had belonged to his mother [280–282].

She had saved her liberty and no doubt also her life. But 30 years of uninterrupted work at the *Kaiser Wilhelm Institut* were brutally stopped. In Sweden, she was scientifically isolated and had meager means. She found it difficult to continue her research. A regular correspondence with Hahn allowed her to participate, to some extent, in the life of the laboratory in Berlin. It is thanks to these letters that we can follow, step by step, the work of Hahn during this crucial period.[1]

Otto Hahn and Fritz Strassmann Set Again to Work

In the spring of 1938, Frédéric Joliot and Otto Hahn meet for the first time at the tenth International Congress of Chemistry which is held in Rome on May 15–21. Their encounter is immediately friendly, and they discuss the latest results of Irène Curie. Hahn tells Joliot that she was certainly mistaken and that he was ready to prove her wrong experimentally [284]. However, Irène persists, and she even sends a more detailed paper to the *Journal de Physique et Le Radium* [285]. When Otto Hahn reads her paper, he feels obliged to repeat her experiments. He explains to all that Irène had mixed everything up and that he is now obliged to set things right [286]. The last paper signed jointly by Hahn, Meitner, and Strassmann is sent just before the hasty departure of Lise Meitner. The paper does not mention the work of Irène Curie, and it announces a new radioactive substance produced, yet again, by bombarding uranium with neutrons [287]. The authors fail to find a place for this substance in the decay scheme of "transuranic" elements; they believe that it is most likely a substance with chemical properties similar to iridium, an *eka-iridium*.

Hahn then reconsiders the problem with Strassmann, and in early November, they send a paper to *Naturwissenschaften* in which they refute the ideas of Irène Curie and Pavle Savić and in which they state:

> *While pursuing new research on the chemical properties of transuranic elements we therefore attempted to produce the 3.5-hour substance of Curie and Savitch. We also succeeded in obtaining this substance by the indicated separation and measurement methods [. . .]*

---

[1] These letters are published in *Im Schatten der Sensation, Leben und Wirken von Fritz Straßmann*, by Fritz Krafft [283].

As far as the 3.5-hour substance of Curie et Savitch, we believe that it is a mixture of substances which we have isolated and identified chemically [288].

The paper is published on November 18, 1938. Their refutation of the results of Irène Curie and Savić is in line with their letter dated January 20, 1938. The $R_{3.5h}$ of Irène Curie would be a mixture, the main component of which would be an isotope of uranium, which is chemically close to barium and lanthanum. Now, in order to produce radium from uranium, *four* electric charges, and therefore *two* α-*particles*, need to be emitted. However, this idea raises difficult problems because the extraction of *two* α-particles consumes quite a lot of energy and one does not see where this energy could come from, since the neutron furnishes hardly any energy. Worse still, as mentioned above, a student of Lise Meitner, Gottfried von Droste, had already attempted to detect the α-particles emitted by the uranium bombarded by neutrons, but in vain [270]. Hahn discusses his results with several physicists who all agree: they fail to see how a low-energy neutron could provoke the emission of two successive α-particles. On November 13, 1938, he was even able to discuss it with Niels Bohr and Lise Meitner. Bohr had invited him to Copenhagen to give a talk, and Lise Meitner, who had been informed, awaited him on the platform in the railway station. Intense discussions took place until Hahn left the following day. Because of the political situation in Germany, Hahn did not mention this encounter with Lise Meitner [289]. In any case, both Bohr and Lise Meitner shared the same opinion: it was difficult to imagine the successive emission of two α-particles.

In order to clear this up, Hahn decides then to repeat the experiments with Strassmann.

More and More Disconcerting Results

Hahn and Strassmann carefully identify, yet again, the substances which are produced when uranium is bombarded with neutrons. At first, the experiment seems to be a repetition of the previous one: there does indeed exist a radioactive substance which is created by the neutrons and which, in all chemical reactions to which it is subjected, reacts as barium, the homologous substance to radium. Hahn and Strassmann believe that it is probably a variety of radium, with the difficulty that it has four electric charges less than uranium and that the transmutation of one into the other is far from obvious. Remember that Pierre and Marie Curie had discovered radium by noting that in natural uranium there was a radioactive substance which had chemical properties similar to those of barium. The separation of barium from radium is not an easy task. Hahn, as Marie Curie had done 40 years earlier, uses the method of fractional crystallization, but, to his great surprise, the separation does not occur! The "radium" behaves *exactly* as barium! On December 19, 1938, Hahn writes to Lise Meitner:

Monday evening, 19, at the lab [. . .]

*Indeed something so strange happens with the "radium isotopes"
that for the time being we tell it only to you[. . .] Our radium isotopes
behave like Ba [barium] [. . .] They can be separated from all elements
except barium. All reactions agree[. . .] The fractional crystallization
does not work[. . .] We always arrive at the terrible conclusion: our
radium isotopes do not behave like radium, but rather like barium[. . .]*

*May be you could propose some fantastic explanation. We do know
that it is not really possible that it explodes into barium. We want now
to test if the Ac isotopes resulting from the disintegration of "radium"
behave not like Ac but rather like La. A most ticklish research! But
we must clarify that.*

[. . .]

*Well, think about it, if some possibility can be imagined; may be
some Ba isotope with an atomic weight much higher than 137? In
case you could propose something worth publishing, that would well
be a way of working together all three of us! We don't believe that
we messed about nor that some infection is playing tricks on us [290].*

Hahn finds himself again in the situation of Irène Curie, who was unable to
separate lanthanum from the substance $R_{3.5h}$! He thinks he should repeat the
experiment of Irène Curie. What could have happened? How could a simple and
slow neutron produce barium? Recall that the isotope 238 of uranium (with atomic
number 92) is by far the most abundant isotope. It has 92 protons and 146 neutrons,
whereas barium (with atomic number 56) has only 56 protons and between 76 and
82 neutrons, if one considers only the isotopes occurring in nature. How could 36
protons and 60 neutrons be extracted in one go? It is so unbelievable that Hahn
and Strassmann speak to none except Lise Meitner about this. Did they fear to be
considered madmen, did they fear to seem ridiculous, was there possibly an error in
their experiments?

Finally, Hahn and Strassmann decide to publish their results in a paper sent to
Naturwissenschaften on December 22, 1938. The title is rather cold: "On the display
and the behavior of alkaline earth metals which are produced by the irradiation of
uranium by neutrons" [291].

They begin by recalling their previous results, with minor changes. The various
observed radioactivities are explained in terms of the formation of four isotopes of
radium, called Ra I, Ra II, Ra III, and Ra IV and originating from the decay of a
uranium nucleus which has absorbed a neutron, after emitting two α-particles. Then,
two thirds down, the paper proceeds in an unexpected direction:

*We must now discuss some more recent research which we publish
only with some hesitation because of the oddity of the results. In
order to identify with indisputable certainty the chemical nature of
the initial members of the radioactive series which we separated with
barium, we performed a fractional crystallization of the active barium
salts using the well known method in order to concentrate (or dilute)
the radium in the barium salts.*

And after stating carefully the method which was used, Hahn and Strassmann continue thus:

> We come to the conclusion that our isotopes of radium have the properties of barium; as chemists we must really claim that the new substances are not radium but barium, because elements other than radium and barium are out of question.

They also give the first results of experiments in which they repeated the experiments of Irène Curie and Pavle Savić. Hahn then receives the answer of Lise Meitner who replies in a letter dated December 21, 1938:

> Your results on radium are really mind-boggling. A reaction produced by slow neutrons and which, according to you, leads to barium! [. . .] For the time being the breaking up of such a magnitude appears difficult to swallow, but we have witnessed so many surprises in nuclear physics that we cannot claim from the outset: it is impossible. By the way, is the hypothesis of heavier transuranic elements discarded [290]?

Thus, Lise Meitner, who is his reference in physics, does not consider the idea to be absurd! He feels very encouraged. Hahn makes a phone call to his friend Paul Rosbaud, who is in charge of the journal *Naturwissenschaften*, asking him to add a paragraph which he probably did not dare write at first:

> As far as the "transuranic" elements are concerned, they are certainly chemically related to their lighter homologous elements, rhenium, osmium, iridium, platinum, without being identical to them. As far as the question to know if they identical to even lighter homologues, masurium, ruthenium, rhodium, palladium, that has not yet been ascertained. This was unimaginable before. For example, the sum of the mass numbers of Ba+Ma for example, that is 138+101 yields 239!

The idea is not stated explicitly but it transpires: if the final fragment is really barium, why could the other fragment not be masurium,[1] since the addition of their masses does indeed yield the mass of uranium (provided the right isotopes are chosen)? Could it be that the uranium nucleus splits up into two large pieces? Hahn and Strassmann consider this to be an extreme possibility without daring to state it openly.

---

[1] At the time, *Masurium* was the name given to the element 43, which two young German chemists, Ida Tacke and Walter Noddack (who were to be married in 1926) thought they had identified in 1925 in a niobium mineral and to which they gave the name *masurium*, in honor of the region where Walter Noddack was born. However, their identification was put into doubt because the experiment could not be repeated. Since the experiments of Segrè and Perrier, which showed that the element 43, which they produced by bombarding molybdenum with deuterons, has *no stable isotopes* [292], we know that the isotope, whose radioactive half-life is the longest (61 days) is the isotope 95. It therefore cannot occur in nature. In 1949, the international union of pure and applied chemistry (IUPAC, which was formed in 1919 by chemists from industry and academia) decided to call it *technetium* to remind that it is an artificial element.

The Word Is Finally Uttered

It is Christmas, and Lise Meitner is invited to spend a week in Kungälv, near Göteborg, by her friend Eva von Bahr-Bergius, a Swedish physicist she had met in Berlin before the First World War. Otto Robert Frisch, the nephew of Lise and provisionally exiled in Copenhagen [293, 294], is also invited. When Frisch arrives, he finds Lise pondering over a letter she had received from Hahn and which she shows him.[1] The results which Hahn describes are so astonishing that Frisch is somewhat skeptical at first: isn't there a mistake somewhere? Lise assures him that not. Hahn is one of the best chemists in the world, and if he claims that he cannot separate chemically the substance from barium, then the substance must surely be barium!

Lise Meitner and Otto Frisch continue the discussion as they walk in the snow. A single slow neutron cannot just cut a nucleus into two pieces. It simply does not have enough energy to extract a big chunk from the nucleus. That is, unless:

> *We walked up and down in the snow, I on skis and she on foot (she said and proved that she could get along just as fast that way), and gradually the idea took shape that this was no chipping or cracking of the nucleus but rather a process to be explained by Bohr's idea that the nucleus was like a liquid drop; such a drop might elongate and divide itself [293].*

Is that possible? The uranium nucleus is in fact a fragile nucleus subject to opposing forces. The nuclear forces acting between neutrons and protons try to keep the uranium nucleus bound, that is, they try to maintain the nucleons together. But the repulsion between the positively charged protons tries to pull them apart. In light nuclei, which have fewer charges, the attractive nuclear forces win out. However, they have a short range, and they act only between nucleons which are in contact so to speak. As a result, as the nucleus gets larger, the number of protons increases and the electric repulsion increases faster than the nuclear attraction. The reason for this is that each proton is attracted only to its immediate neighbors by the short-range nuclear force, but it is repelled by all the other protons by the long-range Coulomb interaction. This is why no stable nucleus exists which is heavier than bismuth. In the uranium nucleus, the balance between the attractive and repulsive forces is so critical that the absorption of a single neutron can throw it out of equilibrium and make the resulting nucleus unstable! Frisch and Lise Meitner conceive the nucleus as a liquid drop which acquires an increasingly elongated shape which leads into its fragmentation into two droplets. As soon as the droplets begin separating from each other, the nuclear attractive forces can no longer hold the pieces together and the strong electric repulsion drives the two pieces apart with very large velocities, exceeding 13,000 km/s which corresponds to an energy of 200 MeV. Frisch and Lise Meitner estimate this energy while sitting on a tree trunk in the forest.

---

[1] It is the letter, dated December 19, 1938, mentioned above on page 360.

It is necessary to check if the energy is conserved during such a process. The energy (i.e., the mass) of the uranium nucleus plus the energy of the neutron must be greater than the sum of the energies of the fragments. It turns out that the difference of these energies is just the required 200 MeV required for the observed fragmentation.

In a book of souvenirs, Frisch describes how this was realized:

Fortunately Lise Meitner remembered the empirical formula for computing the masses of nuclei and worked out that the two nuclei formed by the division of a uranium nucleus together would be lighter that the original uranium nucleus by about one-fifth the mass of a proton. Now whenever mass disappears energy is created, according to Einstein's formula $E = mc^2$, and one-fifth of a proton mass was just equivalent to 200 MeV. So here was the source for that energy; it all fitted [295]!

After the Christmas vacation, Frisch returns to Denmark, eager to discuss this with Bohr, who is about to embark to the United States. Bohr reacts strongly:

I had hardly begun to tell him when he smote his forehead with his hand and exclaimed: "Oh what idiots we all have been! Oh but this is wonderful! This is just as it must be! Have you and Lise Meitner written a paper about it?" "Not yet," I said, "but we would at once"; and Bohr promised not to talk about it before the paper was out [296].

On January 3, 1939, Frisch writes to Lise Meitner [297]. On January 7, 1939, Bohr leaves. On the railway platform, Frisch gives him a first version of the paper which he and Lise Meitner plan to submit to *Nature* [298].

The News Spreads to the United States

Niels Bohr had planned to spend a semester in the Institute for Advanced Study in Princeton where he would find physicists such as Einstein, Wigner, the mathematician von Neumann, as well John Wheeler, a young American physicist he knew well because Wheeler had spent a sabbatical in Copenhagen in 1934. Bohr embarks with Léon Rosenfeld, a 34-year-old Belgian physicist, who was professor at the University of Liège. Bohr held Rosenfeld in high esteem, and he obtained a scholarship for him in order to join him in Princeton. On January 7, the *Drottningholm* leaves the port of Gothenburg to cross the Atlantic in 9 days. During the trip, Bohr tells Rosenfeld of the extraordinary discovery, forgetting to say that he had promised Frisch to keep the news secret until he and Lise Meitner had published their paper [297]. When they reach New York, Fermi, his wife Laura, and John Wheeler are waiting for him on the quay. Bohr spends a few days in New York, and Rosenfeld goes straight to Princeton with Wheeler. Rosenfeld tells about the experiments of Hahn and Strassmann as well as the interpretation of Frisch and Lise Meitner. The news spreads like a bombshell in the United States much to the annoyance of Bohr who did not keep his mouth shut as he had promised. The American physicists rush to their laboratories in order to confirm the observations of Hahn and Strassmann. Bohr is impatient to see the paper of Lise Meitner and Otto Frisch published.

Frisch is not particularly in a hurry. He sends a first draft to his aunt, Lise Meitner, and he discusses it with George Placzek who reacts with skepticism, as usual. He then decides to set up a new experiment: he bombards uranium with neutrons, and indeed, his ionization chamber reveals strong electric impulses which he interprets as being produced by nuclei such as barium. Finally, on January 16, after a long telephone call with his aunt, Frisch submits two papers: the first, "Disintegration of Uranium by Neutrons: a New Type of Nuclear Reaction," signed by him and by Lise Meitner [299], is published on February 11, 1939.

Frisch searched for the right wording. He asked a colleague, the biochemist William Arnold, what was the word used for a the division of a cell into two parts. Arnold replied: "Fission." The word was used by Lise Meitner and Frisch to describe division of uranium into two nuclei. It was immediately adopted by all. The second paper [300] was published on February 18. Frisch describes the results of the experiment which confirmed the phenomenon. He placed a thin layer of uranium in an ionization chamber, and when he brought a neutron source nearby, he detected electric impulses which are characteristic of rare earth nuclei with energies of about 100 MeV, which he had estimated beforehand.

The novelty of the phenomenon, the enormous amount of energy liberated by *a single atom*, impressed both professionals and laymen. On January 28, the headline of the New York Times stated: "An atomic explosion liberates 200,000,000 V." Nuclear fission had entered into the life of humanity.

Confirmations

After being awarded the Nobel Prize in 1935, Frédéric Joliot and Irène Curie continued their research at the *Institut du Radium*. Frédéric then became assistant professor at the Sorbonne, and on January 10, 1937, he became professor at the *Collège de France* where he held the chair of nuclear chemistry, which had replaced the previous chair of mineral chemistry. Finally, he could direct a laboratory. So far, all the important discoveries in nuclear physics had used radioactive sources, and the *Institut du Radium* had particularly intense sources, thanks to Marie Curie. But Frédéric Joliot believed that particle accelerators, such as the cyclotron of Lawrence, would become indispensable to remain competitive. Several accelerators were being constructed in the United States and Europe (in Copenhagen, Liverpool, and Leningrad). Joliot uses his prestige to obtain funds for an electrostatic accelerator in Ivry, near Paris, and a cyclotron at the *Collège de France*. He is helped by a favorable circumstance: the "Front Populaire" government is in power, and it creates a Secretariat for Research which is first entrusted to Irène Curie.[1] A few months later, according to a preestablished plan, she resigns in favor of Jean Perrin who had been struggling for years to establish a real scientific research policy in France and who endeavors to carry it out. The construction of the cyclotron is approved, and Joliot immediately sends the young physicist, Maurice Nahmias, to Berkeley, where Lawrence always gave generous help to cyclotron constructors throughout the world. Furthermore, Joliot invites Hans von Halban to join his team in the Collège de France. Indeed, upon returning from Copenhagen where he had worked on neutron physics with Otto Frisch, Halban had obtained a position in the CNRS.[2] Halban applies for French citizenship which is granted in 1939. Joliot also invites a young and somewhat untypical Russian, Lew Kowarski, whom Bertrand Goldschmidt described as follows:

He was a huge and heavy fellow who would reply by a grunt to your greetings [. . .] He was heavy, displayed finesse and possessed a memory of an elephant; the great intelligence and finesse of his mind, stuffed into his huge body, progressively grew into full bloom as he became successful in this nuclear adventure [301].

Lew Kowarski was born in 1907 in Saint Petersburg. His father was a Jewish publisher and his mother an opera singer. A the age of 11, he fled the Russian revolution with his father. With very meager means, he studied in Gand and later in Lyon where, in 1928, he obtained his diploma as a chemical engineer. He then completed his PhD on the growth of crystals under the direction of Jean Perrin, while working part-time as an engineer in the firm *Le Tube Acier*.

He then entered the *Institut du Radium* without particularly distinguishing himself. When he accepts to become the scientific secretary of Joliot, he is 30 years old and already a confirmed physicist.

On January 16, 1939, Kowarski shows Joliot the paper of Hahn and Strassmann, which had just reached the *Collège de France*. Joliot immediately realizes that it is an important discovery. He works for a few days in his office seeking a way to confirm nuclear fission in a physics experiment in order to complete the purely chemical method used by Hahn and Strassmann. Although he does not know that Frisch is about to send a paper to *Nature,* he reckons that many physicists must have begun working on this [302]. On January 26, he sets up the experiment which immediately gives the result he had hoped for. The experiment is particularly simple and beautiful: Joliot uses a thin copper tube (20 mm in diameter and 5 cm long) the outside of which is coated with a thin layer of uranium oxide. The tube is surrounded with a Bakelite cylinder, the inside surface of which is at a distance of 3 mm from the surface of the copper tube. Joliot places a neutron source inside the tube and, after the irradiation, he examines the radioactivity of the Bakelite with a Geiger counter:

---

[1] Irène Curie became the first woman with the rank of Minister in France, while not even having the right to vote!

[2] The CNRS (*Centre National de Recherche Scientifique*) is still today the major research institution in France.

On finds that the interior surface of the bakelite cylinder receives a complex mixture of radioactive atoms the time evolution of which, as detected by the counter, is similar to that of artificial radioelements formed in the uranium [303].

What has occurred? The neutrons have no trouble to pass through the copper tube, and they bombard the uranium contained in the outer uranium coating. When a fission occurs, the two fragments fly off in opposite directions with high velocities. Those which fly outwards at more or less right angles to the surface pass through the uranium coating, through the 3 mm of air and are stopped at the surface of the Bakelite. When Joliot measures the radioactivity of the Bakelite with the Geiger counter, he finds the radioactivity which had been attributed to the "transuranic" elements. But that was an illusion since they are pieces of the fissioned uranium nucleus.

Joliot then attempts to confirm these results in another way. He tries to detect one of the fast-moving fragments which are produced in the fission process. For this, he uses the Wilson cloud chamber, and he takes a thousand photographs before finding the track of a fission fragment. He reports on this result [304] on February 27.

Neither he or Irène thought of fission in order to explain the experiments of Irène Curie and of Pavle Savić. Bertrand Goldschmidt describes the discussions held in the *Collège de France*:

The work of Irène Joliot and of Savić was discussed during seminars attended by the researchers of the Curie laboratory and of the Collège de France. I participated to each one and none of us suspected the revolutionary solution of the enigma [305].

Indeed, nobody thought of fission. And yet, Ida Noddack had mentioned the possibility of the breaking up of a uranium nucleus into several large pieces.[1] It remained however a theoretical speculation with no experimental backing. Think of the reaction of Bohr ("What idiots we have been!") in order to appreciate that physicists were far from imagining that such a process could occur. Irène was within reach of the discovery: the $R_{3.5h}$, which appeared similar to lanthanum, *was* lanthanum! The fact that Hahn was a pure chemist certainly helped. Having identified chemically the presence of barium, he wrote and published that the neutrons produced barium, while adding that he wrote this "with some hesitation" because it was "in contradiction with all the nuclear physics experiments."

Soon, further confirmations were published: in her paper sent to *Nature*, Lise Meitner had suggested an experiment similar to the one performed by Joliot. She performed it with Otto Frisch, and she published the result [306] on March 18, 1939. Philip Abelson, a young graduate student working in the Radiation Laboratory in Berkeley, directed by Ernest Lawrence, also came very close to the discovery of fission. Thanks to the cyclotron, he was able to use a far more intense source of neutrons than had been available so far and he began to characterize the

---

[1] See p. 352.

"transuranic" elements by the energies (or the wavelengths) of the observed X-rays. Remember that Charles Barkla and Henry Moseley had shown that X-rays emitted by various elements could be grouped into families which he called K and L. The energies of X-rays belonging to a given family increased regularly with the atomic number.[1] The L spectral line always had a higher energy than the K line. And because the atomic number of the presumed "transuranic" elements was higher than the atomic number of uranium, Abelson expected to observe higher energy X-rays *and he found them,* assuming the observed K spectral line was emitted by the transuranic element 95 line, but which was, in fact, the L spectral line of iodine. When the paper of Hahn and Strassmann was published, he checked his measurements and realized he had been mistaken. On February 3, 1939, he sent a short letter to the Physical Review in which he concluded:

This seems to be an unambiguous and independent proof of Hahn's hypothesis of the cleavage of the uranium nucleus [307].

When Niels Bohr had revealed the discovery of fission, Enrico Fermi had recently become professor at the University of Columbia. With a graduate student who was preparing his PhD thesis, he immediately started an experiment, similar to the one being performed by Joliot, aimed at proving that the uranium nucleus splits into two fragments and on February 16, 1939, he sent a letter [308] to the editor of *Physical Review.* A few months later at the University of Chicago, Aristid von Grosse, Eugene Booth, and John Dunning showed that protactinium (the element 91) could also undergo fission when bombarded with neutrons [309]. The Cavendish could not be left behind. Egon Bretscher and Leslie Cook also observed fission and published their results [310] on April 1, 1939.

Fission, a phenomenon so unexpected that a few weeks earlier no one had thought of it, had become an integral part of the nuclear physics landscape. Physicists all over the world were thrilled by it. In January 1940, the American physicist Louis Turner, from the University of Princeton, wrote a review paper which contained no less than 140 references concerning nuclear fission, all written in a single year, namely, 1939.

Niels Bohr: The Theory of Fission, Uranium 235

Niels Bohr had come to the United States with the intention of working in Princeton, but certainly not on fission. However, that is precisely where he made fundamental contributions to the theory of fission. On February 7, 1939, he sends a letter to the editor of *Physical Review* in which he raises a problem which will become very important [311]. Indeed, there was one thing which was difficult to understand: as shown by Lise Meitner, Hahn, and Strassmann in 1938, neutrons

---

[1] See p. 77 and p. 79.

with an energy of about 25 eV (and which therefore travelled at a velocity of about 70 km/s) were strongly absorbed by uranium without however increasing the fission rate, which increased regularly as the neutrons slowed down. The probability of producing fission was maximum for the slowest neutrons, called thermal neutrons, with energies of about 0.025 eV (and velocities of about 2.2 km/s). However, uranium is a mixture of two isotopes, the proportion of which had been recently determined [312]: 99.3% of uranium is the isotope 238 and 0.7% is the isotope 235. Bohr believes that the most abundant isotope, uranium 238, causes the 25 eV resonance but that it does not undergo fission. However, the isotope 235 undergoes fission when exposed to thermal neutrons. Bohr offers a simple explanation of this, based on the binding energy of the neutron.[1] Working with Wheeler, Bohr makes a more complete analysis of fission: in a 25-page paper sent on June 28 to the *Physical Review* [313], Bohr and Wheeler explain fission using the liquid drop model, originally proposed by Gamow and later adopted by Bohr. This model assumes that the forces which bind together the nucleons in the nucleus are similar to those which bind a liquid drop. The forces only act when the nucleons are "in contact" with one another. But in a nucleus, there is an additional force: the electrostatic Coulomb repulsion between the positively charged protons. Bohr and Wheeler study how, in heavy nuclei, the equilibrium between the attractive and repulsive forces can lead to a deformation and finally to the fission of a nucleus into two fragments. The success of their model is due to the fact that, with essentially classical arguments (without appealing to quantum mechanics), they provide a simple explanation for most of the observed facts:

- The observed speed of the two fission fragments
- The way in which the probability of fission varies with the energy of the neutron
- Why, in certain cases, a minimum energy of the neutron is required (a threshold) for fission to take place

Since the publication of this paper in 1939, the description has been improved and made more precise. However, the description provided by Bohr and Wheeler remains the basis of our understanding of the fission process.

---

[1] The "compound nucleus" formed by uranium 235 and a neutron is the uranium 236 nucleus which has an even number of protons (92) and an even number of neutrons (144). Therefore, the binding energy of a neutron is larger for this nucleus: the neutron arrives on top a deeper potential well, and therefore, the 236 nucleus is in a more highly excited state, at an energy where the nuclear states are more densely spaced so that the neutron has a greater probability of finding a "host state" which allows it to form a compound nucleus. The opposite is true for the isotope 238 which, after absorbing a neutron, becomes the isotope 239 and which does not offer the neutron a state with exactly the energy required for it to be absorbed.

The Number of Emitted Neutrons

Several physicists noticed that if uranium broke up into two much lighter pieces, they could not account for the number of emitted neutrons. The fraction of neutrons over protons increases as the nucleus becomes heavier (as the number of nucleons increases). The isotope 236 of uranium, that is, the nucleus formed when the isotope 235 of uranium absorbs a neutron, contains 92 protons which is the atomic number of the element uranium and 144 neutrons. Imagine that this uranium 236 isotope undergoes fission and breaks up into a barium nucleus (with contains 56 protons) and a krypton nucleus (which contains 36 protons). The stable krypton isotope which has the largest number of neutrons has only 50 neutrons. The stable barium nucleus which has the largest number of neutrons has 82 neutrons. The total number of neutrons in the fission fragments is thus $50 + 82 = 132$. Therefore, 12 neutrons are missing! In general, the fission fragments will not consist of stable nuclei but rather of radioactive isotopes (and this is precisely why they could be detected and identified) which have an excess of neutrons and which are transformed into protons by β radioactive decay. But that does not easily account for 12 missing neutrons. Some neutrons may be emitted during the explosive fission process. That is what Joliot meant in his January 30 paper:

> I assume that the excess neutrons of the produced nuclei becomes normal by successive β radioactive decays. But a small number of neutrons may well evaporate first [303].

Such statements are far from being innocent. Indeed, if the fission caused by a single neutron can release several neutrons, *each liberated neutron can in turn cause another fission*, which in turn, can liberate more neutrons thereby causing an even greater number of fission processes, and so on. This "snowball" process would occur in a very short time during which a huge amount of energy would be released. Such an explosive chain reaction, which Joliot mentioned in his Nobel lecture,[1] could occur when the fission process propagates throughout the uranium sample.

However, it was probably not quite so simple; otherwise, an explosion would occur as soon as a uranium sample was exposed to neutrons, and this is not the case. The reason is that the increase of the number of neutrons is a more complicated process. The first requirement is that more than one neutron should be emitted in each fission process. But that is not enough because each emitted neutron does not necessarily induce fission in another nucleus. For example, some of the neutrons can escape from the uranium sample without interacting and others can be absorbed without producing fission. Frédéric Joliot then decides to measure the number of neutrons liberated during a fission process. He asks Hans von Halban to help him because Halban had acquired a considerable experience in neutron physics. Halban suggests to add Lew Kowarski to the team because he had initiated the latter

---

[1] See p. 314.

precisely to this field. They set to work. They make quick progress because the team is composed of physicists with very different characters. But an unexpected event is about to slow down their publications.

Leo Szilard

While Joliot and his team were progressing fast, similar experiments were being performed in the United States, largely due to the insistence of a rather uncommon physicist, a Hungarian newcomer, Leo Szilard.

Born in 1898 in Budapest into a Jewish family [314], Szilard studies engineering first in Hungary and from 1920 onward in Berlin. However, the presence of von Laue, Planck, Schrödinger, Nernst, Haber, Franck, and mainly Einstein attracts him irresistibly to fundamental physics. After completing his PhD in 1922, he continues to work at the *Kaiser Wilhelm Institut* as assistant to Max von Laue and he then becomes *Privatdozent*. By then, he had specialized in theoretical physics which did not prevent him from meddling in several other fields. Together with Einstein, he patented a new electromagnetic pump, the principle of which was used many years later in breeder reactors.

Szilard loved research, but he was also interested in political problems. He was little impressed by hierarchical conventions, and his mind was always overflowing with ideas.

Hitler comes to power. Two months later, Szilard uses the little money he possesses to leave Germany. He goes to Austria and then to England. He has the intention to leave physics and enter biology. But then, he meets an Austrian Jewish student, Maurice Goldhaber, who is preparing his PhD at the Cavendish under the direction of James Chadwick.[1] Discussions with Goldhaber get Szilard interested in nuclear physics.

Szilard then tries to produce radioactive elements for medical use. Working with Thomas Chalmers, a young physicist at the Saint Bartholomew Hospital, he produces neutrons by exposing beryllium to γ-rays. He then tries to produce neutrons with X-rays since γ-rays are nothing but high-energy X-rays. To achieve this, he makes contact with his previous colleagues in Berlin, namely, Arno Brasch and Fritz Lange, who had constructed high-energy X-ray tubes[2] and he asks them to expose beryllium to 1.5 MeV X-rays. A bromine compound is subsequently exposed to the radiation emanating from the beryllium, and the lot is immediately sent to London by airplane. The experiment failed, and today, we understand why. X-rays with energies exceeding 1.66 MeV would have been required to cause the reaction. But this example shows how inventive and clear-sighted Szilard was [315].

---

[1] See p. 260.
[2] See p. 290.

When the paper of Hahn and Strassmann was published, thereby announcing the existence of nuclear fission, Szilard was among the first to think of the possibility of a chain reaction. He was also the first to worry about it.

Is a Chain Reaction Possible?

The huge amount of energy which is stored in an atomic nucleus and which is liberated during radioactive processes led Rutherford and Soddy to suggest, already in 1903, that solar energy might well have the same origin as radioactivity.[1] In 1908, Soddy gave a series of lectures on radioactivity in the University of Glasgow. His lectures became the subject of a book, the third edition of which contained a description of radioactivity "for laymen" [316]. In his book, Soddy explained how radioactive processes would release part of the energy contained in the atom at a rate which could not be controlled by man. In a lyrical style, he foresaw that this gigantic source of energy would be used 1 day:

> *Looking backwards at the great things science has already accom-*
> *plished, and at the steady growth in power and fruitfulness of scientific*
> *method, it can scarcely be doubted that one day we shall come to*
> *break down and build up elements in the laboratory as we now break*
> *down and build up compounds, and the pulses of the world will then*
> *throb with a new source of strength as immeasurably removed from*
> *any we at present control as they in turn are from the natural resources*
> *of the human savage [317].*

Soddy thought, as did Pierre Curie, that the availability of large energy resources would be used to for the benefit of mankind:

> *A race which could transmute matter would have little need to earn*
> *its bread by the sweat of its brow. If we can judge from what our*
> *engineers accomplish with their comparatively restricted supplies of*
> *energy, such a race could transform a desert continent, thaw the*
> *frozen poles, and make the whole world one smiling Garden of Eden.*
> *Possibly they could explore the outer realms of space, emigrating to*
> *more favourable worlds as the superfluous to-day emigrate to more*
> *favourable continents [318].*

Soddy's book impressed the famous writer Herbert George Wells, author of science fiction novels such as *The Time Machine, The Island of Doctor Moreau, The Invisible Man,* and *The War of the Worlds,* published between 1895 and 1898. It inspired him to write the novel *The World Set Free,* written in 1913 and published in 1914. In this novel, he imagines that, in 1933, a physicist succeeds in liberating atomic energy by "inducing radioactivity at will" [319]. The similarity with the

---

[1] See p. 33.

discovery, in 1934, of artificially induced radioactivity is purely accidental, although striking. Wells tells how this new source of energy, which he calls "atomic energy," causes an industrial revolution in the whole world. He also describes how men construct "atomic bombs" (his text states explicitly *atomic bombs*) which have a terrifying destructive power and how these bombs are thrown onto Paris and Berlin during a war in 1956. The remaining part of the novel is a utopian dream as he calls it himself. The destructive power of the "atomic" weapon is so great that it makes war impossible and that it incites the leaders of different countries to form a democratic and pacific world government. A dream indeed.

The young Lew Kowarski read the novel at the age of 10 [320] and was impressed. Leo Szilard read it at the age of 34, and it excited his ever overflowing imagination. It was, of course, only a novel, and the possibility of using the energy stored in the atom belonged to science fiction. On September 11, 1933, Rutherford had delivered a speech to the British Association for the Advancement of Science. The summary, published in *Nature*, ended with the following remark:

> One timely word of warning was issued to those who look for sources of power in atomic transmutations—such expectations are the merest moonshine [321].

And yet, a few months later, in January 1934, Frédéric and Irène Joliot-Curie were discovering artificially induced radioactivity: it became possible to create radioactive nuclei in the laboratory and to "induce radioactivity" as H. G. Wells had predicted. But could this lead to the release of energy? We saw above how Frédéric Joliot explicitly mentioned explosive chain reactions in his Nobel lecture[1] without however mentioning when it could be accomplished.

But Leo Szilard thinks of a way: what if some not too stable nuclei, after being struck by a neutron, could explode while liberating a further neutron? There would then be two neutrons, followed by four, eight, and so on. He has in mind beryllium which might break up into a neutron and two helium nuclei, and indium. On March 17, 1934, he sends the novel of H. G. Wells to Hugo Hirst, the director of a large electricity firm in England. He encloses a letter in which he writes, with a clear allusion to Rutherford:

> Of course, all this is moonshine, but I have reason to believe that in so far as the industrial applications of the present discoveries in physics are concerned, the forecast of the writers may prove to be more accurate than the forecast of the scientists. The physicists have conclusive arguments as to why we cannot create at present new sources of energy for industrial purposes; I am not sure whether they don't miss the point [322].

Szilard makes the experiment: he bombards beryllium with neutrons. The result is negative, but he continues to be obsessed with the idea that the neutrons could

---

[1] See p. 314.

proliferate. Since 1935, he holds a position in the Clarendon Laboratory in Oxford. In 1938, he begins to spend half of his time in the United States, but after the Munich agreements in September 1938, he decides to stay there permanently. In January 1939, he is visiting an old friend, Eugene Wigner, who tells him about the discovery of fission by Hahn and Strassmann. Szilard then immediately thinks that some neutrons could be emitted when fission is produced and that fission may be a way to produce a chain reaction. When he returns to New York, he convinces his friend Walter Zinn, at the University of Columbia, to make an experiment to determine if fission produces neutrons. However, at that time, Szilard holds no position in any laboratory. He has to provide the money for the experiment from his own pocket. Thanks to a loan from a friend, he rents a gram of radium for the few days required to perform the experiment. A block of beryllium is sent from England. It is exposed the γ-rays of radium. The beryllium nucleus breaks up into three pieces, two helium nuclei, which are α-particles, and one neutron. With Zinn, he then, independently, performs the same experiment as Halban, Joliot, and Kowarski,[1] except that the neutrons are detected by the protons which are projected by the neutrons into an ionization chamber:

> Once we had the radium and the beryllium it took just one afternoon to see the neutrons. Mr. Zinn and I performed the experiment. [On March 3, 1939] everything was ready and all we had to do was to turn a switch, lean back, and watch the screen of a television tube. If flashes of light appeared on the screen, that would mean that the large-scale liberation of atomic energy was around the corner. We turned the switch and we saw the flashes. We watched them for a little while and then we switched everything off and went home. That night there was very little doubt in my mind that the world was headed for grief [323].

The "grief" evoked by Szilard was the fact that such a large number of neutrons would make an explosive chain reaction possible and open the way to the possibility of making terrifying bombs. The world was on the brink of war. The possibility that Hitler might possess such a bomb worried Szilard. Recall that German science was then the best in the world. If such a bomb was feasible, Germany was in an excellent position to make one.

Most seriously worried about this possibility, Szilard pressed physicists in the United States, England, and France to refrain from publishing papers concerning fission, in order to prevent German physicists from being informed about possible progress. In spite of the reservations of Fermi, who did not believe a chain reaction was possible, Szilard succeeded in convincing some physicists working on fission in the United States and England. He wrote a letter to Joliot. The request of Szilard was a delicate matter for the team in the *Collège de France* who was leading in the race. But when Joliot learned that an American agency had published, on February 24, a summary of the French experiments, he sent a cable to Szilard telling him that

---

[1] See p. 375.

he thought it was time for him to publish. Clearly, the initiative of Szilard, which was based solely on the good will of physicists, was bound to fail. In fact, it was the war which stopped all publications.

The Last Publications Before the War

Joliot decided to publish the results obtained by the team at the *Collège de France*. On March 8, Kowarski went personally to the Bourget airport to post the paper to *Nature*, in order not to lose time. The experiment is quite simple in principle. A neutron source is placed in a solution of uranyl nitrate, a well-known uranium salt. The neutrons are detected by placing small dysprosium sheets at various locations. Dysprosium becomes radioactive when it absorbs a neutron, as Fermi had shown a few years earlier.[1] After exposing the dysprosium sheets, their radioactivity is measured and this provides a measure of the number of neutrons which were incident on the sheets. One can then deduce the number of neutrons which were present in the uranium nitrate solution. Joliot, Halban, and Kowarski compare the result obtained with the uranium nitrate with that obtained with an ammonium nitrate, which contains no uranium and which serves as a reference. The result is unambiguous: more neutrons are detected in the presence of uranium. But the experiment does not say *how many* more. Their paper makes the following conclusion:

> The interest of the phenomenon observed, as a step towards the production of exo-energetic transmutation chains, is evident. However, in order to establish such a chain, more than one neutron must be produced for each neutron absorbed. This seems to be the case [324].

The team makes another experiment, suggested by Kowarski. Since neutrons produced during a fission process are apparently faster than the neutrons which initiate the fission process, Kowarski suggests to use a source of slow neutrons, which consists of radium which emits γ-rays which in turn produce neutrons when they are incident on beryllium (as shown earlier by Leo Szilard and Thomas Chalmers [325]). The source is surrounded by uranium so as to allow the neutrons to produce fission. The fast neutrons are detected by immersing the whole thing in carbon sulfide, which acts as a detector, because the neutrons incident on sulfur produce a reaction in which the neutron replaces a proton thereby producing radioactive phosphorous-32. It is in fact a selective detector because the reaction can only take place if the energy of the neutrons is larger than about 2 MeV. Thus, if radioactive phosphorous is produced, it cannot be due to the neutrons of the source and it must be due to the neutrons produced by the fission.

---

[1] See p. 324.

The team is helped by the chemist Maurice Dodé who lets the neutrons act for a week before analyzing the solution: he finds radioactive phosphorous-32. Therefore, *neutrons with energies above* 2 MeV *are produced*, and they can only be emitted by the fission of uranium. However, the number of neutrons remains undetermined, and this makes it difficult to decide whether a chain reaction is possible [326].

At about the same time, Szilard publishes the results he obtained with Zinn [327]. They estimate that two neutrons are liberated by the fission of a uranium nucleus. But this is only a rough estimate which makes it difficult to decide whether a chain reaction is possible.

Meanwhile, after a first nonconclusive experiment [308], Fermi also tackles the problem with a young student, Herbert Anderson. He proceeds with an experiment similar to the one of Halban, Joliot, and Kowarski. He places a neutron source inside a cylinder filled with water. Inside, he also places a sheet of rhodium, which becomes radioactive when it is exposed to neutrons. By changing the position of the rhodium sheet, he can estimate the number of neutrons at different positions in the cylinder, both in the presence and in the absence of uranium. As the French team did, he finds a 6% increase of neutrons when the cylinder contains uranium. This proves that neutrons are indeed produced, but he is also unable to tell how many [328]. Fermi, Szilard, and Anderson then set out to make a more precise measurement. This time, they use a 540-l bath of magnesium sulfate. The magnesium will detect the slow neutrons. In this bath, they deposit 200 kg of uranium oxide and in the middle a slow neutron source, consisting of 2 g of radium and 250 g of beryllium. They find a 10% increase in the number of neutrons, which again shows that the neutrons are emitted during the fission process. But to make these neutrons produce further fission processes with a reasonable probability, they should be slowed down because it was an experimental fact that only slow neutrons produce fission. To slow them down, one might attempt to make the neutrons pass through a hydrogenated substance, as Fermi used to do in Rome.[1] However, they do not think this would work because the hydrogen nuclei (protons, in fact) can capture some neutrons so as to form deuterons, and the captured neutrons would be lost for the chain reaction [329].

It is difficult to estimate the number of neutrons because the observed increase in the number of neutrons is the result of two effects acting in opposite directions. The neutrons are slowed down by bouncing off hydrogen nuclei. While they are being slowed down, they can also be captured by hydrogen. Furthermore, as they are slowed down, when they reach the critical velocity of 70 km/sec, they can be captured by the uranium 238, as had been shown in 1937 by Lise Meitner, Hahn, and Strassmann [266]. This reduces the number of neutrons which are sufficiently slowed down to produce fission.

Halban, Joliot, and Kowarski try nonetheless to estimate the number of neutrons liberated by the fission process. They take all these phenomena into account and use the observed probabilities. In a paper sent to *Nature* on April 7, 1939, they

---

[1] See p. 321.

estimate that, on the average, 3.5 neutrons are emitted during the fission of a uranium 235 nucleus. They believe that their estimate may have an error of about 0.7 neutrons [330]. This result had a huge impact: it meant that a chain reaction may well take place. Was the anticipation of H. G. Wells about to take place?

A few papers continued to be published on chain reactions until the summer of 1939. Siegfried Flügge, a theorist from the *Kaiser Wilhelm Institut für Chemie* in Berlin-Dahlem, discusses a practical application of the chain reaction to energy production. He concludes that he cannot conclude, until better data become available, and that it is a quite likely development [331]. In August, Szilard and Zinn publish a paper in which they describe another method to measure the number of neutrons emitted during fission. They obtain a smaller number than estimated by the team of Joliot: 2.3 neutrons on the average [332]. Finally, Joliot, Halban, and Kowarski, joined by Francis Perrin, summarize the situation in a paper sent to the *Journal de Physique et le Radium* in September 1939. They show that, in a sphere containing water and uranium oxide, the number of detected neutrons is more than twice the number of neutrons produced by a source of neutrons placed at the center of the sphere [333]. They conclude that a chain reaction took place in the sphere. However, it is a *convergent* chain reaction, meaning that it ends spontaneously: the number of produced neutrons is insufficient to sustain the chain reaction, either because an insufficient number of neutrons are liberated or because the liberated neutrons are absorbed too fast by the hydrogen in the water, or because they escape from the sphere.

Francis Perrin and the Critical Mass

In the 1930s, the "Perrin tea party" was a real institution. Each Monday afternoon, in his laboratory of chemical physics, next to the *Institut du Radium*, Jean Perrin would receive his guests. In 1967, a vivid description of the tea party was published by Camille Marbo, a pseudonym for Marguerite Appell, the wife of Émile Borel:

> After receiving the Nobel prize[1], the tea parties in the Perrin laboratory became a real Parisian institution. Periodically, external visitors would come and mix with the local scientists. Nine Choucroun and Yvette Cauchois, "workers" of the master, would serve tea to a most varied crowd in widened test tubes and the glass rods, serving as spoons. Paul Valéry and André Maurois would be seen next to Niels Bohr, Paul Langevin or Paul Painlevé. Einstein once appeared among elegant ladies, wives of university professors and students in white overalls. Jean Perrin would flicker from one to another like a flame, between washing tables, vats, gladioli or carnations. Occasionally a scientific

---

[1] Jean Perrin was awarded the Nobel Prize in physics in 1926 "for his work on the discontinuous structure of matter, and especially for his discovery of sedimentation equilibrium".

theme became the object of the party and it was often followed by a colloquium [334].

Francis Perrin, the son of Jean Perrin, worked in the nearby *Institut Henri Poincaré*. He was a natural *habitué* of the tea parties, as well as physicists of the *Institut du Radium*, such as Irène Curie, and physicists of the *Collège de France*, such as Frédéric Joliot. Francis Perrin was well informed about the work of Irène Curie, and from 1939 onward, the work of Joliot on fission. He then made a particularly interesting calculation which introduced a crucial concept for the chain reaction. In a block of uranium, all the liberated neutrons do not necessarily produce further fission because some can be absorbed by other nuclei and others may escape from the uranium block. It is reasonable to expect that fewer will escape if the size of the block is increased and its shape optimized, that is, spherical. Then what is the minimum mass of a spherical uranium block required for a self-sustaining chain reaction to take place? Such a chain reaction would produce an increasing number of neutrons and an increasing amount of energy. With the data available to him at the time, he estimates that a self-sustaining chain reaction would require a 12-ton block of uranium: that is the first time that a *critical mass* is mentioned. The evaluation of Francis Perrin is a very rough estimate because the data he needed were badly known at the time. Furthermore, he calculated the critical mass of natural uranium in spite of the fact that it was known, since the February 1939 paper of Bohr, that only uranium 235 underwent fission. At the time, nobody thought of separating the isotopes 235 and 238 of uranium because it appeared too difficult, indeed practically impossible if an appreciable amount of the isotope 235 was required.

French Patents

Between April 30 and May 4, 1939, Halban, Joliot, Kowarski, and Francis Perrin filed an application for three patents in favor of *Caisse Nationale de la Recherche Scientifique*, an institution created by Jean Perrin and which will soon become the *Centre National de la Recherche Scientifique* (CNRS). The first patent concerns *a device to produce energy*, that is, for what today we call a *nuclear reactor*:

One knows that the absorption of a neutron by a uranium nucleus can cause the latter the break up while energy is emitted as well as new neutrons the average number of which is greater than one. Among the neutrons thus emitted, some can, in turn, cause further uranium nuclei to break-up, and the number of broken uranium nuclei can thus increase, following a geometric progression, with a considerable release of energy [. . .]

But one immediately faces an essential difficulty: since these chains can branch out indefinitely, the reaction can become explosive, and this would considerably reduce the possibility of using the uranium mass as a controlled industrial energy source.

We endeavored to control the emission of energy so as to prevent it from becoming explosive [335].

The problem being thus stated, the authors describe the basic elements of every present nuclear reactor:

- The fuel, namely, natural uranium.
- A substance able to slow down the neutrons, which today we call a *moderator*. For that purpose, they propose to use light substances such as hydrogen, deuterium, beryllium, carbon, or oxygen.
- A substance able to absorb the neutrons, in order to control the chain reaction. They propose to use cadmium.

The second patent concerns a different method of stabilizing the chain reaction and which, in fact, has never been used. The third patent concerns the construction of an *explosive device*:

As a consequence of the present invention, we have attempted to make a practical application of this explosive reaction, not only for mining and public works, but also to construct war devices and, more generally any device which requires an explosive force [336].

What follows is the description of a device which is quite similar to an "atomic bomb" except for the fact that they consider only the use of natural uranium, with which it is not possible to make an explosive chain reaction, a feature which the team of Joliot did not know at the time. The authors asked the patent to remain secret.

References

1. Gamow, G., "Zur Quantentheorie der Atomkerne", *Zeitschrift für Physik* **51**, 204–212, 1928.
2. Gurney, R. W. and Condon, E. U., "Wave mechanics and radioactive decay", *Nature* **122**, 439, September 22, 1928.
3. Rosenblum, S., "Structure fine du spectre magnétique des rayons α du thorium C", *Comptes Rendus de l'Académie des Sciences* **188**, 1401–1403, session of May 27, 1929.
4. Cotton, A., "Le grand électro-aimant de l'Académie des Science", *Comptes Rendus de l'Académie des Sciences* **187**, 77–89, session of July 9, 1928.
5. Cotton, A. and Dupouy, G., "Champs magnétiques donnés par le grand électro-aimant de Bellevue", *Comptes Rendus de l'Académie des Sciences* **190**, 544–47, session of February 10, 1930.
6. Rosenblum, S., "Structure fine du spectre magnétique des rayons α", *Comptes Rendus de l'Académie des Sciences* **188**, 1549–1550, session of June 10, 1929, and **190**, 1124–1127, session of May 12, 1930.
7. Rutherford, E., Chadwick, J. and Ellis, C. D., *Radiations from Radioactive Substances*, Cambridge University Press, London, 1931, note p. 47.
8. Rutherford, E., Ward, F. A. B. and Wynn-Williams, C. E., "A New Method of Analysis of Groups of Alpha-Rays (1) The Alpha-Rays from Radium C, Thorium C, and Actinium C", *Proceedings of the Royal Society, London* **A129**, 211–234, 1930.
9. Rutherford, E., Ward, F. A. B. and Lewis, W. B., "Analysis of the Long Range α-Particles from Radium C", *Proceedings of the Royal Society, London* **A131**, 684–703, 1931.
10. Rutherford, E., Wynn-Williams, C. E. and Lewis, W. B., "Analysis of the α-Particles Emitted from Thorium C and Actinium C", *Proceedings of the Royal Society, London* **A133**, 351–366, 1931.
11. Lewis, W. B. and Wynn-Williams, C. E., "The Ranges of the α-Particles from the Radioactive Emanations and "A" Products and from Polonium", *Proceedings of the Royal Society, London* **A136**, 349–363, 1932.
12. Rutherford, E., Wynn-Williams, C. E. and Bowden, B. V., "Analysis of α-Rays by an Annular Magnetic Field", *Proceedings of the Royal Society, London* **A139**, 617–637, 1933.
13. Rutherford, E. and Ellis, C. D., "Origin of the γ-Rays", *Proceedings of the Royal Society, London* **A132**, 667–688, 1931.
14. *Convegno di Fisica Nucleare, Ottobre 1931*. Reale Accademia d'Italia, 1932.
15. Goudsmit, S. A., "Present difficulties in the theory of hyperfine structure", *Ibid*, pp. 33–49.
16. Pauli, W., "Zur Frage der theoretischen Deutung der Satelliten einiger Spektrallinien und ihrer Beeinflussung durch magnetische Felder", *Naturwissenschaften* **12**, 741–743, September 12, 1924.
17. Bothe, W., "α-Strahlen, künstliche Kernumwandlung und -anregung, Isotope", in *Convegno di Fisica Nucleare, Ottobre 1931*, pp. 83–106, Reale Accademia d'Italia, 1932.
18. Bothe, W. and Becker, H., "Eine γ-Strahlung des Poloniums", *Zeitschrift für Physik* **66**, 307–310, December 3, 1930.
19. Bothe, W., "Erzwungene Kernprozesse", *Physikalische Zeitschrift* **32**, 661–662, September 1, 1931.
20. Birge, R. T., "Mass defects of C^{13}, O^{18}, N^{15} from band spectra and the Relativity relation of mass and energy", *Physical Review* **37**, 841–842, April 1, 1931.
21. Urey, H. C., Brickwedde, F. G. and Murphy, G. M,, "An Hydrogen Isotope of Mass 2", *Physical Review* **39**, 164–165, January 1, 1932.
22. Urey, H. C., Brickwedde, F. G. and Murphy, G. M., "An Hydrogen Isotope of Mass 2 and its Concentration.", *Physical Review* **40**, 1–15, April 1, 1932.
23. Stuewer, R. H., "The naming of the neutron", *American Journal of Physics* **54**, 206–218, March 1986.
24. Murphy, G. M. and Johnston, H., "The Nuclear Spin of Deuterium", *Physical Review* **46**, 95–98, July 1934.

25. Charles Weiner, *Sir James Chadwick, oral history,* American Institute of Physics, 1969; cited by Andrew Brown, *The neutron and the bomb,* p. 56.

26. Brown, A., *The Neutron and the Bomb*, p. 80.

27. Biquard, P., *Frédéric Joliot-Curie et l'énergie atomique*, Seghers, Paris, 1961.

28. Goldsmith, M., *Frédéric Joliot-Curie, a biography*, Lawrence and Wishart, London, 1976.

29. Pflaum, R., *Grand Obsession: Madame Curie and her world*, Doubleday, New York, 1989.

30. Pinault, M., *Frédéric Joliot-Curie*, Odile Jacob, Paris, 2000.

31. Biquard, P., *Frédéric Joliot-Curie*, p. 27.

32. Loriot, N., *Irène Joliot-Curie*, Presses de la Renaissance, Paris, 1991.

33. Joliot, F., "Étude électrochimique des radioéléments. Applications diverses", *Journal de Chimie Physique* **27**, 119–162, 1930.

34. Curie, I. and Joliot, F., "Préparation des sources de polonium de grande densité d'activité", *Journal de Chimie Physique* **28**, 201–205, 1931.

35. Joliot, F., "Sur l'excitation des rayons γ nucléaires du bore par les particules α. Énergie quantique du rayonnement γ du polonium", *Comptes Rendus de l'Académie des Sciences* **193**, 1415–1417, December 28, 1931.

36. Curie, I., "Sur le rayonnement γ nucléaire excité dans le glucinium et dans le lithium par les rayons α du polonium", *Comptes Rendus de l'Académie des Sciences* **193**, 1412–1414, session of December 28, 1931.

37. Curie, I. and Joliot, F., "Émission de protons de grande vitesse par les substances hydrogénées sous l'influence des rayons γ très pénétrants", *Comptes Rendus de l'Académie des Sciences* **194**, 273–275, session of January 18, 1932.

38. Chadwick, J., "Some Personal Notes on the Search for the Neutron", in *Actes du dixième congrès international d'Histoire des Sciences*, edited by Guerlac, H., pp. 159–162, Hermann, Paris, 1962.

39. Six, J., *La découverte du neutron (1920–1936)*, Éditions du CNRS, Paris, 1987.

40. Chadwick, J., "Possible existence of a neutron", *Nature* **129**, 312, February 27, 1932.

41. Chadwick, J., letter to Niels Bohr, dated 24 february 1932, in A. Brown,*The Neutron and the Bomb*, p. 365.

42. Chadwick, J., "The Existence of a Neutron", *Proceedings of the Royal Society, London* **A136**, 692–708, 1932.

43. Curie, I. and Joliot, F., "La complexité du proton et la masse du neutron", *Comptes Rendus de l'Académie des Sciences* **197**, 237–238, session of July 17, 1933.

44. Bainbridge, K. T., "The Mass of Be^9 and the Atomic Weight of Beryllium", *Physical Review* **43**, 367–368, March 1, 1933.

45. Curie, I. and Joliot, F., "Mass of the Neutron", *Nature* **133**, 721, May 12, 1934.

46. Chadwick, J. and Goldhaber, M., "A 'Nuclear Photoeffect' : Disintegration of the Diplon by γ-Rays", *Nature* **134**, 237–238, August 18, 1934.

47. Chadwick, J. and Goldhaber, M., "The Nuclear Photoelectric Effect", *Proceedings of the Royal Society, London* **A151**, 479–493, September 2, 1935.

48. Iwanenko, D., "The Neutron Hypothesis", *Nature* **129**, 798, May 28, 1932.

49. Heisenberg, W., "Über den Bau der Atomkerne I", *Zeitschrift für Physik* **77**, 1–11, 1932.

50. Heitler, W. and London, F. W., "Wechselwirkung neutraler Atome und homöopolare Bindung nach der Quantenmechanik", *Zeitschrift für Physik* **44**, 455–472, 1927.

51. Heitler, W., "La théorie quantique des forces de valence", *Annales de l'Institut Henri Poincaré* **4**, 237–272, 1933.

52. Heisenberg, W., "Über den Bau der Atomkerne II", *Zeitschrift für Physik* **78**, 156–164, 1932.

53. Heisenberg, W., "Über den Bau der Atomkerne III", *Zeitschrift für Physik* **80**, 587–596, 1933.

54. Amaldi, E., "Ettore Majorana, man and scientist", in *Strong and weak interactions, 1966 International School of Physics "Ettore Majorana", Erice, June 19 – July 4*, edited by Zichichi, A., pp. 10–77, Academic Press, New York, 1966.

55. Segrè, Emilio, *A Mind Always in Motion. The Autobiography of Emilio Segre,* University of California Press, Berkeley, 1993

56. Fermi, E., "Un metodo statistico per la determinazione di alcune proprietà dell'atomo", *Atti della R. Accademia Nazionale dei Lincei. Rendiconti della Classe di scienze fisiche, matematiche e naturali* **6**, 602–607, session of December 4, 1927.

57. Fermi, E., "Eine statistische Methode zur Bestimmung einiger Eigenschaften des Atoms und ihre Anwendung auf die Theorie des periodischen Systems der Elemente", *Zeitschrift für Physik* **48**, 73–79, 1928.

58. Thomas, L. H., "The calculation of atomic fields", *Proceedings of the Cambridge Philosophical Society* **23**, 542–548, session of November 22, 1926.

59. Fermi, E., "État actuel de la physique du noyau atomique", in *Comptes rendu de la première section du congrès international d'électricité, Paris 1932*, edited by de Valbreuze, R., Vol. 1, pp. 789–807, Gauthier-Villars, Paris, 1932.

60. Majorana, E., "Über die Kerntheorie", *Zeitschrift für Physik* **82**, 137–145, 1933.

61. Mehra, J., "Eugene Paul Wigner: a biographical sketch", in *The Collected Works of Eugene Paul Wigner*, Vol. 1, pp. 3–14, Springer, Berlin, 1993.

62. Pais, A., "Eugene Wigner", in *The Genius of Science. A Portrait Gallery*, pp. 331–351, Oxford University Press, Oxford, 2000.

63. Heisenberg, W., "Über die Spektra von Atomsystem mit zwei Elektronen", *Zeitschrift für Physik* **39**, 499–518, 1926.

64. Wigner, E. P., "Über nicht Kombinierende Terme in der neueren Quantenmechanik. Erster Teil", *Zeitschrift für Physik* **40**, 492–500, 1927.

65. Wigner, E. P., "Über die elastischen Eigenschweigungen symmetrischer Systeme", *Nachrichten der Gesellschaft der Wissenschaften zu Göttingen Mathematisch-Physikalische Klasse* pp. 133–146, 1930.

66. Bouckaert, L. P., Smoluchowski, R. and Wigner, E. P., "Theory of Brillouin Zones and Symmetry Properties of Wave Functions in Crystals", *Physical Review* **50**, 58–67, July 1936.

67. Wigner, E. P., "On the Mass Defect of Helium", *Physical Review* **43**, 252–257, February 1933.

68. Wigner, E. P., "Über die Streuung von Neutronen an Protonen", *Zeitschrift für Physik* **83**, 253–258, 1933.

69. Harkins, W. D. and Wilson, E. D., "Energy Relations involved in the Formation of Complex Atoms", *Philosophical Magazine* **30**, 723–734, November 1915.

70. Harkins, W. D., "Isotopes : Their Number and Classification", *Nature* **107**, 202–3, April 14, 1921.

71. Harkins, W. D., "The Periodic System of Atomic Nuclei and the Principle of Regularity and Continuity of Series", *Physical Review* **38**, 1270–1288, October 1931.

72. Bartlett, J. H., "Structure of Atomic Nuclei", *Physical Review* **41**, 370–371 (L), August 1932.

73. Bartlett, J. H., "Structure of Atomic Nuclei. II", *Physical Review* **42**, 145–146 (L), October 1932.

74. Elsasser, W. M., "Sur le principe de Pauli dans les noyaux", *Journal de Physique et Le Radium* **4**, 549–556, October 1933.

75. Elsasser, W. M., *Memoirs of a Physicist in the Atomic Age*, Science History Publications & Adam Hilger, New York & Bristol, 1978.

76. Guggenheimer, K., "Remarques sur la constitution des noyaux atomiques I", *Journal de Physique et Le Radium* **5**, 253–356, June 1934.

77. Elsasser, W. M., "Sur le principe de Pauli dans les noyaux II.", *Journal de Physique et Le Radium* **5**, 389–397, 1934.

78. Heisenberg, W., "Die Struktur der leichten Atomkerne", *Zeitschrift für Physik* **96**, 473–484, 1935.

79. Hartree, D. R., "The Wave Mechanics of an Atom with a Non-Coulomb Central Field. Part I. – Theory and Methods", *Proceedings of the Cambridge Philosophical Society* **24**, 89–110, session of November 21, 1928.

80. Hartree, D. R., "The Wave Mechanics of an Atom with a Non-Coulomb Central Field. Part II. – Some Results and Discussion", *Proceedings of the Cambridge Philosophical Society* **24**, 111–132, session of November 21, 1928.

81. Hartree, D. R., "The Wave Mechanics of an Atom with a non-Coulomb Central Field. Part III. – Term Values and Intensities in Series in Optical Spectra", *Proceedings of the Cambridge Philosophical Society* **24**, 426–437, session of July 23, 1928.

82. Bush, V., "The Differential Analyser. A New Machine for Solving Differential Equations", *Journal of the Franklin Institute* **212**, 447–488, October 1931.

83. Feenberg, E. and Wigner, E. P., "On the Structure of the Nuclei between Helium and Oxygen", *Physical Review* **51**, 95–106, January 1937.

84. Brueckner, K. A., Campbell, C. E., Clark, J. W. and Primakoff, A., "Eugene Feenberg 1906–1977", *Nuclear Physics* **A317**, i–vii, 1979.

85. Fock, V., "Näherungsmethode zur Lösung des quantenmechanischen Mehrkörperproblems", *Zeitschrift für Physik* **61**, 126–148, 1930.

86. Hund, F., "Symmetrieeigenschaften der Kräfte in Atomkernen und Folgen für deren Zustände, insbesondere der Kerne bis zu sechszehn Teilchen", *Zeitschrift für Physik* **105**, 202–228, 1937.

87. Wilson, C. T. R., "Condensation of Water Vapour in the Presence of Dust-free Air and other Gases", *Transactions of the Royal Society* **A189**, 265–307, session of April 8, 1897.

88. Geitel, H., "Über die Elektrizitätszerstreuung in abgeschlossenen Luftmengen", *Physikalische Zeitschrift* **2**, 117–119, November 24, 1900.

89. Coulomb, C., *De la quantité d'électricité qu'un corps isolé perd dans un temps donné, soit par le contact de l'air plus ou moins humide, soit le long des soutiens plus ou moins idio-électriques. Troisième mémoire sur l'électricité, 1785*, Gauthier-Villars pour la Société Française de Physique, Paris, 1884.

90. Wilson, C. T. R., "On the leakage of Electricity through dust-free air", *Proceedings of the Cambridge Philosophical Society* **11**, 32, session of November 26, 1900.

91. Wilson, C. T. R., "On the Ionisation of the Atmospheric Air", *Proceedings of the Royal Society, London* **68**, 151–161, session of March 14, published in September 1901.

92. Rutherford, E. and Allen, S. J., "Excited Radioactivity and Ionization of the Atmosphere", *Philosophical Magazine* **4**, 704–723, December 1902.

93. Hess, V. F., "Beobachtung der durchdringenden Strahlung bei sieben Freiballonfahrten", *Sitzungsberichte der Kaiserlichen Akademie der Wissenschaften, Vienne* **121**, Abteilung IIa, 2001–2032, session of October 17, 1912.

94. Kolhörster, W., "Messungen der durchdringenden Strahlung im Freiballon in grösseren Höhen", *Verhandlungen der deutschen physikalischen Gesellschaft* **16**, 1111–1116, November 15, 1913.

95. Kolhörster, W., "Messungen der durchdringenden Strahlung im Freiballon in grösseren Höhen", *Physikalische Zeitschrift* **14**, 1153–1156, November 15, 1913.

96. Millikan, R. A., "High Frequency Rays of Cosmic Origin", *Proceedings of the National Academy of Sciences of the United States of America* **12**, 48–55, January 15 1926.

97. Bothe, W. and Kolhörster, W., "Das Wesen der Höhenstrahlung", *Zeitschrift für Physik* **56**, 751–777, 1929.

98. Blackett, P. M. S. and Occhialini, G. P. S., "Photography of Penetrating Corpuscular Radiation", *Nature* **130**, 363, September 3 1932.

99. Blackett, P. M. S. and Occhialini, G. P. S., "Some Photographs of Penetrating Radiation", *Proceedings of the Royal Society, London* **139**, 699–718, 1933.

100. Skobelzyn, D., "Die Intensitätsverteilung in dem Spektrum der γ-Strahlen von RaC", *Zeitschrift für Physik* **43**, 354–378, 1927.

101. Skobelzyn, D., "Über eine neue Art sehr schneller β-Strahlen", *Zeitschrift für Physik* **54**, 686–702, 1929.

102. Segrè, E., *From X-rays to quarks*, p. 192.

103. Anderson, C. D., "The Apparent Existence of Easily Deflectable Positives", *Science* **76**, 238–239, September 9, 1932.

104. Anderson, C. D., "The Positive Electron", *Physical Review* **43**, 491–494, March 1933.

105. Dirac, P. A. M., "A Theory of Electrons and Protons", *Proceedings of the Royal Society, London* **A126**, 360–365, January 1, 1930.

106. Dirac, P. A. M., "The Proton", *Nature* **126**, 605–606, October 18, 1930.

107. Oppenheimer, J. R., "On the Theory of Electrons and Protons", *Physical Review* **35**, 562–563, March 1930.

108. Bird, K. and Sherwin, M. J., *American Prometheus, The Triumph and Tragedy of J. Robert Oppenheimer*, Alfred A. Knopf, New York, 2005.

109. Dirac, P. A. M., "Quantized Singularities in the Electromagnetic Field", *Proceedings of the Royal Society, London* **A133**, 60–72, September 1, 1931.

110. Dirac, P. A. M., "Théorie du positon", in *Structures et propriétés des noyaux atomique, rapports et discussions du septième conseil de Physique Solvay, 22 au 29 octobre 1933*, pp. 205–212, Gauthier-Villars, Paris, 1934.

111. Curie, I. and Joliot, F., "Sur la nature du rayonnement pénétrant excité dans les noyaux légers par les particules α", *Comptes Rendus de l'Académie des Sciences* **194**, 1229–1232, session of April 11, 1932.

112. Chadwick, J., Blackett, P. M. S. and Occhialini, G. P. S., "New Evidence for the Positive Electron", *Nature* **131**, 473, April 1, 1933.

113. Meitner, L. and Philipp, K., "Die bei Neutronenanregung auftretenden Elektronenbahnen", *Naturwissenschaften* **21**, 286–287, April 14, 1933.

114. Curie, I. and Joliot, F., "Contribution à l'étude des électrons positifs", *Comptes Rendus de l'Académie des Sciences* **196**, 1105–1107, session of April 10, 1933.

115. Curie, I. and Joliot, F., "Sur l'origine des électrons positifs", *Comptes Rendus de l'Académie des Sciences* **196**, 1581–1583, May 22, 1933.

116. Meitner, L. and Philipp, K., "Die Anregung positiver Elektronen durch γ-Strahlen von ThC''''", *Naturwissenschaften* **21**, 468, June 16, 1933.

117. Joliot-Curie, I. and Joliot, F., "Électrons positifs de transmutation", *Comptes Rendus de l'Académie des Sciences* **196**, 1885–1887, session of June 19, 1933.

118. Rutherford, E., "Address of the President, Sir Ernest Rutherford, O. M., at the Anniversary Meeting, November 30, 1927.", *Proceedings of the Royal Society, London* **A117**, 300–316, 1928.

119. Livingston, M. S. and Blewett, J. P., *Particle Accelerators*, McGraw-Hill, New York, 1962.

120. Livingston, M. S. (editor), *The Development of High Energy Accelerators*, Dover, New York, 1966. A reprint of 28 historical articles.

121. Grinberg, A. P., "History of the invention and development of accelerators (1922–1932)", *Soviet Physics, Uspekhi* **18**, 815–831, 1975. Uspekhi fizicheskikh nauk **117**, 333–362, October 1975.

122. Hoffmann, D., "Fritz Lange, Klaus Fuchs, and the Remigration of Scientists to East Germany", *Physics in Perspective* **11**, 405–425, 2009.

123. Brasch, A. and Lange, F., "Experimentelle-technische Vorbereitungen zur Atomzerströmmerung mittels hoher elektrischer Spannungen", *Zeitschrift für Physik* **70**, 10–37, 1931.

124. Marx, E., "Verfahren zur Schlagprüfung von Isolatoren und anderer elektrischen Vorrichtungen", German patent #455933, October 12, 1923.

125. Schenckel, M., "Eine neue Schaltung für die Erzeugung hoher Gleichspannungen", *Elektrotechnische Zeitschrift* **40**, 333–334, July 10, 1919.

126. Greinacher, H., "Über eine Methode, Wechselstrom mittels elektrischer Ventile und Kondensatoren in hochgespannten Gleichstrom umzuwandeln", *Zeitschrift für Physik* **4**, 195–205, 1921.

127. Cockcroft, J. D. and Walton, E. T. S., "Disintegration of Lithium by Swift Protons", *Nature* **129**, 649, April 30, 1932.

128. Vollrath, R. E., "A High Voltage Direct Current Generator", *Physical Review* **42**, 298–304, October 15, 1932.

129. van de Graaff, R., Compton, K. T. and van Atta, L. C., "The Electrostatic Production of High Voltage for Nuclear Investigations", *Physical Review* **43**, 149–57, February 1933.

130. van Atta, L. C., Northrup, D. L., van Atta, C. M. and van de Graaff, R., "The Design, Operation, and Performance of the Round Hill Electrostatic Generator", *Physical Review* **49**, 761–776, May 1936.

131. van Atta, L. C., Northrup, D. L., van de Graaff, R. and van Atta, C. M., "Electrostatic Generator for Nuclear Research at M.I.T.", *Review of Scientific Instruments* **12**, 534–545, November 1941.

132. Tuve, M. A., Hafstad, L. R. and Dahl, O., "Nuclear Physics Studies Using the Van de Graaff Electrostatic Generator", *Physical Review* **43**, 1055 (A), June 15, 1933.

133. Tuve, M. A., Hafstad, L. R. and Dahl, O., "Disintegration Experiments on Elements of Medium Atomic Number", *Physical Review* **43**, 942 (L), July 1, 1933.

134. Ising, G., "Prinzip einer Methode zur Herstellung von Kanal-Strahlen hoher Voltzahl", *Arkiv för matematik, astronomi och fysik* **18**, 1–4, 1924.

135. Walton, E. T. S., "The Production of High Speed Electrons by Indirect Means", *Proceedings of the Cambridge Philosophical Society* **25**, 469–481, session of July 29, 1929.

136. Wideröe, R., "Über ein neues Prinzip zur Herstellung hoher Spannungen", *Archiv für Elektrotechnik* **21**, 387–406, 1928.

137. Lawrence, E. O., McMillan, E. and Thornton, R. L., "The Transmutation function for Some Cases of Deuteron-Induced Radioactivity", *Physical Review* **48**, 493–499, September 15, 1935.

138. Lawrence, E. O. and Edlefsen, N. E., "On the Production of High Speed Protons", *Science* **72**, 376–377, October 10, 1930.

139. Sloan, D. H. and Lawrence, E. O., "The Production of Heavy High Speed Ions Without the Use of High Voltages", *Physical Review* **38**, 2021–2032, December 1, 1931.

140. Coates, W. M. and Sloan, D. H., "High-velocity mercury ions", *Physical Review* **43**, 212–13 (A), February 1, 1933.

141. Sloan, D. H. and Coates, W. M., "Recent Advances in the Production of Heavy High Speed Ions Without the Use of High Voltages", *Physical Review* **46**, 539–542, October 1934.

142. Lawrence, E. O. and Livingston, M. S., "The Production of High Speed Protons Without the Use of High Voltages", *Physical Review* **38**, 834 (L), August 15, 1931.

143. Lawrence, E. O. and Livingston, M. S., "The Production of High Speed Light Ions without the use of High Voltages", *Physical Review* **40**, 19–35, April 1, 1932.

144. Lawrence, E. O., Livingston, M. S. and White, M. G., "The Disintegration of Lithium by Swiftly Moving Protons", *Physical Review* **42**, 150–151, October 1, 1932.

145. Henderson, M. C., Livingston, M. S. and Lawrence, E. O., "The Transmutation of Fluorine by Proton Bombardment and the Mass of Fluorine 19", *Physical Review* **46**, 38–42, July 1934.

146. Henderson, M. C., Livingston, M. S. and Lawrence, E. O., "Artificial Radioactivity Produced by Deuton Bombardment", *Physical Review* **45**, 428–429 (L), March 1934.

147. Lawrence, E. O. and Livingston, M. S., "The Multiple Acceleration of Ions to Very High Voltages", *Physical Review* **45**, 608–612, May 1934.

148. Lawrence, E. O. and Cooksey, D., "On the Apparatus for the Multiple Acceleration of Light Ions to High Speeds", *Physical Review* **50**, 1131–1140, December 1936.

149. Lawrence, E. O., Alvarez, L. W., Brobeck, W. M., Cooksey, D., Corson, D. R., McMillan, E., Salisbury, W. W. and Thornton, R. L., "Initial Performance of the 60-Inch Cyclotron of the William H. Crocker Radiation Laboratory, University of California", *Physical Review* **56**, 124 (L), July 1939.

150. Heilbron, J. L. and Seidel, R. W., *Lawrence and his laboratory. Vol. 1*, University of California Press, Berkeley, 1989.

151. Livingston, M. S., *Particle Accelerators : A Brief History*, Harvard University Press, Cambridge, 1969.

152. Feenberg, E. and Knipp, J., "Intranuclear forces", *Physical Review* **48**, 906–912, December 1935.

153. Wells, W. H., "The Scattering of Protons on Protons", *Physical Review* **47**, 591–596, April 1935.

154. White, M. G., "Scattering of High Energy Protons in Hydrogen", *Physical Review* **49**, 309–316, February 1936.

155. Tuve, M. A., Heydenburg, N. P. and Hafstad, L. R., "The Scattering of Protons by Protons", *Physical Review* **50**, 806–825, November 1936.

156. Breit, G., Condon, E. U. and Present, R. D., "Theory of Scattering of Protons by Protons", *Physical Review* **50**, 825–845, November 1936.
157. Amaldi, E. and Fermi, E., "Sull'assorbimento dei neutroni lenti.– III.", *Ricerca Scientifica* **7**, 56–59, 1936.
158. *Structure et propriétés des noyaux atomiques, rapports et discussions du septième Conseil de Physique Solvay, Bruxelles, 22 au 29 octobre 1933*, Gauthier-Villars, Paris, 1934.
159. Bensaude-Vincent, B., *Langevin, science et vigilance*, Belin, Paris, 1987.
160. Joliot, F. and Curie, I., "Rayonnement pénétrant des atomes sous l'action des rayons α", in *Structure et propriétés des noyaux atomiques. Rapports et discussions du septième Conseil de Physique Solvay, Bruxelles, 22 au 29 octobre 1933*, pp. 121–156.
161. Guerra, F., Leone, M. and Robotti, N., "The Discovery of Artificial Radioactivity", *Physics in Perspective* **14**, 33–58, 2012.
162. Curie, I. and Joliot, F., "Sur les conditions d'émission des neutrons par action des particules α sur les éléments légers", *Comptes Rendus de l'Académie des Sciences* **196**, 397–399, session of February 6, 1933.
163. Meitner, L., "In *Structure et propriétés des noyaux atomiques. Rapports et discussions du septième Conseil de Physique Solvay, p. 176*".
164. Joliot-Curie, I. e. F., "La découverte de la radioactivité artificielle", *Atomes* pp. 9–12, January 1951.
165. Radvanyi, P. and Bordry, M., *La radioactivité artificielle et son histoire*, Seuil/CNRS, Paris, 1984.
166. Curie, I. and Joliot, F., "Un nouveau type de radioactivité", *Comptes Rendus de l'Académie des Sciences* **198**, 254–256, session of January 15, 1934.
167. Curie, I. and Joliot, F., "Artificial production of a new kind of radio-element", *Nature* **133**, 201–202, February 10, 1934.
168. Lauritsen, C. C., Crane, H. R. and Harper, W. W., "Artificial Production of Radioactive Substances", *Science* **79**, 234, March 9, 1934.
169. Crane, H. R., Lauritsen, C. C. and Soltan, A., "Artificial Production of Neutrons", *Physical Review* **44**, 514, September 1933.
170. Crane, H. R., Lauritsen, C. C. and Soltan, A., "Artificial Production of Neutrons", *Physical Review* **45**, 507–512, April 1934.
171. Rutherford, E., letter to F. and I. Joliot-Curie, dated January 29, 1934, Archives Joliot-Curie, Musée Curie, Paris.
172. Cockcroft, J. D., Gilbert and Walton, E. T. S., "Production of Induced Radioactivity by High Velocity Protons", *Nature* **133**, 328, March 3, 1934.
173. Neddermeyer, S. H. and Anderson, C. D., "Energy Spectra of Positrons Ejected by Artificially Stimulated Radioactive Substances", *Physical Review* **45**, 498–499 (L), April 1934.
174. Frisch, O. R., "Induced Radioactivity of Sodium and Phosphorus", *Nature* **133**, 721–722, May 12, 1934.
175. Joliot, F. and Curie, I., "Artificially produced radio-elements", in *International Conference on Physics, London, 1934. A joint Conference organized by the International Union of Pure and Applied Physics and the Physical Society*, Vol. I, pp. 78–89, The Physical Society, London, 1935.
176. Paneth, F. A. and von Hevesy, G., "Über Versuche zur Trennung des Radiums D von Blei", *Sitzungsberichte der Kaiserlichen Akademie der Wissenschaften, Vienne* **122**, Abteilung IIa, 993–1000, session of April 24, 1913.
177. Paneth, F. A. and von Hevesy, G., "Über Radioelemente als Indikatoren in der analytischen Chemie", *Sitzungsberichte der Kaiserlichen Akademie der Wissenschaften, Vienne* pp. Abteilung IIa, 1001–1007, session of April 24, 1913.
178. Coster, D. and von Hevesy, G., "On the Missing Element of Atomic Number 72", *Nature* **111**, 79, January 20, 1923.
179. von Hevesy, G., "The Absorption and Translocation of Lead by Plants. A contribution to the application of the method of radioactive indicators in the investigation of the change of substance in plants.", *Biochemical Journal* **17**, 439–445, 1923.

180. Christiansen, J. A., de Hevesy, G. and Lomholt, S., "Recherche, par une méthode radiochim-ique, sur la circulation du bismuth dans l'organisme", *Comptes Rendus de l'Académie des Sciences* **178**, 1324–26, session of April 24, 1924.

181. Joliot, F., "Les grandes découvertes de la radioactivité", in *Textes choisis*, Éditions sociales, Paris, 1959, p. 66. This written text was planned to be broadcast on April 23, 1957 but it was censored by the French government because the conclusion attracted attention to the dangers of atomic explosions and to the necessity of stopping them.

182. Joliot, F., "Chemical Evidence of the Transmutation of Elements", in *Nobel Lectures, Chemistry 1922–1941*, Elsevier Publishing Company, Amsterdam, 1966.

183. Fermi, E., *Collected Papers (Note e memorie)*, The University of Chicago Press/Accademia Nazionale dei Lincei, Chicago/Rome, 1962.

184. Segrè, E., *Enrico Fermi Physicist*, The University of Chicago Press, 1970, p. 70.

185. *Ibid.* p. 72.

186. Fermi, E., "Versuch einer Theorie der β-Strahlen", *Zeitschrift für Physik* **88**, 161–177, 1934.

187. Sargent, B. W., "Energy distribution curves of the disintegration electrons", *Proceedings of the Cambridge Philosophical Society* **28**, 538–553, 1932.

188. Sargent, B. W., "The Maximum Energy of the β-Rays from Uranium X", *Proceedings of the Royal Society, London* **A139**, 659–673, 1933.

189. Reines, F. and Cowan, C. L., "The Neutrino", *Nature* **178**, 446–449, September 1, 1956.

190. Fermi, E., "Radioattività indotta da bombardamento di neutroni. — I.", *Ricerca Scientifica* **5**, 283, 1934.

191. Fermi, E., "Radioattività provocata da bombardamento di neutroni.— II.", *Ricerca Scientifica* **5**, 330–331, 1934.

192. Fermi, E., "Radioactivity induced by neutron bombardment", *Nature* **133**, 757 (L), May 19, 1934.

193. Rutherford, E., letter to E. Fermi, dated April 23, 1934, in *Enrico Fermi Collected Papers*, p. 641.

194. Amaldi, E., d'Agostino, O., Fermi, E., Rasetti, F. and Segrè, E., "Radioattività ≪ beta ≫ provocata da bombardamento di neutroni.– III.", *Ricerca Scientifica* **5**, 452–453, 1934.

195. Amaldi, E., d'Agostino, O., Fermi, E., Rasetti, F. and Segrè, E., "Radioattività provocata da bombardamento di neutroni.– IV.", *Ricerca Scientifica* **5**, 652–653, 1934.

196. Amaldi, E., d'Agostino, O., Fermi, E., Rasetti, F. and Segrè, E., "Radioattività provocata da bombardamento di neutroni.– V.", *Ricerca Scientifica* **5**, 21–22, 1934.

197. Amaldi, E., D'Agostino, O. and Segrè, E., "Radioattività provocata da bombardamento di neutroni. — VI", *Ricerca Scientifica* **5**, 381, 1934.

198. Fermi, E., Amaldi, E., d'Agostino, O., Rasetti, F. and Segrè, E., "Artificial radioactivity produced by neutron bombardment", *Proceedings of the Royal Society, London* **A146**, 483–500, 1934.

199. Fermi, E., "Artificial Radioactivity Produced by Neutron Bombardment", in *International Conference on Physics, London 1934*, Vol. I, pp. 75–77, The Physical Society, London, 1935.

200. Fermi, E., Rasetti, F. and D'Agostino, O., "Sulla possibilità di produre elementi di numero atomico maggiore di 92", *Ricerca Scientifica* **5**, 536–537, 1934.

201. Fermi, E., "Possible Production of Elements of Atomic Number Higher than 92", *Nature* **133**, 898–899, June 16, 1934.

202. Fermi, L., *Atoms in the family*, p. 91.

203. Hahn, O. and Meitner, L., "Über die künstliche Umwandlung des Urans durch Neutronen", *Naturwissenschaften* **23**, 37–38, January 11, 1935.

204. Fermi, L., *Atoms in the family*, p. 97.

205. Fermi, E., Amaldi, E., Pontecorvo, B., Rasetti, F. and Segrè, E., "Azione di sostanze idrogenate sulla radioattività provocata da neutroni.– I.", *Ricerca Scientifica* **5**, 282–283, 1934.

206. Fermi, E., Pontecorvo, B. and Rasetti, F., "Effetto di sostanze idrogenate sulla radioattività provocata da neutroni. — II.", *Ricerca Scientifica* **5**, 380–81, 1934.

207. Amaldi, E., D'Agostino, O., Fermi, E., Pontecorvo, B., Rasetti, F. and Segrè, E., "Artificial Radioactivity Produced by Neutron Bombardment.– II.", *Proceedings of the Royal Society, London* **A149**, 522–558, April 10, 1935.
208. Curie, I., Joliot, F. and Preiswerk, P., "Radioéléments créés par le bombardement de neutrons. Nouveau type de radioactivité", *Comptes Rendus de l'Académie des Sciences* **198**, 2089–2091, session of June 11, 1934.
209. Moon, P. B. and Tillman, J. R., "Evidence on the Velocity of 'Slow' Neutrons", *Nature* **135**, 904, June 1, 1935.
210. Artsimovitch, L., Kourtschatov, I., Miççovskiï, L. and Palibin, P., "Au sujet de la capture des neutrons lents par un noyau", *Comptes Rendus de l'Académie des Sciences* **200**, 2159–2162, session of June 24, 1935.
211. Tillman, J. R. and Moon, P. B., "Selective Absorption of Slow Neutrons", *Nature* **136**, 66–67, July 13, 1935.
212. Bjerge, T. and Westcott, C. H., "On the Slowing Down of Neutrons in Various Substances Containing Hydrogen", *Proceedings of the Royal Society, London* **A150**, 709–728, July 1, 1935.
213. Amaldi, E. and Fermi, E., "Sull'assorbimento dei neutroni lenti.– I.", *Ricerca Scientifica* **6**, 344–347, 1935.
214. Fermi, E. and Amaldi, E., "Sull'assorbimento dei neutroni lenti.– II.", *Ricerca Scientifica* **6**, 443–437, 1935.
215. Amaldi, E. and Fermi, E., "Sul cammino libero midio dei neutroni nella paraffina", *Ricerca Scientifica* **7**, 223–225, 1936.
216. Amaldi, E. and Fermi, E., "Sui gruppi dei neutroni lenti", *Ricerca Scientifica* **7**, 310–315, 1936.
217. Amaldi, E. and Fermi, E., "Sulle proprietà di diffusione dei neutroni lenti", *Ricerca Scientifica* **7**, 393–395, 1936.
218. Szilard, L., "Absorption of Residual Neutrons", *Nature* **136**, 950–951, December 14, 1935.
219. Rasetti, F., Segrè, E., Fink, G., Dunning, J. R. and Pegram, G. B., "On the Absorption Law for Slow Neutrons", *Physical Review* **49**, 104 (L), January 1936.
220. Amaldi, E. and Fermi, E., "Sopra l'assorbimento e la diffusione dei neutroni lenti", *Ricerca Scientifica* **7**, 454–503, 1936.
221. Amaldi, E. and Fermi, E., "On the Absorption and the Diffusion of Slow Neutrons", *Physical Review* **50**, 899–928, November 15, 1936.
222. Beyerchen, A. D., *Scientists under Hitler : politics and the physics community in the third reich*, Yale University Press, New Haven, 1977.
223. Guérout, S., *Science et politique sous le Troisième Reich*, Ellipses, Paris, 1992.
224. Klein, O. and Nishina, Y., "Über die Streuung von Srahlung durch freie Elektronen nach der neuen relativistischen Quantendynamik von Dirac", *Zeitschrift für Physik* **52**, 853–868, 1929.
225. Yukawa, H., *"Tabibito" (the traveler)*, World Scientific, London, 1982. Translated by L. Brown and R. Yoshida.
226. Yukawa, H., "On the Interaction of Elementary Particles", *Proceedings of the Physico-Mathematical Society of Japan* **17**, 48–57, session of November 17, 1935.
227. Tamm, I. E., "Exchange Forces between Neutrons and Protons, and Fermi's Theory", *Nature* **133**, 981, June 30, 1934.
228. Yukawa, H., *"Tabibito" (The Traveler)*, p 202–203.
229. Wick, G. C., "Range of Nuclear Forces in Yukawa's Theory", *Nature* **142**, 993–994, December 3, 1938.
230. Neddermeyer, S. H. and Anderson, C. D., "Note on the Nature of Cosmic-Ray Particles", *Physical Review* **51**, 884–886, May 15, 1937.
231. Street, J. C. and Stevenson, E. C., "New Evidence for the Existence of a Particle of Mass Intermediate Between the Proton and Electron", *Physical Review* **52**, 1003–1004 (L), November 1937.
232. Anderson, C. D. and Neddermeyer, S. H., "Mesotron (Intermediate particle) as a Name for the New Particles of Intermediate Mass", *Nature* **142**, 878, November 12, 1938.

233. Bhabha, H. J., "The Fundamental Length Introduced by the Theory of the Mesotron (Meson)", *Nature* **143**, 276–277, February 18, 1939.
234. Bethe, H. A., "Theory of Disintegration of Nuclei by Neutrons", *Physical Review* **47**, 747–759, May 1935.
235. Perrin, F. and Elsasser, W. M., "Théorie de la capture sélective des neutrons lents par certains noyaux", *Comptes Rendus de l'Académie des Sciences* **200**, 450–452, session of February 4, 1935.
236. Breit, G. and Wigner, E. P., "Capture of slow neutrons", *Physical Review* **49**, 519–531, April 1936.
237. Bohr, N., "Neutron Capture and Nuclear Constitution", *Nature* **137**, 344–348, February 29, 1936.
238. Bohr, N. and Kalckar, F., "On the transmutation of atomic nuclei by impact of material particles. I. General theoretical remarks", *Det Kongelig Danske Videnskabernes Selskab, Matematisk-fysiske Meddleser* **14**, No. 10, 1–40, 1937.
239. Feenberg, E., *Shell theory of the nucleus*, Princeton University Press, Princeton, 1955.
240. Feather, N., *Lord Rutherford*, Priory Press, London, 1973, p. 189.
241. *Ibid.*, p. 184.
242. Brown, A., *The Neutron and the Bomb*, p. 158.
243. Eve, A. S., Chadwick, J., Thomson, J. J., Bragg, W. H., Bohr, N., Soddy, F. and Smith, F. E., "The Right Hon. Lord Rutherford of Nelson, O. M., F. R. S.", *Nature* **140**, 746–754, October 30, 1937.
244. Fermi, E., "Tribute to Lord Rutherford", *Nature* **140**, 1052, December 18, 1937.
245. Bethe, H. A. and Bacher, R. F., "Nuclear Physics. A. Stationary States of Nuclei", *Reviews of Modern Physics* **8**, 82–229, April 1936.
246. Bethe, H. A., "Nuclear Physics. B. Nuclear Dynamics, Theoretical", *Reviews of Modern Physics* **9**, 69–244, April 1937.
247. Livingston, M. S. and Bethe, H. A., "Nuclear Physics. C. Nuclear Dynamics, Experimental", *Reviews of Modern Physics* **9**, 245–390, July 1937.
248. Schweber, S. S., *In the Shadow of the Bomb : Bethe, Oppenheimer, and the Moral Responsibility of the Scientist*, Princeton University Press, Princeton, 2000.
249. Pollard, E., "Nuclear Potential Barriers: Experiments and Theory", *Physical Review* **47**, 611–620, 1935.
250. von Weizsäcker, C. F., "Zur Theorie der Kernmasse", *Zeitschrift für Physik* **96**, 431–458, 1935.
251. Schüler, H. and Schmidt, T., "Über Abweichungen des Atomkerns von der Kugelsymmetrie", *Zeitschrift für Physik* **94**, 457–468, 1935.
252. Bethe, H. A. and Placzek, G., "Resonance Effects in Nuclear Processes", *Physical Review* **51**, 450–484, March 1937.
253. Amaldi, E., "George Placzek", *Ricerca Scientifica* **26**, 2038–2042, July 1956.
254. von Grosse, A. and Agruss, M. S., "Fermi's Element 93", *Nature* **134**, 773, November 17, 1934.
255. Hahn, O. and Meitner, L., "Die Muttersubstanz des Actiniums, ein neues radioaktives Element von langer Lebensdauer", *Physikalische Zeitschrift* **19**, 208–218, May 15, 1918.
256. Meitner, L., "Über das Protactinium", *Naturwissenschaften* **6**, 324–326, May 31, 1918.
257. Meitner, L., "Über die von I. Curie und F. Joliot entdeckte künstliche Radioaktivität", *Naturwissenschaften* **22**, 172–174, March 16, 1934.
258. Meitner, L., "Wege und Irrwege zur Kernenergie", *Naturwissenschaftliche Rundschau* **16**, 167–169, May 1963.
259. Shea, W. R., editor, *Otto Hahn and the rise of nuclear physics,* D. Reidel, Dordrecht, 1983.
260. Meitner, L., "Einige Erinnerungen an das Kaiser-Wilhelm-Institut für Chemie in Berlin-Dahlem", *Naturwissenschaften* **41**, 97–99, March 1954.
261. Noddack, I., "Das periodiches System der Elemente und seine Lücken", *Angewandte Chemie* **47**, 301–305, May 19, 1934.

262. Hahn, O. and Meitner, L., "Über die künstliche Umwandlung des Urans durch Neutronen. II. Mitteil", *Naturwissenschaften* **23**, 230–231, April 5, 1935.
263. Hahn, O., Meitner, L. and Strassmann, F., "Einige weitere Bemerkungen über die künstlichen Umwandlungsprodukte beim Uran", *Naturwissenschaften* **23**, 544–545, August 2, 1935.
264. Meitner, L. and Hahn, O., "Neue Umwandlungsprozesse bei Bestrahlung des Uran mit Neutronen", *Naturwissenschaften* **24**, 158–159, March 6, 1936.
265. Sime, R. L., *Lise Meitner*, p. 352.
266. Meitner, L., Hahn, O. and Strassmann, F., "Über die Umwandlungsreihen des Urans, die durch Neutronbestrahlung erzeugt werden", *Zeitschrift für Physik* **106**, 249–270, 1937.
267. Hahn, O., "Über eine neue radioaktive Substanz im Uran", *Berichte der deutschen chemischen Gesellschaft* **B 54**, 1131–1142, June 11, 1921.
268. Hahn, O. "Über das Uran Z und seine Muttersubstanz", *Zeitschrift für physikalische Chemie* **103**, 461–481, 1923.
269. von Weizsäcker, C. F., "Metastabile Zustände der Atomkerne", *Naturwissenschaften* **24**, 813–814, December 18, 1936.
270. von Droste, G., "Über Versuche eines Nachweiss von α-Strahlen während der Bestrahlung von Thorium und Uran mit Radium + Beryllium-Neutronen", *Zeitschrift für Physik* **110**, 84–94, 1938.
271. Goldschmidt, B., *Pionniers de l'atome*, Stock, Paris, 1987, p. 27.
272. Goldschmidt, B., "Hans Halban (1908–1964)", *Nuclear Physics* **79**, 1–11, 1966.
273. Curie, I., von Halban, H. and Preiswerk, P., "Sur la création artificielle des éléments d'une famille radioactive inconnue lors de l'irradiation du thorium par les neutrons", *Comptes Rendus de l'Académie des Sciences* **200**, 1841–1843, session of May 27, 1935.
274. von Halban, H. and Preiswerk, P., "Sur l'existence de niveaux de résonance pour la capture de neutrons", *Comptes Rendus de l'Académie des Sciences* **202**, 133–135, session of January 13, 1936.
275. von Halban, H. and Preiswerk, P., "Recherches sur les neutrons lents", *Journal de Physique et Le Radium* **8**, 29–40, January 1937.
276. Curie, I. and Savitch, P., "Sur les radioéléments formés dans l'uranium irradié par les neutrons (I)", *Journal de Physique et Le Radium* **8**, 385–387, October 1937.
277. Hahn, O. and Meitner, L., letter to Irène Curie, dated January 20, 1938, Archives Joliot-Curie, Musée Curie, Paris.
278. Curie, I. and Savitch, P., "Sur le radioélément de période 3,5 heures formé dans l'uranium irradié par les neutrons", *Comptes Rendus de l'Académie des Sciences* **206**, 906–908, session of March 21, 1938.
279. Curie, I. and Savitch, P., "Sur la nature du radioélément de période 3,5 heures formé dans l'uranium irradié par les neutrons", *Comptes Rendus de l'Académie des Sciences* **206**, 1643–1644, session of May 30, 1938.
280. Sime, R. L., "Lise's Meitner escape from Germany", *American Journal of Physics* **58**, 262–67, 1990.
281. Sime, R. L., *Lise Meitner, a life in Physics*, pp. 184–209.
282. Rife, P., *Lise Meitner and the Dawn of Nuclear Age*, pp. 160–177.
283. Krafft, F., *Im Schatten der Sensation, Leben und Wirken von Fritz Strassmann*, Verlag Chemie, Weinheim, 1981.
284. Joliot, F., *Textes choisis*, p. 35.
285. Curie, I. and Savitch, P., "Sur les radioéléments formés dans l'uranium irradié par les neutrons (II)", *Journal de Physique et Le Radium* **9**, 355–359, September 1938.
286. Cook, L. G., "Personal reminiscences of the *Kaiser Wilhelm Institut*, Berlin, 1937–38 and the nuclear project in Canada, 1944–45.", in *50 years with nuclear fission, 25–28 April 1989, Gaitherzburg*, edited by Behrens, J. W., American Nuclear Society, La Grange Park, 1989.
287. Hahn, O., Meitner, L. and Strassmann, F., "Ein neues langlebiges Umwandlungsprodukt in den Trans-Uranreihen", *Naturwissenschaften* **26**, 475–476, July 22, 1938.

288. Hahn, O. and Strassmann, F., "Über die Entstehung von Radiumisotopen aus Uran durch Bestrahlen mit schnellen und verlangsamten Neutronen", *Naturwissenschaften* **26**, 755–756, November 18, 1938.

289. Sime, R. L., *Lise Meitner*, p. 227.

290. Krafft, F., *Im Schatten der Sensation*, p. 264.

291. Hahn, O. and Strassmann, F., "Über den Nachweis und das Verhalten bei der Bestrahlung des Urans mittels Neutronen entstehenden Erdalkalimetalle", *Naturwissenschaften* **27**, 11–15, January 6, 1939.

292. Perrier, C. and Segrè, E., "Some Chemical Properties of Element 43", *Journal of Chemical Physics* **5**, 712–716, September 1937.

293. Frisch, O. R. and Wheeler, J. A., "The discovery of fission", *Physics Today* **20**, 43–48, November 1967.

294. Frisch, O. R., *What Little I Remember*, Cambridge University Press, Cambridge, 1979, p. 115.

295. *Ibid.*, p. 116.

296. *Ibid.*

297. Stuewer, R. H., "Bringing the news of fission to America", *Physics Today* **38**, 48–56, October 1985.

298. Sime, R. L., *Lise Meitner*, p. 243.

299. Meitner, L. and Frisch, O. R., "Disintegration of Uranium by Neutrons: a New Type of Nuclear Reaction", *Nature* **143**, 239–240, February 11, 1939.

300. Frisch, O. R., "Physical Evidence for the division of Heavy Nuclei Under Neutron Bombardment", *Nature* **143**, 276, February 18, 1939.

301. Goldschmidt, B., *Pionniers de l'atome*, p. 102.

302. Biquard, P., *Frédéric Joliot-Curie*, p. 55.

303. Joliot, F., "Preuve expérimentale de la rupture explosive des noyaux d'uranium et de thorium sous l'action des neutrons", *Comptes Rendus de l'Académie des Sciences* **208**, 341–43, session of January 30, 1939.

304. Joliot, F., "Observation par la méthode de Wilson des trajectoires de brouillard des produits de l'explosion des noyaux d'uranium", *Comptes Rendus de l'Académie des Sciences* **208**, 647–649, session of February 27, 1939.

305. Goldschmidt, B., *Pionniers de l'atome*, p. 39.

306. Meitner, L. and Frisch, O. R., "Products of Fission in the Uranium Nucleus", *Nature* **143**, 471–472, March 18, 1939.

307. Abelson, P., "Cleavage of the Uranium Nucleus", *Physical Review* **55**, 418 (L), February 1939.

308. Anderson, H. L., Booth, E. T., Dunning, J. R., Fermi, E., Glasoe, G. N. and Slack, F. G., "The Fission of Uranium", *Physical Review* **55**, 511–512 (L), March 1939.

309. von Grosse, A., Booth, E. T. and Dunning, J. R., "The Fission of Protactinium (Element 91)", *Physical Review* **56**, 382 (L), August 1939.

310. Bretscher, E. and Cook, L. G., "Transmutations of Uranium and Thorium Nuclei by Neutrons", *Nature* **143**, 559–560, April 1, 1939.

311. Bohr, N., "Resonance in Uranium and Thorium Disintegrations and the Phenomenon of Nuclear Fission", *Physical Review* **55**, 418–419, February 1939.

312. Nier, A. O., "The Isotopic Constitution of Uranium and the Half-Livres of the Uranium Isotopes. I.", *Physical Review* **55**, 150–153, January 15 1939.

313. Bohr, N. and Wheeler, J. A., "The Mechanism of Nuclear Fission", *Physical Review* **56**, 426–450, September 1939.

314. Lanouette, W., *Genius in the shadows. A biography of Leo Szilard : the man behind the bomb*, C. Scribner's Sons, New York, 1992.

315. Brasch, A., Lange, F., Waly, A., Banks, T. E., Chalmers, T. A., Szilard, L. and Hopwood, F. L., "Liberation of Neutrons from Beryllium by X-Rays: Radioactivity Induced by Means of Electron Tubes", *Nature* **134**, 880, December 8, 1934.

316. Soddy, F., *The Interpretation of Radium*, John Murray, London, 1912.

317. *Ibid.*, p. 238.
318. *Ibid.*, p. 251.
319. Wells, H. G., *The World Set Free*, E. P. Dutton & Company, London, 1914.
320. Weart, S., *Scientists in power*, Harvard University Press, Cambridge, 1979, p. 67.
321. Rutherford, E., "Atomic Transmutations", *Nature* **132**, 432–433, September 16, 1933.
322. Weart, S. and Weiss Szilard, G., *Leo Szilard : His Version of the Facts*, p. 38.
323. *Ibid.*, p. 55.
324. von Halban, H., Joliot, F. and Kowarski, L., "Liberation of neutrons in the nuclear explosion of uranium", *Nature* **143**, 470–471, March 18, 1939.
325. Szilard, L. and Chalmers, T. A., "Detection of Neutrons Liberated from Beryllium by Gamma Rays: a New Technique for Inducing Radioactivity", *Nature* **134**, 494–495, September 29, 1934.
326. Dodé, M., von Halban, H., Joliot, F. and Kowarski, L., "Sur l'énergie des neutrons libérés lors de la partition nucléaire de l'uranium", *Comptes Rendus de l'Académie des Sciences* **208**, 995–997, session of March 27, 1939.
327. Szilard, L. and Zinn, W. H., "Instantaneous Emission of Fast Neutrons in the Interaction of Slow Neutrons with Uranium", *Physical Review* **55**, 799–800, April 15, 1939.
328. Anderson, H. L., Fermi, E. and Hanstein, H. B., "Production of Neutrons in Uranium Bombarded by Neutrons", *Physical Review* **55**, 797–798 (L), April 1939.
329. Anderson, H. L., Fermi, E. and Szilard, L., "Neutron Production and Absorption in Uranium", *Physical Review* **56**, 284–286, August 1, 1939.
330. von Halban, H., Joliot, F. and Kowarski, L., "Number of neutrons liberated in the nuclear fission of uranium", *Nature* **143**, 680, April 22, 1939.
331. Flügge, S., "Kann der Energieinhalt der Atomkerne technisch nutzbar gemacht werden?", *Naturwissenschaften* **27**, 402–410, June 9, 1939.
332. Zinn, W. H. and Szilard, L., "Emission of Neutrons by Uranium", *Physical Review* **56**, 619–624, October 1939.
333. Halban, H., Joliot, F., Kowarski, L. and Perrin, F., "Mise en évidence d'une réaction nucléaire en chaîne au sein d'une masse uranifère", *Journal de Physique et Le Radium* **10**, 428–434, October 1939.
334. Marbo (Berthe Borel), C., *A travers deux siècles, souvenirs et rencontres : 1883–1967*, Grasset, Paris, 1967, p. 183.
335. Halban, H. v., Joliot, F., Kowarski, L. and Perrin, F., "Dispositif de production d'énergie", Patent #976.541, taken out by the "Caisse Nationale de la Recherche Scientifique" on May 1, 1939. In Frédéric and Irène Joliot-Curie, *Œuvres scientifiques complètes*, p. 678–683.
336. Halban, H. v., Joliot, F., Kowarski, L. and Perrin, F., "Perfectionnements aux charges explosives", Patent #971.324, taken out by the "Caisse Nationale de la Recherche Scientifique" on May 4, 1939. In Frédéric and Irène Joliot-Curie, *Œuvres scientifiques complètes*, p. 687–691.

The Upheavals of the Second World War

Had I known that the Germans would not succeed in producing an atomic bomb, I never would have lifted a finger.

Albert Einstein, to *Newsweek* magazine, March 10, 1947. In Alice Calaprice, *The expanded quotable Einstein.*

A Chronology

Two refugees, one German, the other Austrian, show the English the extraordinary destructive power which a pure uranium 235 bomb would have. After a tedious start, the United States devote immense resources, both financial and human, in order to build a uranium or plutonium "atomic" bomb. Hiroshima and Nagasaki are destroyed by nuclear fire.

The history of the first nuclear reactor and of the first atomic bomb are not the subject of this book which describes the development and the understanding of the physics of atomic nuclei. However, these events changed the status of research, nuclear physics research in particular, to such an extent that one cannot ignore them. The following is a brief chronology of the making of the first atomic weapons [1]:

- *April 1939, England, the Tizard Committee*: Sir Henry Tizard, president of Imperial College and president of the Air Defense Committee incites the British government to investigate the possibility of making a uranium bomb [2, 3].

B. Fernandez and G. Ripka, *Unravelling the Mystery of the Atomic Nucleus: A Sixty Year Journey 1896 — 1956*, DOI 10.1007/978-1-4614-4181-6_6, © Springer Science+Business Media New York 2013

He consults George Paget Thomson (the son of J. J. Thomson), who is professor at Imperial College, and William Lawrence Bragg, who succeeded to Rutherford at the Cavendish. Both are Nobel Prize laureates in physics. Their conclusion is rather negative. However, Tizard believes that so much is at stake that he asks Thomson and Mark Oliphant in Birmingham to perform further experiments before reaching a conclusion.

- *August 24, 1939*: the German–Soviet pact.
- *September 3, 1939, the war*: German troops have invaded Poland. France and Great Britain declare war to Germany.
- *October 11, 1939*: Roosevelt receives a letter from Einstein [4]. More than any other physicist, Szilard worries about the possibility that Hitler might eventually possess an atomic bomb. After failing to get the American Army interested, he convinces Einstein to write a letter to the president Franklin Roosevelt in order to warn him. The letter, dated August 2, 1939 has become famous [5]. It is only handed to Roosevelt on October 11 by his private economic advisor Alexander Sachs, with whom Szilard succeeded in getting in touch. Roosevelt creates an Advisory Committee on Uranium, directed by Lyman Briggs, the director of the Bureau of Standards. The committee includes Szilard, Wigner, Sachs, Teller, Adamson, lieutenant colonel in the Army, and Hoover, a Navy officer. In spite of the opposition of Adamson, a $6,000 budget is attributed to continue research on the chain reaction. But the decision does not materialize.
- *February 1940*: the fission of pure uranium 235 is measured in the Columbia University by Alfred Nier, Eugene Booth, John Dunning, and Aristid von Grosse. The probability that a neutron colliding with a uranium 235 nucleus should produce fission is indeed very large [6].

Mark Oliphant admits to his laboratory in Birmingham two physicists who are Jewish refugees: Otto Frisch and Rudolf Peierls. Otto Frisch and his aunt Lise Meitner had explained the experiments of Hahn and Strassmann.[1] Rudolf Peierls, born in 1907, made a brilliant career as a theorist in Munich with Sommerfeld and later as assistant of Pauli in Hamburg [7]. Whereas all the English physicists are working on radar, Frisch and Peierls are excluded from this because they come from enemy countries and the research is top secret. However, they are free to work on nuclear fission. Peierls has already published a paper on the critical mass required to sustain a chain reaction [8]. Frisch asks him what would happen if one had pure uranium 235. They set to work, and in a few weeks, they obtain an amazing result. With about 1 kg of uranium 235, one could make a terrifying bomb against which there could be no defense.[2]

Their conclusions are written in a report known as the *Frisch–Peierls memorandum* [3] which they hand to Oliphant, who in turn delivers it to the

---

[1] See p. 363.

[2] This first estimate was based on incomplete data, and later, it was found that a larger mass was required. About 15 kg of uranium 235 are required to make an atomic bomb.

Tizard Committee. In this memorandum, they give a broad outline of what the manufacture of an atomic bomb could be:

A moderate amount of ^{235}U would indeed constitute an extremely efficient explosive [...].
One might think of about 1 kg as suitable size for a bomb [...].
Owing to the spreading of radioactive substances with the wind, the bomb could probably not be used without killing large numbers of civilians, and this may make it unsuitable as a weapon for use by this country.

They consider the case where Germany would succeed in making such a bomb, and they propose a strategy which later will be called deterrence:

The most effective reply would be a counter-threat with a similar bomb. Therefore it seems to us important to start production [of uranium 235] as soon and as rapidly as possible, even if it is not intended to use the bomb as a means of attack.

The memorandum made such an impression that a committee was formed for this purpose, the "MAUD Committee." However, at the time, the separation of the 235 and 238 isotopes of uranium appeared to be a most difficult, almost utopian task, especially if kilograms of pure uranium 235 had to be produced. But Frisch and Peierls insist that the problem has to be tackled with great urgency.

– *February 14, 1940*: in a five-page report to the French government, Joliot, Halban, and Kowarski propose two methods to obtain an energy producing chain reaction [9]. Joliot succeeds in getting Raoul Dautry, the French Minister of Armament, interested. After an incredible operation of the French Secret Services, the young officer Jacques Allier succeeds in getting hold of the world stock (185 l) of heavy water[1] which was produced in Norway by a hydroelectric power plant [10].

Joliot immediately sets to work. He leads the race for energy production, but he no longer believes in the military use of uranium because he thinks that the separation of uranium 235 is beyond reach.

– *March 1940*: in the United States, the $6,000, which the Briggs Committee had promised to Fermi and Szilard, is finally granted, thanks to the unrelenting efforts of Szilard. They will enable the purchase of uranium and ultrapure graphite in order to determine whether, yes or no, a chain reaction can be produced with natural uranium.

– *April 9, 1940*: the German troops invade Denmark and Norway.

– *April 27, 1940*: second meeting of the Briggs Committee which decides to wait for the results of the ongoing experiments before taking further action.

---

[1]Heavy water is water in which both hydrogen atoms are replaced by deuterium atoms (D_2O instead of H_2O).

- *May 1940*: two American physicists, Edwin McMillan and Philip Abelson, observe the element 93, the first real transuranic element (see below p. 425).
- *May 10, 1940*: German troops invade the Netherlands, Belgium and France.
- *June 18, 1940*: as the German troops advance, Joliot, Halban, and Kowarski retreat to Clermont-Ferrand, then to Bordeaux. From there, Halban and Kowarski go to England with the intention of pursuing research on the chain reaction. They take with them the world stock of heavy water. Joliot returns to Paris and he joins the *Resistance* and will become one of its leaders as president of the *Front national de lutte pour la libération et l'indépendance de la France*.[1] Frédéric Joliot and Irène Curie stop all nuclear research during the war. On the radio in London, General de Gaulle calls for resistance to the invader.
- *June 1940*: Roosevelt creates the National Defense Research Committee which will supervise the scientific and technological war effort and which is under the aegis of the Office of Scientific Research and Development (OSRD), directed by a talented engineer and electrician, Vannevar Bush. Bush presides the new committee of which the Uranium Committee is a subbranch.
- *August 1940*: a group of English physicists visits Canada and the United States with the intention of finding a research site which is less exposed to the bombing of the German *Luftwaffe*. They inform Fermi of the Frisch–Peierls memorandum. Fermi is very skeptical about the possibility of making a bomb, as most physicists are at the time. He informs them about research aiming at enriching natural uranium so as to increase the abundance of the 235 isotope from 0.7% to 3% or 4%. The aim is not to make a bomb but to build a nuclear reactor which would use natural water as a moderator, to slow down the neutrons. Meanwhile the British physicists try to verify the theoretical hypotheses of Frisch and Peierls, while Halban and Kowarski are cordially welcomed in England where they can pursue their research with natural uranium and with the heavy water which they brought with them from France.
- *January 1941*: in Berkeley, Glenn Seaborg, Edwin McMillan, Joseph Kennedy, and Arthur Wahl discover the element 94, which they propose to call *plutonium*.[2]
- *October 1941*: the British send to the United States a copy of the report which confirms the validity of the outline proposed in the Frisch–Peierls memorandum. The report has a great impact: Vannevar Bush believes that work on the uranium bomb must be pushed forward. A report of the American Academy of Sciences, although cautious, points to the same direction. Roosevelt agrees to an exchange of data between English and American scientists, and he even favors a common project. But England, assuming to be ahead of the United States, is in no hurry to do so.
- *December 6, 1941*: the United States decide to accelerate the research program aimed at making the bomb. Vannevar Bush modifies the organization: the Uranium Committee becomes an autonomous committee, still presided by Briggs, but its effective direction is entrusted to James Conant. He studies

---

[1] The National Front for the liberation and the independence of France.
[2] See p. 427.

the separation of uranium 235 by gaseous diffusion, magnetic deviation, and centrifugal processes.

- *December 7, 1941*: the Japanese Navy attacks Pearl Harbor without declaring war.

- *December 11, 1941*: Germany and Italy declare war to the United States.

- *December 18, 1941*: in a meeting of the new committee called "Section S1" of the OSRD, Conant presents the new policy of the American government, which has decided to make an all-out effort to make an atomic bomb.

- *April 1942*: in January 1942, Compton, who is the responsible scientist of the project, decides to regroup the research in Chicago. In order not to attract attention, the created laboratory is called the Metallurgical Laboratory, the Met. Lab. in short. Fermi moves there in March, and he begins the construction of a nuclear reactor, the "pile" as he calls it. Indeed, it consists of a stack of ultrapure graphite and uranium blocks. It is constructed in a squash court situated under the terraces of the football stadium, the Stagg Field of the University of Chicago.

- *September 1942*: Henry Stimson, the War Secretary, assigns general Leslie Groves to coordinate all the war effort aimed at making a uranium bomb. The project is given the harmless name the *Manhattan District*. It will be known as the *Manhattan Project*. Groves asks Robert Oppenheimer, a young and talented theorist from Berkeley, to be the scientific director of the project.

- *Fall, 1942: the* Manhattan District *purchases* a 60 000-acre ground in a scarcely populated valley of the Tennessee river, which flows from the north of Alabama, through the state of Tennessee, into Ohio. On this ground, on the riverbank of the Clinch, which flows into the Tennessee, a new city is created, called Oak Ridge. It is an industrial center with the sole aim of extracting uranium 235 from natural uranium. By the end of the war, Oak Ridge will have 75 000 inhabitants. The network of workshops was attributed the innocent name *Clinton Engineering Works*. It enriched uranium by gas diffusion and magnetic separation, in fact by a combination of both. By the spring of 1945, about 60 kg of enriched uranium was delivered by Oak Ridge.

- *November 1942*: to work on the bomb, Groves and Oppenheimer decide to create a new laboratory in an isolated place, relatively cut off from the world. They choose the plateau of Los Alamos in New Mexico, at an altitude of 2000 m and about 50 km north of Santa Fé. A city is born there, and by 1945, about 7000 people, physicists, and technicians with their families will be living there in rather austere conditions.

- *December 2, 1942*: constructed under the direction of Fermi and Leo Szilard in the University of Chicago, the first nuclear reactor *diverges*. A chain reaction is said to *diverge* if it sustains itself as it does in a nuclear reactor and to *converge* if it stops by itself. Compton sends the following message to Conant: "The Italian navigator has just landed in the new world."

- *December 28, 1942*: President Roosevelt decides to launch a grand-scale project to separate uranium 235 and to produce plutonium with the aim of constructing an atomic bomb.

- *Early 1943*: in order to produce plutonium in large quantities, the Manhattan District purchases a 586-square-mile plot of land on the Columbia river in the state of Washington. The population, which originally consisted of the 500 inhabitants of the Hanford and Richland villages, will reach 60 000 inhabitants in 1944. Three nuclear reactors are constructed to produce plutonium. The first begins to work in September 1944 and the third in the summer of 1945. Factories for chemical extraction of plutonium are also constructed in Hanford.
- *November 1943*: the best experts in England are sent to Los Alamos under the direction of Chadwick: Otto Frisch, Rudolf Peierls, George Placzek, Philip Moon, James Tuck, Egon Bretscher, and Klaus Fuchs.
- *April 12, 1945*: Franklin Roosevelt dies at the age of 63 from a cerebral hemorrhage. The vice-president Harry Truman becomes president at the age of 61.
- *May 8, 1945*: Germany capitulates.
- *July 16, 1945*: the first plutonium atomic explosion takes place in Alamogordo, in the New Mexico desert. The uranium bomb will not be tested before it is dropped on Hiroshima.
- *July 24, 1945*: uranium is delivered to Los Alamos and the construction of a uranium atomic bomb is immediately started.
- *August 6, 1945 at 9:15 am*: the bomber Enola Gay drops the first atomic bomb, called Little Boy on Hiroshima from an altitude of 31 600 feet (about 9600 m). It destroys the city completely within a radius of 2 km of the point of impact, kills 71 000 people and injures 68 000.
- *August 9, 1945*: the plutonium bomb, called Fat Man is dropped on Nagasaki from an altitude of 1 950 feet (about 600 m) killing 35 000 people and injuring 60 000.
- *August 14, 1945*: Japan capitulates. End of World War II.

The New Face of Physics After the War

Talented physicists, industrial and financial power, and refugees from Europe, ruined by the war, make science dominate in the United States. The fact that nuclear physics can only develop in large laboratories accentuates the American supremacy in the field.

At the end of World War II, most of the physicists who had worked for the construction of the atomic bomb wanted to pursue the study of the atomic nucleus. But this proved to be difficult because the American army wanted to control nuclear physics and keep it secret. Indeed, General Groves incited the War Department to prepare a law which would restructure nuclear research in the United States. The project was supported by the democratic senator Edwin C. Johnson and the democratic congressman Andrew Jackson May. This was the May-Johnson Bill which they tried to pass through Congress quickly, in order to avoid excessive information and discussion. According to the bill, all nuclear research would be placed under the authority of the War Department, and heavy sanctions (including imprisonment) would threaten any scientist who would not comply to secrecy. This caused a rebellion among the scientists who had worked for the Manhattan Project, led (again) by Leo Szilard. Politicians, senators, and congressmen were alerted. The bill was finally given up at the end of 1945 [11–13]. This was also the time when an association of scientists, called Atomic Scientists of Chicago, was created as well as the *Bulletin of the Atomic Scientists of Chicago,* which in 1946 became the *Bulletin of the Atomic Scientists.* Finally, the McMahon Bill was voted in 1946. It placed civil (nonmilitary) research under the authority of the Atomic Energy Commission (AEC) the civil directors of which were named by the president of the United States. David Lilienthal, an industrialist who had presided the Tennessee Valley Authority, became its first director on January 1, 1947. General Groves regretted that a large number of physicists should leave Los Alamos which survived only as a military research center. The English scientists returned to their home country. Fermi became professor at the University of Chicago and Hans Bethe at the University of Cornell. James Franck returned to the University of Chicago where he had been professor since 1938 as well as Herbert Anderson. Edward Condon became the director of the National Bureau of Standards, Eugene Wigner returned to Princeton where he taught mathematical physics. However, Leo Szilard was obliged to abandon nuclear physics following a F.B.I. investigation which was initiated by General Groves who never forgave him to have sunk the May-Johnson Bill [14]. Szilard turned to biology, his youthful love, and he ended his career in this field. Robert Oppenheimer resigned from his position as scientific director in Los Alamos. For some time, he taught quantum mechanics in Berkeley, and in 1947, he became the director of the Institute for Advanced Study in Princeton.

Big Science: Physics on a Large Scale

Research did not simply take off from where it had stopped during the war. The nuclear physicists who had worked in Los Alamos were an extremely brilliant group, and they had lived through a unique experience. They had considerable means at their disposal, and their labs were very well equipped. At the end of the war, they had acquired a considerable prestige. They seemed able to make anything, as long as they were provided with adequate means, almost as magicians. After the war, they had no trouble in obtaining funds to equip large laboratories devoted to fundamental research.

Experimental physics required far greater means than those available before 1940. In the 1930s, a good laboratory had some general apparatus set in a proper building, a photographic laboratory, perhaps also a Wilson chamber, and some Geiger counters. It is with such equipment that the atomic nucleus was discovered as well as the neutron, artificial radioactivity, neutron physics, and fission, to mention but a few of the great discoveries made in the years 1932–1938. The arrival of the first accelerators was soon to change the landscape and the laboratory in Berkeley was a foremost example.

In 1932, Lawrence had claimed he could use "normal equipment" for his first cyclotron. But his subsequent cyclotrons became larger and larger and required increasingly large research teams devoted to their construction, repair, and running. After 1945, large accelerators were constructed in the United States, naturally in Berkeley, but also in Rochester, Harvard, Columbia, Chicago, etc. They were financed at first by the army and then by the AEC. From then on, *big science*,[1] that is, science on a large scale, prevailed. In contrast to *little science*, which had been practiced until 1940, *big science* kept laboratory directors in contact with politicians, thereby enabling them to secure increasingly important funding. Between 1945 and 1971, the budget devoted to research had increased by a factor of 20 in the United States. This corresponds to an average 12% increase each year [16]. The number of physicists also grew rapidly, the best students being naturally attracted by the prestige acquired by nuclear physics and by the most eminent physicists who were now in the United States.

Team Work

Another important transformation: research is now performed in teams composed of increasing numbers of physicists [17, 18]. Until 1940, research was performed either by a single physicist or by small teams, often consisting of two physicists. One of the largest research teams at that time was the one of Fermi in Rome: their

---

[1]So called by Alvin Weinberg, who had participated in the Manhattan Project in Los Alamos and had become the director of the laboratory in Oak Ridge [15].

papers bore signatures of five or six physicists. The Joliot–Halban–Kowarski team formed in 1940 was also an innovation: it consisted of three people with different specialities and characters who joined efforts to reach a common goal. As in Fermi's team, one member is without doubt the leader, but each member has a role which is far from simply executing orders. One can also mention the team formed by Lise Meitner, Otto Hahn, and Fritz Strassmann, the research of which led to the discovery of nuclear fission.

After the war, research in experimental nuclear physics could no longer be performed by a single physicist. Accelerators usually functioned 24 h a day. Detection equipment became increasingly complex and required extensive competence which could not be handled by a single person. Today, papers are signed by regularly increasing numbers of authors. This evolution is even greater in particle physics, which is devoted to the study of the inner structure of neutrons, protons, and other elementary particles and not to the way in which they form nuclei. The new field requires higher energies, experiments are performed on an even larger scale, and papers are signed by an even greater number of physicists. The number reaches hundreds today and has reached a thousand at the Large Hadron Collider in Geneva, which began to function in 2008.

The study of nuclear structure does not require such large accelerators because the energies involved are of the same order of magnitude as the energies of neutrons and protons in a nucleus, namely, tens of MeV. Even today, accelerators devoted to the study of nuclear structure do not exceed tens or perhaps hundreds of MeV.

The H-Bomb: Political and Military Implications

After the atomic bomb had been constructed, nuclear physics became a strategic issue for the governments of all countries. As a result, it was supplied with generous funding, but there were also constraints. In order to maintain its monopoly on nuclear arms for as long as possible, the United States kept secret a large amount of nuclear data which could be used to construct a nuclear reactor and, *a fortiori*, to construct an atomic bomb. Indeed, the Cold War had begun even before the Second World War was over. It became official, so to speak, after the famous speech delivered at Westminster College in Fulton, Missouri on March 5, 1946 by Winston Churchill, in which he declared: "From Stettin in the Baltic to Trieste in the Adriatic an iron curtain has descended across the Continent." From then on, the potential enemy of the United States and Western Europe became the Soviet Union.

The basic nuclear physics required to make a bomb had been known for several years so that fundamental research, aimed at understanding the structure of the nucleus, no longer had much incidence on nuclear arms. In fact, research in basic nuclear physics was not subject to orders coming from the military who, anyway, would have been unable to direct research in the field. Nevertheless, basic research and research applied to the construction of nuclear reactors remained linked to military research because a new weapon was being devised, the *hydrogen bomb* (H-bomb) which was potentially far more powerful than the *atomic bomb* (A-bomb).

Indeed, as early as 1942, Edward Teller had proposed to build a bomb which relied on the fusion of hydrogen or deuterium nuclei instead of the fission of uranium nuclei. The power of such a bomb is, in principle, unlimited. But the technical difficulties which needed to be overcome convinced Oppenheimer and General Groves to give priority to the construction of the A-bomb. However, Teller did not give up. Born in Budapest in 1908, Edward Teller [19] obtained his PhD in Leipzig in 1930. He then obtained a position in Göttingen, but he left after being expelled by the Nazi regime in 1933. He emigrated to the United States and became an American citizen in 1941. After 1942 he pursued, essentially alone, a study aimed at building an H-bomb.

After the war, an intense debate opposed those who were in favor or against the construction of a H-bomb. The former included Teller, Lawrence, and Lewis Strauss, a self-made man, a banker, who became a commissioner for the Atomic Energy Commission (AEC). A large number of physicists opposed the project, Oppenheimer to begin with as well as Fermi, Szilard, Bethe, and Franck. At first, the opponents seemed to have the upper hand. The United States seemed to have the monopoly of the "simple" A-bomb so that there seemed to be no hurry to construct an H-bomb. But the Cold War became increasingly threatening and communism seemed to spread: the civil war in Greece, the Berlin blockade (from May 1947 to June 1948) which the United States countered by an air lift, the Prague Putsch in February 1948, the victory of Mao Zedong in China in mid-1949, and finally the invasion of South Korea in June 1950, to which the United States responded immediately and which ended in 1953 with a return to the situation of before 1950. Furthermore, on August 29, 1949, the Soviet Union exploded its first atomic bomb (A-bomb). This led to an arms race and the supporters of the hydrogen bomb (H-bomb) won [13].

In November 1952, the first H-bomb exploded on an island in the Pacific Ocean. It was 700 times more powerful than the A-bomb which had destroyed Hiroshima. However, it provided the United States only with a short-lived monopoly. On August 12, 1953, the Soviet Union exploded its first H-bomb. It is during this mounting tension that the republican senator Joseph McCarthy launched in 1950 a violent assault against so-called communist infiltrations in the American administration. One of the victims of this witch hunt was Oppenheimer, who was subject to 3 weeks of hearings, from April 12 to May 6, 1954, by the Personnel Security Board, which decided that he was a "security risk." Oppenheimer was denied access to nuclear secret data, and he was expelled from the Consulting Committee of the AEC. He was paying for the left wing opinions he had held in his youth but mostly for his opposition to the H-bomb [20, 21].

The American Supremacy

After the Second World War, a large number of scientists, among the best, are living in the United States where they find optimal means and facilities to work.

And students pour in. A measure of the supremacy of the United States in the field is provided by the published papers: most of them appear in the Physical Review, an American journal. For 10 or 20 years, the funds provided for experiments are concentrated in the United States. This provides an essential advantage to experimentalists but also to theorists who can maintain personal and quick contact with their experimental colleagues who are obtaining new data.

Europe and Japan After the War

Great Britain

In spite of being a privileged ally of the United States, Great Britain is considerably weakened at the end of the war. The country had not been invaded, and it continued to function. Great Britain was an almost full participant to the Manhattan Project, and it could measure the strategic importance of possessing an atomic bomb as well as nuclear energy. For this purpose, it launched an ambitious nuclear program with the creation in 1946 of an Atomic Energy Research Establishment (AERE) in Harwell, near Oxford, on an airfield used during the war by the Royal Air Force. The first British nuclear reactor is built there, and it functions in August 1947. Britain also constructs accelerators devoted to basic research. In 1954, the AERE is incorporated into the United Kingdom Atomic Energy Authority (UKAEA). Most of the British physicists who had participated in the Manhattan project returned home after the war so that Great Britain maintained most of the quality of its research except for the fact that it could not compete with the means available to the laboratories and to the powerful industry of the United States. However, the contributions of the British theoreticians continue to be important. In Birmingham, the theoretical physics school, led by Rudolf Peierls, is among the most brilliant. However, the United States continue to attract a large number of British physicists, among whom the best, Maurice Pryce and Jeffrey Goldstone, for example.

France

After the German occupation and the subsequent liberation, France is in a worse state than after World War I in spite of the considerably smaller number of lost lives. Indeed, the industrial infrastructure and means of transport are either destroyed or obsolete. Food rationing, which begun in 1941, continues until January 1949. The supply of towns is difficult because railway tracks and railway stations have been destroyed.

In this context, scientific research was poor. One should add however that even before the war it only appeared to thrive because of a handful of personalities such as Frédéric and Irène Joliot-Curie, Paul Langevin, Alexandre Proca, or Léon Brillouin. The *Centre National de Recherche Scientifique* (CNRS) was but a small structure, created by Perrin in 1939 in order to compensate for the inability of universities to reform and to create real research laboratories [22].

During the war, Frédéric Joliot remained in Paris and led a double life as director of a laboratory and leader of a Resistance movement. His laboratory in the *Collège de France* was placed under German military surveillance, but luckily, the German officer in charge of the laboratory was no other than Wolfgang Gentner, a friend of Joliot and an anti-Nazi. As early as 1941, Joliot joined and became the president of the *Front national de lutte pour la libération et l'indépendance de la France.*[1] This movement was launched by the Communist Party and it aimed at uniting all the resistance against the German occupation. It will join the *Conseil National de la Résistance*[2] which in turn nominates Joliot to become the director of the CNRS, a task he undertakes in August 1944 as soon as Paris is liberated [23]. Joliot proceeds to a thorough reform of this institution which will become the spearhead of French research. He is granted important funds by the government. Between September 1944 and July 1945, the number of researchers working in the CNRS increases from 600 to 970 and the number of technicians from 450 to 570. The budget increases from 300 million francs in 1944–1945 to 632 million in 1946 [24].

Once the CNRS got going, Joliot returned to nuclear energy. He convinced de Gaulle to create an organism designed to be in charge of nuclear energy in order to enable France to catch up with the United States. The *Commissariat à l'Énergie Atomique*[3] (CEA) is created on October 18, 1945. On January 2, 1946, Joliot becomes the High Commissioner of the CEA, and he begins this adventure with a small team composed of physicists and chemists of the *France libre* who had worked in Canada and in England: Lew Kowarski, Pierre Auger, Bertrand Goldschmidt, Jules Guéron, and Francis Perrin. The challenge consists in constructing, before the end of 1948, an experimental nuclear reactor similar to the one Fermi had made in 1942. In spite of the state of dilapidation of French industry, the project succeeds, and on December 15, 1948, the reactor ZOE[4] "diverges."

In the newly formed CEA, Joliot created teams which had the task of preparing the future experimental nuclear reactor ZOE but also to pursue fundamental research. A small but brilliant team of physicists was formed initially around Jacques Yvon: Jules Horowitz, Claude Bloch, Michel Trocheris, Anatole Abragam.[5] Following the advice of Kowarski, they were systematically sent abroad to the best laboratories: Cal Tech, Copenhagen, Oxford, etc. These "Four Musketeers," soon joined by Albert Messiah, will form the theoretical physics laboratory of the CEA. They will attract the best young students, and in a few years, the laboratory becomes

---

[1] National Front for the fight for the liberation and independence of France.

[2] National Resistance Council.

[3] Atomic Energy Commission.

[4] The acronym ZOÉ was coined by Lew Kowarski: "Z" and "É" stand for "zero energy" (*zéro énergie* in French) and "O" stands for "oxide," because the reactor used uranium oxide.

[5] The autobiography of Anatole Abragam, *De la physique avant toute chose?* contains an interesting account of theoretical physics in France before the war.

one of the best theoretical laboratories in the world. Albert Messiah gave the first quantum mechanics course in France and subsequently published a book which became a reference in the whole world. This is also where physicists such as Claude Cohen-Tannoudji or Pierre-Gilles de Gennes were formed. Thanks to this small team and in the wake of enthusiasm which followed the liberation and reconstruction of the country, a French theoretical physics school was born. Indeed, it is a birth and not a renaissance because theoretical physics hardly existed in France before the war. "We had no seniors, but we followed passionately the seminar of Proca," Albert Messiah once told us, referring to the only theoretical physics seminar which was animated by Proca from 1946 to 1955, the year he died. Theoretical physics was also developed in the CNRS, in the universities of Paris, Marseille, and Strasbourg. Teams were initiated by young physicists, such as Maurice Lévy or Philippe Nozières, who were formed abroad, a practice which did not exist before the war.

Experimental nuclear physics in France started seriously only in 1950, when several accelerators were built in the Centre d'Etudes Nucléaires[1] in Saclay. The laboratory of the University of Orsay was initiated by Irène Joliot-Curie in 1954 and inaugurated by Frédéric Joliot-Curie in 1958. Other laboratories were built in Strasbourg, Grenoble, Bordeaux, etc.

Germany

Of all the European countries, Germany suffered the most extensive destructions. Its economy was ruined, the laboratories destroyed by bombing or placed in the hands of the Allies. Daily life and even survival became difficult. But German science also felt the effect of another disaster: the forced exodus of Jewish scientists which began in 1933.[2]

Should German science be reconstructed? The clear-sighted Churchill had understood that the Versailles Treaty, which followed the First World War and which was meant to punish Germany, had in fact helped Hitler to come to power. Thus, when Churchill met Roosevelt in 1941 (even before the United States entered the war), he defended the idea that a similar error should not be repeated after the victory of the Allies. It is thanks to Great Britain that German science was reborn. One of the leading laboratories, the *Kaiser Wilhelm Gesellschaft*, could function again in 1946, led by Max Planck and called the *Max Planck Gesellschaft zur Förderung der Wissenschaften in der Britischen Zone*. As specified by its name, its activities were limited to the British occupation zone. After the death of Planck on October 4, 1947, the institution was replaced by the *Max Planck Gesellschaft*, created on February 26, 1948, and presided by Otto Hahn. It was extended throughout all West Germany in 1949.

---

[1] Center of Nuclear Studies, a section of the CEA (the French Atomic Energy Commission)
[2] See p. 327–331.

Under the impetus of researchers who had behaved well during the war, universities also resumed research activities. Among the physicists whom we have met previously, we can mention Walther Bothe, Wolfgang Gentner, Max von Laue, Otto Hahn, Friedrich Hund, Siegfried Flügge, Fritz Strassmann, and Walther Gerlach. The case of Heisenberg is special: he had been the director of the *Uranverein*, a uranium club which was in charge of nuclear research and which made a great effort to build a nuclear reactor without however succeeding before the end of the war. In 1942, he delivered a report to Albert Speer, *Reichsminister für Rüstung und Kriegsproduktion* (Minister of Armament and War Production) stating the possibility of making a nuclear bomb and concluding that it would be confronted with the greatest difficulties in the foreseeable future. As a result, the construction of a uranium bomb did not become a priority for the prevailing regime, and contrary to the fears of the Americans and the British, no military research was initiated. Did Heisenberg deliberately undermine a military nuclear project or was he voicing his sincere opinion at the time? Had he thought of storing the plutonium produced in a nuclear reactor for a future military use? It is a subject of controversy to this day, and we shall not dwell upon it. It is true that Heisenberg was never a member of the Nazi Party and that he tried, as much as possible, to defend Jewish scientists so that nobody accused him of anti-semitism. His high scientific reputation allowed him to resume his scientific career and to work on the reconstruction of the ruined German science. He became the director of the *Max Planck Institut für Physik* in Göttingen. He tried to dispel the isolation which befell upon German physicists when the Nazis took over and to have them again admitted in the international community. After the war, the United States considered nuclear physics to be an essential military and strategic feature, and they were not ready to allow Germany to resume research, not even fundamental research, in this field. But with the onset of the Cold War, the United States faced a different enemy. On May 5, 1955, in Paris, the Allies made an agreement to abolish all restrictions on research pursued in West Germany, which could then resume research in nuclear physics and nuclear energy. Shortly later, Chancellor Adenauer was tempted to establish a study of nuclear weapons. However, the German scientists, foremost of whom were Weizsäcker, Hahn, and Heisenberg, voiced a strong opposition to the project, which was given up.

Japan

During the first half of the twentieth century, scientific research in Japan had attained the topmost level. Hantaro Nagaoka became known for his "saturnal" model of the atom, and he continued to publish regularly in the *Philosophical Magazine* and in *Nature*. The following generation of scientists included talented experimentalists: Kenjiro Kimura, Shin'ichi Aoyama, or Takeo Hori, who were formed in Europe, particularly in Copenhagen. The best known physicists were doubtlessly theoreticians such as Yoshio Nishina, Hideki Yukawa, the inventor of the meson, and Sin-Itiro Tomonaga, who was awarded the Nobel Prize in physics in 1965 for his work on quantum electrodynamics. After the war, Japan was ruined and famished. The American occupation army destroyed five cyclotrons and threw them into the port of Tokyo on the grounds that they were used for nuclear physics and could potentially produce nuclear weapons. But again, when the enemy facing the Unites States had changed, research could slowly resume. Yukawa was invited to the United States in 1948, and in 1949, he was awarded the Nobel Prize in physics for having predicted the existence of the meson. By 1950, Japanese scientists were again part of the international scientific community. The funds available to experiments came more slowly, following the growth of the Japanese economy.

Is "Big Science" Really the Result of the War?

Was it solely the war which led governments to discover the importance of research and which led to the expansion of "big science"? In a book, published in 1963, the physicist and science historian Derek de Solla Price claims that this is far from obvious [25]. He presents convincing evidence pointing to the fact that the growth of research had been going on for a very long time (he measures it in units of centuries). This applies to the allotted funds, to the number of scientific journals, to the number of published papers, and to the number of researchers. He claims that it has always been an *exponential* growth, meaning that it involved each year a given fraction of the existing size. Price estimates the growth to be about 4.7% each year. It therefore doubles in 15 years. While research involved a small number of people and small funds, the increase was hardly noticeable. But when it began to involve a non-negligible fraction of public resources, the growth appeared to undergo a transition. This is precisely what happened in the United States in the 1940s. We saw that the appearance of accelerators had begun before the war, and it was bound to modify the way in which nuclear physicists worked.

The construction of large telescopes in the United States is another example of "big science" which has nothing to do with the war [26]. The Mount Wilson Observatory was founded at the beginning of the twentieth century under the leadership of the American astronomer George Hale and the funds were provided by the Carnegie Institution in Washington. The 1.52-m telescope was built in 1908 and the 2.5-m one in 1917. The construction of the giant 5-m telescope at Mount Palomar in Southern California began in 1928, thanks to six-million dollars granted by the Rockefeller Foundation. It began functioning in 1948.

It is true that the budgets of nuclear physics laboratories increased considerably in the years that followed the war, especially when they emphasized potential military applications of their research. It is also true, however, that advances in fundamental nuclear physics had little incidence on nuclear arms, which raised technological problems. But it took some time for governments to realize this.

References

1. Rhodes, R., *The making of the atomic bomb*, Simon and Schuster, New York & London, 1986.
2. Gowing, M., *Britain and atomic energy, 1939-1945*, St Martin's Press, New York, 1964.
3. Szasz, F. M., *British Scientists and the Manhattan Project. The Los Alamos Years*, MacMillan, London, 1992.
4. Cantelon, P. L., Hewlett, R. G. and Williams, R. C. (editors), *The American Atom. A documentary History of Nuclear Physics from the Discovery of Fission to the Present*, University of Pennsylvania Press, Philadelphia, 1984.
5. Lanouette, W., *Genius in the shadows*, p. 205.
6. Nier, A. O., Booth, E. T., Dunning, J. R. and von Grosse, A., "Nuclear Fission of Separated Uranium Isotopes", *Physical Review* **57**, 546 (L), March 15, 1940.
7. Peierls, R., *Bird of Passage. Recollections of a physicist*, Princeton University Press, Princeton, 1985.
8. Peierls, R., "Critical conditions in neutron multiplication", *Proceedings of the Cambridge Philosophical Society* **35**, 610–615, 1939.
9. Joliot, F., "Confidential report to the French government", Musée Curie, Paris, February 13, 1940.
10. Goldschmidt, B., *Pionniers de l'atome*, Stock, Paris, 1987, p. 100.
11. Goldschmidt, B., *The atomic complex: a worldwide political history of nuclear energy*, American Nuclear Society, La Grange Park, 1982. Translated from *Le complexe atomique*, Fayard, Paris, 1980.
12. Lanouette, W., *Genius in the shadows*.
13. Herken, G., *Brotherhood of the Bomb. The Tangled Lives and Loyalties of Robert Oppenheimer, Ernest Lawrence and Edward Teller*, Henry Holt and Co., New York, 2002.
14. Lanouette, W., *Genius in the shadows*, p. 305-313.
15. Weinberg, A. M., "Impact of Large-Scale Science and the United States", *Science* **134**, 161–164, July 21, 1961.
16. Kowarski, L., "Psychology and structure of large-scale physical research", *Bulletin of the Atomic Scientists* **V**, 186–191, June-July 1949.
17. Kowarski, L., "Nuclear Research Centres", *OEEC Publications* **2**, 77–81, 1958.
18. Kowarski, L., "New Forms of Organisation in Physical Research after 1945", in *Storia della fisica del XX secolo, Scuola internazionale di fisica ≪ Enrico Fermi ≫*, edited by Weiner, C., pp. 370–401, Academic Press, New York & London, 1977.
19. Goodchild, P., *Edward Teller : the real Dr. Strangelove*, Harvard University Press, Cambridge, Mass., 2004.
20. McMillan, P. J., *The Ruin of J. Robert Oppenheimer and the Birth of the Modern Arms Race*, Viking, New York, 2005.
21. Bird, K. and Sherwin, M. J., *American Prometheus, The Triumph and Tragedy of J. Robert Oppenheimer*, Alfred A. Knopf, New York, 2005.
22. Day, C. R., "Science, applied science and higher education in France 1870–1945, an historiographical survey since the 1950", *Journal of Social History* **26**, No. 2, 367–384, Winter 1992. This review article contains many references on Science in France and its decline between 1870 and 1939.
23. Weart, S., *Scientists in power*, Harvard University Press, Cambridge, 1979.
24. Pinault, M., *Frédéric Joliot-Curie*, p. 310.
25. de Solla Price, D. J., *Little science, big science*, Columbia University Press, New York, 1963.
26. Lankford, J. and Slavings, R. L., "The Industrialization of American Astronomy, 1880-1940", *Physics Today* pp. 34–40, January 1996.

The Time of Maturity

"It's a Snark!" was the sound that first came
to their ears,
    And seemed almost too good to be true.
Then followed a torrent of laughter and cheers:
    Then the ominous words, "It's a Boo –"

Lewis Carroll, *The Hunting of the Snark*

New Experimental Means

*A variety of new accelerators are aimed at creating beams of
increasingly energetic particles. New means of detecting and analyzing particles are invented, mostly due to the development of
electronic devices which become numerous, varied and flexible,
the fruit of apparently unlimited imagination.*

The nuclear physics lab in the 1950s bears little likeness to the one in the 1930s.
As soon as the Second World War ended, physicists resumed their work, first in the
United States and then in the rest of the world. Particle accelerators were developed
and diversified as well as systems designed to detect particles of various sorts. From
then on, every laboratory required its own accelerator which served as a kind of
projector able to "illuminate" the tiny nuclei with projectiles whose wavelengths
were much shorter than the nuclei and which therefore required high energies. As
they were perfected, the instruments became "eyes" of physicists enabling them
to observe the detailed structure of nuclei. They made nature "speak" and they
became the source of theoretical progress, which in turn led to the development of

B. Fernandez and G. Ripka, *Unravelling the Mystery of the Atomic Nucleus:*
A Sixty Year Journey 1896 — 1956, DOI 10.1007/978-1-4614-4181-6_7,
© Springer Science+Business Media New York 2013

novel instruments. An excellent example is the invention of the transistor. Physics progresses both when new data allow theory to make progress and when progress in theory leads to further data. This chapter is an overview of the main developments in nuclear instrumentation, involving both accelerators and detectors. The reader who is impatient to learn how the story continues may proceed directly to the section Data accumulate.

New Accelerators Have Ever Increasing Energies

Cyclotrons started to develop before the war, first in the United States and then in Europe and in Japan, but fundamental research stagnated during the war. Between 1934 and 1939, 11 cyclotrons were built in the United States, in laboratories other than Berkeley. Two small cyclotrons were built in 1937 in Tokyo and Leningrad, followed by further ones in Cambridge and Copenhagen (1938); in Liverpool, Osaka, Paris, and Stockholm (1939); in Heidelberg (1943); and by a second more powerful one in Tokyo after 1941. Most of these cyclotrons, in particular the ones in Liverpool, Paris, Stockholm, and Copenhagen, benefited from the experience gained by physicists formed in Berkeley [1]. Then a new generation of accelerators appeared after 1945.

The Synchro-Cyclotron

In 1939, the largest cyclotron was the 60-inch cyclotron in Berkeley. It was 1.5 m in diameter and was able to accelerate protons to an energy of 10 MeV, deuterons[1] to an energy of 21 MeV, and α-particles (which are helium nuclei consisting of two protons and two neutrons) to an energy of 42 MeV. As early as 1939, Lawrence had obtained the money required to set up a cyclotron 4.67 m in diameter, the famous 184-inch cyclotron. But during the war, the huge electric magnet, which was to become part of it, was used for the magnetic separation of the 135 uranium isotope. When peace returned, it was soon realized that the new accelerator could not function as an ordinary cyclotron, because of relativistic effects, which were somewhat hastily thought easy to control: indeed, the mass of a particle increases with its velocity,[2] and the latter increases more slowly due to this relativistic effect. Recall that the trajectory of a proton in a cyclotron is a spiral: each time the proton circles once in the accelerator, it increases its velocity and the radius of its orbit also increases. The increase of the trajectory radius compensates exactly the increase of the particle velocity so that the particle takes exactly the same time to make

---

[1] The deuteron is the heavy hydrogen isotope consisting of a proton and a neutron; see p. 249.

[2] According to the Einstein formula $E = mc^2$, the mass of a proton, which is accelerated to an energy of 10 MeV, increases by about 1%.

each turn. When the particle velocity is small (compared to the velocity of light), relativistic effects are negligible. But when the accelerated proton reaches energies of about ten million electron volts (10 MeV), the increase of its mass prevents it from increasing sufficiently its velocity, and it takes more time to make one turn. This upsets the synchronization of its passing from one dee to the next with the alternating electric current, which will only accelerate the particle if it changes in phase with it. Otherwise, it slows it down!

The solution to this problem consisted in varying the frequency of the oscillating electric field in order to keep it in phase with the accelerated particle. That was not an easy task. Indeed, the time it takes for a proton to reach its maximum energy is about one thousandth of a second, and during this time, it makes about ten million turns in the accelerator. And there is a price to pay: only about 1% of the electrically charged particles are accelerated; the rest is lost. This is due to the necessity of first increasing and then decreasing the frequency of the electric field. Indeed, the acceleration can only begin when the frequency reaches its maximum value in order to achieve the right synchronization at low velocity. The decreasing frequency then accompanies a bunch of particles to its maximum energy, and during that time, no other bunch of particles is accelerated. The accelerator no longer produces a continuous beam but a so-called "bunched" bream. Edwin McMillan called this new accelerator a *synchro-cyclotron*. It was made possible by his discovery of the "stability phase" [2] in 1945, not knowing that the Russian Vladimir Veksler had made the same discovery [3] in 1944.

It is this stability which makes the accelerator work: particles belonging to the same bunch remain bunched together as they are accelerated. One might have feared that the bunch would spread out as it spirals in the accelerator. However, the particles remain bunched as the frequency is gradually changed. The principle was tested on the 37-inch cyclotron in Berkeley [4], and, already in 1946, the large 184-inch synchro-cyclotron [5] accelerated deuterons to an energy of 190 MeV and α-particles to 380 MeV. Other synchro-cyclotrons were soon to follow. Six years later, there were six in the United States, one in Harwell (England), one in Amsterdam, and one at McGill in Montreal (Canada). These were followed by synchro-cyclotrons in Uppsala, Liverpool, Dubna, and CERN in Geneva [6]. But in the meantime, a new idea appeared.

The Proton Synchrotron

Nothing in principle prevented building synchro-cyclotrons able to accelerate protons and other particles to any desired energy, except practical and financial considerations. However the 4.70-m electric magnet was already so large that it became difficult to imagine increasing its size in order to increase the energy of the accelerated particles. But in 1943, the Australian physicist Mark Oliphant, who was professor in Birmingham, had thought of another method. Instead of using one big magnet, why not set up a ring of successive smaller electric magnets. The trajectory of the accelerated particles remains fixed, and in order to be in phase with their increasing velocity and energy, the magnetic fields must increase. It takes but a few seconds to accelerate a bunch of particles after which one begins anew. The idea was however only published 2 years after the war [7, 8]. This ring-shaped accelerator, called a synchrotron, had the advantage of involving considerably smaller and lighter electric magnets which guided the protons. It was the ancestor of present high-energy accelerators. The first such synchrotron was built in Brookhaven, on Long Island, not far from New York. It was called a *cosmotron* and it began working in 1952. It accelerated protons to an energy of 3000 MeV, or 3 GeV (a GeV is one thousand MeV or one billion electron volts). This accelerator was soon followed by a similar 1-GeV synchrotron in Birmingham in 1953, a 6.4-GeV Bevatron in Berkeley in 1954, a 10-GeV *synchro-phasotron* in Dubna (USSR) in 1957, a 2.5-GeV *Saturne* synchrotron in Saclay (France) in 1958, and more [6].

Electron Accelerators

The Betatron

The first cyclotrons were able to accelerate protons, deuterons, and α-particles. At first, the accelerator inventor Rolf Wideröe attempted to accelerate electrons, but he failed.[1] His idea of building a "radiation transformer" was taken up later by Donald Kerst who built a "magnetic induction accelerator" which was able to accelerate electrons to an energy of 2.3 MeV which in turn produced X-rays upon impinging on a tungsten target [9, 10]. In 1941, in the laboratories of the *General Electric Company*, Kerst constructed a new 20-MeV accelerator [11] for the University of Illinois. In the same laboratory, a new 100-MeV accelerator was constructed in 1945. These accelerators were mainly used to produce X-rays by bombarding, for example, tungsten targets with the accelerated electrons (the precise nature of the target is not crucial; it needs only to be heavy enough and to have a high enough melting point in order to resist high temperatures). The X-rays were radiated by the high-energy electrons which were suddenly slowed down by the target. But in order to be used for physics, considerably higher energies were required. This was achieved 10 years later by Kerst with the *betatron*, which produced 300-MeV electrons [12]. The small size of the betatron (2.44 m in diameter) was both an advantage and a disadvantage. Indeed, the electrons being light particles soon reach the velocity of light, and if they are constrained to such a small orbit, they radiate energy, just as an antenna does. The acceleration therefore becomes less efficient since an increasing fraction of the energy gained is lost in radiation. This sets a limit of about 1000 McV to such accelerators.

---

[1] See p. 295.

The Electron Synchrotron

The process by which protons are accelerated in the synchro-cyclotron, described above, can also be used to accelerate electrons. Since the ring is considerably larger than that of the betatron, it does not suffer from the limitations of the latter. The first electron synchrotron was constructed in Berkeley in January 1949, under the direction of Edwin McMillan. It could accelerate electrons to an energy of 300 MeV. Within a few years, this kind of accelerator proliferated: in 1950, five were functioning in the United States and one in Glasgow. Less than 10 years later, an energy close to 1000 MeV was reached at Cal Tech, at the University of Cornell, in Stockholm, Rome, as well as in Tokyo [13].

The Linear Electron Accelerator

It is also Rolf Wideröe who constructed the first device capable of accelerating particles to a given energy by accelerating them *in successive steps.*[1] The particles travel in a straight line, whence the name *linear accelerator.* For protons and heavier particles (deuterons, α-particles) the linear accelerator was soon supplanted by the cyclotron in which particles have spiral-shaped trajectories. David Sloan had constructed a 1.14-m long linear accelerator, but it was limited to rather slow particles.[2] The idea of a linear accelerator for electrons had already been contemplated in the 1930s by the American physicist William Hansen, at the University of Stanford. However, no high-frequency generator with sufficient power existed at that time [14]. The development of the *magnetron* (for radar) during the war, and of the *klystron*,[3] made it possible to construct increasingly powerful linear electron accelerators in Stanford: 6 MeV in 1947, 35 MeV in 1950, and mainly the famous *Mark III* accelerator [15] which reached 700 MeV in 1954 and 1000 MeV in 1960. It was an impressive machine, 90 m long, consisting of ten sections, and built entirely underground in order to shield the environment from its radiation. Similar accelerators will be built in Orsay (France), at the Massachusetts institute of technology (MIT), and in Saclay (France).

---

[1] See p. 295.

[2] See p. 297.

[3] The klystron was invented in 1937 by Russel and Sigurd Varian to improve the radar technologies. It is a special vacuum tube used as an amplifier at microwave and radio frequencies. The magnetron is a high-powered vacuum tube that generates higher-frequency microwaves using the interaction of a beam of electrons with a constant magnetic field. It was invented in 1940 by the British physicists John Randall and Harry Boot. It allowed to locate airplanes by radar with a better precision.

Electrostatic Accelerators

In 1931, Robert Van de Graaff constructed the first electrostatic accelerator endowed with a belt.[1] The first accelerators of this kind were already operating in 1933, and they could accelerate protons to an energy of about 1 MeV. The greatest danger lurking in such accelerators is the huge spark, a real lightning, which can occur between the "terminal" set to several million volts et the remainder which is grounded. The accelerator constructed at the MIT in 1950 could reach 4 million volts after patient technical developments [16]. Then MIT launched a project designed to reach 12 million volts, but it never exceeded 9 million volts, nor did a similar project in Los Alamos.

Hundreds of electrostatic accelerators were built throughout the world. The highest voltages reached about 5 million volts in Brookhaven, Los Angeles, Harwell in England, and Saclay in France.

As early as 1947, Robert Van de Graaff, his colleague John Trump, and several others started a business, the *High Voltage Engineering Corporation* (HVEC), which sold electrostatic accelerators with increasing energies reaching 12 million volts in 1965. They enjoyed a quasi monopoly and sold dozens of "Van de Graaff" accelerators throughout the world.

The last technical progress, but not the least, consisted in doubling the energy of the accelerated particles while maintaining the same maximal voltage. This was not a new idea. It had been suggested in 1936 by two American physicists, Willard Bennett and Paul Darby [17]. In a normal electrostatic accelerator, the protons, for example, are produced in the electrode which is maintained at a high voltage, by ionizing hydrogen, that is, by tearing off the electrons which are normally bound to the protons. Repelled by the high positive voltage, the proton is accelerated to the earthed potential. If the positive voltage is 1 million volts, the proton acquires an energy of 1 MeV (one million electron volts). Bennett and Darby observed that in certain instances, a second electron could bind itself to a proton which already had one electron. Such an atom is in fact a *negative* ion. Indeed, it is a very fragile ion: as soon as it collides (e.g., with a molecule in the gas), it can lose its extra electron or even both of its electrons if the collision is sufficiently violent. They had the idea of producing such negative ions, not at the high-voltage end but in the earthed part of the accelerator. The negative ions are then accelerated to the positive voltage part, thereby acquiring a 1 MeV energy (in our example). Then, after passing through a vessel containing a gas, *they lose their two electrons*, thereby becoming positive ions, and they are *again accelerated* to the earthed part of the accelerator, thereby gaining an extra 1 MeV. With such a trick, a voltage of 1 million volts can accelerate protons to 2 MeV! Such accelerators are called *tandems*. They were built and sold by Robert Van de Graaff in the 1960s throughout the world. The energy imparted to particles by such electrostatic accelerators was admittedly lower than by accelerators such as cyclotrons and synchrotrons. In spite of this, it was the

---

[1] See p. 293.

"Van de Graaff" accelerators which furnished most of the experimental data on nuclei. Indeed, they had distinct advantages: their energies were well adapted to fathom the structure of nuclei, their energies could easily be modified in a continuous manner, they could accelerate particles to very precise energies with a dispersion of the order of one in a thousand, and finally, they could furnish a really continuous beam, a great advantage for coincidence measurements. Their golden era lasted from 1950 to about 1970.

New Detectors, New Measuring Instruments

The Parallel Plate Spark Chamber

In 1934–1935, Heinrich Greinacher, who had introduced electronic amplification for ionization chambers, proposed a new invention: the "hydraulic" counter. It consisted of a fine metallic pin placed very close to a jet of water while a 2,000-V voltage was maintained between them. When a particle passed close to the pin, the electric perturbation deviated the water jet, thereby allowing the particle to be detected. But in spite of numerous improvements, the counter remained unstable and slow and it was therefore not adopted [18–20].

A more successful counter was imagined by Salomon Rosenblum, who was then a refugee in Princeton, and by Chang, namely, the *spark counter* [21]. A thin platinum wire (0.2 mm in diameter) is stretched out parallel to a polished brass plate at a distance of 1.5 mm and maintained at 3,000 V relative to the plate. When a particle passes close to the wire, a spark is generated which can be observed and even heard. The counter is insensitive to electrons and to γ rays, but it responds well to α-particles. The sensitive area remains very close to the wire. Rosenblum constructed such a counter, consisting of 10 wires, 5 mm apart. He used the counter to detect α-particles emerging from a magnetic spectrometer. This was the forerunner of multi-wire counters which underwent a phenomenal development and for which Georges Charpak was awarded the Nobel Prize in 1992.

The *parallel plate* counter, a descendant of the spark counter, was introduced in 1948 by two American physicists, Leon Madansky and Robert Pidd [22]. They were searching for a fast detector with which they could measure the lifetimes of *mesons*, particles which will be described shortly, and the lifetime of which is about 2.5 hundred millionth (2.5×10^{-8}) of a second. The particles are again detected by the avalanche of electrons caused by the few electrons set free when they are ripped off from gas molecules by a particle passing by, exactly as in a spark chamber or a Geiger-Müller counter. But in the parallel plate counter, there is no wire. Instead, two circular copper plates, 5 cm in diameter and 1 mm apart, are maintained at a relative voltage ranging from 900 to 3000 V. The proximity of the plates creates a sufficiently strong electric field, which accelerates the ripped off electrons to an energy sufficiently high to cause an avalanche and a corresponding strong and rapid signal. The plates are immersed in a mixture of argon and butane gas at a pressure of 2 atmospheres. The proper choice of gas is crucial to make the counter work properly. This was the ancestor of the present *parallel plate counters* which are often much larger, and the plates of which are thin organic films which become electric conductors when coated with a thin aluminum layer.

The Return of Scintillators Observed by *Photomultipliers*

After having been the key to major discoveries (the discovery of a nucleus in the heart of the atom to begin with[1]), scintillation detectors were given up, partly due to the controversy between the *Cavendish* and the *Institut für Radiumforschung* in Vienna.[2] They were replaced by new electric Geiger counters deemed more objective than the human eye. In its original version, the scintillation method consists in observing visually with a microscope the weak light emitted by a particle passing through a thin sheet of zinc sulfide. In 1941, the Hungarian physicist Zoltán Bay had thought of using a photoelectric cell, which transfers light energy to electrons and which is coupled to an *electron multiplier* [23]. An isolated photoelectric cell would not have worked because the light emitted by the passing α-particle was too weak. The *electron multiplier* was patented in 1923 by Joseph Slepian, who was performing research in *Westinghouse*, and further perfected by Vladimir Zworykin working in the *Radio Corporation of America* (RCA) [24, 25]. It consists of an evacuated glass tube, the extremity of which is maintained at a very high negative voltage. The electrons which are emitted at the cathode are accelerated towards the anode, at the other end of the tube. But the tube is set up so that, on their way, they hit intermediate electrodes called *dynodes*. The latter produce further electrons so that the small number of electrons produced at the photocathode can be *multiplied by a factor of one billion*. Such an avalanche of electrons is similar to a Geiger counter, the difference however being that the number of produced electrons reflects rather faithfully the number of electrons produced initially at the cathode. This device counted the number of particles in a far more objective fashion than the human eye. It was also considerably faster, and the intensity of the signal was proportional to the energy lost by the particle as it produces light in the scintillating substance.

The coupling of a photocathode to an electron multiplier was developed for radio-electric industry, and it became known as a *photomultiplier*. Commercial catalogues were offering dozens of models after the war.

In 1944, Marietta Blau, a Jewish Austrian physicist who had to flee from the *Institut für Radiumforschung* in Vienna and became a refugee in the United States, proposed to use such a device to detect α-particles [26]. At the same time and independently, two American physicists in Berkeley made a similar proposition [27]. It became common usage to detect charged particles, neutrons,

---

[1] See pages 70, 72, 74, 200.
[2] See p. 203.

and γ-rays using a scintillator coupled to a photomultiplier. Numerous scintillation substances were used and still are. Their choice depends on what needs to be detected. For example, neutrons are well detected using an organic substance which is rich in hydrogen. For the detection of γ-rays, one usually uses sodium iodide doped with thallium (meaning that it contains a minute quantity which is very precisely determined) [28]. A great variety of such substances are available. Photomultipliers underwent an impressive development and are still widely used today.

The Invention of the Transistor and of the *p-n Junction*

Initially, quantum mechanics was formulated in order to understand the dynamics of electrons surrounding the nucleus. But it also applies to chemistry, to the crystalline structure of solids, and to the structure of the nucleus. In the 1930s it was applied to electric conductivity of various substances, and it could explain why some substances are electric conductors, why others are not, and that further substances exist, called *semiconductors*, which are neither proper electric conductors nor proper electric isolators. In conducting metals, some electrons are barely bound to the atomic nuclei, and they can move freely in the metal. However, in isolating substances, each electron is bound to neighboring nuclei and it cannot propagate. In semiconductors, the number of barely bound electrons is very small [29].

At the beginning of the century, one semiconducting substance was used in radio receptors, namely, lead sulfide, called galena. It can indeed be used as a *diode*, which, when it is in contact with a metallic pin, allows an electric current to pass in one direction but not in the other, because the electric resistance of the contact is far greater in one direction than in the other. This was an empirical fact which remained unexplained until the advent of quantum mechanics between 1932 and 1939. The explanation was formulated by Alan Wilson in Great Britain [30], Walter Schottky in Germany [31, 32], and Nevill Mott, who was still in Great Britain [33].

During the war, a great effort was made both in England and in the United States to develop *crystal rectifiers* which were needed for radars. In 1946, the *Bell Telephone Company* launched a research program aiming at a better understanding of such phenomena and at making a semiconductor rectifier, a diode as well as a semiconductor amplifier (equivalent to a *triode* tube). This is how the *transistor* is born in 1948, in the hands of Walter Brattain and John Bardeen [34]. It was a point transistor, the crystal of which was germanium and very difficult to handle [35, 36].

The word *transistor* was coined by John Pierce, another physicist in *Bell* labs (who also wrote science fiction stories signed by the name J. J. Coupling) [37].

Shortly later, in the same lab, William Shockley invented the *p-n junction* transistor which was much easier to use in industrial applications. He replaced the fragile tungsten or phosphorous bronze pins by introducing, in a controlled fashion, impurities into an otherwise very pure silicon crystal. Such transistors quickly spread and replaced tubes in electronic circuits. This transistor had many advantages: it was smaller, it required less energy, it was more reliable, less fragile, and cheaper to manufacture, at least after some initial problems had been solved.

In a paper published in *Physics Today*, the Canadian physicist Kenneth McKay from Bell labs wrote: "Each time a nuclear physicist observes a new effect produced by an atomic particle, he tries to use it to make a counter." [38] This is precisely what happened with isolating crystals such as diamond. Already in 1913, Wilhelm Röntgen noticed that a crystal exposed to radiation becomes slightly conducting [39]. But the current produced by a single particle was far too weak to be detected without amplification. After World War II, several physicists succeeded in detecting particles using a crystal, mostly diamond, but also cadmium sulfide or silver chlorine [40–44]. The crystal detector is based on a process similar to that of an ionization chamber: the particle passing through the counter tears off a few electrons which are attracted by an electric field, thereby causing a transient electric current. In an isolating substance, however, electrons have a tendency to stick to atoms which are missing an electron. This is a *recombination* process which distorts considerably the measurement so that the measured electric current is not a faithful representation of the energy loss of the passing particle. At this stage, crystal detectors were at best in an experimental phase.

However, semiconducting crystals were more promising: electrons in semiconductors are more mobile and have less probability of recombining. If one uses a semiconductor, one can set up a few volts of an "inverse" voltage, meaning in the direction where the current cannot flow. The electrons teared off by a charged particle passing through the crystal can move and create a current for a short time, thereby providing for a *signal* of the particle. Furthermore, the amplitude of the signal reflects the number of liberated electrons and therefore the energy lost by the particle in the crystal. This is in fact a solid ionization chamber which is smaller and much denser than a gas ionization chamber. The first works on the subject were published in 1951 by Kenneth McKay [45] and in 1956 by James Mayer and Ben Gossick [46]. Possibly put off by the setbacks of isolating crystal counters, physicists were at first somewhat wary of crystal semiconductor counters. Furthermore, the manufacture of *p-n* silicon junctions raised several serious technical problems. One had to learn to make rather large and very pure *monocrystals*. Crystalline substances such as crystallized sugar, kitchen salt, or quartz are consist usually of small crystals which can be detected by their planar and brilliant surfaces. But the counters required crystals with atoms maintaining their orientation over distances of the order of 1 cm, with no breaks or faults. This implied an alignment of hundreds of millions of atoms! Impurities, meaning atoms other than silicon, had to be very scarce, less than one in ten billion! The sample also required to be "doped" with impurities in a controlled fashion so as to increase or decrease the number of electrons. The proportion of such controlled impurities was about one for ten million silicon atoms... This is why it took 10 years before the first papers were published describing physics experiments using semiconductor detectors [47]. But thereafter they developed rapidly. A paper published in 1962 and reviewing the use of such detectors had no less than 231 references [48]!

Semiconductor counters were simple, reliable, and endowed with an excellent resolution. They could distinguish particles with energies differing no more than a few 10 000 V.

As all detectors today, they require electronic equipment to collect, amplify, and sort out the signals.

The Growing Presence of Electronics

Until 1939, electronic equipment had a modest place in the experimental arsenal of physicists. It was used to detect automatically the "simultaneous" firing of two Geiger-Müller counters[1] and also to amplify the weak electric signals of ionization chambers and to make a rough estimate of the energy of the detected particle. But such instruments were laboratory makeshifts, using amplifiers extracted from radio receivers. For example, for the discovery of artificial radioactivity, amplifiers of commercial radio receivers were used.[2] A great progress occurred after the war: laboratories employed teams of electronic engineers in order to develop instrumentation adapted to their experiments. In experimental halls, the size of electronic equipment kept growing. Electronics furnished more sensitive detectors for more precise measurements. Some crucial measurements, such as the time separating the firing of two counters, were not conceivable without electronics. For example, this is how the velocity of neutrons is measured. Electronic equipment also permits the selection of rare events, in which the energy of a given particle lies within a definite range, which was called a *channel*. *Multichannel selectors* were developed which were able to analyze the distribution of detected particles according to their energies in different channels, the channel number revealing immediately the particle energy. Selectors began involving ten, one hundred, and then 256, 1027, 4096... channels.

And what about computers? For the time being, they play no role in experimental physics, in spite of the fact that the first computer was born in 1945. They remain absent until the end of the 1950s. Until then, calculations were performed with slide rules (Enrico Fermi always had a small slide rule in his pocket) or with electric mechanical calculators. Computers begin to play an increasingly important role in physics only in the 1960s, first for computations and only later in instrumentation.

Photographic Emulsions: A Special Case

Becquerel discovered radioactivity because it left a mark on photographic plates which were soon replaced by electric counters. In 1911, the German chemist Max Reinganum used a microscope to observe the traces left on a photographic plate by

---

[1] See p. 219.
[2] See p. 308.

α-particles. It was soon realized that the tracks consisted of a number of black spots which form an image of the path of the particle [49].

In 1925, Marietta Blau succeeded in observing much fainter tracks produced by protons. However, the interest for physics remained rather meager [50,51], except in one field, where photographic plates became very useful: the detection and study of cosmic rays. They had been first discovered in an ionization chamber and further studied in a Wilson chamber, which displayed a shower of particles, probably created by particles colliding with oxygen or nitrogen nuclei in the atmosphere. Since this radiation originated in outer space, physicists sought to detect it at high altitude, on the top of mountains or in balloons which could reach greater altitudes, as Victor Hess and Werner Kolhörster had done.[1] They aimed at observing the "primary" radiation coming from outer space and not only the "secondary" radiation reaching the ground, due to collisions in the atmosphere. For this purpose, photographic plates were ideal detectors. They were light and easy to handle and they could record particles passing through. After being developed, they could be kept for a long time before being analyzed. The first such experiment was performed in the United States in 1935. The balloon named *Explorer II* reached an altitude of 22 000 m and remained above 21 000 m for over 2 hours. One of the embarked photographic plates did indeed display tracks of cosmic rays [52,53].

With the help of Victor Hess, Marietta Blau managed to expose photographic plates on top of mount Hafelekar (2334 m high, in the Austrian Tyrol) for a period of 5 months. She observed the first tracks of a nuclear reaction produced by a cosmic ray [54,55]. The photograph consisted of several tracks, all emanating from one point, suggesting the explosion of a nucleus hit by a seemingly very-high-energy cosmic ray. She called the picture a *star*. Numerous other observations followed.

One of the major difficulties in using photographic plates is their weak sensitivity. Several physicists attempted to produce more sensitive emulsions, and in 1945, the British firm Ilford sold emulsions which they boasted to have "nuclear quality" and which indeed were better adapted. For the following few years, photographic plates remained a major experimental tool to study cosmic rays. Several elementary particles were discovered this way.

---

[1] See p. 280.

Data Accumulate

Protons and neutrons, similar and yet so different, are finally recognized as belonging to the family of nucleons. Real transuranic elements are discovered, which require a revision of the periodic table of Mendeleev. The radioactive lifetime of the neutron is measured. The scattering of high energy electrons allows one to observe the distribution of electric charge inside the nucleus.

Protons and neutrons are very similar, except for the fact than only protons carry an electric charge. They have similar masses and also almost identical nuclear interactions: this property is called the *charge independence* of nuclear forces.[1] Their similarity led the Danish physicist Christian Møller to give both protons and neutrons a common name, *nucleons* [56]. Thus, for example, the uranium 238 nucleus contains 238 *nucleons*. The name stuck and is still used today.

The Papers of Bethe

After the war, the understanding of the structure of nuclei was much the same as it was in 1939. The three 1936–1937 papers of Bethe[2] were still considered to be a standard reference for nuclear physicists, and they were even published a second time in *Reviews of Modern Physics*. Except for transuranic nuclei, which had thoroughly been studied for military purposes, nothing much had changed. But the change came with the advent of accelerators and new detectors.

Real Transuranic Nuclei

Since the experiments of Lise Meitner, Hahn, and Strassmann in 1937, it was known that uranium could absorb 25-keV neutrons in a resonant fashion, that is, with great efficiency. The resulting uranium isotope 239 would disintegrate by emitting a β-ray (an electron) with a half-life of about 23 min. The emission of the β-ray is due to the transformation, in the nucleus, of a neutron into a proton so that *the resulting substance is necessarily composed of the element 93*, a real transuranic element.

---

[1] See p. 303.
[2] See p. 343.

Unfortunately, every attempt to identify this new element precisely, using chemical methods, had failed. In a paper published in June, 1939, Emilio Segrè thought that the question had not been settled:

> *The necessary conclusion seems to be that the 23-minute uranium decays into a very long-lived element 93 and that transuranic elements have not yet been observed [57].*

What made it so difficult to observe this element? The fact that a very small quantity was produced because of the weak intensity of the neutron source and that, in addition, the radioactive half-life of the produced element was rather long (known to be 2.35 days) so that the number of decays per second was very small.

However, the cyclotron in Berkeley could be used as an intense source of neutrons. It was sufficient to accelerate deuterons and to aim the beam onto any target. Indeed, a deuteron consists of a proton and a loosely bound neutron. The collision of a deuteron with a nucleus leaves them unbound. It is as though the proton had carried the neutron along and then let it loose. The cyclotron acted effectively as a neutron source *several million times more intense* than that which was available to Otto Hahn or Enrico Fermi, who bombarded pulverized beryllium by α-particles emitted by a radioactive substance.

Neptunium

Edwin McMillan and Philip Abelson decide to expose uranium to neutrons using a cyclotron. They produce a sufficient amount of the element 93 to study its chemical properties. They are surprised to discover that *the chemical properties are not similar to those of rhenium*, of which everyone expected it to be the homologue. Indeed, in the Mendeleev table, radium appeared in the same column as barium, of which is the element 56. Then, for each new element, one had to pass on to the next column. This placed the element 93 into the same column as rhenium, the element 75.[1] And yet it did not seem to follow this rule:

> *It is interesting to note that the new element has little if any resemblance to its homolog rhenium [. . .]. This fact, together with the apparent similarity to uranium, suggests that there may be a second "rare earth" group of similar elements starting with uranium [58].*

McMillan and Abelson suggested that the new element could belong to a group of elements similar to rare earths, which are elements ranging from lanthanum (the element 57) to lutetium (the element 71) which all have very similar chemical properties and which all fit into a single cell of the Mendeleev table, a property

---

[1]This part of the Mendeleev table, as it was conceived in 1938, is shown on page 357.

which Bohr explained in 1921 in terms of the filling of electronic shells.[1] Was this happening with the element 93? It was a bit early to tell, but there was definitely something in the wind.

The element 93 was the first element beyond uranium, therefore *the first real transuranic element*. McMillan and Abelson proposed to call it *neptunium*, because the planet Neptune is the first planet beyond Uranus, which had given a name to uranium.

Plutonium

On Monday, June 3, 1940, Glenn Seaborg, a young chemist in Berkeley, read the paper of McMillan and Abelson, whereupon he wrote in his diary:

> The descendant $^{239}94$ has a lifetime which is too long to be observed. I know about this work while it is going on and I am in a hurry to work in this exciting field [59].

What Seaborg called a "descendant" is what was left of the neptunium nucleus (the element 93) after it β-decayed. It emitted an electron which carried one unit of negative charge, and it had therefore to increase its electric charge by one unit, thereby becoming the element 94. This still remained to be proved.

Born in Michigan in 1912, Glenn Seaborg spent the first 10 years of his life in Ishpeming, a small mining town where his parents worked. They had emigrated from Sweden. The family then settled in California where Seaborg studied and where he obtained his Ph.D. in 1937. He first became the assistant to the great American chemist Gilbert Lewis, then instructor in 1939, assistant professor in 1941, and finally professor in 1945. Together with Joseph Kennedy, another chemist in Berkeley, Seaborg begins to study the radioactivity of the element 93, which becomes the subject of the Ph.D. thesis of his student Arthur Wahl. They bombard uranium with the deuterons of the cyclotron in Berkeley, and they attempt to use radiochemistry to determine the substances thus produced. They quickly realize that the observed α-particles are emitted neither by uranium nor by the element 93. They are probably emitted by the element 94. On January 28, 1941, they send a paper to the *Physical Review* just a few days after the United States entered the war. The authors ask to postpone its publication [60]. On February 25, 1941, they obtain a chemical proof that they had indeed discovered the element 94, which they propose to call *plutonium*, because Pluto is the second planet beyond Uranus [61, 62].

As they had suspected, the new element 94 has chemical properties similar to those of uranium and not of osmium. The idea that there may exist a series of elements with similar chemical properties, as rare earth elements, was progressively emerging.

---

[1] See p. 116.

The discovery of this new element had important military implications. Basing his argument on the theory of Bohr, Fermi had shown that this element might fission in the same fashion as uranium 235. Provided one could produce a sufficient amount, it might turn out to be easier to make a plutonium bomb, rather than separating uranium 235 from uranium 238. This is why they kept their discovery secret until 1946.

Actinides

The idea that there might exist a new family of elements, which have very similar chemical properties, as the rare earth elements do, was not really new. As early as 1921, Bohr had explained the similarity of chemical properties of lanthanides by the fact that the outer electron orbit, with "six quanta" and responsible for most of the chemical properties, contained two electrons, whereas the inner orbits, with 4 or 5 quanta, were filled progressively as one went from lanthanum to lutetium.[1] In the table of elements which he presented at the famous *Bohr Festival* in Göttingen in 1922,[2] Bohr had drawn a (dotted line) frame indicating a possible existence of a new family of very similar substances, without however insisting. In his Nobel speech on November 11, 1922, he restated this possibility, adding that nothing could be claimed with certainty, because too few elements in that region were known at the time. In 1935, the German chemist Aristid von Grosse, who had emigrated to the United States, mentioned explicitly this possibility [63]. The results obtained by McMillan, Abelson, and Seaborg certainly pointed in that direction. One should also mention the theoretical calculation of the German physicist Maria Goeppert-Mayer, who also emigrated to the United States. In a paper published in 1941, she showed that the existence of another series of "rare earth" elements (as McMillan and Abelson called them) was theoretically not only possible but also very likely. According to her calculations, this series of new elements could begin even before uranium, starting with the element 91 (praseodymium) [64]. In a secret document sent to the uranium committee in 1942 and published only in 1948, Seaborg gives a detailed list of the chemical properties of neptunium and plutonium. He concludes, as McMillan and Abelson did, that there could well exist a new family of "rare earths," the first element of which could be uranium, thorium, or even actinium [62]. In 1945, he became convinced that the family started with actinium, and he proposed to call the elements *actinides*, as had been done for *lanthanides* [65]. A few years later, he made a detailed review of the new periodic table in which the *actinides* had acquired their position [66]. The discovery of *actinides* caused a considerable modification of the periodic table[3] in the region of

---

[1]Recall that the "number of quanta," also called the principal quantum number n, determines the size and the energy of the electron orbit. See p. 126 and p. 116 for the theory of lanthanides.

[2]See p. 123.

[3]The periodic table, or table of Mendeleev, in its present form, is shown at the end of this book.

heavy elements: *actinium, thorium, praseodymium, uranium, neptunium, plutonium, and all the elements which had not yet been observed up to the element 103 were placed in the same cell of the periodic table*, in the same fashion as the elements 57–71 are all in the cell of lanthanum. From then on, physicists and chemists could predict the chemical properties of the new elements and this helped to discover them. Thus, in 1949, they observed the element 95, called *americium*, and the element 96, called *curium*; in 1950, they observed the element 97, called *Berkelium*, and the element 98, called *californium*; in 1955, the elements 99 and 100, called *Einsteinium* and *Fermium*.

In 1955, Edwin McMillan and Glenn Seaborg were awarded the Nobel Prize "for their discoveries in the chemistry of transuranic elements." Their search however extended over a long time. The element 101 was observed in 1955, the element 102 in 1958, and the element 106 in 1974. The latter were all observed in Berkeley. Meanwhile, other laboratories joined the race, namely, Dubna in the Soviet Union and Darmstadt in Germany. The last identified element is 116, discovered in Dubna in the year 2000.

The Lifetime of the Neutron

As soon as it became certain that the mass of the neutron exceeds the mass of the proton, it was surmised that the neutron is radioactive: that it could decay into two particles (plus a neutrino the mass of which is very small, possibly zero). In the issue of *Physical Review*, dated May 1, 1950, two papers received within one week confirmed the radioactivity of the neutron and gave an estimate of the radioactive lifetime, which is the time required for half of a sample of neutrons to decay. In the first paper, Arthur Snell, from Oak Ridge, Tennessee, together with several collaborators, confirmed that the neutron decays into a proton and an electron, which they detected *in coincidence*, and they estimated the radioactive half-life to be between 10 and 30 min [67]. In the second paper [68], John Robson, working in the Canadian laboratory in Chalk River, quotes a lifetime between 9 and 25 min. A year later, he describes more precise measurements: the radioactive lifetime of the neutron is 12.8 min with a possible uncertainty of 2.5 min [69]. This is compatible with today's estimate of 885.7 s, that is, 14 min and 45.7 s, plus or minus 0.8 s.

Note that this is the lifetime of a free neutron and not the lifetime of a neutron which is bound to a nucleus. For neutrons in a nucleus, the balance of energies and masses is more complicated. If a neutron decays in a proton, it is the nucleus which is transformed in the corresponding β-decay. The new nucleus has one neutron less and one proton more. But such a process will only occur if it *does not* require the absorption of energy from an external source. Otherwise, the nucleus is stable and will not β-decay. Indeed, this is what we observe: the nuclei of the atoms which surround us have neutrons and yet they are stable; they do not decay.

Electron Scattering and the Electric Charge Distribution in Nuclei

As early as 1934, the English physicist Nevill Mott had calculated the elastic scattering of electrons on a nucleus [70], using the relativistic Dirac equation.[1] He formulated in a modern way what Rutherford had expressed in 1911.[2] Mott assumed that the nucleus was a point positive electric charge, which is reasonable from the point of view of a low-energy electron. Indeed, quantum mechanics shows that the size of a nucleus will modify the scattering of the electron only if the wavelength of the electron has a size comparable to the size of the nucleus. It is for the same reason that a microscope cannot resolve details of objects smaller than about 1 μ, which is roughly the size of the wavelength of visible light. But the postwar electron accelerators—in Stanford, California, at first—made increasingly energetic electron beams available.[3] The electrons had correspondingly small wavelengths. The electron beam energies were 6 MeV in 1947, 35 MeV in 1950, 700 MeV in 1954, and 1000 MeV in 1960. The corresponding electron wavelengths were 190 fm, 35 fm, 1.8 fm, and 1.2 fm[4]. The effect of the nuclear size on the elastic scattering of electrons had been studied by the Austrian physicist Eugen Guth [71]. In 1947, an American physicist, Morris Rose, working in the Oak Ridge Laboratory (which, at the time, still depended on the Manhattan Project) studied elastic electron scattering and showed that the electric charge distribution inside the nucleus could be deduced directly from the distribution of elastically scattered electrons [72]. The electron scattering reveals the "Fourier transform" of the electric charge, a mathematical operation well known to electricians. This is only true within an approximation, known as the "Born approximation." A more precise relation requires a more complex analysis of the data, but the essential result is clear: the angular distribution of elastically scattered electrons provide a kind of "snapshot" of the electric charge inside the nucleus, of what we call the *charge distribution* of the nucleus. Later, more precise calculations were performed by the British physicist Lewis Elton [73], using a more precise theory which required a considerably longer computation time. He used the so-called phase-shift method, and he was only able to perform his calculations when the first computers became available in the 1950s.

The charge distribution, which is roughly the distribution of protons in the nucleus, is a fundamental property: it was discovered that the charge distribution was approximately constant in the interior and decreased rapidly at the nuclear surface. But is the charge really uniformly distributed in the nucleus? What is the surface thickness, that is, the distance within which the density varies from zero to the value it has in the interior?

---

[1] See p. 146.
[2] See p. 74.
[3] See p. 417.

The first experiment, which revealed the effect of the nuclear size on the scattering of electrons, was performed in 1951 at the University of Illinois, on aluminum, copper, silver, and gold nuclei. The impinging electron beam had an energy of 15.7 MeV. This energy was however too small to provide a measurement of the nuclear size [74]. Two years later, the electron synchrotron in the University of Michigan, in Ann Arbor, provided a beam of 34-Mev electrons. The analysis of the elastic scattering of the electrons revealed that the target nuclei had a radius about 30% smaller than what had been estimated by other means [75] (the scattering of neutrons and the α-decay of radioactive nuclei).

At about the same time, the research team led in Stanford by Robert Hofstadter measures the angular distribution of 116-MeV electrons which undergo elastic scattering on beryllium, tantalum, and lead [76]. His data do not reveal a precise picture of the charge distribution in the nuclei, but they suggest a smooth surface, the charge density gradually falling from a constant density inside to a vanishing density outside. In 1954, the electron beam in Stanford reaches an energy of 186 MeV, thereby allowing the team to draw a more precise picture of the charge density. In an important paper, published in 1956 in *Reviews of Modern Physics* [77], Hofstadter summarizes the results of the measurements: the radius of several nuclei have been measured as well as what Hofstadter calls the "nuclear surface thickness," which is the distance in which the charge density drops from 90% to 10% of its central value: it is about 1 fm in all nuclei. Thus, electrons "see" the nucleus as an object with a diffuse surface about 1 fm[1] thick, which is roughly one third of the radius of the oxygen nucleus and 15% of the radius of the lead nucleus. The experiments also confirm that the volume of the nucleus is proportional to the number of nucleons. But the observed radius is smaller than that which had been deduced from other means. This is not surprising because the electrons "see" the charge distribution of the nucleus, whereas the other experiments, namely, the scattering of α-particles, of protons or neutrons, as well as radioactive α-decay, suggest a nuclear size which is deduced from the distance at which the interaction of the nucleus with the impinging particle begins to be felt. In 1955, Robert Hofstadter and Robert McAllister measured the elastic scattering of 188- and 236-MeV electrons from hydrogen. Their results revealed that the charge distribution of the proton has a finite size, with a radius of the order of 0.7 fm [78, 79]. In 1961, Robert Hofstadter will be awarded the Nobel Prize "for his pioneering studies of electron scattering in atomic nuclei and for his thereby achieved discoveries concerning the structure of the nucleons."

---

[1] fm denotes a *femtometer*, that is, of 10^{-15} m. It is a natural unit on the nuclear scale: the radius of the oxygen nucleus is about 3 fm, and that of lead about 6.5 fm.

The "Shell" Structure of Nuclei

> *A paradoxical idea, which had long been disputed by the wise Niels Bohr, is finally confirmed: yes, protons and neutrons do form successive shells in the nucleus, just as electrons do in the atom. This is an irreducibly quantum effect: the protons and the neutrons interact strongly in the nucleus, and yet they pass by barely noticing one another!*

How are the protons and neutrons arranged inside the nucleus? The question had been raised ever since the discovery of the neutron in 1932. In 1933, the two German physicists exiled in Paris, Walter Elsasser and Kurt Guggenheimer, suggested that the neutrons and protons in the nucleus might form successive "shells" (Elsasser used the word "envelopes") in a similar fashion as the electrons do in the atom but at a different scale.[1] Guggenheimer based his argument on the separation energy of neutrons and protons, that is, the energy required to extract them from the nucleus. By scanning through the table of elements, he noticed that the separation energy differed little from one nucleus to the next in the table, except for sudden discontinuities for certain numbers of neutrons or protons. This suggested the "quantum effect" which is observed in atoms when electronic shells become filled. Elsasser had a more theoretical starting point. He claimed that the Pauli exclusion principle has to be applied to protons and neutrons in the nucleus and that they have to be treated on the same footing (e.g., with no preformed α-particles in the nucleus). In the nuclear interior, each nucleon feels an *average* force due to the other nucleons. Because the other nucleons exert forces in all directions, the average force should vary slowly inside just as the force exerted on water molecules does inside a water droplet. The force however becomes important at the nuclear surface where the nucleon feels a strong attraction tugging it back towards the center of the nucleus. He applied this idea by assuming the nucleus to be a simple *potential well* $^\diamond$ in which the force felt by each nucleon is similar to that felt by a golf ball at the bottom of a hole, with a flat bottom. It moves freely along the bottom, but a lot of energy is required to extract it from the hole. The calculations of Elsasser predicted some of the numbers of protons and neutrons required to form closed shells, but the numbers did not all correspond to the ones observed by Guggenheimer.

---

[1] See p. 272.

A Model of Quasi-independent Particles?

Recall the devastating argument put forth by Bohr against this model: neutrons and protons are tightly packed inside the nucleus. One cannot imagine that a proton or a neutron can move in a well-defined orbit (the classical equivalent of a state with "well-defined quantum numbers"). As soon as it enters such an orbit, it collides with the other nucleons and is pushed into another orbit (the classical equivalent to a transition to a state with different quantum numbers). In spite of his objection, some physicists still adhered to the "shell model" of the nucleus, namely, Wigner and Feenberg, who had made some encouraging calculations based on this model.[1]

The Symmetries and Supermultiplets of Wigner et Feenberg

For nuclei heavier than oxygen, the "independent particle" model, in other words the shell model, did not give good results. This was hardly surprising especially if one argued as Bohr did. While admitting that the shell model was not as valid for nuclei as it was for atoms, Wigner attempted to account for the sudden discontinuities in the separation energies of nucleons, using arguments based on symmetry [80]. The idea of Wigner and of Friedrich Hund, as mentioned above,[2] consisted in exploring the consequences of two basic hypotheses:

- The force acting between two nucleons is the same, independently of whether they are both neutrons, protons, or a neutron and a proton. If they are both protons, one must add the electric repulsion between charged particles, but the nuclear force is assumed to be the same. This is called the *charge independence* of the nuclear force.[3]
- The force between two nucleons does not depend on the orientation of their spins.[4]

Thus, the wavefunction of a given nucleus should be almost identical[5] to the wavefunction obtained by replacing a proton by a neutron or that obtained by modifying the orientations of the nucleon spins. Wigner called "supermultiplets" the set of wavefunctions thus obtained. Of course, when a proton replaces a neutron, nothing guarantees that the resulting nucleus will be in its ground state. Wigner tried to identify states of neighboring nuclei which had such parentage, and he found

---

[1] See p. 276.

[2] See p. 277.

[3] See p. 303.

[4] But recall that, even if the force does not depend on the orientation of the spins, the Pauli exclusion principle will favor certain spin orientations (see p. 264). Quantum mechanics is most surprising.

[5] But not completely identical because the repulsive Coulomb force acts on protons but not on neutrons

some. Together with Eugene Feenberg, he developed these ideas in an important paper published in 1941 in the British journal *Reports on Progress in Physics* [81].

Arguments Put Forth by Maria Goeppert-Mayer

We already encountered the young German physicist Maria Goeppert-Mayer, who suggested the probable existence of elements similar to the rare earths, close to uranium, and called actinides today.[1]

Maria Goeppert was born in 1906 in Silesia which, today, is part of Poland. From the age of 4, she lived in Göttingen, where she studied mathematics and then turned to physics. She obtained her Ph.D. in 1930 and her thesis advisor was Max Born. In that same year, she married a young American chemist, Joseph Mayer, an assistant of James Franck. They go to the United States. Joseph becomes professor in the John Hopkins University in Baltimore, and Maria works without a salary, collaborating with her husband in theoretical chemistry. In 1939, the couple moves to New York where Joseph Mayer obtains a position in the University of Columbia. Maria then works with Harold Urey, the chemist who discovered deuterium.[2] She is helped and encouraged by Fermi. It is at this time that she publishes her work on elements which did not yet bear the name "actinides." During the war, she joins the Manhattan Project. She works with Urey on the separation of uranium 235. In 1946, she obtains her first position as professor in the University of Chicago. In 1960, she moves to California with her husband. She dies in 1972.

In 1948, Maria Goeppert-Mayer takes a second look at the data which had become more numerous and precise: she examines the masses of nuclei, their spins, and even their magnetic moments. The data acts both as a constraint to which models of nuclei have to comply and as a source of inspiration. Adopting the method of Elsasser, Maria Goeppert-Mayer also notices disturbing regularities. Nuclei possessing 20, 50, or 82 protons or 20, 50, 82, or 126 neutrons have remarkable properties [82]:

- Nuclei which have 20 or 50 protons have numerous isotopes:

  - Calcium (20 protons) has five isotopes, and their number of neutrons ranges from 20 to 28.
  - Tin (50 protons) is the element which has the largest number of isotopes, namely, 10.

Thus, 20 protons can form stable nuclei with 20 to 28 neutrons, and 50 protons can form stable nuclei with 62 to 74 neutrons:

---

[1] See p. 428.
[2] See p. 249.

- The same phenomenon occurs when the number of neutrons is equal to 20, 50, 82, or 126.
- The probability of absorbing a neutron in a collision is smaller than average for nuclei having 20, 50, 82, or 126 neutrons.
- A new argument: when uranium 235 undergoes fission, it does not generally split into two equal pieces. Maria Goeppert-Mayer notices that the most probable pieces have 50 or 82 neutrons.
- Maria Goeppert-Mayer also notices, as Guggenheimer had done, a discontinuity in the slow increase of binding energy when nuclei have 82 neutrons.

Bohr could well claim that nucleons could not form shells in nuclei. And yet the data assembled by Maria Goeppert-Mayer seemed strong evidence to the contrary: a "shell" appeared to become closed, meaning that it would become filled with either neutrons or protons, or both, whenever the number of neutrons or protons reached 20, 50, or 82 and when the number of neutrons reached 126 (no known nucleus possessed 126 protons).

Maria Goeppert-Mayer did not generalize, but she pointed out that the calculations made by Wigner and Feenberg[1] in 1937 had indeed predicted *a shell closure* when the number of protons or neutrons was equal to 20. These calculations were based on the assumption that each nucleon could be treated as a particle which moves independently of the others and that it only feels the presence of the other nucleons by way of the average potential which they create. A careful study of the binding energies was made by Walter Barkas [83] in 1939. They seemed to confirm the calculations.

The trouble was that the same calculations gave some wrong results for nuclei with masses larger than about 40, that is, beyond calcium. So what was wrong? What was so special about the numbers 20, 50, 82, and 126? To Eugene Wigner, they had "magical virtues." The term "magic numbers" prevailed!

The Spin-Orbit Interaction

Such was the state of affairs when one fine day Maria Goeppert-Mayer had a discussion with Fermi, which she recalled later:

> *At that time Enrico Fermi had become interested in the magic numbers. I had the great privilege of working with him, not only at the beginning, but also later. One day as Fermi was leaving my office he asked: "Is there any indication of spin-orbit coupling?" Only if one had lived with the data as long as I could one immediately answer: "Yes, of course and that will explain everything." Fermi was skeptical, and left me with my numerology [. . .].*

---

[1] See p. 276.

In ten minutes the magic numbers were explained, and after a week, when I had written up the other consequences carefully, Fermi was no longer skeptical. He even taught it in his class in nuclear physics [84].

What is the *spin-orbit coupling*? It describes a simple effect. The proton (as well as the neutron) has an intrinsic spin, which one can visualize as a rotating top. Consider now a proton describing an orbit in the nucleus. Fermi was asking if the nucleus was more stable when the proton spun around itself in the same direction as it rotated along its orbit or in the opposite direction. The difference could be caused by a so-called spin-orbit coupling which had already been observed in the orbits of electrons in the atom.

One may be surprised to see classical descriptions of "orbiting nucleons" constantly used by nuclear physicists. It is not because they fail to realize that the classical description is wrong but because they simply use such terms to describe *the corresponding quantum state*. The term *orbit* is often used to describe the quantum wavefunction of the particle.

Maria Goeppert-Mayer set to work, and she quickly discovered that the spin-orbit interaction had miraculous effects: it predicted that nuclei possessing 20, 50, or 82 protons and/or 20, 50, 82, and 126 neutrons were more strongly bound and had therefore a more solid structure than other nuclei. In the letter she sent on February 4, 1949, to the *Physical Review*, she uses the terms *magic numbers* [85].

Independently, a German physicist, Hans Jensen, a year younger than she, obtained the same result.

Johannes Hans Daniel Jensen

Hans Jensen was born in 1907 in Hamburg and studied there. In 1941, he became professor of theoretical physics in Hannover, and in 1949, he became professor at the University of Heidelberg. Two of his colleagues, Otto Haxel in Göttingen and Hans Suess, a nuclear chemist in Hamburg [86], drew his attention to the disturbing regularities and to the magic numbers of neutrons or of protons which make nuclei particularly stable. But, just like Maria Goeppert-Mayer, he was unable to account for these numbers. He was invited to give a talk in Copenhagen, and he discovered there her first paper [82], in which she noted, as he did, the anomalies which occur for certain numbers of neutrons or protons, without however giving an explanation. This gave him the courage to speak about it at a conference which was attended by Niels Bohr:

I shall never forget that afternoon. Niels Bohr listened very attentively and threw in questions which became more and more lively. Once he remarked: "But that is not in Mrs. Mayer's paper!"; evidently Bohr had already carefully read and pondered about her work. The seminar turned into a long and lively discussion. I was very much impressed by the intensity with which Niels Bohr received, weighted, and compared these empirical facts, facts that did not at all fit into his own picture of nuclear structure. From that hour on I began to consider seriously the possibility of a "demagification" of the "magic numbers" [87,88].

Jensen set himself the task of "demagifying" the magic numbers, that is, to release them from their incomprehensible magic spell and to give them a physical interpretation. When he returned to Heidelberg, he discussed this with Otto Haxel and Hans Suess. They had the idea of trying out a strong spin-orbit interaction. Just as Maria Goeppert-Mayer in Chicago did, they realized that it could explain the famous "magic numbers." Their first paper was refused by a journal for the excellent reason that it "did not really deal with physics, but it only played with numbers." Jensen, Haxel, and Suess then sent two short papers to *Naturwissenschaften* [89,90], and they wrote a full paper in English which they sent to Victor Weisskopf, a Jewish Austrian physicist who had emigrated to the United States in 1937. The paper, dated April 1949, was published on June 1, 1949, as a letter to the *Physical Review* [91]. Jensen, Haxel, and Suess could not have known the paper of Maria Goeppert-Mayer, dated February 4, but which was only published on June 15, 1949. A year later, they published a more detailed paper [92] in *Zeitschrift für Physik*, and in 1952, they wrote a review of the very impressive results of the new shell model [93].

A Paradoxical Model

The publications of Maria Goeppert-Mayer and of Haxel, Jensen, and Suess had an explosive effect. No one so far had expected such a strong spin-orbit interaction, possibly because Bethe, in his famous 1936 paper, considered almost as a bible by nuclear physicists, had suggested that the spin-orbit interaction was probably quite weak [94].

However, several physicists were convinced that the nucleus had a shell structure because nuclei containing 2, 8, 10, 20, 50, and 82 neutrons or protons and 126 neutrons displayed special properties. But they were unable to explain the "magic numbers" beyond 20. It must be admitted that the state of nucleons in a nucleus differs considerably from that of electrons in an atom, where they form shells around the central positively charged nucleus. The shells bear quantum numbers. The principal quantum number determines the size of the electron orbit (to speak in classical terms) and therefore also the sequence of the shells, formed by a set of electron orbits. But in the nucleus, the nucleons are more closely packed and the principal quantum number no longer determines the sequence of the shells in the presence of a strong *spin-orbit interaction.* Shells are still formed by sets of orbits. But their succession is altered by the spin-orbit interaction. That is what made it so difficult to interpret the magic numbers.

However, the arguments put forth by Bohr against a shell structure of the nucleus had not been proved wrong. They were simply ignored, a somewhat embarrassing situation. Without really understanding why, experiments seemed to indicate that the nucleus has a shell structure, as the atom does, in spite of their big differences. In a first approximation, the nucleons appear to move independently of one another, each one interacting with the others by means of the average potential created by the others. This indicates that each proton and neutron can travel over a distance comparable to the size of the nucleus without colliding with another nucleon. How can that be? Bohr's argument rests on the estimate that the average distance between collisions is considerably smaller than the nuclear size. Where is the error? In the nuclear physics course, which he was giving in the University of Chicago, from January to June 1949, and the lectures notes of which were written up and published by three of his students, Fermi proposed an interesting explanation of why nucleons might move almost freely in the nucleus. Imagine the nucleus in its ground state, that is, in its lowest energy state. In the shell model, each proton (or neutron) is described by four quantum numbers, and each proton must differ from another by at least one quantum number. If the proton collides with another nucleon, the only thing which quantum mechanics allows it to do is to change its quantum numbers. But its new quantum numbers must be different from the quantum numbers of the other nucleons present in the nucleus. This means that the colliding nucleon can only be excited into orbits which are not already occupied by another nucleon. The probability it should do so is reduced by the fact that the empty orbits have a higher energy, so that it remains in its original orbit. It appears to pass through the collisions without noticing the other nucleons [95]! It is the Pauli exclusion principle which is acting here: a purely quantum effect with no classical analogue.[1]

Fermi's argument could explain why the mean free path of nucleons in a nucleus may be considerably larger than what Bohr had estimated, but it remains qualitative. It is worth noting that the argument put forth by Fermi in his course was given *before* he knew of the large spin-orbit interaction discovered by Maria Goeppert-Mayer and Hans Jensen. Fermi's argument was later restated by Weisskopf [96]. Today, the shell model is still the basis of our understanding of the structure of the nucleus.

Meanwhile, and this was perhaps its best justification, this simple model of the nucleus, in which the neutrons and the protons fill successive shells, was able to account for a considerable amount of experimental observations. For example:

- *The problem of isomers.* We already mentioned, in another context, the existence of certain nuclei which could remain for a lapse of time of the order of seconds or even days, in an excited state, which is a state with an energy higher than the energy of their groundstate.[2] In almost all cases, a nucleus which is in an excited state makes a quick transition to its ground state by emitting a photon (γ-decay).

---

[1]This shows how careful one must be when using classical terms to describe quantum systems. The Pauli principle, which forbids two identical particles (e.g., two neutrons or two protons) to be in the same state, or, equivalently, to bear the same quantum numbers, has no analogue in classical mechanics. Classical mechanics cannot even account for the experimentally verified fact that there exist identical composite objects, such as, for example, identical atoms. Classical electron orbits of two atoms would never be exactly the same!

[2]See p. 353.

Weizsäcker had explained the abnormally long lifetime of the excited states of some nuclei by the fact that there was a large difference between the spins of the excited state and of the ground state. But why did some states have such a large spin? The spin is an intrinsic angular momentum of the nucleus. It results from adding the intrinsic spins of the nucleons to the angular momentum of their motion in the nucleus. If this motion were chaotic, how could they acquire a large angular momentum? Furthermore, it had been noted that nuclei with large spin occur near closed shells, that is, in nuclei with proton or neutron numbers close to the magic numbers. The shell model could explain this, as Feenberg noted [97] in 1948, before the publications of Maria Goeppert-Mayer and Hans Jensen.

- *The magnetic moments of nuclei.* As early as 1937, the German physicist Theodor Schmidt [98] noticed a relation between the magnetic moments of nuclei and their spin.[1] Nuclei which have an *odd number of protons* and an even number of neutrons have a magnetic moment which increases with their spin, although there are important variations. It seemed natural to assume that their magnetic moment was caused by the rotation of their protons, which create a magnetic field. Nuclei which have an even numbers of protons but an *odd number of neutrons* have magnetic moments which do not appear to depend on their spin. Schmidt tried to explain this in terms of a simple model, which gave a qualitatively correct result: he assumed that the magnetic moment of a nucleus was due to the motion of the last odd nucleon, the other nucleons forming a spinless core with a zero magnetic moment. This implied that the last odd nucleon could move in an orbit which is independent of the other nucleons. This was, in fact, an anticipation of the shell model. There again, Feenberg published a paper in the same issue of the *Physical Review* as the paper of Maria Goeppert-Mayer. He reviewed the evidence in favor of the shell model et developed further the calculations of Schmidt [99]. That same issue of *Physical Review* contained a paper on the same subject by Lothar Nordheim [100], a German Jewish physicist who emigrated to the United States after spending some time in France in 1933.

The shell model was an immediate success and became the subject of an increasing number of publications. Hans Jensen was invited to the United States where he worked with Maria Goeppert-Mayer. Together, they wrote a book, *The Elementary Theory of Nuclear Shell Structure*, which became a standard reference for all nuclear physicists during decades [101]. The shell model was winning in spite of it being paradoxical and lacking solid justification. The situation was similar to that of the Bohr atom in 1913. How could it be? The experimental facts had spoken, leaving physicists puzzled.

---

[1]Nuclei, which have an even number of protons and an even number of neutrons, have a spin zero and a zero magnetic moment.

Elastic Scattering and the "Optical Model"

The mean free path of nucleons in the nucleus appears to be surprisingly large. Elastic scattering on nuclei is accounted for by a very simple model analogous to a light wave incident on a semi opaque crystal. After certain refinements and the use of the new computers, this "optical model" becomes most successful.

The mean free path of a nucleon in the nucleus, that is, the average distance a nucleon travels in the nucleus before colliding with another nucleon, was a problem which needed to be understood if the shell model was to be explained. Was the mean free path smaller than the nuclear size, as Bohr had surmised, or was it large enough to justify the shell model of Maria Goeppert-Mayer and Hans Jensen? The problem was tackled from another angle by a team in Berkeley which included the Canadian physicist Leslie Cook, the American Edwin McMillan, and two students, Jack Peterson and Duane Sewell. They measured the absorption of neutrons by about 15 nuclei ranging from hydrogen to uranium [102]. Robert Serber, the lab theoretician, showed that their results could be understood if one assumed that nuclei were partially "transparent" to neutrons which had a mean free path of the order of the nuclear radius [103], that is, 4–5 femtometers.[1] This meant that about half of the neutrons could pass through the nucleus without colliding, and it contradicted the Bohr model of the compound nucleus. On the contrary, it favored the shell model of Maria Goeppert-Mayer and Hans Jensen, which assumed that nucleons could move freely inside the nucleus, independently of the other nucleons.

However, a doubt remained: the measurements were made for neutrons with an energy of 90 MeV which is considerably higher than the energy, roughly 20 MeV, of either protons or neutrons in the nucleus. It was not obvious that the mean free path of neutrons in the nucleus was the same as that of high-energy neutrons.

Born in 1909, Robert Serber studied in Wisconsin. He was a pupil and became a friend of Robert Oppenheimer, who assigned to him important work in the Manhattan Project. At the end of the war, he became professor in Berkeley. Serber was a respected theorist and known to maintain good contact with experimentalists, whose work he always followed most attentively.

To calculate the absorption of neutrons in the nucleus, Serber used an idea which Bethe had proposed in 1940. Bethe introduced an absorption coefficient which appeared in the Schrödinger equation for the neutron as an imaginary potential [104]. He compared this to the absorption of light which penetrates a refringent substance such as, for example, glass (or our cornea). Lenses of corrective

---

[1] A femtometer (fm) is equal to 10^{-15} m or 10^{-13} cm. Nuclear sizes vary from 1 fm for the proton, the nucleus of hydrogen, to about 7 fm for the radius of the lead nucleus.

sunglasses are a typical example. Light is deviated by the lenses (so as to make a correction) and, at the same time, it is partially absorbed. In optics, the refraction is taken into account by an index of refraction and the absorption by an *imaginary index of refraction*.[1] Bethe obtained an a priori unexpected result: the probability of a neutron getting absorbed is larger if the absorption increases *gradually* when the neutron penetrates into the refringent nucleus.

The Nucleus Is Like a Cloudy Crystal Ball

Two years later, while working with two students in Berkeley, Sidney Fernbach and Theodore Taylor, Serber described in detail the collision of a neutron with a nucleus [105]. The interaction of the neutron with the nucleus is described in terms of an attractive potential$^\diamond$. It has the shape of a sphere inside which, as Bethe had formulated, the potential has an imaginary absorptive part. The particle thus has a probability of being absorbed, in which case it does not emerge from the nucleus. This is similar to a light wave passing through a semi-transparent (or, equivalently, semiopaque) sphere: upon entering the sphere, the light wave is refracted, meaning deviated, as it is upon emerging from the sphere. The model of Serber, Fernbach and Taylor accounts globally for the observed neutron scattering and absorption. The semitransparency of the nucleus thus appears to be naturally derived from the data. The model was branded the *cloudy crystal ball* model and also simply as the "optical model" because of the analogy put forth by Bethe.

"Optical" Attempts

So far, and for some time yet, only "elastic scattering" was observed, by measuring the average widening of a beam of particles passing through a sheet of matter. One could only make a rough guess at how many particles were deviated at a given angle [106]. An interesting information was however obtained: the nucleus is partially transparent to neutrons. But without more precise data, it was difficult to say more.

However, physicists gradually improved the experiments and became able to measure more and more accurately how many neutrons or protons are scattered elastically at a given angle. This is called the *angular distribution* of the scattered particle. At first, the elastic scattering of 340-MeV protons was observed in Berkeley [107] and analyzed in terms of the optical model with moderate success [108]. Then came the measurement of elastic scattering of 18-MeV protons,

---

[1]The word "imaginary" is meant here in the mathematical sense where a complex number has a real and imaginary part.

by Piet Gugelot [109], who was able to display a real angular distribution, although only with angles which increased in steps of 30°. The data was analyzed by David Saxon, a theoretician in the University of California in Los Angeles, and his student Robert Le Levier. The result was encouraging although still only qualitative [110]. Le Levier and Saxon based their calculations on an optical model similar to the one used by Serber.

Much more precise measurements of elastic scattering of about 20-MeV protons were published successively by Bernard Cohen and Rodger Neidigh [111] in Oak Ridge[1] and by Irving Dayton in Princeton [112]. It became obvious that the calculations of Le Levier and Saxon did not yield satisfactory results nor did those of the two physicists in Princeton, David Chase and Fritz Rohrlich, who included the effect of the Coulomb repulsion on the protons [113].

The Woods-Saxon "Optical" Potential

On July 15, 1954, the *Physical Review* publishes a letter [114] signed by two physicists from Los Angeles, David Saxon and his student Roger Woods (who did his thesis on the subject). Their work is based on a hardly original idea, and yet it changed completely the situation. Woods and Saxon represent the interaction between a proton and the nucleus by a potential well$^\diamond$ with a *diffuse surface*, so that the attraction felt by the proton increases gradually (instead of abruptly) over a distance of about 1 femtometer, which is a fraction of the nuclear radius. It is an "optical" potential similar to the one of Bethe [104] in that it contains both a real and an absorptive imaginary potential (imaginary in the mathematical sense) which both increase gradually and remain constant inside the nucleus. To model the nuclear surface, Woods and Saxon use a simple formula which has become so famous that we quote it here:

$$U(r) = \frac{V + i\,W}{1 + e^{\frac{r - r_0}{a}}}$$

In this expression, V represents the depth of the real potential in the center of the nucleus, W represents the imaginary absorptive part of the potential, r is the distance separating the proton from the center of the nucleus, r_0 is the radius of the nucleus, and a the surface thickness, that is, the distance over which the potential grows.

Why were Woods and Saxon the first to calculate the scattering of protons in such a potential? It is because *they were the first to use a computer for the calculation*. Woods wrote the computer program. Until then calculations were made by hand on paper with the possible help of office adding machines, slide rules,

---

[1]The Oak Ridge National Laboratory in Oak Ridge, Tennessee, had been created during the war, in order to separate uranium 235, required to make an atomic bomb.

logarithmic tables, and tables of other mathematical functions. To study scattering, the Schrödinger equation had to be solved analytically in terms of functions listed in numerical tables (such as sines, cosines, exponentials, and a few other functions). Every physicist had a thick book on his desk, containing tables of frequently used mathematical functions. The solution of the Schrödinger equation for a proton scattered by a potential such as the Woods-Saxon potential above was impossible without a computer. Worse, the physicist is not interested in only one solution. He wishes to see how the solution depends on the parameters of the potential. This would be an easy task if an analytical solution were known, but if the solution is only numerical, one needs to repeat the calculation with different sets of parameters. Dozens of solutions need to be performed. Woods and Saxon state that it took their computer 15–20 min to calculate an angular distribution of protons with a given set of parameters (today's computers could do the job in a fraction of a second). They were able to make the calculation with various sets of parameters.

The results obtained were most impressive and fitted well the experimental data. Woods and Saxon showed that it is essential to assume a diffuse nuclear surface, which in their potential has a thickness equal to the parameter a.

The Computer: A Decisive Instrument

The first computer was conceived by the mathematician John von Neumann and by two American physicists, J. Presper Eckert and John Mauchly, who analyzed in 1945 the limitations of the first electronic *calculator*, the ENIAC (*Electronic Numerical Integrator And Computer*) the construction of which had secretly begun during the war at the *Moore School of Electrical Engineering* of the university of Pennsylvania [115]. On June 30, 1945, after discussing with Eckert and Mauchly, von Neumann wrote a famous text entitled *First Draft of a Report on the EDVAC*, in which he proposed the construction of a real *computer*, called the *Electronic Discrete Variable Computer* (EDVAC). It became operational in 1952 by which time von Neumann had left the *Moore School*. Eckert and Mauchly set up their own company because they believed computers had a future. In 1951, after certain difficulties, they began producing the first commercial computer, the UNIVAC (*Universal Automatic Computer*). At the *Institute for Advanced Study* in Princeton, von Neumann pursued his own research. Other important research was pursued in Great Britain by the mathematician Alan Turing, who, during the war, had succeeded to decipher messages encrypted by the Germans with their famous machine *Enigma*.

Woods and Saxon used a SWAC computer, an acronym for the *Standards West Automatic Computer*. It belonged to the first generation of computers. It began running on August 17, 1950. It was the first computer built on the West Coast of the United States at the *Institute for Numerical Analysis* in Los Angeles, an institute which was part of the *National Bureau of Standards*.[1]

What distinguishes a real computer from a calculator, be it as powerful as the ENIAC, is the following: in a calculator, the successive operations were read from punched cards or paper tape, or by hardwired connections. There was no internal memory, hence no stored program. The idea of von Neumann, Eckert, and Mauchly was to write the instructions to process data into the memory of the computer and to endow it with a system of internal commands which control the succession of mathematical operations, in other words, to write a *computer program.*

---

[1]The *National Bureau of Standards* is a governmental agency which, since its foundation in 1901, has the role of "working with industry in order to develop and apply technology, measurements and standards" in the interest of the nation. In 1988, it became the *National Institute of Standards and Technology* (NIST).

Direct Nuclear Reactions

Physicists discover that nuclear reactions, which they no longer call transmutations, do not always proceed by a preliminary formation of an intermediate compound nucleus, as predicted by the theory of Niels Bohr. Such reactions yield precious information on the structure of the nucleus. The computer becomes an indispensable tool of the physicist.

Ever since the first observation of a nuclear reaction, made by Rutherford in 1919, physicists have tried to learn more about the nucleus by bombarding it with all sorts of projectiles and by observing the nuclear reactions thus produced.

The physical process giving rise to the nuclear reaction remained however to be understood. In 1936, Bohr proposed a description of a nuclear reaction[1] which was immediately adopted by nuclear physicists. During 10 years, they believed that they could thus explain all the observed nuclear reactions, which they first called "transmutation reactions" because they involved a transformation of the nucleus and therefore a transmutation of the corresponding element. In the model proposed by Bohr, the incident particle is first absorbed by the nucleus, thereby forming a so-called "compound nucleus" which disintegrates later, after a time which appears infinitely short to us but which is much longer than the time required for the nucleons to rearrange themselves in the nucleus. In that lapse of time, the nucleons can move from one end of the nucleus to another hundreds, even thousands of times. Their trajectories become random and governed by laws of probability. It is easy to understand why, after this time and under such conditions, particles are emitted from the compound nucleus in directions which are no longer determined by the direction of the incident particles.

However, in a paper he wrote with Rudolf Peierls and George Placzek [116], Bohr also mentioned another possibility: when the incident particles have a large velocity, the energy, which they exchange with the nucleus during the collision, could be localized close to the point of impact, in which case they could eject a nucleon from the nucleus without a preliminary formation of a compound nucleus.[2] The new experimental results soon confirmed this scenario. They showed that in certain cases nuclear reactions can occur during which, for example, a proton or a neutron is deposited in the surface of a nucleus while only slightly perturbing the other nucleons. Another nucleus can be formed in this manner.

---

[1] See p. 336.
[2] See p. 336.

Such reactions were called *direct reactions*. They made it possible to learn how an extra nucleon can be added to the nucleus, into which shell it falls, and so forth. Direct reactions will constitute a large body of data upon which theory will feed.

The Stripping of a Deuteron

Berkeley Shows How to "Undress" a Deuteron

In 1947, three physicists in Berkeley, A. Helmholtz, Edwin McMillan, and Duane Sewell, measure the neutrons emitted by eight elements which are bombarded by deuterons [117]. The bombarded elements range from beryllium to uranium, and the deuterons are incident with an energy of 190 MeV. Recall that the deuteron is a nucleus composed of a proton loosely bound to a neutron. An unexpected result is observed: the neutrons are ejected mainly in the direction of the incident deuterons and deviate only little from this direction. For the Berkeley physicists, this suggests that the nuclear reaction does not proceed as in the model proposed by Bohr because, if it did, the neutrons would be emitted in all possible directions. Rather, the experiments suggest that when the deuteron approaches the nuclear target, it splits into its components: the proton is captured whereas the neutron continues moving in the same direction, without being significantly perturbed by its small (roughly 2 MeV) binding energy to the proton. They publish their results in the *Physical Review* and, in the same issue of that journal, Robert Serber gives a theoretical interpretation of this nuclear reaction which he calls *stripping* [118], because the target nucleus strips off a nucleon from the incident particle (a deuteron in this case). The stripped deuteron ends up naked having lost its proton partner and emerging as a simple neutron. The calculation of Serber accounts successfully for both the observed cross section, which is the probability that such a process should occur, and for the fact that the neutrons are emitted in the direction of the incident deuterons.

One can understand the model proposed by Serber as follows: the velocity of the incident deuteron is much larger than the relative velocity of the neutron and the proton in the incident deuteron so that when the deuteron passes close to the nucleus its proton can be captured in a span of time so short that the neutron has no time to react. The neutron simply continues its forward motion, and *in this case, no compound nucleus is formed*. The proton is captured *directly*. The deuteron is nonetheless slightly deviated as it approaches the target nucleus, and this deviates slightly the trajectory of the emerging neutron. The process described by Serber holds only if the deuterons are incident with a high velocity. But it was later also successfully applied by a Ph.D. student of Weisskopf in MIT, David Peaslee at much lower energies, ranging from 2–15 MeV [119]. However, these experiments and calculations concerned neutrons emitted at all angles and at all possible energies.

The Angular Distributions Measured in Birmingham and the Butler Theory

The cyclotron, built by James Chadwick in Liverpool, delivered a beam of 8-MeV deuterons. Three physicists, Hannah Burrows, William Gibson, and Josef Rotblat, use this beam to bombard oxygen 16 and to detect with great precision the emitted protons [120]. This time, the neutron is captured by the oxygen 16 nucleus, which becomes oxygen 17. The latter can be formed in different configurations which have different energies: a ground state, which has the lowest energy, and so-called "excited" states. The energies of the oxygen 17 nucleus can be deduced from the observed energies of the emitted protons which carry away the energy of the incident deuteron which has been stripped of its neutron, deposited into one of the states of oxygen 17. They used the track left on a photographic plate[1] to measure both the energy of the protons and the angle at which they are scattered: the track left on the photographic plate confirms that it is the track of a proton (as witnessed by the thickness of the track), and it also indicates the direction in which the proton is emitted after the nuclear reaction. The results are surprising: the *angular distribution* of the emitted protons, that is, the distribution of the angles at which they are emitted, depends on whether the oxygen 17 nucleus is formed in its ground state or in one of its excited states. Furthermore, the angular distributions are oscillatory. When the oxygen 17 nucleus is formed in its ground state, a peak of the angular distribution is observed at 34°, a minimum at about 85° and a smaller peak at about 120°. If a large photographic plate had been placed facing the emerging protons, they would have produced a track of concentric circles. It is a young Australian physicist who makes the first theory of the reaction in which a neutron is stripped off the deuteron and who calculates the angular distribution of the emerging protons [121].

Born in 1926, Stuart Butler obtained a scholarship to prepare his Ph.D. in theoretical physics under the direction of Rudolf Peierls. Butler makes an important discovery: the angular distribution of the emitted protons *is characteristic of the angular momentum of the newly formed oxygen 17 nucleus*. In its ground state, the target oxygen 16 nucleus has a zero intrinsic angular momentum, or spin, as do all *even–even* nuclei (which have even numbers of neutrons and protons). According to the simple shell model, the incident deuteron skims past the oxygen 16 target, and its neutron is stripped off and captured into one of the empty shell model orbits of the oxygen 16 nucleus. In this orbit, the neutron acquires an angular momentum which it must remove from the incident deuteron, and therefore, the remaining proton must modify its trajectory accordingly. Butler immediately applies his theory to the experiments of Hannah Burrows, Gibson, and Rotblat mentioned above as well as to the experiments performed by two other British physicists, John Holt et C. T. Young, who had measured somewhat earlier the angular distributions of protons emitted when aluminum is bombarded by deuterons [122].

---

[1]See p. 423.

This development had far reaching consequences: by bombarding an even–even nucleus, one could determine the quantum numbers of the states of the neighboring nucleus, which was the isotope with one extra neutron. Soon after, Butler gave a more detailed account of his calculations [123]. As almost always in physics, he was obliged to make approximations in order to complete his calculation: he neglected the Coulomb force acting between the target nucleus and the incident deuteron (as well as the emerging proton), and he also assumed that the protons followed straight line trajectories, which in quantum mechanics are represented by plane waves propagating in straight lines as waves do on the surface of the sea. He also assumed that the transition between the initial state, consisting of a deuteron plus the target nucleus, to the final state, consisting of a proton and the newly formed nucleus, take place instantaneously. He finally assumed that the deuterons and the protons do not penetrate the nucleus.

The Success and Development of the Theory of Butler

The theory of Butler opened a path towards the determination of the states of a nucleus, foremost its spin, which is formed after a stripping nuclear reaction. It became most successful. For example, Joseph Rotblat used it to determine the spin of two states of the carbon 13 nucleus which is a carbon 12 nucleus with an added neutron [124]. These were the first steps towards *nuclear spectroscopy*.

At the same time, several theoretical works studied the approximations made by Butler, with the hope of improving them. Two American physicists, Paul Daitch and Bruce French [125], showed that the approximation made by Butler was in fact equivalent to the well-known approximation made by Max Born in his work on the collision of an electron with a nucleus.[1] Born considered the incoming and outgoing electron trajectories to be straight lines and the interaction with the nucleus to be a small perturbation. The approximation became known as the *Born approximation*.

Two young French physicists, Jules Horowitz and Albert Messiah, also tackle the problem, and they show that the approximation can be improved by assuming that the trajectories are not straight lines but trajectories which are bent by the target nucleus. In quantum mechanical language, this means that the deuteron and the emerging proton are not described by plane waves but by a wave which is distorted by the target nucleus. They assume that the deuteron does not penetrate inside the nucleus. They also neglect the electric Coulomb repulsion [126,127]. The advantage of their theory lies in the fact that they are able to clearly identify the approximations which are made, and this makes it easy, so to speak, to improve them, as we shall see shortly. Two American physicists, Norman Francis and Kenneth Watson, developed a similar theory. They improved the Born approximation by replacing the straight line trajectories, which are plane waves in the language of quantum

---

[1]See p. 140.

mechanics, by waves which they calculated in an *optical potential* which in this case was made completely absorptive in the nuclear interior. They proposed to call their approximation an "optical model of stripping reactions." [128] But, just like Horowitz and Messiah, they do not make numerical calculations.

The idea of Horowitz and Messiah is pursued further by an American physicist, William Tobocman, who proposes to modify the Born approximation by including the Coulomb interaction in addition to the nuclear optical potential. He forbids the access to the interior of the nucleus by a somewhat different mathematical procedure [129]. He adds that the calculation requires a computer, and he immediately proceeds to use one with the help of Malvin Kalos [130]. Together, they make the first complete calculation of a stripping reaction using the improved Born approximation. The incident and emerging particles are represented by waves which are distorted in the neighborhood of the nucleus. They account very well for the observed stripping reactions and not only qualitatively as the initial calculations of Butler did. Tobocman and Kalos used the first computer constructed by IBM for the department of numerical calculations in the University of Cornell. It was a card-programmed computer (CPC), where both data and instructions were punched on cards. The approximation was soon referred to as the *distorted-wave Born approximation* (DWBA), a name first used by Thomas [131] and rapidly adopted by everyone.

The DWBA and the Computer: Indissoluble Partners

Why was it necessary to wait until 1954 before such calculations were performed? The idea had already been formulated in the 1930s in order to calculate the collision of an electron with an atom and the subsequent stripping of one of its electrons. For such reactions, the original Born approximation, which used straight line trajectories, gave worse results than the method of "distorted waves" which took into account the distortion of the trajectories by the Coulomb force exerted by the electrically charged nucleus [132].

In the second of his famous 1936–1937 papers, Bethe had written:

It is, in the case of atoms, far superior to the often used Born approximation in which the average interaction between atom and incident electron is neglected. However, the method of the distorted wave functions is not at all applicable in nuclear physics [133].

Bethe presented two arguments against this method:

- The argument presented by Bohr against the idea of an average mean field: indeed, a particle penetrating the nucleus is subject to many collisions and its trajectory is chaotic.
- The low energy of the incident particles prevented one to consider the interaction between the nucleus and the particle to be a weak perturbation of its trajectory.

The first argument lost much of its power when the shell model of the nucleus proved to be so successful. The model describes the nucleus as a configuration of nucleons filling successive shells, each one moving independently of the others. The second argument was no longer valid in 1950, as compared to 1937, because new accelerators were now furnishing particles, such as deuterons, with far greater energy.

But what made the DWBA so useful was the *possibility of performing the calculation*, thanks to computers. The aim of physical theory is to discover equations which describe physically observed quantities, such as the position of a planet at a given time. The theory requires both basic principles and solvable equations. An equation which is completely unsolvable is of little use. Faced with an equation which is too difficult or impossible to solve exactly, the art of the physicist is to discover a good approximation to the solution. Before the 1950s means were sought to perform calculations by hand, using only paper, a pencil, and numerical tables of common functions, such as logarithms, sines, cosines, exponentials, and so forth. The appearance of computers suddenly opened the way to solving equations numerically and to perform much more difficult calculations in a reasonable time. The DWBA, together with the optical potential, is the first example of a theory which could not have been be verified without a computer.

Computers grew quickly more powerful. For example, a home computer today is at least one million times faster (in the number of operations performed per second); it has a "memory" at least a million times larger than the SWAC which was considerably more powerful than the ENIAC!

Today, the computer has become as important for the physicist as water or electricity. He uses the computer to make calculations, to analyze experimental data, to correspond and maintain contact with the whole world, to write his papers, prepare his talks, and so forth.

Direct Reactions and Reactions Which Proceed Though the Formation of a Compound Nucleus

The stripping of a deuteron is the first example of what was called a *direct* reaction, as opposed to reactions which proceed in two steps: the formation of a compound nucleus followed by the decay of the compound nucleus. The two processes are very different. A deuteron which is stripped loses its neutron (or proton) upon skimming past the nucleus. The process takes the time required to travel a distance comparable to the size of the nucleus. In contrast, it takes about a thousand times more time to form a compound nucleus and for the latter to decay, although the time involved (it has been measured to be about 10^{-18} s) may seem tiny to us.

When a particle (a proton, neutron, deuteron, α-particle,...) collides with a nucleus, the collision can be either head on or peripheral. Each of these possibilities occurs with a certain probability. Direct reactions occur in peripheral collisions, and a compound nucleus is formed in head-on collisions. Since such collisions last for very different times, their probabilities can be evaluated independently. Usually, a direct reaction occurs when the emitted particle is observed to move in the same direction as the incident particle, and a compound nucleus has been formed if the particle emerges at a very different angle. Several direct reactions were progressively discovered:

- *Inelastic scattering* during which the scattered particle remains the same while inducing a modification of the internal structure of the target nucleus, which is left in an excited state after the collision. An *excited state* of the nucleus is a higher energy state in which the neutrons and the protons have a different configuration than in the ground state, that is, the state which has the lowest energy. A nucleus in an excited state decays to its ground state within a more or less short time, while emitting the energy it had to absorb in order to become excited and which had been transmitted by the incident particle. It decays as an unstable scaffolding would do until all the pieces have nowhere lower to fall. When the nucleus reaches its ground state, it becomes stable because there exists no lower energy state it can decay into.
- There exist different kinds of stripping reactions: for example, an incident α-particle (which is the nucleus of the helium 4 atom, consisting of two protons and two neutrons) can lose part of its nucleons during a collision.
- Inverse reactions also occur: a proton passing by a nucleus can *pick up* a neutron stolen from the nucleus and emerge as a deuteron. Many other possibilities can happen. For example, a deuteron can pick up a proton and emerge as a helium 3 nucleus, which is an isotope of helium 4.

In the ensuing years, the study of nuclear reactions became an important subject because the information extracted from the observed reactions revealed the interior structure of the nucleus, but only insofar as the theory of the nuclear reaction was well understood.

A Collective Behavior

> *The nucleons in a nucleus can move together coherently. The nucleus displays vibrations in which the protons and neutrons move in opposite directions. The shape of some nuclei is deformed and vibrations of the nuclear shapes are also observed.*

While the shell model was becoming successful in explaining an increasing amount of experimental observations, further data was revealing the existence of "collective" excitations of nuclei which were understood not as individual excitations of single nucleons but as vibrations of the average potential in which the nucleons were moving.

Photonuclear Reactions

For a long time, physicists had bombarded nuclei with γ-rays in order to "excite" them, that is, in order to transmit sufficient energy to modify their ground state configuration and to subsequently observe how they decay back into their ground state. Among the first to do so systematically was Walther Bothe, who was known for having invented the method of coincidence counting[1] and for the experiments which eventually led to the discovery of neutrons.[2] As early as 1934, together with W. Horn and Wolfgang Gentner who were students in his laboratory in Heidelberg, Bothe studied how γ-rays with sufficiently high energy were deviated by aluminum and lead nuclei [134, 135]. This experiment was similar to the one made by Compton[3] and to the one which he had made with Hans Geiger,[4] except for the fact that the γ-rays had a considerably higher energy. A similar experiment had been performed by James Chadwick who observed how the deuterium nucleus broke up into a neutron and a proton after being hit by a γ-ray. This enabled him to make a precise measurement of the mass of the neutron.[5]

For such experiments, the source of γ-rays consisted in radioactive materials. This was a handicap because of the low number of γ-rays emitted per second and because the energy of the γ-rays was limited and produced few nuclear reactions. These disadvantages disappeared with the advent of electron accelerators, notably

---

[1] See p. 193 and p. 219.
[2] See p. 248.
[3] See p. 124.
[4] See p. 193.
[5] See p. 260.

the *betatrons*, which were precisely designed to produce high-energy X-rays by bombarding a heavy target such as tungsten, for example, which is well suited because it melts only at a high temperature, thereby allowing the samples to get hot. The maximum energy of the emitted X-rays increases with the energy of the incident electrons, and it reaches values at which the X-rays are preferably called γ-rays.[1] In 1945, the University of Illinois in the United States had a betatron accelerator[2] which could accelerate electrons to an energy of 20 MeV. When passing through a metal plate, the electrons produce photons of various energies, reaching almost 20 MeV.

George Baldwin and William Koch used these photons to produce artificial radioactivity by exposing to the photon radiation samples ranging from carbon to silver. By increasing progressively the energy of the incident photons, they determined the threshold at which artificial radioactivity is triggered. They also attempted to measure the photofission of uranium, that is, fission of the uranium nucleus caused by a photon instead of a neutron [136].

Giant Resonances

When the 100 MeV betatron begins functioning, Baldwin resumes his experiments, this time with Stanley Klaiber [137]. They notice that 18-MeV photons are the most efficient to produce fission. A kind of "resonance" effect seems to occur: photons endowed with 10 or 22 MeV energies have an efficiency[3] which is about five times smaller. But the measurement is difficult and their conclusion remains qualitative. Another team from the same laboratory make similar experiments using the 22-Mev betatron and 23 targets, ranging from deuterium to bismuth [138]. Among these targets, tantalum appears to display a resonant effect similar to the one observed by Baldwin and Klaiber.

An interpretation of this phenomenon is soon proposed by Maurice Goldhaber, who had measured the photodisintegration of the deuteron[4] and who emigrated to the United States in 1938, as well as by Edward Teller, the father of the hydrogen bomb.[5] The idea of Goldhaber and Teller is that the resonant effect is due to a particular oscillatory mode of the nucleus:

---

[1] The term *X-rays* is usually used to denote photons (particles of light) with energies which range from a dozen to hundreds of thousands of electron volts (eV). The term γ-rays denotes photons with higher energies. The exact frontier is not clearly defined, and one often speaks of hard or soft X-rays. In the entry *photon* of the glossary, a table is given listing the names given to various radiations according to their energy.

[2] See p. 416.

[3] More precisely, a cross section$^\diamond$.

[4] See p. 260.

[5] See p. 404.

We propose to interpret these frequencies as resonances, somewhat different from those caused by definite nuclear levels, and analogous to the "reststrahl frequencies" of polar crystals. We assume that the γ-rays excite a motion in the nucleus in which the bulk of the protons move in one direction while the neutrons move in the opposite direction. We shall call this motion the "dipole vibration" [139].

The basic idea is that the γ-ray photons interact mostly with the electrically charged protons and less with the neutrons, so that a γ-ray photon will have a tendency to "push" the protons, leaving the neutrons unaffected. However, there is a strong attraction between neutrons and protons, and the former begin to oscillate against the latter.

This "giant resonance" becomes an important subject: the interpretation of Goldhaber and Teller remains undisputed. This vibrational mode has an energy which varies slowly from one nucleus to another and which decreases slowly as the nucleus gets heavier. It is certainly a very "collective" behavior of the nucleons, quite different from the single particle excitations observed in stripping and direct reactions. In fact, another very collective phenomenon was known to exist, namely, fission!

Are All Nuclei Spherical?

Atoms are considered to be spherical. But what does this geometrical attribute mean for an atom? Simply that it appears to be same, no matter what angle it is looked at. The reason is that the attractive force exerted by the nucleus on the electrons is the same in all directions.

But the interior of the nucleus is different. There is nothing in the center of the nucleus to guide the nucleons, as Rutherford had initially assumed. Even in the shell model, nucleons move independently in an average potential produced by the other nucleons. What produces a spherical shape is the *surface tension*: the same phenomenon occurs in a liquid drop. The molecules which are at the surface of the drop are not equally attracted in all directions because, close to the surface, there are more molecules attracting them towards the interior than in the opposite direction. The result is that the molecules at the surface form a kind of skin which tries to contract as the skin of an inflated toy balloon does, and it contracts most when the drop adopts a spherical shape, because that is the shape which minimizes the surface area. However, in a nucleus, the electric Coulomb repulsion attempts to repel protons from one another. As the charge of the nucleus grows, *the electric repulsion increases faster than the surface tension*. The equilibrium between these opposing effects is broken when the nucleus becomes too heavy. This is why there exist no stable nuclei heavier than bismuth (which has 83 protons and therefore 83 units of elementary charge). Fission is an illustration of such an instability, as shown first by Lise Meitner and Otto Frisch,[1] and later by Niels Bohr and John Wheeler.[2]

The Quadrupole Moment: An Indicator of Nuclear Deformation

The quadrupole moment of the electric charge distribution in a nucleus is a quantity which is zero when the charge is distributed spherically, positive when the charge has an elongated shape, and negative when it has a flattened shape.[3] It specifies the shape of a nucleus. In 1935, the German physicists Hermann Schüler and Theodor Schmidt had already suggested that certain nuclei may not have spherical shapes.[4] They deduced this from observations of the hyperfine structure of atomic spectra, which are small perturbations of the spectra caused by the magnetic moment of the nucleus.

In 1949, the American physicist Walter Gordy, a specialist of atomic spectroscopy, published a paper in which he considered all the known quadrupole moments. He stressed a remarkable feature [140]: the quadrupole moment is zero or close to zero for nuclei having "magic numbers" (2, 8, 20, 50, and 82) of protons, and it increases considerably between these numbers. For Gordy, this confirmed the shell model of the nucleus, and it also meant that nuclei which did not have a magic number of protons did not have a spherical shape. Some nuclei were far from spherical: assuming that the nucleus has an ellipsoidal shape, like a rugby football, the major axis would be 12% larger than the minor axis for europium and 24% larger for the isotope 175 of lutetium.[5] Could these observed quadrupole moments be explained by the shell model of Maria Goeppert-Mayer and Hans Jensen, in which nucleons move independently of the others? Yes, but only to some extent, as three American physicists, Charles Townes, Henry Foley, and William Low, noted [141] at the end of 1949. Their calculations yielded zero quadrupole moments for nuclei with magic numbers but far too small quadrupole moments for other nuclei.

James Rainwater and Aage Bohr

In the beginning of 1950, two physicists share an office in the University of Columbia in New York. James Rainwater, 33 years old, graduated from the California Institute of Technology and began his Ph.D. in the University of

---

[1] See p. 363.

[2] See p. 368.

[3] See p. 347.

[4] See p. 347.

[5] By comparison, the major axis of a rugby football is 50% larger than the minor axis.

Columbia when the war broke out. He participated actively in the Manhattan Project. He obtained his Ph.D. only in 1946 after his works were declassified. In 1950, he became instructor and professor in 1952. In 1950, he is an experimentalist who specializes in the physics of slow neutrons. His deskmate in Columbia is Aage Bohr, 28 years old, the fourth son of Niels Bohr. Aage Bohr studied in Copenhagen where in the 1930s, still an adolescent, he mixed with the physicists who were invited by his father: Pauli, Klein, Nishina, Heisenberg, Kramers, and more. He begins his university studies in 1940 just when Denmark is invaded by German troops. In 1943, the Bohr family must flee because of the Jewish mother of Niels Bohr. They seek refuge first in Sweden and then in England. Finally, Niels Bohr goes to the United States and joins the Manhattan Project. His son Aage joins him and serves as his secretary. After the war, Aage Bohr obtains his Ph.D. and goes first to Princeton and then to Columbia, where he stays from January to August 1905. There he meets Rainwater. They discuss a talk given by Charles Townes [141] on recent measurements of quadrupole moments. In April 1950, Rainwater sends a paper to *the Physical Review* in which he states that the observed quadrupole moments could be explained if nuclei were deformed and that this was compatible with the shell model [142]. He argues that each nucleon outside a closed shell would deform the potential created by the other nucleons in the closed shells by exerting a pressure on the surface of the nucleus inside which it is confined. He admits that this is more an idea than a theory, and he publishes nothing more on the subject.

A year later, Aage Bohr publishes a paper in the same journal in which he states:

> *The individual particle model, which describes the stationary state of a nucleus in terms of the motion of the individual nucleons in an average nuclear field, has accounted successfully for a large number of nuclear properties. In the simplest form of this model the nucleons are assumed to move in a field of spherical symmetry and the quantization of angular momenta is similar as in atomic structures.*
>
> *This extreme model meets with the difficulty, however, that nuclei are found to have very large electric quadrupole moments [. . .].*
>
> *However, it is possible to allow for the existence of the large quadrupole moments, and still retain many essential features of the individual particle model by assuming that the average nuclear field in which the nucleons move deviates from spherical symmetry [143].*

Bohr does not explain the origin of the deformation, but he shows that if the nucleons move in a deformed average potential, both the large observed quadrupole moments and the magnetic moments could be explained. He also shows that when one observes the quadrupole moment of a quantum state of a nucleus, one does not measure directly the quadrupole moment of the deformed intrinsic state of the nucleus but the product of the former multiplied by a factor which depends on the spin of the nucleus and which can be small when the nuclear spin is small: for a nucleus with a spin equal to $\frac{3}{2}\hbar$, the factor is 1/10 or 1/5, depending on whether the intrinsic state spins in the same or the opposite direction as the nucleon. This small factor implies that the deformation of the intrinsic nuclear state is even larger than what was believed!

Aage Bohr, the Resolution of a Paradox

Upon his return to Copenhagen, Aage Bohr adopts a more general view of the problem. At first, his father, Niels Bohr, who advocated the liquid drop model of the nucleus, was opposed to a model in which nucleons move independently, as electrons do in an atom. But the indisputable success of the shell model of Maria Goeppert-Mayer and Hans Jensen, according to which the nucleons do move independently of the others, faced physicists with a paradox which made them almost schizophrenic. They were invited to accept one model together with the opposite model: the nucleus behaved as a liquid drop and, at the same time, as a system of nucleons moving almost independently. (It was however realized at the time that they moved *almost* independently in a first approximation.)

Aage Bohr believes that it is necessary to construct a theory which embodies both models. At the end of 1952, he publishes a 40-page paper in the *Communications to the Royal Society of Sciences of Denmark*. He begins by stating his motivations:

> The liquid drop model and the single particle model represent opposite approaches to the problem of nuclear structure. Each refers to essential aspects of nuclear structure, and it is to be expected that features of both models must be taken into account simultaneously in a detailed description of nuclear properties [144].

Aage Bohr considers the observed quadrupole moments which appear to confirm both models:

- The quadrupole moment vanishes in "magic" nuclei, a feature which confirms the shell model.
- But the observed quadrupole moments of nuclei which are far from closed shells are too large to be accounted for by nucleons moving in a spherical potential. This is what led Rainwater to propose that the nucleons should move in a deformed potential, in which case the quadrupole moment results from a contribution of many nucleons.

Aage Bohr aims at making a synthesis of the two models:

> The necessity of combining the two models is clearly indicated by the observed behavior of nuclear quadrupole moments. On the one hand [...] the quadrupole moments give definite evidence for shell structure [...]. On the other hand, for many nuclei, the magnitude of the quadrupole moments is too large to be accounted for in terms of individual nucleons and suggests that the equilibrium shape of the nucleus itself deviates from spherical symmetry.
>
> The behavior of the quadrupole moments finds a simple explanation if one considers the motion of the individual particles in a deformable nucleus. Due to the centrifugal pressure exerted by the particles on the nuclear walls, the nucleus may acquire a considerable deformation [144].

Aage Bohr suggests an independent particle model, similar to the shell model, but rather a deformed shell model, to which he adds the possibility for a nucleus to undergo *collective rotation and vibration*.

A Unified Model of the Nucleus

Two physicists working in the Theoretical Physics Institute in Copenhagen reconcile the independent motion with the collective motion of nucleons in the framework of a generalized shell model, in which the potential can be deformed. A hard core is discovered in the nucleon-nucleon interaction, which means that at short distance nucleons repel each other.

Aage Bohr returns to Copenhagen in 1950. He meets a young American physicist, Ben Mottelson, who plans to spend a year there with a postdoctoral position.

Ben Mottelson

Born in 1926, Mottelson obtained his Ph.D. in the University of Harvard under the direction of Julian Schwinger, the physicist who will be awarded the Nobel Prize in 1965 for his work on electrodynamics (together with the Japanese physicist Sin-Itiro Tomonaga and the American physicist Richard Feynman). After his Ph.D., Mottelson chooses to go to the Institute of Theoretical Physics in Copenhagen. He is attracted to this prestigious and international laboratory where quantum theory had been developed and where Niels Bohr is still active. Aage Bohr and Ben Mottelson soon become friends, and they begin a collaboration which will last all their life. They will publish jointly their most important work. An additional American grant allows Mottelson to extend his stay in Copenhagen for two extra years. Finally, he decides to settle there and he obtains Danish citizenship in 1971. The collaboration of Aage Bohr and Ben Mottelson will leave a definite mark on the theory of nuclear structure.

Their first publication is a contribution to the international conference on nuclear reactions held in Amsterdam on September 1–6, 1952. They show that the shell model explains certain complexities of radioactive β-decay, namely, why some transitions between an unstable state and another take place very fast, while others are much slower. They note however that certain phenomena most likely involve several nucleons and cannot be explained in the strict framework of the shell model:

In order to give a more comprehensive description of nuclear properties, one is therefore led to consider a model which recognizes collective motion as well as single particle motion as the fundamental types of motion in the nucleus. [145].

New Data, New Confirmations

Until then, the hypothesis of nuclear deformation relied almost exclusively on the observed quadrupole moments. But in the years 1951–1953 additional experimental evidence became available from:

* *The first excited states of even–even nuclei.*[1] In a paper devoted to the classification of isomers, Maurice Goldhaber and Andrew Sunyar note that the first excited state of a large number of even–even nuclei (19 are mentioned in their paper) have a spin equal to 2 and that *the excited state decays very fast to the ground state* while emitting a γ-ray, as much as 100 times faster than what would be expected from a single nucleon which would make a transition from one orbit into another as in the shell model [146]. Recall that *excited states* of nuclei are configurations or internal arrangements of neutrons and protons, which have a higher energy than the ground state configuration. A nucleus in an excited state decays back into its ground state, while the energy of the excited state is carried away by a γ-ray. The fact that the decay takes place very quickly suggests that the excited state may be quite similar to the ground state. Bohr and Mottelson believe that the quickly decaying excited state may be a rotation or a vibration of the nucleus. Goldhaber and Sunyar relate this phenomenon to the large observed quadrupole moments. At the same conference, Gertrude Scharff-Goldhaber[2] reports on 54 similar occurrences [147]. The physicists in Berkeley made another observation concerning α-decay: they found that the first excited states of heavy nuclei such as the isotopes of uranium, radium, and thorium had small energy spacings which varied slowly with the number of neutrons [148]. Again, this could be explained by the rotation of a deformed nucleus, several nuclei having similar deformations.
* *Rotational bands*: the nitrogen molecule looks something like a small dumbbell, because the nuclei of the molecule are much heavier than the surrounding electrons and they maintain a fixed distance between each other. The molecule can however rotate as a whole, and the angular momentum of the rotating molecule is quantized by quantum mechanics. It can take the values 0, 2, 4, 6, and so on. The same applies to the rotation of even–even nuclei. The states with angular momentum (or spin) 0, 2, 4, 6,... form what is called a *rotational band*. When such a spectrum is observed, it becomes a strong indication that it corresponds to the rotation of a deformed state. The energy spacings between members of the rotational band also indicate how strong the deformation is. Rotational bands appeared to occur in the spectra of radium 226 and ionium [149, 150] (ionium is the isotope 230 of thorium; see p. 165), of hafnium isotopes [151] and of the isotope 150 of samarium [152].

---

[1] The nuclei which have an even number of both protons and neutrons are called *even–even nuclei*.

[2] Gertrude Scharff is a German physicist born in Mannheim in 1911. She obtained her Ph.D. in Munich in 1935 after which she was obliged to flee Nazi Germany because she was Jewish. She emigrated to England where she met Maurice Goldhaber, an exiled Austrian Jew. They married, and in 1939, they emigrated to the United States where they both had exceptional careers.

Bohr and Mottelson: The Key to Nuclear Spectra

In November 1952 and March 1953, Aage Bohr and Ben Mottelson send two letters to the editor of *Physical Review*. In the first letter [153], they show that the observed rapid transitions to the ground state can be explained quite simply if one assumes a deformed nucleus. They even estimate the expected quadrupole moments which are roughly equal to the observed ones. In the second letter [154], they insist on the rapid accumulation of experimental data which confirm their hypothesis. At the end of 1953, they publish a 173-page paper in the *Mathematics and Physics Communications of the Royal Society of Sciences of Denmark*. The paper describes in detail a unified model of nuclear structure [155]. The paper will become a reference for a generation of nuclear physicists all over the world.

Bohr and Mottelson aim at formulating a model which can explain the properties of all nuclei: the spins and energies of their ground and excited states, the probability of exciting the nucleus, their quadrupole moments, and more. They adopt the shell model of Maria Goeppert-Mayer and Hans Jensen. The nucleons move almost independently in the nuclear interior, and they form successive shells as electrons do in the atom. A nucleus in its ground state may be excited in roughly two ways. Either a nucleon makes a transition to another orbit or the nucleus as a whole can vibrate or rotate if it is deformed. But how do the individual nucleons react to the collective rotation of the nucleus? How are the nucleon orbits modified? The answer to these questions will yield a real *nuclear spectroscopy* which will allow to predict if a nucleus is deformed or not, if an excited state is due to a collective vibration or rotation, or if it is due to the motion of a single nucleon in the deformed average potential:

> In the present paper, we consider the further development of such a unified nuclear model incorporating collective and individual-particle features, and pursue its consequences, especially for the nuclear properties pertaining to the ground state and the low energy region of excitation. The available empirical evidence is analyzed in an attempt to ascertain to what extent a comprehensive interpretation is possible on the basis of such a description of the nucleus.

The authors begin by defining what they call *collective variables:* they are parameters which define the deformation and the vibrations of the nucleus. They then examine in detail how the motion of the individual nucleons is modified by the rotation of the nucleus and how the two are *coupled*: the rotation affects the motion of the nucleons which in turn affects the rotation. The simplest case is when the rotation is much slower than the motion of the individual nucleons. In that case, the rotation is quite independent of the motion of the nucleons.

Bohr and Mottelson pursue by analyzing in detail the available data on quadrupole moments, magnetic moments, the energies of the excited states, and the transition probabilities, that is, the average time required for a nucleus to decay back into its ground state, possibly passing through an intermediate lower energy excited state. The model of Bohr and Mottelson becomes a framework which will allow physicists to analyze their experimental observations. Nuclear spectroscopy is born.

Within a few years after the publication of the paper of Bohr and Mottelson, a proliferation of theoretical developments took place. We mention only two examples, the so-called Nilsson orbits and Coulomb excitation.

The Nilsson Orbits

A young Swedish physicist, Sven-Gösta Nilsson, published a paper in 1955, in which he calculated how the energies of nucleon orbits were modified by the nuclear deformation [156]. He identified the quantum numbers of orbits in a deformed potential. He calculated that some orbits could considerably lower their energies as the deformation grew, thereby upsetting the order in which the orbits are filled. The paper of Nilsson also became an indispensable tool of nuclear physicists.

Coulomb Excitation

How can one provoke a "collective excitation" the nucleus, that is, how can one cause it to vibrate or rotate? Already in 1938, Victor Weisskopf had proposed to bombard nuclei with low-energy charged projectiles, so that the Coulomb repulsion would maintain the projectile at a distance beyond the range of the nuclear interactions. The projectile could however interact with the nucleus by means of the long-range Coulomb interaction which would act on the nucleus as a whole, rather than with an individual nucleon [157]. But Weisskopf gave only a global formula. In 1952, the idea was taken up by the Russian physicist Karen Ter-Martirosyan who formulated the first real theory of the process [158]. A year later, this "Coulomb excitation" was developed by the Danish physicist Aage Winther and the Swiss physicist Kurt Alder [159]. At the same time, experiments were performed by the Danish physicist Torben Huus and the Yugoslavian physicist Črtomir Zupančič. The experiments confirmed that nuclei could be excited this way and they confirmed the theory. The experiments were able to measure the deformation of the tungsten nucleus [160]. Further experiments showed what a useful tool Coulomb excitation was. In 1956, a definitive theory of Coulomb excitation was published by Alder, Bohr, Huus, Mottelson, and Winther [161].

Coulomb excitation became a reliable method of studying collective excitations of nuclei, their vibrations, and rotations. The experiments consisted in measuring carefully inelastic scattering, that is, scattering during which some energy is transferred to the target nucleus. Various projectiles were used with rather low energies. It soon became apparent that inelastic scattering at higher energies, at which nuclear interactions would also occur, could also excite collective modes. However, the advantage of Coulomb excitation lies in the fact that the process is well understood theoretically because it involves the well-known electromagnetic interaction. The information about the target nucleus obtained this way was more reliable.

The Birth of Nuclear Spectroscopy

Atomic physics was born from the pressing need of understanding atomic spectra, their beautiful lines with rainbow colors, and immutable spacings for each atom. The atomic spectra remained incomprehensible for a long time, until Niels Bohr suggested in 1913 that *the spectral lines corresponded to transitions* between two states of the atom. They were light rays emitted by an electrons which "jumped" from one orbit to a lower energy orbit. What became known as atomic *spectroscopy* consisted in determining the states of the atom and their properties which were determined by their *wavefunction*.

Nuclear spectroscopy began only in the 1950s. It also aimed at determining the various *states* of a nucleus, their properties, which amounts to a determination, as far as possible, of the wavefunction of the state. It is more difficult to achieve this than in the case of the atom because the interactions which bind the nucleons to the nucleus are less well known. However, the shell model of Maria Goeppert-Mayer and Hans Jensen, extended to deformed nuclei by Bohr and Mottelson, is a solid theoretical model of the structure of the nucleus. To achieve this understanding, the experimental tools, which become increasingly perfected, can be roughly grouped into the following families:

- Inelastic scattering of protons or α-particles can determine *collective excitations*.
- *Stripping* reactions, in which a neutron (or a proton) is "deposited" into a spherical or a deformed nucleus, allows one to identify the final state of the target nucleus, the structure of which is similar to the motion of independent nucleons with possible effects of collective motion on the nucleons.
- The detection of γ-rays emitted when one of the excited states gets rid of its energy by making a transition to another state, either the ground state or an intermediate state. As in the case of atoms, the energy is emitted in the form of light rays which are γ-rays which have roughly a million times higher energies than the photons of visible light and which are detected by different means. The intensity of emission of this γ radiation yields information concerning the relation between the initial and final states of the transition. For example, two successive members of a "rotational band" are almost identical because they correspond to the rotation of the same state at two different angular velocities. Naturally, the *coincidence* detection of two photons is a better determination of successive decays, when, for example, a nucleus progressively decays into successively slower rotating members of a rotational band.
- Natural and artificial α and β radioactivity.

Experimentation involves hundreds of physicists. We will not dwell on the numerous experiments, each one of which makes a contribution to the overall picture. The observations are summarized in published nuclear data sheets. The following illustrates how the data accumulates: in 1944, Glenn Seaborg published a "table of isotopes" in *Reviews of Modern Physics*. It is a 32-page paper which contains all the available information on the isotopes known at the time [162]. The table was reedited several times by many authors. In 1948, 1953, and 1958, the required number of pages grew to 83, 183, and 320 [163–165].

Nobel Awards

In 1963, the Nobel Prize in physics was shared by three physicists:

- Eugene Wigner "for his contributions to the theory of the atomic nucleus and of elementary particles, in particular for the application of fundamental principles of symmetry"
- Maria Goeppert-Mayer and Hans Jensen for "their discovery of the shell structure" of nuclei

Twelve years later, in 1975, James Rainwater, Aage Bohr, and Ben Mottelson were also awarded the Nobel Prize in physics "for their discovery of the connection between collective motion and independent particle motion in atomic nuclei and for their theory of nuclear structure based on this connection."

These Nobel prizes recognized the major developments of the theory of the atomic nucleus in the period 1948–1953.

The Nuclear Force

The heaven sent cosmic rays finally reveal the π meson, the massive quantum predicted by Yukawa. It is also produced in abundance by the accelerator in Berkeley. Three π mesons are discovered, one bearing a positive charge, one neutral and one negatively charged. A multitude of other mesons are discovered. Heisenberg's initial intuition is confirmed: the interaction between nucleons is short ranged and attractive, but strongly repulsive at very short distances. The dream of a simple and elegant picture fades away and a complex reality becomes apparent.

When Anderson and Neddermeyer discovered in cosmic rays a particle with a higher mass than the electron and a lower mass than the nucleon, they immediately thought it could be the famous *massive quantum* which had been predicted by Yukawa. Indeed, it appeared to have just about the right mass.[1] But after further observations had been made, the cosmic meson appeared to differ more and more from the quantum Yukawa had predicted. To begin with, its mass was about two times too small to account for the observed scattering of two protons [166]. Worse, its radioactive lifetime was too long, about 1 μs, whereas theoretical estimates predicted it to be 100 times shorter [167–169]. Worse still, its absorption by nuclei appeared to be absurd: the positively charged mesons, after being slowed down by the photographic emulsion, should be repelled by nuclei and therefore decay spontaneously, while the negatively charged mesons should be attracted and absorbed by nuclei [170]. Three Italian physicists, Marcello Conversi, Ettore Pancini, and Oreste Piccini, who had been working since 1941, amid the war, on the absorption of "mesons" by matter, made a series of observations which were in complete disagreement with these expectations [171]. The analysis of their data made by Fermi, Teller, and Weisskopf showed that the negatively charged cosmic "mesons" interacted far more weakly with nuclei (by a factor of about 10^{12}, that is, a thousand billion times less) than the Yukawa particle.

The Discovery of the π Meson

It was then thought that *possibly two different mesons may exist*. If the Yukawa meson interacts strongly with a nucleus and if its lifetime is very short, it may not

---

[1]See p. 333.

reach the surface of the earth, because it may either be absorbed by a nucleus in the atmosphere *or decay into a different meson*. It might be this *secondary meson* that is detected in the laboratory. This idea had been put forth by the Japanese physicists Shoichi Sakata and Takesi Inoue [172], and also by Yasutaka Tanikawa [173], well before the observations of Conversi, Pancini, and Piccini. It had first been discussed in September 1943 during a symposium on the theory of mesons. However, because of the war, it was only published in 1946 and 1947 in the new Japanese journal *Progress of Theoretical Physics*, which was not read much in the United States in spite of the articles being written in English. During a discussion which took place at a meeting on quantum mechanics at Shelter Island near New York, on June 4–7, 1947, the American physicist Robert Marshak proposed the same idea [174].

In the issue of Nature, dated May 24, 1947, which had not yet reached the United States on June 4, three physicists from the University of Bristol, Cecil Powell, Cesare Lattes, and Giuseppe Occhialini, reported an important observation made at high altitude:

> The experimental data suggest that there exists two kinds of mesons with different masses [175].

The same team confirmed this observation by exposing photographic plates at an altitude of 2800 m at the Pic du Midi in France and at 5500 m at Chacaltaya in the Andes in Bolivia [176]. Thus, two distinct particles exist with masses between the electron and nucleon masses. For this reason, they continue to be called *mesons*.[1] Lattes, Occhialini, and Powell further observe that the heavier primary mesons decay into secondary mesons. They propose to call the former π *mesons* and the latter μ *mesons*. They are cautious in estimating their masses which they believe to be in a ratio of 1.5–2. One could not dream of a better confirmation of Yukawa's massive quantum and of the idea put forth by Tanikawa, Sakata, and Inoue, as well as by Marshak.

The next step was soon taken at the new synchrotron[2] in Berkeley, which provided a beam of 380-MeV α-particles. This allowed Eugene Gardner and Cesare Lattes to produce a large number of the π mesons of Powell, and they attributed to them a mass about 300 times the mass of the electron [177], while Reginald Richardson estimated its lifetime to be 8×10^{-9} s, that is, 8 billionths of a second [178]. This agreed with the theoretical estimates.

The π^0 Completes the Pion Trio

The number of π mesons produced in Berkeley was about a hundred million times greater than those observed in cosmic rays. This made it possible to discover their

---

[1] See p. 334.
[2] See p. 415.

properties with greater precision. The first major discovery in Berkeley was the existence of a *neutral meson*, called the π^0 meson. This completed the family of π mesons: there were three, one positively charged, the other negatively, and the third was neutral [179]. The neutral meson had been predicted in 1938 in order to account for the interaction between two neutrons or two protons which could not exchange a charged meson [180–182]. The properties of π mesons were soon discovered. They have a mass of about 139 MeV, roughly eight times smaller than the mass of the nucleon. They are zero-spin particles with *negative parity* [183]. Parity is a quantum number related to symmetry under a mirror reflection. The world observed through a mirror is almost identical to the real world, with some differences: the mirror image of your right hand is a left hand, corkscrews turn in the opposite direction... The π mesons are now called *pions*.

Other particles were discovered in the following years, heavier mesons and further fermions.[1] The theory of the nuclear force mediated by meson exchange becomes more complex: nucleons can exchange not only one but two or three heavier mesons, the corresponding range of the force being divided by two or three, respectively. This accounts for the short-range part of the force. All these meson exchanges combine, and the resulting nuclear force is more complicated than the original simple force conceived in the 1930s.

The Hard Core

In order to explain the saturation of nuclear forces, that is, the fact that nucleons are attracted to one another while remaining at a certain distance, Heisenberg had speculated that the nuclear force should become repulsive at very short distance so as to prevent the collapse of the nucleus. The average distance between nucleons should be the one which minimizes the energy.[2]

However, Majorana considered this *complicated and not esthetically satisfying*. His exchange force had the advantage of explaining the saturation and on relying on very general arguments. His model seemed *simple and elegant* and physicists forgot about the short-range repulsion. At a meeting in Minneapolis in 1977, Rudolf Peierls recalls the motivations of physicists at the time:

> We had to find a new law of force. It was reasonable to hope that this would be something as simple and as fundamental as Coulomb's law. After all, Coulomb's law, and indeed the whole content of Maxwell's equations, can be demonstrated by a very limited number of experiments, and with the use of very few parameters. Therefore, we looked for something simple. This was, I think, the background to

---

[1] See p. 147 and p. 229.
[2] See p. 267.

the attitude of most theoreticians, when they rejected the idea of a repulsive core to the forces, which would make nuclear forces act like interatomic forces in molecules or solids [184].

In his venerated 1936 paper, Bethe rejected the idea of a force which would be attractive at "large" distance and repulsive at short distance:

Thus, the repulsive forces which prevent the interpenetration of atoms are, in this case, primarily responsible for the binding energy being proportional to the number of atoms. However, it would seem very unsatisfactory to transfer such a mechanism to nuclei: it would involve the assumption of a force between elementary particles, viz., protons and neutrons, which would be attractive at large distances and repulsive at small distances, an assumption which one would make only very reluctantly [94].

For Bethe, it is a matter of principle:

For particles with internal structure, such as atoms or the α-particle, the assumption of such a force is, of course, not objectionable but results directly from simple assumptions about the forces between elementary particles.

The fundamental law governing elementary particles, which are *void of internal structure*, had to be simple, and protons and neutrons were considered to be elementary particles at that time.

Fifteen years later, this was challenged by experiments made in Berkeley. Owen Chamberlain and Clyde Wiegand measured the elastic proton–proton scattering at 340 MeV, the highest energy available with the accelerator. To their great surprise, they found a clear difference between the scattering and therefore interaction between the proton–proton and proton–neutron force [185]. Did that imply that one should give up the hypothesis of charge independence, which stated that there was a *single strong interaction* which was the same when acting between two protons, two neutrons, or a proton and a neutron? Chamberlain and Wiegand thought so, but Robert Jastrow, a theorist in Princeton, proposed another explanation [186]: charge independence could be maintained, *provided that the force between two nucleons was repulsive at short distances*, as Heisenberg had surmised 19 years earlier. Jastrow exploited the only difference between proton–neutron and proton–proton pairs, namely, Pauli's exclusion principle which forbids certain relative orientations of the nucleon spins. For certain orientations, the nucleons can come close to enough to each other so as to feel the short-range repulsion: the data are then compatible with charge independence.

A last remark: until the end of the 1960s, before the quark model, protons and neutrons were considered as elementary particles with no internal structure. Once that hypothesis was lifted, Bethe's objection fell since it was only valid for elementary particles.

From then on, one had to live with this short-range repulsion which made calculations more difficult.

Nuclear Matter

Physicists accept the challenge of the shell model and attempt to explain the nucleus as a system of N interacting particles. A theoretical justification is found of the shell model extended to deformed nuclei.

In the span of a few years, roughly between 1948 and 1953, the way physicists conceived the nucleus became completely altered by the success of the "shell" model of Maria Goeppert-Mayer and Hans Jensen, and by its extension to deformed shapes by Aage Bohr and Ben Mottelson. The nucleus was viewed as a set of particles moving independently of each other in an average potential created by the same particles. The model was able to explain a vast amount of experimental observations. It remained to understand why the model worked so well.

The Challenge

It should be realized that the shell model came as a surprise to many physicists. Indeed, it contradicted all that was known previously. In their thick book, *Theoretical Nuclear Physics,* published in 1952, and which remained a reference for a long time, Victor Weisskopf and John Blatt write:

> The great success of the shell model of nuclei is most surprising and is not yet understood on the basis of our present knowledge about nuclear forces and nuclear dynamics [187].

For them, the paradox is that each nucleon appears to move throughout the nuclear volume without colliding with another:

> The one requirement for the validity of the independent-particle model which seems to violate most strongly our ideas of the nucleus is the lack of any effective interaction between nucleons. The existence of orbits in the nucleus with well-defined quantum numbers is possible only if the nucleon is able to complete several "revolutions" in this orbit before being perturbed by its neighbors.

They repeat the argument put forth Weisskopf in 1951: it may have something to do with the Pauli principle. But it remained a qualitative argument.

How could this paradox be resolved? Edward Teller and Montgomery Johnson [188] had the following idea: it may not be possible to explain the nucleus in terms of two-body interactions acting between the nucleons. Instead, mesons which mediate these interactions might form a cloud which would create a common potential inside which the nucleons evolve. Unfortunately, experimental data contradicted this hypothesis [189]. The nearly independent motion of nucleons remained a mystery.

Keith Brueckner, Jeffrey Goldstone, Hans Bethe, and a Few Others

In 1954 and 1955, a series of papers were published by a young American physicist, Keith Brueckner, who was already known for his work on the mesonic theory of nuclear interactions. Some of the papers were written together with his collaborators. Born in 1924, Brueckner obtained his Ph.D. in Berkeley in 1950 and then he joined the University of Indiana in Bloomington.

Brueckner simplified the problem by studying what takes place in the central part of a heavy nucleus, far enough from the surface, so that surface effects can be ignored. This amounts to consider the inside of a nucleus as a chunk of "infinite" nuclear matter of uniform density, much more extended than the range of nuclear forces. He further simplified the problem by assuming that this "nuclear matter" is composed of equal numbers of protons and neutrons and by neglecting the Coulomb repulsion between the protons. Such nuclear matter has a density equal to 0.17 nucleons per cubic femtometer.[1] This implies that the average distance between the nucleons is approximately 1.5 femtometer, which is larger than the range of the short-range repulsion. In addition, the binding energy of nuclear matter is equal to 15 MeV per nucleon. Brueckner assumed a nuclear interaction close to the one which was observed, namely, attractive at short range and with a "hard core," that is, repulsive at very short range, which prevents the nucleons from penetrating one another.

Physicists often make estimates using perturbation theory. One starts with a problem which one can solve and which is close to the real problem, and then one treats the difference as a small perturbation. This is how the trajectories of planets around the sun are calculated: one first calculates the trajectory of a lonely planet around the sun in the absence of the other planets. Then one calculates the small perturbation on the trajectory caused by the other planets. Unfortunately, this method could not be used in the case of idealized nuclear matter, mainly because of the intensity of the short-range hard repulsive core between the nucleons. This is where Brueckner makes a decisive contribution: he is able to make a mathematical transformation which allows him to proceed by successive approximations: he shows that one can rigorously replace the complicated interaction acting between the nucleons by a pseudo-potential which can be calculated in terms of the nucleon-nucleon interaction and the average potential in which the nucleons evolve. The theory of Brueckner is successful: he calculates [190] a density of nucleons equal to 0.16 nucleons per cubic femtometer and a binding energy of 12 MeV. This was a remarkable result for a first calculation. He progressively improves his

---

[1] A heavy nucleus, such as lead 208, contains 208 nucleons within a sphere with a radius equal to about 6.6 femtometers (6.6×10^{-13} cm). The volume is thus 1200 cubic femtometers, and the average density of nucleons is about $\frac{208}{1200} = 0.17$ nucleons per cubic femtometer.

calculation during a collaboration with an English physicist, Richard Eden, visiting Bloomington [189, 191, 192].

Richard Eden returns to Cambridge at the Clare College, and he meets Hans Bethe, who came to spend a sabbatical year at the Cavendish Laboratory. Personal contacts between scientists are very important for research which does not rely exclusively on published scientific papers but which is constantly irrigated by visiting scientists from other laboratories.

Hans Bethe is very interested in the work of Brueckner, and he attempts to formulate the theory on a firmer basis. In March 1956, he publishes a 38-page paper in the *Physical Review* in which he acknowledges the contribution of Brueckner before formulating a complete nuclear theory. It is interesting to note his introduction:

> *Nearly everybody in nuclear physics has marveled at the success of the shell model [...]. While the success of the shell model has thus been beyond question for many years, a theoretical basis for it has been lacking. Indeed, it is well established that the forces between two nucleons are of short range, and of very great strength, and possess exchange character and probably repulsive cores. It has been difficult to see how such forces could lead to any over-all potential and thus to well-defined states for the individual nucleons [193].*

It is in Cambridge that Bethe met the student Jeffrey Goldstone (born in 1933) who informed him about the theory of Brueckner. Goldstone improved and considerably simplified the theory [194] by formulating it in terms of diagrams similar to the diagrams which Richard Feynman had introduced a few years earlier in quantum electrodynamics [195].

Solid Foundations

With what became known as the Brueckner-Bethe-Goldstone theory, the description of nuclei acquired a solid foundation. The intuition of Fermi and Weisskopf was confirmed quantitatively. It is the Pauli exclusion principle which allows nucleons to move almost independently in the interior of the nucleus without interacting otherwise than with an average potential: since most of the orbits *are already occupied by the other nucleons*, a collision would oblige the nucleons to make a transition to considerably higher energy orbits. Although a short-range repulsion can do this, the quantum theory predicts only a small probability for the process to occur. As a result, the most probable event is that the nucleons remain in their orbits, unperturbed by collisions.

From then on, the shell model of the nucleus (be it spherical or deformed) will prevail. Naturally, many problems remain to be solved which require ingenuity, persistence, and imagination. Indeed, the calculations are quite complex and they remain difficult in spite of the dazzling development of computers.

Quantum mechanics does not exclude *completely* the possibility of nucleons in the nucleus to collide and to be excited into higher energy orbits. It simply makes such collisions less probable and less frequent. This means that the nucleus can be described by the shell model to a first—and surprisingly good—approximation. Corrections to the shell model can and have been progressively calculated.

And What About Niels Bohr's Original Objection?

Recall that Niels Bohr remained opposed for a long time to the idea that nucleons could move almost independently in the nucleus. His objection relied on his belief that the mean free path of a nucleon in the nucleus was very much smaller than the nuclear size. It is with this hypothesis that he formulated his theory of the compound nucleus. He assumed that a neutron which penetrates a nucleus immediately shares its energy with the other nucleons.[1] Did the success of the shell model prove that his model of the compound nucleus wrong? That appeared unlikely because the compound nucleus explained many experimental observations. Are we faced with a contradiction?

In fact, the Pauli principle is mostly "felt" by nucleons in the nucleus when the latter is in its ground state because it is in this state that the nucleons occupy all the lowest energy orbits. It is in such a state that the Pauli principle makes it difficult for them to make transitions into other orbits. However, the compound nucleus which is formed when an external neutron enters the nucleus is not in its ground state. It is in an excited state at about 8 MeV above the ground state. At such an excitation energy, there are many available configurations for nucleons to make transitions into, and these configurations are not excluded by Pauli principle. This allows many nucleons in the nucleus to share the energy of the incoming neutron, the mean free path of which is thus considerably reduced as compared to the mean free path of a neutron in the ground state of the nucleus. This is what Niels Bohr had not understood.

The End of an Era

Three international conferences on nuclear physics were held between 1956 and 1958: one in Amsterdam [196] on July 2–7, 1956, devoted to nuclear reactions; another in Rehovoth in Israel [197] on September 8–14, 1957, devoted to nuclear structure; and a third in Paris [198] on July 7–12, 1958, devoted to all nuclear physics. Aage Bohr, Ben Mottelson, Hans Jensen, Keith Brueckner, and Victor Weisskopf were among the prominent speakers, together with a growing generation from several countries.

---

[1] See p. 336.

Frédéric Joliot-Curie, who presided the organization committee of the Paris conference, delivered the welcome address on July 7, 1958. He died on August 14. He survived his wife Irène by only 2 years. She died of a violent leukemia on March 17, 1956. Wolfgang Pauli died in Zurich on December 15, 1958, from a cancer in the pancreas, 4 months after Frédéric Joliot-Curie. Enrico Fermi, one of the greatest physicists of all time, died on November 29, 1954, from a stomach cancer, and Einstein died a few months later, on April 18, 1955. Erwin Schrödinger died on January 4, 1961, and Niels Bohr on November 18, 1962. Otto Hahn disappeared on July 28, 1968, soon followed by Lise Meitner on October 1968, Max Born on June 5, 1970, James Chadwick on July 24, 1974, and Heisenberg on February 1, 1976... The pioneers had paved the way for a new generation.

References

1. Heilbron, J. L. and Seidel, R. W., *Lawrence and his laboratory. Vol. 1*, University of California Press, Berkeley, 1989.
2. McMillan, E., "The Synchrotron—A Proposed High Energy Particle Accelerator", *Physical Review* **68**, 143–144 (L), September 1, 1945.
3. Veksler, V. I., "A new method for acceleration of relativistic particles", *Comptes Rendus (Doklady) de l'Académie des Sciences de l'URSS* **43**, 329–331, 1944.
4. Richardson, J. R., MacKenzie, K. R., Lofaren, E. J. and Wright, B. T., "Frequency Modulated Cyclotron", *Physical Review* **69**, 669–670 (L), June 1, 1946.
5. Brobeck, W. M., Lawrence, E. O., MacKenzie, K. R., McMillan, E., Serber, R., Sewell, D. C., Simpson, K. M. and Thornton, R. L., "Initial Performance of the 184-Inch Cyclotron of the University of California", *Physical Review* **71**, 449–450, April 1947.
6. Livingston, M. S. and Blewett, J. P., *Particle Accelerators*, McGraw-Hill, New York, 1962.
7. Oliphant, M. L., Gooden, J. S. and Hide, G. S., "The acceleration of charged particles to very high energies", *Proceedings of the Physical Society* **59**, 666–677, 1947.
8. Gooden, J., Jensen, H. H. and Symonds, J. L., "Theory of the proton synchrotron", *Proceedings of the Physical Society* **59**, 677–693, 1947.
9. Kerst, D. W., "Acceleration of Electrons by Magnetic Induction", *Physical Review* **58**, 841 (L), November 1940.
10. Kerst, D. W. and Serber, R., "Electronics Orbits in the Induction Accelerator", *Physical Review* **60**, 53–58, July 1941.
11. Kerst, D. W., "A New Induction Accelerator Generating 20 Mev", *Physical Review* **61**, 93–94 (L), January 1942.
12. Kerst, D. W., Adams, G. D., Koch, H. W. and Robinson, C. S., "Operation of a 300-Mev Betatron", *Physical Review* **78**, 297 (L), May 1950.
13. Livingston, M. S. and Blewett, J. P., *Particle Accelerators*, p. 397.
14. *Ibid.*, p. 328.
15. Chodorow, M., Ginzton, E. L., Hansen, W. W., Kyhl, R. L., Neal, R. B. and Panofsky, W. K., "Stanford High-Energy Linear Electron Accelerator (Mark III)", *Review of Scientific Instruments* **26**, 134–204, February 1955.
16. van de Graaff, R., Trump, J. G. and Buechner, W. W., "Electrostatic Generators for the Acceleration of Charged Particles", *Reports on Progress in Physics* **11**, 1–18, 1948.
17. Bennett, W. H. and Darby, P. F., "Negative Atomic Hydrogen Ions", *Physical Review* **49**, 97–99, January 1936.
18. Greinacher, H., "Über einen hydrolischen Zähler für Elementarschtrahlen", *Helvetica Physica Acta* **7**, 360–367, 1934.
19. Greinacher, H., "Über den hydrolischen Zähler für Elementarstrahlen (II. Mitteilung). Messung des elementaren Photoeffekts an Wasser", *Helvetica Physica Acta* **7**, 514–519, 1934.
20. Greinacher, H., "Über einen weiteren hydroelektrischen Zähler für Elementarstrahlen und Photo-Elektronen", *Helvetica Physica Acta* **8**, 89–96, 1935.
21. Chang, W. Y. and Rosenblum, S., "A Simple Counting System for Alpha-Ray Spectra and the Energy Distribution of Po Alpha-Particles", *Physical Review* **67**, 222–227, April 1945.
22. Madansky, L. and Pidd, R. W., "Characteristics of the Parallel-Plate Counter", *Physical Review* **73**, 1215–1216, May 1948.
23. Bay, Z., "Electron Multiplier as an Electron Counting Device", *Review of Scientific Instruments* **12**, 127–133, March 1941.
24. Zworykin, V. K., Morton, G. A. and Malter, L., "The Secondary Emission Multiplier—A New Electronic Device", *Proceedings of the Institute of Radio Engineers* **24**, 351–375, March 1936.
25. Zworykin, V. K. and Rajchman, J. A., "The Electrostatic Electron Multiplier", *IRE Proceedings* **27**, 558–566, September 1939.

26. Blau, M. and Dreyfus, B., "The Multiplier Photo-Tube in Radioactive Measurements", *Review of Scientific Instruments* **16**, 245–248, September 1945.
27. Curran, S. C. and Baker, W. R., "Photoelectric Alpha-Particle Detector", *Review of Scientific Instruments* **19**, 116, February 1948.
28. Mcintyre, J. A. and Hofstadter, R., "Measurement of Gamma-Ray Energies with One Crystal", *Physical Review* **78**, 617–619, June 1950.
29. Wilson, A. H., "The Theory of Electronic Semi-Conductors", *Proceedings of the Royal Society, London* **A133**, 458–491, October 1, 1931.
30. Wilson, A. H., "A Note on the Theory of Rectification", *Proceedings of the Royal Society, London* **A136**, 487–498, June 1, 1932.
31. Schottky, W., "Zur Halbleitertheorie der Sperrschicht- und Spitzengleichrichter", *Zeitschrift für Physik* **113**, 367–414, May 1939.
32. Schottky, W., "Vereinfachte und erweiterte Theorie der Randschicht-gleichrichter", *Zeitschrift für Physik* **118**, 539–592, February 1942.
33. Mott, N. F., "The Theory of Crystal Rectifiers", *Proceedings of the Royal Society, London* **A171**, 27–38, May 1, 1939.
34. Bardeen, J. and Brattain, W. H., "The Transistor, A Semi-Conductor Triode", *Physical Review* **74**, 230–231, July 15, 1948.
35. Riordan, M. and Hoddeson, L., *Crystal fire : the invention of the transistor and the birth of the information age*, Norton, New York, 1997.
36. Riordan, M., Hoddeson, L. and Herring, C., "The invention of the transistor", *Reviews of Modern Physics* **71**, S336–S345, March 1999.
37. Riordan, M. and Hoddeson, L., "*Crystal fire*, p. 159".
38. McKay, K. G., "The Crystal Conduction Counter", *Physics Today* **6**, 10–13, May 1953.
39. Röntgen, W. C. and Joffé, A., "Ueber die Elektrizitätsleitung in einiger Kristallen und über den Einfluss der Bestrahlung darauf", *Annalen der Physik, Leipzig* **41**, 449–498, 1913.
40. van Heerden, P. J., "The Crystal Counter. A New Apparatus in Nuclear Physics for the Investigation of β and γ-rays. Part I", *Physica* **16**, 505–516, June 1950.
41. Wooldridge, D. E., Ahearn, A. J. and Burton, J. A., "Conductivity Pulses Induced in Diamond by Alpha-Particles", *Physical Review* **71**, 913 (L), June 15, 1947.
42. Curtiss, L. F. and Brown, B. W., "Diamond as a Gamma-Ray Counter", *Physical Review* **72**, 643 (L), October 1, 1947.
43. Hofstadter, R., Milton, J. C. D. and Ridgway, S. L., "Behavior of Silver Chloride Crystal Counters", *Physical Review* **72**, 977–978, November 15, 1947.
44. Chynoweth, A. G., "Conductivity Crystal Counters", *American Journal of Physics* **20**, 218–226, April 1952.
45. McKay, K. G., "Electron-Hole Production in Germanium by Alpha-Particles", *Physical Review* **84**, 829–832, November 1951.
46. Mayer, J. W. and Gossick, B. R., "The Use of Au-Ge Broad Area Barrier as Alpha-Particle Spectrometer", *Review of Scientific Instruments* **27**, 407–408 (L), June 1956.
47. McKenzie, J. M. and Bromley, D. A., "Observation of Charged-Particle Reaction Products", *Physical Review Letters* **2**, 303–305, April 1959.
48. Miller, G. L., Gibson, W. M. and Donovan, P., "Semiconductor Particle Detectors", *Annual Review of Nuclear Science* **12**, 189–220, 1962.
49. Reinganum, M., "Streuung und photographische Wirkung der α-Strahlen", *Physikalische Zeitschrift* **12**, 1076–1077, December 1, 1911.
50. Blau, M., "Über die photographische Wirkung natürlicher H-Strahlen", *Sitzungsberichte der Akademie der Wissenschaften in Wien* **134**, 427–436, session of July 9, 1925.
51. Blau, M., "Die photographische Wirkung von H-Strahlen aus Paraffin und Aluminium", *Zeitschrift für Physik* **34**, 285–295, 1925.
52. Rumbaugh, L. H. and Locher, G. L., "Neutrons and Other Heavy Particles in Cosmic Radiation of the Stratosphere", *Physical Review* **49**, 855 (L), June 1, 1936.
53. Wilkins, T. R. and St. Helens, H., "Direct Photographic Tracks of Atomic Cosmic-Ray Corpuscles", *Physical Review* **49**, 403 (L), March 1936.

54. Blau, M. and Wambacher, H., "Vorläufiger Bericht über photographische Ultrastrahlunter-suchungen nebst einigen Versuchen über die 'spontane Neutronemission'. Auftreten von H-Strahlen ähnlichen Bahnen entsprechend mehreren Metern Reichweite in Luft", *Sitzungs-berichte der Akademie der Wissenschaften in Wien* **146**, 469–477, session of July 1, 1937.

55. Blau, M. and Wambacher, H., "Disintegration Processes by Cosmic Rays with the Simultaneous Emission of Several Heavy Particles", *Nature* **140**, 585, October 2, 1937.

56. Møller, C., "On the theory of mesons", *Det Kongelig Danske Videnskabernes Selskab, Matematisk-fysiske Meddleser* **18**, No. 6, 1941.

57. Segrè, E., "An Unsuccessful Search for Transuranic Elements", *Physical Review* **55**, 1104–1105, June 1939.

58. McMillan, E. and Abelson, P. H., "Radioactive Element 93", *Physical Review* **57**, 1185–1186, June 1940.

59. Seaborg, G. T., *The Plutonium story. The journals of professor Glenn T. Seaborg, 1939–1946*, Battelle Press, Columbus, 1994.

60. Seaborg, G., McMillan, E. M., Kennedy, J. W. and Wahl, A. C., "Radioactive Element 94 from Deuterons on Uranium", *Physical Review* **69**, 366–367 (L), April 1, 1946.

61. Seaborg, G., Wahl, A. C. and Kennedy, J. W., "Radioactive Element 94 from Deuterons on Uranium", *Physical Review* **69**, 367 (L), April 1946.

62. Seaborg, G. and Wahl, A. C., "The Chemical Properties of Elements 94 and 93", *Journal of the American Chemical Society* **70**, 1128–1134, 1948.

63. von Grosse, A., "The Chemical Properties of Elements 93 and 94", *Journal of the American Chemical Society* **57**, 440–441, March 1935.

64. Goeppert Mayer, M., "Rare-Earth and Transuranic Elements", *Physical Review* **60**, 184–187, August 1941.

65. Seaborg, G., "The Chemical and Radioactive Properties of the Heavy Elements", *Chemical and Engineering News* **23**, 2190–93, December 10, 1945.

66. Seaborg, G., "Place in the Periodic System and Electronic Structure of the Heaviest Elements", *Nucleonics* **5**, 16–36, November 1949.

67. Snell, A. H., Pleasonton, F. and McCord, R. V., "Radioactive Decay of the Neutron", *Physical Review* **78**, 310–311, May 1950.

68. Robson, J. M., "Radioactive Decay of the Neutron", *Physical Review* **78**, 311–312 (L), May 1950.

69. Robson, J. M., "The Radioactive Decay of the Neutron", *Physical Review* **83**, 349–358, July 15, 1951.

70. Mott, N. F., "The Scattering of Fast Electrons by Atomic Nuclei", *Proceedings of the Royal Society, London* **A124**, 425–442, June 4, 1929.

71. Guth, E., "Über die Wechselwirkung zwischen schnellen Elektronen und Atomkernen", *Anzeiger der Akademie der Wissenschaften in Wien* **24**, 299–306, session of November 22, 1934.

72. Rose, M. E., "The Charge Distribution in Nuclei and the Scattering of High Energy Electrons", *Physical Review* **73**, 279–284, February 1948.

73. Elton, L. R. B., "The Effect of Nuclear Structure on the Elastic Scattering of Fast Electrons", *Proceedings of the Physical Society, London* **A63**, 1115–1124, October 1950.

74. Lyman, E. M., Hanson, A. O. and Scott, M. B., "Scattering of 15.7-Mev Electrons by Nuclei", *Physical Review* **84**, 626–634, November 15, 1951.

75. Pidd, R. W., Hammer, C. L. and Raka, E. C., "High-Energy Electron Scattering by Nuclei", *Physical Review* **92**, 436–437, October 1953.

76. Hofstadter, R., Fechter, H. R. and McIntyre, J., "Scattering of High-Energy Electrons and the Method of Nuclear Recoil", *Physical Review* **91**, 422–423, July 15, 1953.

77. Hofstadter, R., "Electron Scattering and Nuclear Structure", *Reviews of Modern Physics* **28**, 214–254, 1956.

78. Hofstadter, R. and McAllister, R. W., "Electron Scattering from the Proton", *Physical Review* **98**, 217–218, April 1955.

79. McAllister, R. W. and Hofstadter, R., "Elastic Scattering of 188-MeV Electrons from the Proton and the Alpha Particle", *Physical Review* **102**, 851–856, May 1956.

80. Wigner, E. P., "On the Structure of Nuclei beyond Oxygen", *Physical Review* **51**, 947–958, June 1937.

81. Wigner, E. P. and Feenberg, E., "Symmetry properties of nuclear levels", *Reports on Progress in Physics* **8**, 274–317, 1941.

82. Goeppert Mayer, M., "On Closed Shells in Nuclei", *Physical Review* **74**, 235–239, August 1948.

83. Barkas, W. H., "The Analysis of Nuclear Binding Energies", *Physical Review* **55**, 691–698, April 1939.

84. Goeppert Mayer, M., "The Shell Model", in *Nobel Lectures, Physics 1963–1970*, Elsevier, Amsterdam, 1972.

85. Goeppert Mayer, M., "On Closed Shells in Nuclei. II", *Physical Review* **75**, 1969–1970 (L), June 1949.

86. Waenke, H. and Arnold, J. R., "Hans Suess 1909–1993", *Biographical Memoirs of the National Academy of Sciences* **87**, 3–20, 2005.

87. Jensen, J. H. D., "Glimpses at the History of the Nuclear Structure Theory", in *Nobel Lectures, Physics 1963–1970*, Elsevier, Amsterdam, 1972.

88. Jensen, J. H. D., "The History of the Theory of Structure of the Atomic Nucleus", *Science* **147**, 1419–1423, March 19, 1965.

89. Suess, H. E., Haxel, O. and Jensen, J. H. D., "Zur Interpretation der ausgezeichneten Nucleonenzahlen im Bau der Atomkerne", *Naturwissenschaften* **36**, 153–155, July 1949.

90. Jensen, J. H. D., Sueß, H. E. and Haxel, O., "Modellmäßige Deutung der ausgezeichneten Nucleonenzahlen im Kernbau", *Naturwissenschaften* **36**, 155–156, July 1949.

91. Haxel, O., Jensen, J. H. D. and Suess, H. E., "On the "Magic Numbers" in Nuclear Structure", *Physical Review* **75**, 1766, June 1, 1949.

92. Haxel, O., Jensen, J. H. D. and Suess, H. E., "Modellmäßige Deutung der ausgezeichneten Nukleonenzahlen im Kernbau", *Zeitschrift für Physik* **128**, 295–311, 1950.

93. Haxel, O., Jensen, J. H. D. and Suess, H. E., "Das Schalenmodell des Atomkerns", *Ergebnisse der Exakten Naturwissenschaften* **26**, 244–290, 1952.

94. Bethe, H. A. and Bacher, R. F., "Nuclear Physics. A. Stationary States of Nuclei", *Reviews of Modern Physics* **8**, 82–229, April 1936.

95. Fermi, E., *Nuclear physics*, The University of Chicago Press, Chicago, 1949, p. 168.

96. Weisskopf, V. F., "Nuclear Models", *Science* **113**, 101–102, January 26, 1951.

97. Feenberg, E., "Nuclear Shell Structure and Isomerism", *Physical Review* **75**, 320–22 (L), January 15, 1949.

98. Schmidt, T., "Über die magnetischen Momente der Atomkerne", *Zeitschrift für Physik* **106**, 358–361, 1937.

99. Feenberg, E. and Hammack, K. C., "Nuclear Shell Structure", *Physical Review* **75**, 1877–1893, June 15, 1949.

100. Nordheim, L. W., "On Spins, Moments, and Shells in Nuclei", *Physical Review* **75**, 1894–1901, June 1949.

101. Goeppert-Mayer, M. and Jensen, J. H. D., *Elementary theory of nuclear shell structure*, John Wiley, New York, 1955.

102. Cook, L. G., McMillan, E. M., Peterson, J. M. and Sewell, D. C., "Total Cross Sections of Nuclei for 90-Mev Neutrons", *Physical Review* **72**, 1264–1265, December 1947.

103. Serber, R., "Nuclear Reactions at High Energies", *Physical Review* **72**, 1114–1115, December 1947.

104. Bethe, H. A., "A Continuum Theory of the Compound Nucleus", *Physical Review* **57**, 1125–1144, June 15, 1940.

105. Fernbach, S., Serber, R. and Taylor, T. B., "The Scattering of High Energy Neutrons by Nuclei", *Physical Review* **75**, 1352–1355, May 1949.

106. Burkig, J. W. and Wright, B. T., "Survey Experiment on Elastic Scattering", *Physical Review* **82**, 451–452, May 1951.

107. Richardson, R. E., Ball, W. P., Leith, C. E. and Moyer, B. J., "Elastic Scattering of 340-Mev Protons", *Physical Review* **83**, 859–860 (L), August 1951.
108. Gatha, K. M. and Riddell, Jr., R. J., "An Investigation into the Nuclear Scattering of High Energy Protons", *Physical Review* **86**, 1035–1039, June 1952.
109. Gugelot, P. C., "Some Data on the Elastic Scattering of 18.3-Mev Protons", *Physical Review* **87**, 525–526, August 1952.
110. Le Levier, R. E. and Saxon, D. S., "An Optical Model for Nucleon-Nuclei Scattering", *Physical Review* **87**, 40–41, July 1952.
111. Cohen, B. L. and Neidigh, R. V., "Angular Distributions of 22-Mev Protons Elastically Scattered by Various Elements", *Physical Review* **93**, 282–287, January 15, 1954.
112. Dayton, I. E., "The Elastic Scattering of 18-Mev Protons by Al, Fe, Ni, and Cu", *Physical Review* **95**, 754–758, August 1954.
113. Chase, D. M. and Rohrlich, F., "Elastic Scattering of Protons by Nuclei", *Physical Review* **94**, 81–86, April 1954.
114. Woods, R. D. and Saxon, D. D., "Diffuse Surface Optical Model for Nucleon-Nuclei Scattering", *Physical Review* **95**, 577–578, July 1954.
115. Williams, M. R., *A history of computing technology*, Prentice-Hall, Englewood Cliffs, N.J., 1985.
116. Bohr, N., Peierls, R. and Placzek, G., "Nuclear Reactions in the Continuous Energy Region", *Nature* **144**, 200–201, July 29, 1939.
117. Helmholz, A. C., McMillan, E. M. and Sewell, D. C., "Angular Distribution of Neutrons from Targets Bombarded by 190-Mev Deuterons", *Physical Review* **72**, 1003–1007, December 1947.
118. Serber, R., "The Production of High Energy Neutrons by Stripping", *Physical Review* **72**, 1008–1016, December 1947.
119. Peaslee, D. C., "Deuteron-Induced Reactions", *Physical Review* **74**, 1001–1013, November 1948.
120. Burrows, H. B., Gibson, W. M. and Rotblat, J., "Angular Distributions of Protons from the Reaction $O^{16}(d, p)O^{17}$", *Physical Review* **80**, 1095, December 15, 1950.
121. Butler, S. T., "On Angular Distributions from (d, p) and (d, n) Nuclear Reactions", *Physical Review* **80**, 1095–1096 (L), December 15, 1950.
122. Holt, J. R. and Young, C. T., "The Angular Distribution of Protons from the Reaction $^{27}Al(d,p)^{28}Al$", *Proceedings of the Physical Society* **63**, 833–838, August 1950.
123. Butler, S. T., "Angular distributions from (d,p) and (d,n) nuclear reactions", *Proceedings of the Royal Society, London* **A208**, 559–579, September 24, 1951.
124. Rotblat, J., "The Spins and Parities of the 3.7-3.9-Mev Doublet in C^{13}", *Physical Review* **83**, 1271–1272, September 1951.
125. Daitch, P. B. and French, J. B., "The Born Approximation Theory of (d, p) and (d, n) Reactions", *Physical Review* **87**, 900–901 (L), September 1952.
126. Horowitz, J. and Messiah, A. M. L., "The Mechanism of Stripping Reactions", *Physical Review* **92**, 1326–1327, December 1, 1953.
127. Horowitz, J. and Messiah, A. M. L., "Sur les réactions (d, p) et (d, n)", *Journal de Physique et Le Radium* **14**, 695–706, December 1953.
128. Francis, N. C. and Watson, K. M., "The Theory of the Deuteron Stripping Reactions", *Physical Review* **93**, 313–317, January 15 1954.
129. Tobocman, W., "Theory of the (d, p) Reaction", *Physical Review* **94**, 1655–1663, June 1954.
130. Tobocman, W. and Kalos, M. H., "Numerical Calculation of (d, p) Angular Distributions", *Physical Review* **97**, 132–136, January 1, 1955.
131. Thomas, R. G., "Collision Matrices for the Compound Nucleus", *Physical Review* **97**, 224–237, January 1955.
132. Mott, N. F. and Massey, H. S. W., *The Theory of atomic collisions*, Clarendon Press, Oxford, 1933. p. 100.
133. Bethe, H. A., "Nuclear Physics. B. Nuclear Dynamics, Theoretical", *Reviews of Modern Physics* **9**, 69–244, April 1937.

134. Bothe, W. and Horn, W., "Die Sekundärstrahlung harter γ-Strahlen", *Zeitschrift für Physik* **88**, 683–698, 1934.

135. Bothe, W. and Gentner, W., "Die Streu- und Sekundärstrahlung harter γ-Strahlen", *Naturwissenschaften* **24**, 171—172, March 13, 1936.

136. Baldwin, G. C. and Koch, H. W., "Threshold Measurements on the Nuclear Photo-Effect", *Physical Review* **67**, 1–11, January 1945.

137. Baldwin, G. C. and Klaiber, G. S., "Photo-Fission in Heavy Elements", *Physical Review* **71**, 3–10, January 1, 1947.

138. McElhinney, J., Hanson, A. O., Becker, R. A., Duffield, R. B. and Diven, B. C., "Thresholds for Several Photo-Nuclear Reactions", *Physical Review* **75**, 542–554, February 1949.

139. Goldhaber, M. and Teller, E., "On Nuclear Dipole Vibrations", *Physical Review* **74**, 1046–1049, November 1, 1948.

140. Gordy, W., "Relation of Nuclear Quadrupole Moment to Nuclear Shell Structure", *Physical Review* **76**, 139–140 (L), July 1949.

141. Townes, C. H., Foley, H. M. and Low, W., "Nuclear Quadrupole Moments and Nuclear Shell Structure", *Physical Review* **76**, 1415–1416 (L), November 1949.

142. Rainwater, J., "Nuclear Energy Level Argument for a Spheroidal Nuclear Model", *Physical Review* **79**, 432–434 (L), August 1950.

143. Bohr, A., "On the Quantization of Angular Momenta in Heavy Nuclei", *Physical Review* **81**, 134–138, January 1951.

144. Bohr, A., "The Coupling of Nuclear Surface Oscillations to the Motion of Individual Nucleons.", *Det Kongelig Danske Videnskabernes Selskab, Matematisk-fysiske Meddleser* **26**, No. 14, 1–40, 1952.

145. Bohr, A. and Mottelson, B. R., "Beta-Decay and the Shell Model, and the Influence of Collective Motion on Nuclear Transitions", *Physica* **18**, 1066–1078, December 1952.

146. Goldhaber, M. and Sunyar, A. W., "Classification of Nuclear Isomers", *Physical Review* **83**, 906–918, September 1, 1951.

147. Scharff-Goldhaber, G., "Excited States of Even-Even Nuclei", *Physica* **18**, 1105–1109, December 1952.

148. Asaro, F. and Perlman, I., "First Excited States of Even-Even Nuclides in the Heavy Element Region", *Physical Review* **87**, 393–394, July 1952.

149. Curie, I., "Étude du rayonnement γ de l'ionium", *Journal de Physique et Le Radium* **10**, 381–386, December 1949.

150. Rasetti, F. and Booth, E. C., "Gamma-Ray Spectrum of Ionium (Th^{230})", *Physical Review* **91**, 315–318, July 1953.

151. Goldhaber, M. and Hill, R. D., "Nuclear Isomerism and Shell Structure", *Reviews of Modern Physics* **24**, 179–239, July 1952.

152. Scharff-Goldhaber, G., "Excited States of Even-Even Nuclei", *Physical Review* **90**, 587–602, May 1953.

153. Bohr, A. and Mottelson, B. R., "Interpretation of Isomeric Transitions of Electric Quadrupole Type", *Physical Review* **89**, 316–317 (L), January 1953.

154. Bohr, A. and Mottelson, B. R., "Rotational States in Even-Even Nuclei", *Physical Review* **90**, 717–719, May 1953.

155. Bohr, A. and Mottelson, B. R., "Collective and Invidual Aspects of Nuclear Structure", *Det Kongelig Danske Videnskabernes Selskab, Matematisk-fysiske Meddleser* **27**, No. 16, 1–175, 1953.

156. Nilsson, S. G., "Binding States of Individual Nucleons in Strongly Deformed Nuclei", *Det Kongelig Danske Videnskabernes Selskab, Matematisk-fysiske Meddleser* **29**, No. 16, 1–68, 1955.

157. Weisskopf, V. F., "Excitation of Nuclei by Bombardment with Charged Particles", *Physical Review* **53**, 1018 (L), June 1938.

158. Ter-Martirosyan, K. A., "Excitation of Nuclei by the Coulomb field of Charged Particles", *Zhurnal Eksperimental' noi i Teoreticheskoi Fiziki (Journal of Experimental and Theoretical Physics)* **22**, 284–296, 1952. English translation in Kurt Alder and Aage Winther, eds., *Coulomb excitation, a collection of reprints*, pp. 15–19.

159. Alder, K. and Winther, A., "The Theory of Coulomb Excitation of Nuclei", *Physical Review* **91**, 1578–1579 (L), September 1953.
160. Huus, T. and Zupančič, Č., "Excitation of Nuclear Rotational States by the Electric Field of Impinging Particles", *Det Kongelig Danske Videnskabernes Selskab, Matematisk-fysiske Meddleser* **28**, No. 1, 1–19, 1953.
161. Alder, K., Bohr, A., Huus, T., Mottelson, B. R. and Winther, A., "Study of Nuclear Structure by Electromagnetic Excitation with Accelerated Ions", *Reviews of Modern Physics* **28**, 432–542, October 1956.
162. Seaborg, G. T., "Table of Isotopes", *Reviews of Modern Physics* **16**, 1–32, 1944.
163. Seaborg, G. T. and Perlman, I., "Table of Isotopes", *Reviews of Modern Physics* **20**, 585–667, 1948.
164. Hollander, J. M., Perlman, I. and Seaborg, G. T., "Table of Isotopes", *Reviews of Modern Physics* **25**, 469–651, 1953.
165. Strominger, D., Hollander, J. M. and Seaborg, G. T., "Table of Isotopes", *Reviews of Modern Physics* **30**, 585–904, 1958.
166. Hoisington, L. E., Share, S. S. and Breit, G., "Effects of Shape of Potential Energy Wells Detectable by Experiments on Proton-Proton Scattering", *Physical Review* **56**, 884–890, November 1939.
167. Yukawa, H. and Sakata, S., "Mass and Mean Life-Time of the Meson", *Nature* **143**, 761–762, May 6, 1939.
168. Nordheim, L. W., "Lifetime of the Yukawa Particle", *Physical Review* **55**, 506 (L), March 1939.
169. Bethe, H. A. and Nordheim, L. W., "On the Theory of Meson Decay", *Physical Review* **57**, 998–1006, June 1940.
170. Tomonaga, S.-I. and Araki, G., "Effect of the Nuclear Coulomb Field on the Capture of Slow Mesons", *Physical Review* **58**, 90–91, July 1940.
171. Conversi, M., Pancini, E. and Piccioni, O., "On the Disintegration of Negative Mesons", *Physical Review* **71**, 209–210 (L), February 1947.
172. Sakata, S. and Inoue, T., "On the Correlations between Mesons and Yukawa Particles", *Progress of Theoretical Physics*. **1**, 143–149, November-December 1946.
173. Tanikawa, Y., "On the Cosmic-Ray Meson and the Nuclear Meson", *Progress of Theoretical Physics* **2**, 220–221, November–December 1947.
174. Marshak, R. E. and Bethe, H. A., "On the Two-Meson Hypothesis", *Physical Review* **72**, 506–509, September 1947.
175. Lattes, C. M. G., Muirhead, H., Occhialini, G. P. S. and Powell, C. F., "Processes Involving Charged Mesons", *Nature* **159**, 694–697, May 24, 1947.
176. Lattes, C. M. G., Occhialini, G. P. S. and Powell, C. F., "Observation of the Tracks of Slow Mesons in Photographic Emulsions", *Nature* **160**, I : pp. 453–456, October 4; II : pp. 486–492, October 11, 1947.
177. Gardner, E. and Lattes, C. M. G., "Production of Mesons by the 184-Inch Berkeley Cyclotron", *Science* **107**, 270–271, March 12, 1948.
178. Richardson, J. R., "The Lifetime of the Heavy Meson", *Physical Review* **74**, 1720–1721, December 1948.
179. Steinberger, J., Panofsky, W. K. and Steller, J., "Evidence for the Production of Neutral Mesons by Photons", *Physical Review* **78**, 802–805, June 15, 1950.
180. Kemmer, N., "The Charge-Dependence of Nuclear Forces", *Proceedings of the Cambridge Philosophical Society* **34**, 354–364, session of May 16, 1938.
181. Kemmer, N., "Quantum theory of Einstein-Bose particles and nuclear interactions", *Proceedings of the Royal Society, London* **A166**, 127–153, 1938.
182. Frölich, H., Heitler, W. and Kemmer, N., "On the nuclear forces and the magnetic moment of the neutron and the proton", *Proceedings of the Royal Society, London* **A166**, 154–177, 1938.
183. Panofsky, W. K., Aamodt, R. L. and Hadley, J., "The Gamma-Ray Spectrum Resulting from Capture of Negative π-Mesons in Hydrogen and Deuterium", *Physical Review* **81**, 565–574, February 1951.

184. Peierls, R., "The Development of our Ideas on the Nuclear Forces", in *Nuclear Physics in retrospect*, edited by Stuewer, R. H., pp. 183–211, University of Minnesota Press, 1979.
185. Chamberlain, O. and Wiegand, C., "Proton-Proton Scattering at 340 Mev", *Physical Review* **79**, 81–85, July 1950.
186. Jastrow, R., "On the Nucleon-Nucleon Interaction", *Physical Review* **81**, 165–170, January 1951.
187. Blatt, J. M. and Weisskopf, V. F., *Theoretical Nuclear Physics*, John Wiley & Sons, New York, 1952, p. 777.
188. Johnson, M. H. and Teller, E., "Classical Field Theory of Nuclear Forces", *Physical Review* **98**, 783–787, May 1955.
189. Brueckner, K. A., Eden, R. J. and Francis, N. C., "High-Energy Reactions and the Evidence for Correlations in the Nuclear Ground-State Wave Function", *Physical Review* **98**, 1445–1455, June 1955.
190. Brueckner, K. A., Levinson, C. A. and Mahmoud, H. M., "Two-Body Forces and Nuclear Saturation. I. Central Forces", *Physical Review* **95**, 217–228, July 1954.
191. Brueckner, K. A., Eden, R. J. and Francis, N. C., "Nuclear Energy Level Fine Structure and Configuration Mixing", *Physical Review* **99**, 76–87, July 1955.
192. Brueckner, K. A., Eden, R. J. and Francis, N. C., "Theory of Neutron Reactions with Nuclei at Low Energy", *Physical Review* **100**, 891–900, November 1955.
193. Bethe, H. A., "Nuclear Many-Body Problem", *Physical Review* **103**, 1353–1390, September 1956.
194. Goldstone, J., "Derivation of the Brueckner many-body theory", *Proceedings of the Royal Society, London* **A239**, 267–279, February 26, 1957.
195. Feynman, R. P., "The Theory of Positrons", *Physical Review* **76**, 749–759, September 15, 1949.
196. *Amsterdam Nuclear Reactions Conference, Physica* **22**, 941–1123, 1956.
197. Lipkin, H. J. (editor), *Proceedings of the Rehovoth Conference on Nuclear Structure held at the Weizmann Institute of Science, Rehovoth, september 8–14, 1957*, North Holland, Amsterdam, 1958.
198. Gugenberger, P. (editor), *Comptes rendus du congrès international de physique nucléaire, Interactions nucléaires aux basses énergies et structure des noyaux, Paris, 2–7 juillet 1958*, Dunod, Paris, 1959.

Where the Narrative Ends

Jeg hørte ham mumle ind i Edderkoppens Spind.
— ""Du flinke, lille Væver! Du lærer mig at holde
ud! rives itu dit Spind, begynder Du frofra igjen
og fuldender! atter itu — og ufortrøden tager Du
igjen fat, forfra! — forfra! det er det man skal! og
det lønnes.""

H. C. Andersen, *Vinden fortæller om Valdemar
Daae og hans Døttre*

I heard him as he thus spoke; he was looking at
a spider's web, and he continued, "Thou cunning
little weaver, thou dost teach me perseverance.
Let any one tear thy web, and thou wilt begin
again and repair it. Let it be entirely destroyed,
thou wilt resolutely begin to make another till it
is completed. So ought we to do, if we wish to
succeed at last."

Translated by H. P. Paull (1872)

At the end of the 1950s, nuclear physics is in full bloom. The structure of nuclei
is beginning to be understood, and increasingly powerful experimental means
probe the nucleus as never before. The economic prosperity as well as strategic
considerations, royally endow nuclear physics both humanly and financially. The
early years of teeming activity are followed by a period of intensive work devoted
to the consolidation and the deepening of our understanding of nuclei.

B. Fernandez and G. Ripka, *Unravelling the Mystery of the Atomic Nucleus:*
A Sixty Year Journey 1896 — 1956, DOI 10.1007/978-1-4614-4181-6_8,
© Springer Science+Business Media New York 2013

The structure of nuclei is basically understood but further experimental data reveal further properties and the interactions between nuclei. In the 1930s there existed a hope of constructing an ideal theory: one would start with an interaction between the constituents of the nucleus. It was expected that the interaction would be as simple as gravitation or as electromagnetic interactions which govern the structure of the atom. This interaction would then be used to calculate and predict the properties of every nucleus. After all, it is with such a theory that we understand the motion of planets in our solar system as well as the state of electrons in atoms. The electric interaction between electrons is simple and quantum mechanics describes accurately the properties of atoms, even if, in practice, calculations can be quite complex.

But gradually this turned out to be more complicated. To begin with, the hope of discovering a simple interaction between the nucleons was given up. Meson exchange made it more complicated. Furthermore, it was discovered in the late 1960s that protons and neutrons are not elementary particles. Instead they are composed of *quarks* which interact by exchanging *gluons,* the underlying theory of which is called quantum chromodynamics (QCD).

The nucleus is composed of a large number of interacting nucleons so that, in almost all cases, we are only able to make approximate calculations of their structure. The same problem is encountered in other fields of physics, in chemistry, for example, even more so in biology. In spite of knowing the interactions between the constituents we are far from being able to predict the chemical properties and the structure of molecules.

This is also the reason why experiments are required to discover the properties of nuclei. Surprising experimental discoveries often point out how a theory, which is able to account for some properties, all too often fails to *predict* them. For example, certain nuclei were discovered to have exceptionally deformed shapes with a ratio of three between the major and minor axes. They had not been predicted. Another example is the discovery in the 1960s of the so-called double fission barrier: some fissioning nuclei undergo a strong deformation and spend an unusually long time before breaking up into two pieces. Both of these examples were later explained in terms of "shell effects" which favor energetically certain seemingly outlandish configurations. But occasionally theory *is* predictive. For example, it predicted the existence of some elements around a proton magic number between 114 and 126 and a neutron magic number around 126. These elements could have an exceptionally long half-life. They were named *superheavy elements.* The heaviest element which has been produced and identified so far has 116 protons and 175 neutrons! This again is a shell effect.

Today, physicists explore exotic and metastable nuclei. They are able to produce nuclei with a much larger neutron excess than normally stable nuclei and also nuclei which have exceptionally high spin. Research alternates between great discoveries and patient, obstinate work. In her book *Pierre Curie,*[1] written in 1923, Marie Curie wrote on p. 104–105:

---

[1] Marie Curie-Skłodowska, *Pierre Curie,* Macmillan, New York, 1923.

A great discovery never pops out of the brain of a scientist, ready made, as Minerva did, fully armed, out of the brain of Jupiter. It is the fruit of preliminary and accumulated work. Between days of fertile production are days of uncertainty during which nothing seems to work and nature itself appears to be hostile. That is when one has to resist becoming discouraged.

Rutherford made a similar description of the evolution of physics in a talk delivered in 1936 in Cambridge, entitled "Forty years of physics"[1]:

Science goes step by step, and every man depends on the work of his predecessors. When you hear of a sudden unexpected discovery—a bolt from the blue as it were—you can always be sure that it has grown up by the influence of one man on another, and it this mutual influence which makes the enormous possibility of scientific advance. Scientists are not dependent on the ideas of a single man, but on the combined wisdom of thousands of men, all thinking of the same problem, and each doing his little bit to add to the great structure of knowledge which is gradually being erected.

*   *

*

Dear reader, this is where we interrupt the story relating how a few black spots detected on a photographic plate led some men—and women—to conceive instruments as well as theories able to reveal secrets of nature. Have I succeeded in communicating to you my admiration of the towering knowledge acquired in the span of barely 60 years due to the unrelenting and imaginative work of so many physicists, some with and some without fame, all bound to this adventure?

---

[1] See p. 72.

Glossary

The entries of this glossary are marked in the text by a small diamond-shaped exponent$^\diamond$.

Absolute zero, absolute temperature. Absolute zero is the lowest temperature which a body can reach. It is equal to $-273.15\,°C$. The absolute temperature of a body is its temperature measured relative to the absolute zero temperature. It is obtained by adding 273.15 to the temperature expressed in degrees Celsius.

In the kinetic theory of gases, the molecules are assumed to be in constant disordered motion. The average kinetic energy of the molecules is proportional to the absolute temperature of the gas. When the temperature is lowered, the gas liquefies, in which case the molecules are in contact. If the temperature is further reduced, the liquid solidifies and becomes a crystal. The molecules remain then at fixed distances one from the other; however, they still vibrate around their mean position. When the temperature reaches absolute zero, the molecules become completely motionless. This limit can never be reached in practice.

Action. In classical mechanics, the action is the quantity which integrates, along the path of a moving particle, such as an atom or a planet, the product of its momentum, that is, its mass multiplied by its velocity. The action has a remarkable property discovered by the French mathematician, philosopher and physicist Pierre Louis Moreau de Maupertuis (1698–1759): the classical equations of motion are such that, among all the possible paths, the one which the particle follows is the one for which the action is smallest. This is the famous *principle of minimum action*.

In quantum mechanics, the *quantum of action*, which was discovered by Max Planck and which is equal to Planck's constant h, is the smallest value which an action can take. The action can take values equal to $h,\ 2h,\ 3h, \ldots, nh$. In the

B. Fernandez and G. Ripka, *Unravelling the Mystery of the Atomic Nucleus:*
A Sixty Year Journey 1896 — 1956, DOI 10.1007/978-1-4614-4181-6,
© Springer Science+Business Media New York 2013

scale of daily life (but not in the atomic scale), n is a very large integer because Planck's constant h is very small so that the action appears to be continuous. This is no longer the case in the scale of atoms and nuclei.

Angular momentum. To the motion of a massive body, we can assign a velocity or, better, a *momentum*, which is equal to its velocity multiplied by its mass . Consider a body rotating around an axis, such as a planet orbiting around the sun, or a hand-held sling which is spun around, or still a spinning top. To the circular motion of the body, we can assign an *angular momentum* which is equal to the momentum of the body multiplied by the distance separating the body from its rotation axis. For the planet, the latter would be the distance to the sun, whereas for the sling it would be the distance between the sling and the hand. For the spinning top, the angular momentum is a bit harder to calculate because various parts of the top are at different distances from its rotation axis. An angular momentum has a definite direction in space and it is represented by a vector which is parallel to the rotation axis.

Angular momentum is a useful concept because when a system, such as a planet and the sun, does not undergo external perturbations, its angular momentum remains constant in time. When a top is set spinning fast, it would continue spinning around its axis indefinitely were it not for the friction exerted on its tip which touches the ground. It is on this same principle that modern gyrocompasses are built on boats, enabling them to orient themselves.

Angular momentum is measured in the same units as the *action*. In quantum mechanics, the angular momentum is quantized in units of an elementary angular momentum. The angular momentum of a particle orbiting around a center (such as an electron around a nucleus) is called *orbital* angular momentum and, in quantum mechanics, it can only take discrete values which are multiples of $\hbar$, which is Planck's constant h divided by 2π (the usual notation is $\hbar = h/2\pi$). However, particles, such as electrons, neutrons, and protons, for example, have an intrinsic angular momentum which is nonvanishing even when the particle is a rest. This intrinsic angular momentum is called the *spin* of the particle. It does not correspond to a rotation of matter and it cannot be stopped. In quantum mechanics, it can take either integer or half-integer multiples of $\hbar$ ($= h/2\pi$). For example, the π meson has a spin equal to zero. The electron, the proton, and the neutron have spins $\frac{1}{2}$, that is, equal to $\frac{1}{2}\hbar$ (see *spin*). The photon has a spin 1.

Atomic mass, atomic weight. In the nineteenth century, the Avogadro number (see *Avogadro*) was not known so that it was not possible to determine the weight of an atom. However, it was possible to compare atomic weights. It was observed that, when they were expressed in units of the atomic weight of hydrogen, the atomic weights of other elements were almost integers. It was therefore decided to determine the relative atomic weights. In fact, for practical reasons, it was decided to use another unit, namely, 1/16 of the atomic weight of oxygen. This only made a small difference. Today, we prefer to refer to masses (see *mass*) rather than to weights. See *mass and weight.*

Avogadro's number. At the beginning of the nineteenth century, Dalton showed how the *ratios* of the masses of atoms of various elements could be determined from the chemical formulas of composite substances. However, it is only at the end of the century that the *number of hydrogen atoms* contained in 1 g of hydrogen became known. As soon as this number N, which Jean Perrin proposed to call *Avogadro's number*, was known, the masses of individual atoms could be determined (see *Brownian motion*). Today, Avogadro's number is known to a precision of one part in twenty million. It is equal to $N = 6.022\,141\,79 \times 10^{23}$ with an uncertainty of $0.000\,000\,3 \times 10^{23}$.

Bar. Unit in terms of which pressure is measured. See *pascal*.

Binding, binding energy, bound systems. The sun and the earth which is orbiting around the sun constitute what is called a *bound system* because they cannot separate without being given the energy to do so. A grounded space rocket is obviously bound to the earth. If it is sent into outer space at a sufficient speed, it can become an orbiting satellite in which sense it remains bound to the earth. The rocket must be given further energy to leave the earth for good, and even more energy to leave the solar system. The velocity, which the rocket needs to reach in order to be separated from the earth or from the sun, is called the *escape velocity*. Similarly, the atom is a bound system consisting of a nucleus and electrons. The energy required to extract an electron from the atom (thereby transforming the atom into a positive ion) is called the *binding energy* of the electron in the atom. The inverse process, consisting of the capture of an electron by an atom, liberates (in the form of light quanta) an energy equal to the binding energy of the electron.

Black-body radiation. If you bring your hand close to a hot iron, you are quickly convinced that it radiates heat. The heat is emitted by electromagnetic waves which are identical to light waves, but which have a long wavelength so that they are not visible by the eye. This is called infrared radiation. When a piece of iron is heated up to about 500 °C, it emits a weak deep red glow. As its temperature rises, it becomes more luminous and it changes color. For example, the tungsten wire of an electric lamp reaches 2,600 °C and it emits a white light. In general, a body which is heated up to a certain temperature emits a radiation which is a mixture of different wavelengths (colors) and the mixture changes when the temperature changes. What we call "white light" is light the different colors of which are similar to the light emitted at the surface of the sun at a temperature of about 5,600 °C.

Every substance has an absorption coefficient which depends on the nature of its surface: a brilliant or white surface reflects a greater proportion of incident light. In 1860, Kirchhoff introduced the concept of a "blackbody" which is a substance which ideally absorbs 100 % of the incident radiation. A material which is painted in matt black is similar to the "blackbody" of physicists. A closer approximation to the blackbody is a closed cavity with absorptive walls and with a small hole through which the interior can be observed, somewhat like a baker's oven. Indeed, the radiation which enters through the hole has a small probability of escaping and is thus almost completely absorbed.

The concept of blackbody made it possible to define fundamental physical laws deduced from thermodynamics. For example, the Stefan-Boltzmann law

Fig. 1 The distribution of
emitted energy versus
wavelength according to the
Planck formula for the
black-body radiation at
5800 K

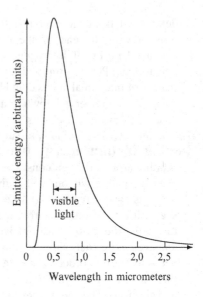

(1879–1885) states that the radiated energy is proportional to the absolute temperature raised to the fourth power.

The consequences of these laws are important. For example, if the temperature of a glowing wire in a lamp is raised from 2600 °C to 3000 °C (as in tungsten halogen lamps), the emitted energy increases by 68 % and the emitted light has less infrared and more visible light: the lighting efficiency is almost doubled.

Is it possible to calculate the distribution of wavelengths (colors) of light emitted by a blackbody at a certain temperature, using the laws of thermodynamics? At the end of the nineteenth century, this was the problem facing physicists such as Wilhelm Wien, Lord Rayleigh, James Jean, and Max Planck. Planck was able to explain the long wavelength part of the distribution by assuming that the radiation was not emitted in a continuous fashion by rather as indivisible packages, which became known as quanta of light. Figure 1 shows the distribution of energy radiated by a black body at the temperature of the surface of the sun. The maximum radiated energy is for visible light with wavelengths in the range of 0.4–0.8 μm.

Brownian motion. By observing in a microscope small particles of pollen suspended in water, the botanist Robert Brown noticed in 1827 that they were animated with a random motion. He realized that this motion had nothing to do with the biological nature of the particles nor with the motion of the water. The origin of this motion remained a mystery for a long time. In 1860 the Italian Giovanni Cantoni and in 1877 the Belgian Jesuits Joseph Delsaux and Ignace Carbonelle suggested that the motion could be due to collisions of the pollen particles with water molecules. In 1865, the Austrian physicist Joseph Loschmidt gave the first estimate of the size of molecules. However, the size of the pollen particles was much greater than the size of the molecules and the collisions were far too frequent (about a thousand billion collisions per second) for individual collisions to be observed. In a paper published in 1905, Einstein showed that the observed displacements of the particle were due to fluctuations of the collisions: at a given time, the particle does not undergo exactly the same number of collisions in every direction, so that it is set in motion in a chaotic fashion and it slowly drifts away. The same phenomenon occurs when a drop of wine is deposited in still water: the drop slowly diffuses. Einstein was able to relate the observed displacement of the particles to Avogadro's number. By measuring the speed of the displacement, Jean Perrin succeeded in estimating Avogadro's number.

Calorie, gram-calorie. One calorie (which used to be called the small calorie or the gram-calorie) is the amount of heat (of energy) required to raise the temperature of 1 g of water by 1 °C. The large calorie, also called the kilogram-calorie, is equal to 1,000 small calories. Today, we use a universal unit of energy, namely the joule (J): 4.184 J are required to raise the temperature of 1 g of water by 1 °C. This means that it takes 4.184 s for a 1-W water heater to do this. Nowadays, calories are only used to label food products which are endowed with a "calorific value." It usually refers to kilogram-calories (kcal) and to their equivalent in joules (or kilojoules, *i.e.,* 1000 J).

Cathode rays. In 1675, Jean-Felix Picard (1620–1682), a French astronomer and priest, gave the first description of an extraordinary luminous phenomenon. When he transported his barometer at night, he observed that, whenever the mercury was abruptly shaken, a bluish light appeared in the vacuum above it. The experiment was repeated several times at the *Académie des Sciences*, in particular by Daniel Bernoulli (1700–1782), a Dutch-Swiss mathematician, but no reasonable explanation could be found. Seventy years later, the phenomenon was linked to electricity by the German physician Christian Ludolf (1707–1763), who showed in 1744 that *"the luminous barometer is made perfectly electrical by the motion of the quicksilver; first attracting, and then repelling bits of paper, etc. suspended by the side of the tube."*

For a long time, luminous phenomena produced by electric discharges in rarefied gases remained curiosities which were displayed in salons and fun fairs. But the almost simultaneous appearance of the Geissler tube and the Ruhmkorff coil lead Julius Plücker (1801–1868), a mathematician in the University of Bonn who became an experimental physicist rather late in life, to discover that the luminous phenomena changed progressively as the vacuum in the tube improved. He noticed in 1857–1959 that "a wonderful green glow" appeared on the glass facing the negative electrode that Faraday had called the cathode. Eleven years later, his pupil Johann Wilhelm Hittorf, a German physicist, noticed that the glow was caused by "rays" which emanated from the cathode and propagated in a straight line. These observations were confirmed by the German physicist Eugen Goldstein (1850–1930) who called them *"Cathodenstrahlen,"* meaning cathode rays.

From then on, physicists throughout all Europe began to study cathode rays. They adhered to two different interpretations. The *materialists*, among whom Joseph John Thomson (1856–1940), the director of the Cavendish Laboratory, believed that the cathode rays are electrically charged material particles. The others considered them to be rays, like electromagnetic waves. Many German physicists adhered to the latter group, namely, Heinrich Hertz (1857–1894) and his student Philipp Lenard (1862–1947), Gustav Wiedemann (1826–1899), and Eugen Goldstein (1850–1930).

The materialist interpretation prevailed. In his thesis work in 1875, Jean Perrin (1870–1942) showed that cathode rays carried negative electric charge. Joseph John Thomson (1856–1940) measured the deviation of cathode rays by magnetic and electric fields. He deduced their velocity as well as their charge/mass ratio. Their velocity was barely 10 % of the velocity of light, which excluded the possibility that they should be electromagnetic waves. The measured charge/mass ratio remained remarkably constant, no matter what material the cathode was made of. Thomson concluded that cathode rays were particles of identical nature. Subsequently, he succeeded in measuring both the mass and the charge of these particles. They carried the smallest electric charge which had ever been observed and which in due time became the fundamental unit of electric charge. The word "electron" was proposed in 1888 by George Johnstone Stoney (1826–1911) to designate the natural unit of electric charge. Eventually, it was used to designate the new particles.

Chain reaction. Chain reactions are at the root of all explosive processes. When city gas accumulates into a space containing oxygen, a single spark can provoke an explosion. The gas molecules can combine with the oxygen molecules thereby emitting heat (in what is called an *exothermic* reaction). But in order to do so, the oxygen-gas mixture needs to be heated up to a certain temperature and this is what the spark does. The spark allows a few molecules to combine and in doing so they emit sufficient heat to allow further molecules to combine. This snowball effect, so to speak, allows the chemical reaction to propagate through the whole volume in a very short time, during which a large amount of heat is produced together with a sudden expansion of the gas: it is an *explosion*. The nuclear chain reaction is based on the same principle. However, the amount of heat which is emitted in more than a million times larger than in chemical reactions. This is why nuclear explosions are colossal.

Cross section. Consider, for example, a flux of neutrons passing through matter. The neutrons can collide and interact with the nuclei of the atoms. Seen from afar, the spherical nuclei appear as tiny disks presenting a surface roughly equal to the surface of a circle. Consider the case where a neutron is absorbed whenever it hits a nucleus, thereby forming a new isotope of the element. The *cross section* of this process is then equal to the surface of the apparent disk. Now assume that, when it hits the nucleus, the neutron has a 50 % probability of being scattered and a 50 % probability of being absorbed. The cross section *for producing a new isotope* is then equal to half of the surface of the apparent disk. It is as if, for this particular process (the absorption of the neutron), the nucleus presents a smaller surface. The measurement of a cross section

Fig. 2 A schematic representation of the diffraction of incoming sea waves by a jetty or a pier in a harbour; the waves approaching from the sea turn around the extremity of the jetty

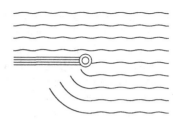

yields the probability for a process to occur. Nuclei are very small: their radii range from 10^{-15} to 6×10^{-15} m. The cross sections for nuclear reactions are therefore of the order of the square of their radii, which is 10^{-28} m^2, a unit which is called a *barn*[1]. Many cross sections are expressed in units of *millibarns*, one millibarn being equal to one thousandth of a barn.

The Curie law and the Curie temperature. It is by studying the magnetization of various substances that Pierre Curie formulated the physical law which bears his name. When placed in a magnetic field, in the vicinity of a magnet, for example, some substances (called diamagnetic) reduce somewhat the magnetic field in their interior, other substances (called paramagnetic) increase it somewhat, and some metals such as iron, nickel, and cobalt increase it very strongly: the magnetic field can become 10,000 times more intense in their interior. Such metals are called ferromagnetic and they become magnets themselves. Pierre Curie showed that, above a certain temperature, a ferromagnetic material becomes paramagnetic.

Diffraction. We have all observed incoming sea waves which are stopped by a pier. At the extremity of the pier, the waves appear to turn somewhat, and the extremity seems to emit waves into the calm waters inside the harbor (Fig. 2). This phenomenon is called diffraction: waves have a tendency to turn around any obstacle they encounter, so that shadows are never perfectly sharp. For this same reason, we can hear people talking in another room, in spite of obstacles which prevent the sound from propagating in a straight line. The waves turn less around obstacles when their wavelength decreases. This is why the direction from which a high pitched musical note comes from is better perceived than the direction of a low pitched note. It is also diffraction which sets a limit to the sharpness of photographs and to the resolution of microscopes: the edges of the lenses deviate somewhat the light waves. Diffraction may also be observed by gazing, for example, at the back light of a car through a fine tissue: several luminous points appear instead of one because light is deviated by the thin threads of the tissue and the luminous source can reach the eye by passing through several close lying holes between the threads.

---

[1]The origin of the term *barn* comes from the American expression *as big as a barn*. It was used jokingly by physicists working in the *Manhattan Project* during the Second World War, as an antiphrasis to designate the low cross sections of neutrons interacting with certain nuclei. (R. D. Evans, *The Atomic Nucleus*, New York, McGraw-Hill, 1955, footnote p. 10).

Ebonite. Ebonite is one of the first substances called "plastic." It is hardened rubber containing between 30 % and 40 % sulfur. In the 1950s, it was used as an electric isolator. It is black, elastic, and brittle.

Electric and magnetic fields. When two parallel conducting plates are connected to the terminals of an electric battery, an electric field is established in the space between the metallic plates. This means that an electric charge, situated in this space, is subject to a force which draws it towards the plate of opposite charge. The field strength is the force exerted on a unit charge. It is measured in volts per meter. The force increases as the separation of the two plates decreases.

A permanent magnet establishes a magnetic field in its vicinity. The magnetic field exerts a force on other magnets as well as on a conducting wire through which an electric current is flowing. More generally, a magnetic field exerts a force on a moving electric charge (but not on a static charge). A magnet has a north and a south pole. The compass is a little magnet which orients itself along the magnetic field of the earth, the latter acting as a huge magnet.

Electric discharges in gases. When a condenser is placed in a gas at low pressure and charged to a voltage of several thousand volts, the gas no longer acts as an electric isolator: an electric current starts flowing between the positive and negative plates of the condenser. This electric discharge can be either weak and continuous or abrupt and violent. It can cause a variety of occasionally spectacular phenomena, depending on the electric potential between the plates and on the pressure of the gas. Lighting is an example of a violent electric discharge. In fluorescent tubes a continuous discharge is maintained the light of which excites fluorescence on the glass wall of the tube (see *cathode rays*).

Electrolysis. Electrolysis was discovered in 1800 by two Englishmen, Anthony Carlisle (1768–1840) and William Nicholson (1753–1815), barely 2 years after the demonstration of Volta. They inserted two electrodes connected to a battery into salted water and they observed that hydrogen bubbles were emitted from the negative electrode (the cathode) and that oxygen bubbles were emitted from the positive electrode (the anode). The electric current passing through the water had broken the water down into its components. The cause of this phenomenon was understood later.

Electron volt (eV), million electron volts (MeV). The electric current, which, for example, makes a light bulb shine, is a current of electrons which stems from the negative terminal and which flows into the positive terminal. The energy supplied by the electric current per unit time is usually expressed in watts, 1 W being equal to 1 J/s. Common batteries, for example, have a voltage of 4.5 V. The power, that is, the energy per unit time, which they supply is equal to the voltage multiplied by the electric current. The electric current is measured in units of amperes, usually called amps. An electric current of one ampere corresponds to a flow of 6.2415×10^{18} electrons per second. Therefore, each electron supplies a tiny energy which is measured in units of 1 eV. The electron volt is the energy which an electron acquires when it passes through a an electric potential of one volt. Its abbreviation is *eV*. The electron volt is a unit which is well suited for the description of electrons in atoms because the energies involved in atomic phenomena and in chemical reactions are of that order of magnitude. But in a nucleus, the energies involved are one million times larger. The natural unit used to describe nuclear phenomena is a million electron volts, the abbreviation of which is *MeV*. For example, the energy required to extract a neutron from a nucleus is about 8 MeV. The energy emitted by a nucleus which undergoes fission is about 200 MeV. A particle accelerator must provide a beam of particles with energies of several MeV in order to produce nuclear reactions.

Elementary electric charge. It is the smallest amount of electric charge which can exist. In units of Coulombs ©, it is equal to $1.60217649 \times 10^{-19}$ C, with an uncertainty of four on the last digit. The proton has a positive elementary charge and the electron a negative elementary charge. An electric current is a flow of electrons. An electric current equal to 1 A corresponds to a flow of 6.241509×10^{18} electrons per second.

Entropy. Entropy is a quantity introduced by Rudolf Clausius. It is an abstract quantity which is defined mathematically and which cannot be measured directly: there exists no "entropymeter" similar to a thermometer. Boltzmann showed that entropy is a measure of the disorder of a physical system, in the sense of the maximal information one can have of the system. For example, a system composed of wine and water which is not mixed has a lower entropy than the corresponding mixture of wine and water. Indeed, less information on the location of the wine and water molecules is available in the mixed solution. Entropy is a fundamental quantity linked to the second law of thermodynamics (see *thermodynamics*).

Esu (electrostatic unit). In the past, electric charge was expressed in "units of electric charge," *esu* in short. The modern unit of electric charge is the Coulomb, C in short, which is equal to 2 997 924 580 esu. The electron has a negative electric charge equal to $1.60217649 \times 10^{-19}$ C $= 4.803204 \times 10^{-10}$ esu. It is the smallest electric charge known to exist and all observed charges are multiples of this elementary charge.

Femtometer (fm). The *femtometer* is equal to 10^{-15} m. It is the typical length scale of nuclear systems. The size of neutrons and protons are of the order of a femtometer (fm). The radii of nuclei vary from 1 to about 6–7 fm. For a long time, nuclear physicists used the term *fermi* to designate the femtometer. The radii of atoms are of the order of an *ångström*, which is equal to 0.1 nm, or 10^{-10} meters, a 100 000 times larger than the femtometer.

Geissler tube. Heinrich Geissler, a technician and glass blower, was born in Thüringen in 1815 and died in Bonn in 1879. He invented the mercury vacuum pump. Before, vacuum pumps were mechanical devices similar to the first pump constructed in 1654 by Otto von Guericke. However, such pumps were unable to produce vacua with pressures below about 1 mb because the gaskets were made with greasy leather and were not sufficiently airtight. Geissler invented a clever method of producing a vacuum in a vessel by filling it first with mercury and then emptying it out using a system of communicating vessels composed of flexible tubes. The pressures obtained this way were a 100 times lower. The quality of such vacua were to become crucial for the study of electrical discharges in rarefied gases. This is also where Geissler made decisive improvements: in 1856, he constructed the "Geissler tube" which was a glass tube from which air was expelled, with a residual pressure of a hundredth of a millimeter of mercury. He succeeded in implanting electrodes into the sealed tube thereby permitting electrical manipulations. This kind of tube was subsequently perfected by William Crookes and it became known as the "Crookes tube."

Gold leaf electroscope. The gold leaf electroscope is a plugged bottle with an electric wire passing through the plug. The wire is connected to two very thin and light gold plates which hang vertically in contact inside the bottle. When the electric wire becomes electrified, it supplies an electric charge to the gold leaves, which become electrically charged. They repel each other and remain separated.

Gram-atom, gram-molecule. See *mole*.

Groups, group theory. Certain operations which one can perform on a physical system, displacements or rotations, for example, have the remarkable property that two successive displacements or rotations are equivalent to a single displacement or rotation. Another example is the modification of the seating arrangement of guests at a dinner table. A modification of the seating arrangement corresponds to what mathematicians call a *permutation* of the guests. Two successive permutations correspond to another permutation. Such properties can be represented mathematically in terms of *groups*. which were invented between 1830 and 1832 by the mathematician Évariste Gallois (1811–1832). There exist *rotation groups, translation groups, permutation groups*, and many others. Group theory is useful to express quantities which do not change when an operation is performed. The invariance of certain quantities when specific operations are performed is called a *symmetry*. For example, the symmetry of a circle can be expressed by the fact that the circle does not change when it is rotated around its center. The rotations of the circle form a group and the symmetry of the circle maintains the shape of the circle. In quantum mechanics, the wavefunction of an atom must not be modified, or rather it should undergo a definite modification, when its electrons are exchanged because electrons are indistinguishable particles. This also corresponds to a given symmetry. In 1927, Wigner showed how group theory can be applied to classify the quantum states of atoms with many electrons.

Ground state. See *state*.

Half-life. See *radioactive half-life*.

Hectopascal. The hectopascal is a unit of pressure, equal to 100 Pa (see: *pascal*).

Ideal gas. An ideal gas is a gas which is composed of infinitely small non-interacting molecules. Real gases at sufficiently low pressure are very similar to ideal gases. In an ideal gas, the absolute temperature T, the pressure P and the volume V are related by a very simple law, namely $PV = RT$ where R is the ideal gas constant. Thus, for example, if the pressure of the gas is doubled, then, at constant temperature, its volume is halved.

Inverse square law. The inverse square law applies, for example, to the manner in which massive systems (e.g., planets) are attracted to each other according to Newton's law of gravitation. The law states that the attractive force between two massive objects decreases in proportion to the square of the inverse distance separating them. If the distance is doubled, the attractive force is diminished by a factor of 4. The same inverse square law applies to the electric force acting between two electrically charged particles.

Ion. The atom consists of a very small and positively charged nucleus surrounded by negatively charged electrons. The number of electrons is equal to the number of positively charged protons inside the nucleus so that the atom is electrically neutral. Thus, the hydrogen atom has one electron because the electric charge of its nucleus (which is a proton) is equal to 1. The carbon atom has six electrons because its nucleus has six protons. The oxygen atom has eight electrons and its nucleus has eight protons, and so on. Each chemical element is determined by the number of electrons of the atom, usually called Z. During a collision, an electron of an atom may be ejected. The atom then possesses a positive electric charge and it is called a *positive ion*. The atom might also capture an extra electron in which case it becomes a *negative ion*. A similar ionization process occurs when kitchen salt (sodium chloride) is dissolved into water. The water molecules dissociate the sodium chloride molecules into negative chloride ions and positive sodium ions.

Kinetic theory of gases. Clausius, Maxwell, Boltzmann, and Gibbs formulated the kinetic theory of gases in the middle of the nineteenth century. The theory describes gases in terms of molecules which move in all directions in a chaotic fashion, colliding with each other and bouncing off the walls of the gas container. The collisions of the molecules with the walls create the pressure exerted by the gas on the walls. The average kinetic energy of the molecules of a gas is proportional to the absolute temperature of the gas.

Line (spectral line). When sunlight is observed through a prism, or better still, through a modern spectroscope, all the colors of a rainbow are displayed together with some dark lines appearing as thin dark shadows. If however we use the same spectroscope to observe the yellow light emitted by kitchen salt thrown into the hot flame of a gas burner, we see a few thin bright lines, called spectral lines. Each element radiates a certain number of characteristic spectral lines, which serve as a kind of fingerprint. This is the phenomenon upon which spectroscopy is based. It was founded by Kirchhoff and Bunsen who showed that the dark spectral lines of the sun are in precise correspondence with the bright white lines of known elements, thereby revealing the presence of these elements on the surface of the sun.

Luminescence and phosphorescence. Certain substances become luminous when they are exposed to either visible or invisible (ultraviolet) light. This physical phenomenon is called luminescence. The light emitted by a luminescent substance has a characteristic and well-defined color or wavelength. In general, the substance becomes luminous in a very short time (about one hundred millionth of a second) after being exposed to the "excitation" radiation. In the interior of a fluorescent lamp a violet light is produced by an electric current flowing through a rarefied gas (see *electric discharge*). The light emitted by the fluorescent lamp is created when this violet light hits the interior surface of the bulb. The wavelength of light emitted by a fluorescent substance is always longer than the incident radiation which causes the luminescence. This is Stoke's law which Einstein explained assuming the existence of light quanta. Certain luminescent substances emit light much more slowly, and they can remain luminous even hours after being irradiated: this phenomenon is called phosphorescence.

Mass and weight. In current language, mass and weight often mean the same thing. They are however distinct quantities. Weight is the force which is exerted when I step onto my weigh scale which indicates how much I weigh. It is the attractive force exerted by the earth on my body. The mass of a body, also called the inertial mass, is directly linked to the quantity of matter the body contains. It is the mass which determines the speed acquired by a body upon which a force is exerted. A body with high mass (a truck) will move slower than a body with a low mass (a bicycle), even if there is no friction between the body and the soil. The mass of an astronaut is the same on the earth as on the moon, but not his weight. His weigh scale will indicate that he is six times lighter on the moon than on the earth. The moon is lighter than the earth and it exerts a smaller attraction on the astronaut. While he is orbiting around the earth, the astronaut floats around weighing nothing. His weigh scale would be of little use since it would float around just like him! And yet the mass of the astronaut remains the same.

Mean free path. The mean free path of the molecules in a gas is the average distance which a molecule travels before hitting another molecule. The mean free path depends on the size of the molecule, and Maxwell showed that it is related to the viscosity of the gas (which is a measure of the rate at which the gas flows through a small hole in the container or through a porous surface). Hence, a measurement of the viscosity allows one to calculate the size of the molecules, as Loschmidt did in 1865. The size of small molecules is of the order of 10^{-7} mm (one ten-millionth of a millimeter). In oxygen at a temperature of $20\,°C$, the average velocity of molecules is 460 m per second. At atmospheric pressure, their mean free path is of the order 5×10^{-5} mm (five hundred thousandth of a millimeter), that is, about 500 times larger than their size. A molecule undergoes about ten billion collisions per second. See *kinetic theory of gases*.

Meson, π meson, μ meson. See *pion* and *muon*.

MeV. The MeV is a unit of energy equal to one million electron volts (see *electron volt*). It is a basic unit in nuclear physics because the energies involved in nuclear processes are of the order of one MeV. By contrast, the energies of electrons in atomic processes are of the order of 1 eV (eV). One eV (electron volt) is equal to $1.602\,177 \times 10^{-19}$ J and 1 Mev a million times more.

Millibar. The millibar is a unit of pressure equal to one thousandth of a bar (see *pascal*).

Mole. One mole, which used to be called a *gram-molecule*, is the amount of matter consisting of a number of *molecules* equal to Avogadro's number $N = 6.022\,142 \times 10^{23}$. A gram-atom is used to designate a substance composed of N *atoms*. One gram of hydrogen contains N atoms, and it is therefore 1 g-atom. One mole of hydrogen has a mass of 2 g because each molecule is composed of two hydrogen atoms. One mole of a gas has a volume equal to 22.4 liters at the atmospheric pressure and at a temperature of 20°, whatever the gas is. The "hypothesis of Avogadro" was that one mole of any gas always consists of the same number (Avogadro's number) of molecules.

Muon. Usual name attributed to the μ meson. See *pion*.

Neutron. The neutron is a constituent (together with the proton) of the atomic nucleus. It has a mass similar to the mass of the proton but it has zero electric charge. The nuclei of given elements have a definite number of protons, but the number of neutrons of a given element may vary, thereby giving rise to isotopes of the element.

Nucleon. Neutrons and protons are called nucleons. They have similar properties such as mass, spin and interactions. The nucleus carbon 12 is composed of six protons and six neutrons: it is composed of 12 nucleons.

Pascal. The pascal is a unit of pressure, equal to the pressure exerted by a force of 1 N on a surface equal to 1 m^2 (a mass of 1 kg has a weight of about 10 N). A pressure equal to 1 Pa is a low pressure, roughly equal to one hundred thousandth of atmospheric pressure. This is why the unit hectopascal (equal to 100 Pa) is frequently used. Another frequently used unit of pressure is the *bar*, which is equal to 100 000 Pa. Normal atmospheric pressure is equal to about 1.013 bars, which is equal to 1013 hPa and to 101 300 Pa. For historical reasons, another unit of pressure is still occasionally used. It is the height, expressed in millimeters, of the mercury column of a barometer, namely 76 cm or 760 mm at normal atmospheric pressure. A pressure equal to 1 mm of mercury, which is equal to 1.33 hPa, has also been used.

Periodic law : Dmitri Mendeleev, Lothar Meyer. At the end of the nineteenth century, chemists were searching for a classification of the numerous elements which had been discovered. The German chemist Lothar Meyer (1830–1895) and the Russian chemist Dmitri Mendeleev (1834–1907) made two interesting observations: if the elements are ordered by increasing mass, elements with similar chemical properties reappear regularly. This is how the periodic table of elements was born. It is displayed at the end of the book in its modern form, where elements belonging to a given column have similar chemical properties. But in order to make it work properly, Mendeleev left several entries empty, assuming that they corresponded to as yet undiscovered elements. He wrote :

We should still expect to discover many unknown simple bodies; for example, those similar to Al and Si [aluminum and silicon], elements with atomic weights of 65 to 75.

Table 1 Range of energies and wavelengths of photons. The various categories sometimes considerably overlap. One micrometer, also called a *micron*, is equal to 10^{-6} m, meaning one millionth of a meter.

Name	Range of energies (eV)	Range of wavelengths
Radiofrequencies	10^{-10} to 10^{-5}	10 km to 10 cm
Microwaves (RADAR, domestic ovens)	10^{-6} to 0.1	1 mm to 1 m
Infrared (domestic oven and iron)	10^{-4} to 1	1 cm to 1 μm
Light visible by the human eye	1.2 to 2.5	0,4 to 0.8 μm
Ultraviolet	2.5 to 2000	0.4 to 0.0005 μm
X-rays	10 to 100 000	0,1 to 10^{-5} μm
Gamma rays	More than 10 000	Less than 10^{-5} μm

The subsequent discovery of the missing elements made Mendeleev and his "table of elements" famous. However, the reason why such regularities existed and why the elements belonging to a given column had similar chemical properties remained a mystery. Lothar Meyer published his table in 1870, whereas Mendeleev published his in 1869. Although he was credited for publishing it first, Mendeleev always took care to acknowledge the work of Meyer.

Photoelectric effect. The photoelectric effect was discovered by Heinrich Hertz (1857–1894) in 1887 while he was performing experiments on the propagation of electromagnetic waves. Electrons are emitted from a metallic surface when it is exposed to light rays of suitable wavelength, or color. In 1902, Phillipp Lenard (1862–1947) showed that the energy of the emitted electrons depended on the wavelength (therefore on the color) of the light rays and not on their intensity. In 1905, Einstein explained the photoelectric effect by assuming the existence of light quanta. It is the photoelectric effect which transforms directly the energy of solar light into electric energy in photovoltaic cells.

Photon. The photon is the quantum of light, that is, the smallest quantity of light (of a given frequency) which can be emitted. Photons travel at the speed of light which is equal to 299 792 458 m/s. The product $\lambda \nu$ of the wavelength λ and the frequency ν of a photon is equal to the speed of light. The energy of the light quantum is determined by its frequency ν according to Planck's formula $E = h\nu$ in which h is Planck's constant. Thus, a high-frequency photon (which has a small wavelength λ) transports more energy than a photon of low frequency (which has a long wavelength λ). Photons are given different names depending on their frequency, as shown in Table 1.

Pion, π meson. The name *meson* was first attributed to a particle discovered in 1937 by Anderson and Neddermeyer (see p. 333) because its mass lay between the masses of the proton and of the electron (in Greek "meso" means "in between"). In 1947, it was discovered that there were two particles of this type, and they were called the π meson and μ meson (see p. 469). Today, the π meson is called a *pion* and the μ meson a *muon*.

Potential. The difference between the potential at a point A and a point B is the potential energy acquired by a unit charge (in the case of an electric field) or a unit mass (in the case of a gravitational field) when it is displaced from A to B. An electron acquires a potential energy of 1 electron volt (eV) when it passes through a potential difference of 1 V. The force acting on an electric charge is related to the variation (the derivative) of the potential: if the potential increases, it acts against the displacement of the particle; if it decreases, it accelerates it. The nuclear force acting on a nucleon in a nucleus is often represented by a potential well, the edges of which are very steep, so that the force acting on a nucleon is very strong at the surface of the nucleus. The bottom of the well is rather flat so that the force acting on a nucleon inside the nucleus is not very strong.

Potential energy. Potential energy designates a latent energy which does not show itself. For example, the water in a mountain lake has a potential energy which can become mechanical energy when the water is allowed to flow into a turbine. A battery cell has a chemical potential energy which can become either heat or light energy if the battery is connected to a light bulb. The fuel of a car has a chemical potential energy which transforms partly into mechanical energy (making the car move) and partly into heat when the fuel is burnt in the cylinders of the car.

Potential well. When an electrically charged particle, such as the α-particle, approaches a nucleus at a certain speed, at first it feels a repulsion because the nucleus has a positive electric charge, as the α-particle does. This repulsion, called the Coulomb barrier, slows down the α-particle transforming its kinetic energy (related to its velocity) into potential energy. A similar phenomenon is observed when a ball is thrown up a hill. When the α-particle gets close enough to the nucleus, it feels a strong attraction due to the nuclear force, somewhat like a ball falling into a hole. This attraction is called a *potential well*. In order to extract the particle from the potential well, the energy which must be transferred to the particle is equal to the depth of the potential well, as measured from its edge.

α-decay is the radioactivity due to an α-particle which escapes from a nucleus. According to classical mechanics, the particle cannot escape the nucleus if its energy is below the Coulomb barrier. However, in quantum mechanics, the particle can, with a small but finite probability, pass through the Coulomb barrier. Figure 3 shows a qualitative shape of the potential well felt by an α-particle which approaches the nucleus ^{40}Ca. The α-particle first feels the slowly rising Coulomb potential and then suddenly it falls into a deep nuclear potential well.

Precision of a measuring instrument. The precision of a measuring instrument is what allows one to give a precise value of a measurement. For example, if I am able to measure the length of a 10-m-long room to within

Fig. 3 The qualitative shape of the potential well felt by an α-particle which approaches a nucleus. The α-particle first feels the slowly rising Coulomb potential and then suddenly it falls into a deep nuclear potential well. The steepness of the slope indicates the strength of the force acting on the particle

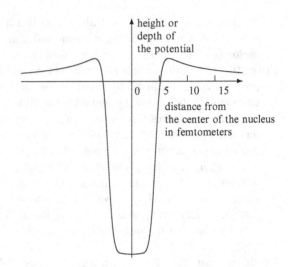

1 mm, I say that I have a precision of 1 mm, that is, of one ten thousandth of the length. The precision of an instrument is not always related to the resolution of the instrument. The resolution is the ability of an instrument to separate two neighboring images. The lens of a camera can have an exceptional sharpness, which corresponds to a good resolution. However, the image may be distorted and this may diminish the precision with which lengths and distances can be evaluated from the image. To measure the height of a tower using a telescope, I need good precision. To distinguish details, I need good resolution.

Proton. The proton (together with the neutron) is a constituent of the nucleus at the center of the atom. The proton carries a positive elementary electric charge. The nucleus of the lightest atom, namely, hydrogen, consists of one proton, although a rare isotope of hydrogen exists, deuterium, which is composed of one proton and one neutron (one atom out of 6,400 hydrogen atoms).

R, constant of ideal gases. See *ideal gases.*

Radioactive half-life. The radioactivity of every radioactive substance decreases in time. The radioactive substance decays by transforming into another substance, which may or not be radioactive itself. The decrease in radioactivity is exponential, meaning that the number of atoms which decay in a given time is a fixed fraction of the total number of atoms in the substance. As the decay progresses, the number of decaying atoms decreases so that the number of decays per second also decreases. For example, if half of the atoms of a substance decay during one day, then each day the radioactivity will be equal to half of what it was the day before. One kilogram of this radioactive substance will reduce to 500 g the next day, to 250 g the following day, to 125 g the third day, and so on.

The time taken for half of the atoms to decay is called the radioactive half-life. Radioactive half-lives can be very different. The longest measured half-lives reach several billion years (several times 10^9 years). Uranium, for example, has a radioactive half-life of 4.5×10^9 years. The shortest half-lives can be fractions of a second. The term "half-life" is sometimes used instead of "radioactive half-life." But beware: two half-lives do not amount to a full lifetime! In the example above, the half-life was 1 day, but after 2 days 1/4 of the radioactive atoms remain.

Refraction. When light passes through a medium such as air, glass, or water, the direction in which it propagates changes: the phenomenon is called refraction. It is refraction which causes a ruler which is half immersed in water to appear bent. Refraction is used in eye glasses, magnifying glasses, microscopes, and so forth. The change in the direction of the propagating light is not the same for all wavelengths (colors). This can be seen by making light propagate through a prism. Red is more deviated than blue. This was discovered by Newton who thus discovered that white light is a superposition of all the colors of a rainbow. See *diffraction*.

Resolution, resolving power. Originally, to resolve meant to "break down a substance into its constituents" (in Latin *resolvere* means to release or detach). The resolving power of the lens of a camera is the ability of the lens to produce an image with sufficient sharpness so that two sufficiently close objects (such as two luminous spots) appear as distinct objects. More generally, the resolving power of an instrument is its ability to distinguish two objects which are separated, depending on the instrument, by a short distance (spatial resolution), or by a short time (time resolution), or by close lying energies (energy resolution). For example, a 40-m resolution of a photograph of the earth taken from a satellite means that objects separated by less than 40 m on the surface of the earth will not appear as distinct objects on the photograph. See *precision*.

Resonance. Resonances occur in systems vibrating with a given frequency (a number of oscillations per second). Energy can be transferred to an oscillating system, provided that it is transferred at a suitable frequency, close to the frequency of the oscillating system. For example, one can increase the motion of a swing by giving it a push at regular time intervals: it is essential that the push should be given at the right moment. A famous example is provided by the bridge which collapsed while a military troop was crossing it because the frequency of the marching steps happened to match exactly the natural oscillation frequency of the bridge. More generally, any vibrating system can increase the amplitude of its vibration by receiving impulses at frequencies equal to its proper frequencies.

Resolving power (of a measuring instrument). See *resolution*.

Rhumkorff coil. The German instrument maker Heinrich Daniel Ruhmkorff was born in Hanover in 1803 and, after several journeys in Europe, he settled in Paris where he died in 1874. He had a little workshop in the *rue Champollion,* where he constructed a device capable of producing very high electric voltages. In fact, he perfected the induction coil of Nicholas Callan, based on the discovery of Faraday (1791–1867): the device was a transformer whose secondary coil had about 100 000 turns. The primary coil was fed by a battery the contact of which was periodically broken. Each time the primary circuit was broken, a very high voltage was induced in the secondary coil, reaching up to 100 000 V! The ignition of spark plugs in engines, for example, works in a similar fashion. Due to his know-how and his admittedly empirical but deep understanding of electric phenomena, his so-called "Ruhmkorff coils" became famous in all of Europe. Faraday, Zeeman, and Röntgen used them. Until that time, there were few other ways of producing high electric voltages. Electric batteries connected in series produced only limited voltages and friction devices produced high voltages but very weak currents.

Scattering. In nuclear physics, the term *scattering* is used to describe the *deviation* of fast particles by the nucleus of the atom, followed by a loss of kinetic energy. The case of a perfectly elastic collision, such as the ideal collision of billiard balls, is called *elastic scattering*, during which the incident particles transfer some velocity to the target, without however modifying the total kinetic energy. If I throw a ball made of soft modeling clay onto a wall, it does not bounce back. Its kinetic energy is transformed into deformation energy and there is a slight increase in the temperatures of the ball and of the wall.

In nuclear physics, if the incident particle or the target nucleus absorb energy during the collision, the process is called *inelastic scattering*.

Spectroscopy. In 1675, Isaac Newton showed that white sunlight passing through a prism was decomposed into what he called a "spectrum" of various colors ranging from red to violet. Physicists made further progress by showing that the spectrum continued in fact beyond the violet (towards the ultraviolet) and beyond the red (towards the infrared). In 1813, Joseph von Fraunhofer (1787–1826) discovered that the solar spectrum contained several dark lines and he published his results in 1817. Using a prism or a diffraction grating (with 4000 spacings over 12 mm, roughly 300 spacings per mm), he constructed the first real spectroscope which made it possible to measure the wavelength of the spectral lines. In his first catalogue, he recorded 576 spectral lines. The dark lines had in fact already been observed in 1802 by William Hyde Wollaston (1766–1826), who however interpreted them as "gaps" separating the different colors of the sun.

In 1859, two professors of the University of Heidelberg, Gustav Kirchhoff (1824–1887) and Robert Bunsen (1811–1899), showed how the observation of spectral lines could provide the means of an efficient chemical analysis. They observed the spectral lines using the famous "Bunsen burner," the flame of which, not very bright by itself, made it easy to observe the spectral lines of elements. Kirchhoff explained the "inversion" of the spectral lines observed by Fraunhofer: the dark lines observed in the solar spectrum correspond to the bright spectral lines observed on earth; they indicate the presence of elements which absorb the radiation on the solar surface. This soon led to the discovery of new elements.

See *line*.

Spin. The term "spin" was originally coined to designate the intrinsic angular momentum of an elementary particle at rest. As such, it is an intrinsic property of a point particle, with no counterpart in classical mechanics, and it is not related to the rotation of matter. An electron at rest has an angular momentum equal to $\frac{1}{2}\hbar$ where $\hbar$ is Planck's constant divided by 2π: it is said to have spin 1/2. The neutrino has spin 1/2 and the photon, which is a quantum of light, has spin 1. The proton and the neutron, when considered to be point particles, have spin 1/2.

However, the term "spin" is also used to designate the angular momentum of a composite particle at rest, such as a nucleus. In this case, the spins and the orbital angular momenta of the nucleons add up (according to quantum mechanical rules) to yield the total spin of the nucleus. The spins of two nucleons pointing in opposite directions can add up to a zero total spin. Nuclei with even numbers of both protons and neutrons have zero spin in their ground state, that is, in their lowest energy state as they are found in nature. Nuclei with odd numbers of neutrons or protons have half-integer spin.

State. In quantum mechanics, atoms and nuclei exist only in certain configurations, called *states*, which have definite energies. If an atom is modified by promoting, for example, an electron into a higher energy state, its binding energy (which is equal to the energy required to extract all the electrons from the atom) is decreased. The largest binding energy occurs for the most stable state of the atom, called the *ground state*. The ground state can only be modified by transferring energy to the atom. When the binding energy of an atom is smaller, the atom is said to be in an *excited state*. It can then decay into its ground state by radiating its excess energy.

Supersaturated water. At a given temperature, air can contain at most a certain amount of humidity which is water vapor. Hygrometers indicate the percentage of maximum humidity presently in the air. When the temperature falls, the maximum humidity is reduced and the excess humidity condenses in the form of condensation on windows, dew in gardens, mist, or clouds. However, a curious phenomenon occurs: if one slowly reduces the temperature of calm and clean air, no condensation occurs when the humidity reaches 100 %. In fact, the temperature may be lowered so as to exceed somewhat 100 % humidity. In this case, the water vapor is said to be *supersaturated*. But supersaturated vapor is an unstable state and condensation is triggered by the presence of impurities such as dust and ions which act as nuclei around which condensation forms. The terrestrial atmosphere always contains dust particles as well as ions which are permanently created by cosmic rays. The cloud chamber of C. T. R. Wilson is based on that phenomenon.

Thermodynamics: the two principles. The two principles of thermodynamics, the science which relates heat to mechanics, were formulated in the nineteenth century. They form the basis of a beautiful theoretical structure with numerous and often nonintuitive consequences, concerning heat engines, the production of cold, etc. What we now call the *second principle* *of thermodynamics* was in fact the first to be formulated, in 1824, by Sadi Carnot (1796–1832), in a short and barely noticed 30-page book *Réflexions sur la puissance motrice du feu*.[1] The principle states that a heat engine can only produce work by extracting heat from a hot source and transmitting heat to a colder source.

Heat and mechanical work became related when the concept of energy was introduced by Robert Mayer (1814–1878) in 1842. Energy was defined more precisely in 1843 by James Joule (1818–1889), and it was given a definitive formulation by Rudolf Clausius (1822–1888), who stated what today we call the *first principle of thermodynamics:*

> *The total energy of an isolated system remains constant.*

According to the first principle, mechanical energy and heat are two forms of energy. Mechanical energy in the form of work can be transformed into heat, and vice versa, but the total energy can neither increase nor decrease. This makes it impossible to construct a perpetual motion machine "of the first kind": energy cannot be produced from nothing.

The *second principle* was formulated by Clausius in 1850 and later, in 1854, by William Thomson (Lord Kelvin 1824–1907). They rediscovered the law of Carnot and proved that it is equivalent to the statement:

> *Heat cannot spontaneously pass from a cold body to a hot body.*

It is however Clausius who, by introducing the concept of entropy (see this word), gave a final formulation to *the second principle:*

> *The entropy of an isolated system can only increase.*

An unexpected and spectacular result of the second principle is the existence of a minimum temperature: the temperature cannot fall below $-273.15°$, the so-called "absolute zero." One can cool a body close to this temperature but it is impossible to reach it exactly.

Wavelength, period and frequency. If one watches a cork floating on top of waves in water, one sees the cork going up and down with the water surface while remaining in the same place. The up and down oscillation of the water surface is transmitted to the water molecules by the neighboring molecules. What appears to move is the wave. This example shows that wave motion is a subtle concept. The waves do not displace the water (except for the up and down displacement). They displace energy. What propagates is the surface displacement which constitutes the wave.

During the time it takes the cork to perform one cycle of up and down motion, the wave has moved a certain distance which is called the *wavelength*. The time it takes to go through one cycle is called the *period* and the number of cycles performed per unit time is called the *frequency*.

---

[1]Reflections on the motive power of fire.

Electromagnetic waves are oscillations of electric and magnetic fields. They propagate extremely fast, at the speed of light (299 792.458 km/s in vacuum). Visible light consists of electromagnetic waves the wavelengths of which span from about 0.4 μm[1] (red light) to about 0.7 μm (violet light). FM (frequency modulation) radio waves have frequencies of the order of 100 MHz (megahertz), meaning one hundred million cycles per second. They propagate at the speed of light, which implies that their wavelength is about 3 m, the wavelength being the distance travelled by the wave during one cycle. AM (amplitude modulation) radio waves called "long waves" have wavelengths of 1–2 km. They therefore have a lower frequency between 300 and 150 kHz (a kilohertz is equal to 1000 Hz, meaning 1000 cycles per second). AM radio waves called "medium waves" have frequencies of 520–1610 kHz, and wavelengths of 600–200 m.

Zeeman effect. The observed spectral lines of an element are modified when the atoms emit radiation while being exposed to a magnetic field. Each spectral line is split up into a group of 2, 3, 4, 5... close lying lines and the distance separating the lines increases with the intensity of the magnetic field. The phenomenon was discovered by Pieter Zeeman in 1897 and it led to the discovery of the spin of the electron.

---

[1]A micrometer is one millionth of a meter, one thousandth of a millimeter.

Bibliography of cited books

1. Aston, F. W., *Isotopes*, Edward Arnold & Co, London, 1922.
2. Aston, F. W., *Mass Spectra and Isotopes*, Edward Arnold & Co, London, 1933.
3. Badash, L., *Rutherford and Boltwood, Letters on radioactivity*, Yale University Press, New Haven, 1969.
4. Barbo, Loïc, *Curie, le rêve scientifique*, Belin, Paris, 1999.
5. Barbo, Loïc, *Les Becquerel, une dynastie scientifique*, Belin, Paris, 2003.
6. Bensaude-Vincent, B., *Langevin, science et vigilance*, Belin, Paris, 1987.
7. Beyerchen, Alan D., *Scientists under Hitler: politics and the physics community in the third reich*, Yale University Press, New Haven, 1977.
8. Biquard, P., *Frédéric Joliot-Curie et l'énergie atomique*, Seghers, Paris, 1961.
9. Bird, Kai and Sherwin, Martin J., *American Prometheus, the triumph and tragedy of J. Robert Oppenheimer*, Alfred A. Knopf, New York, 2005.
10. Birks, John B. Birks, John B., ed. *Rutherford at Manchester*, Heywood & Company Ltd., London, 1962.
11. Blatt, John M. and Weisskopf, Victor F., *Theoretical Nuclear Physics*, John Wiley & Sons, New York, 1952.
12. Bohr, Niels *Collected Works*, North Holland, Amsterdam, 1972–96; Léon Rosenfeld, ed. (vol. 1 to 4) then Erik Rüdinger and Finn Aaserud; vol. 1, 1972: *Early work (1905–1911)* ; vol. 2, 1981: *Work on atomic physics (1912–1917)* ; vol. 3, 1976: *The Correspondence principle (1918–1923)*; vol. 4, 1977: *The Periodic system (1920–1923)*; vol. 5, 1984: *The Emergence of quantum mechanics (mainly 1924–1926)*; vol. 6, 1985: *Foundations of quantum physics I (1926–1932)*; vol. 7, 1996: *Foundations of quantum physics II (1933–1958)*; vol. 8, 1986: *The Penetration of charged particles through matter (1912–1954)*.
13. Born, M., *My Life & my Views*, Charles Scribner's Sons, New York, 1968.
14. Born, M., *My life: recollections of a Nobel Laureate*, Charles Scribner's Sons, New York and Taylor and Francis, London, 1978.
15. Born, M., *The Born-Einstein letters; correspondence between Albert Einstein, and Max and Hedwig Born from 1916 to 1955*, with a foreword by Werner Heisenberg and commentaries by Max Born, Macmillan, London, 1971. Translated by Irene Born from the German edition: *Albert Einstein, Hedwig und Max Born Briefwechsel 1916–1955*, Nymphenburger Verlagshandlung, München, 1969.
16. Bouguer, P., *Traité d'optique sur la gradation de la lumière*, H.L. Guérin et L.F. Delatour, Paris, 1760.
17. Brink, David M., *Nuclear forces*, Pergamon Press, Oxford, 1965.
18. Broglie, Louis de, *Recherches sur la théorie des quanta*, Masson, Paris, 1963.

B. Fernandez and G. Ripka, *Unravelling the Mystery of the Atomic Nucleus:*
A Sixty Year Journey 1896 — 1956, DOI 10.1007/978-1-4614-4181-6,
© Springer Science+Business Media New York 2013

19. Broglie, Maurice de, *Les premiers conseils de physique Solvay et l'orientation de la physique depuis 1911*, Albin Michel, Paris, 1951.
20. Brown, A., *The Neutron and the Bomb. A biography of Sir James Chadwick*, Oxford University Press, Oxford, 1997.
21. Campbell, J., *Rutherford: scientist supreme*, AAS, Christchurch, N.Z., 1999.
22. Cantelon, Philip L., Hewlett, Richard G. and Williams, Robert C., *The American Atom. A documentary History of Nuclear Physics from the Discovery of Fission to the Present*, University of Pennsylvania Press, Philadelphia, 1984.
23. Caroe, G. M., *William Henry Bragg, 1862–1942: man and scientist*, Cambridge University Press, Cambridge and New York, 1978.
24. Cassidy, David C., *Uncertainty: the life and science of Werner Heisenberg*, Freeman, New York, 1992.
25. Charpentier-Morize, M., *Perrin, savant et homme politique*, Belin, Paris, 1997.
26. Crawford, E., *The Beginnings of the Nobel Institution. The Science Prizes, 1901–1905*, Maison des Sciences de l'Homme and Cambridge University Press, Cambridge, 1984.
27. Curie, Ève, *Madame Curie*, Gallimard, Paris, 1938; Folio/Gallimard, 1981.
28. Curie, P., *Œuvres scientifiques*, Gauthier-Villars, Paris, 1908; Éditions des archives contemporaines, Paris and Montreux, 1984.
29. Curie-Skłodowska, M., *Pierre Curie*, MacMillan, New York, 1923. Translated from the first French edition, *Pierre Curie*, Payot, Paris, 1923. It includes an autobiography of Marie Curie, which she forbade to be translated into French. The latest French edition (Odile Jacob, Paris, 1996) also includes a study by Irène Curie of Pierre and Marie Curie's laboratory books, and the Marie Curie's personal diary (1906–1907).
30. Dalton, J., *New System of Chemical Philosophy*. Volume 1, Bickerstaff, Manchester, Part I, 1808; part II, 1810; Volume II, Bickerstaff, London, 1827.
31. Darrigol, O., *From c-Numbers to q-Numbers: The Classical Analogy in the History of Quantum Theory*, University of California Press, Berkeley, 1992. http://ark.cdlib.org/ark:/13030/ft4t1nb2gv/
32. Dirac, Paul A. M., *The principles of quantum mechanics*, Clarendon Press, Oxford, 1930.
33. Einstein, A., *The Collected papers of Albert Einstein*, Princeton University Press: vol. 1, 1987: *The early years, 1879–1902*; vol. 2, 1989: *The swiss years: writings, 1900–1909*; vol. 3, 1993: *The swiss years: writings, 1909–1911*; vol. 4, 1995: *The swiss years: writings, 1912–1914* ; vol. 5, 1993: *The swiss years: correspondence, 1902–1914*; vol. 6, 1996: *The Berlin years: writings, 1914–1917*; vol. 7, 2002: *The Berlin years: writings, 1918–1921*; vol. 8, 1998: *The Berlin years: correspondence, 1914–1918*; vol. 9, 2004: *The Berlin years: correspondence, January 1919–April 1920*.
34. Einstein, A. and Besso, M., *Correspondance avec Michele Besso 1903–1955*, French translation of the German correspondence, notes and introduction by Pierre Speziali, Hermann, Paris, 1972.
35. Einstein, A., *Textes choisis*, translation and notes by F. Balibar, O. Darrigol and B. Jech, Éditions du Seuil/Éditions du CNRS, Paris, 1989–1991.
36. Einstein, A. and Infeld, L., *The Evolution of Physics*, Cambridge University Press, London, 1938.
37. Enz, Charles P., *No time to be brief: a scientific biography of Wolfgang Pauli*, Cambridge University Press, Cambridge, 2002.
38. Elsasser, Walter M., *Memoirs of a Physicist in the Atomic Age*, Science History Publications & Adam Hilger, New York & Bristol, 1978.
39. Eve, Arthur S., *Rutherford*, The University Press, Cambridge, 1939.
40. Farmelo, G., *The Strangest Man. The hidden life of Paul Dirac, quantum genius*, Faber and Faber, London, 2009.
41. Feather, N., *Lord Rutherford*, Glasgow, Blackie & Sons, 1940; Priory Press, London, 1973.
42. Feenberg, E., *Shell theory of the nucleus*, Princeton University Press, Princeton, 1955.
43. Fermi, E., *Collected Papers (Note e memorie)*, E. Segrè, ed., The University of Chicago Press/Accademia Nazionale dei Lincei, Chicago/Rome, 1962.

44. Fermi, E., *Nuclear physics*, The University of Chicago Press, Chicago, 1949. A course given by Enrico Fermi at the University of Chicago; notes compiled by Jay Orear, A.H. Rosenfeld, and R.A. Schluter.

45. Fermi, L., *Atoms in the family. My life with Enrico Fermi*, The University of Chicago Press, Chicago, 1954.

46. Fölsing, A., *Albert Einstein : a biography*, Viking, New York, 1997. Translated by Ewald Osers from *Albert Einstein: eine Biographie*, Suhrkamp, 1993.

47. Frank, P., *Einstein, his life and times*, Alfred A. Knopf, New York, 1947.

48. Frisch, Otto R., *What Little I Remember*, Cambridge University Press, Cambridge, 1979.

49. Ganot, A., *Introductory course of natural philosophy: for the use of schools and academies.* Edited by William G. Peck, New York: A.S. Barnes & Burr, 1862. Translated from *Traité élémentaire de physique expérimentale et appliquée et de météorologie, à l'usage des Établissements d'instruction, des aspirants aux grades des Facultés et des candidats aux diverses écoles du Gouvernement*, chez l'auteur, Paris, 1853.

50. Gerlach, W. and Hahn, D., *Otto Hahn. Ein Forscherleben unserer Zeit*, Wissenschaftliche Verlagsgesellschaft MbH, Stuttgart, 1984.

51. Gillispie, Charles C. and Holmes, Frederic L. and American council of learned societies, *Dictionary of scientific biography* (eight volumes), Charles Scribner's Sons, New-York, 1981–1990.

52. Goeppert-Mayer, M. and Jensen, J. Hans D., *Elementary Theory of Nuclear Shell Structure*, John Wiley & Sons, New York, 1955.

53. Goldschmidt, B., *The atomic complex: a worldwide political history of nuclear energy*, American Nuclear Society, La Grange Park, 1982. Translated by Bruce M Adkins from the French original edition: *Le complexe atomique*, Fayard, Paris, 1980.

54. Goldschmidt, B., *Pionniers de l'atome*, Stock, Paris, 1987.

55. Goldsmith, M., *Frédéric Joliot-Curie, a biography*, Lawrence and Wishart, London, 1976.

56. Goodchild, P., *Edward Teller: the real Dr. Strangelove*, Harvard University Press, Cambridge, Mass., 2004.

57. Gowing, M., *Britain and atomic energy, 1939–1945*, St Martin's Press, New York, 1964.

58. Groves, Leslie R., *Now It Can Be Told: The story of the Manhattan project*, Harper, New York, 1962.

59. Guérout, S., *Science et politique sous le Troisième Reich*, Ellipses, Paris, 1992.

60. Hahn, O., *A scientific autobiography*, Charles Scribner's Sons, New York, 1966. Translated and edited by Willy Ley, with an Introduction by Glenn T. Seaborg, from *Von Radiothor zur Uranspaltung*, Friedrich Vieweg & Sohn, Braunschweig, 1962.

61. Heilbron, John L., *H. G. J. Moseley: the life and letters of an English physicist, 1887–1915*, University of California Press, Berkeley, 1974.

62. Heilbron, John L., *The Dilemmas of an upright man: Marx Planck as spokesman for German science*, University of California Press, Berkeley, 1986.

63. Heilbron, John L. and Seidel, Robert W., *Lawrence and his laboratory. Vol. 1*, University of California Press, Berkeley, 1989.

64. Heisenberg, E., *Inner exile: recollections of a life with Werner Heisenberg*, Bikhäuser, Boston, 1984, with an introduction by Victor Weisskopf, translated by S. Cappellari and C. Morris from *Das politische Leben eines Unpolitischen: Erinnerungen an Werner Heisenberg*, Piper, München und Zürich, 1980.

65. Heisenberg, W., *Physics and beyond: encounters and conversations*, Harper & Row, New York, 1972. Translated from *Der Teil und das Ganze*, R. Piper & Co Verlag, Munich, 1969.

66. Herken, G., *Brotherhood of the Bomb. The Tangled Lives and Loyalties of Robert Oppenheimer, Ernest Lawrence and Edward Teller*, Henry Holt and Co., New York, 2002.

67. Hermann, A., *The Genesis of Quantum Theory (1899–1913)*, M.I.T. Press, Cambridge, 1971. Translated from *Frühgeschichte der Quantentheorie (1899–1913)*, Physik Verlag, Mosbach/Baden, 1971.

68. Hermann, A., *Max Planck in Selbstzeugnissen und Bilddokumenten*, Rowohlt Taschenbuch, Reinbeck bei Hamburg, 1973.

69. Hermann, A., *Werner Heisenberg, 1901–1976*, Inter Nationes, Bonn, 1976. Translated by Timothy Nevill from the German edition: *Heisenberg*, Rowohlt, 1976.

70. Hurwic, A., *Pierre Curie*, Flammarion, Paris, 1995.

71. Jammer, M., *Concepts of Mass in classical and modern physics*, Harvard University Press, Cambridge, 1961.

72. Jammer, M., *The Conceptual Development of Quantum Mechanics*, McGraw Hill, New York, 1996.

73. Jensen, C., *Controversy and consensus : nuclear beta decay 1911–1934*, edited by Finn Aaserud, Birkhäuser, Basel, 2000.

74. Joliot, Frédéric, *Textes choisis*, Éditions sociales, Paris, 1959.

75. Joliot, Frédéric and Curie, Irène, *Œuvres scientifiques complètes*, edited by H. Faraggi, H. Langevin-Joliot, N. Marty and P. Radvanyi, Presses Universitaires de France, Paris, 1961.

76. Kelvin, William T., Baron, and Tait, Peter G., *Elements of natural philosophy*, Clarendon Press, Oxford, 1867.

77. Kelvin, William T., Baron, and Tait, Peter G., *Treatise on Natural Philosophy*, Clarendon Press, Oxford, 1867.

78. Klein, Martin J., *Paul Ehrenfest*, North-Holland, Amsterdam and New York, 1985.

79. Kowarski, L., *Réflexions sur la science, la pensée de Lew Kowarski à travers ses écrits de 1947–1977*, Gabriel Minder, Institut universitaire de hautes études internationales, Genève, 1978.

80. Krafft, F., *Im Schatten der Sensation, Leben und Wirken von Fritz Strassmann*, Verlag Chemie, Weinheim, 1981.

81. Kursunoglu, Behram N. and Wigner, E. Paul, editors, *Reminiscences about a great physicist: Paul Adrien Maurice Dirac*, Cambridge University Press, Cambridge, Mass., 1987.

82. Lanouette, W. and Silard, B., *Genius in the shadows. A biography of Leo Szilard: the man behind the bomb*, C. Scribner's Sons, New York, 1992.

83. Livingston, M. Stanley, ed., *The Development of High Energy Accelerators*, Dover, New York, 1966.

84. Livingston, M. Stanley, *Particle Accelerators: A Brief History*, Harvard University Press, Cambridge, 1969.

85. Livingston, M. Stanley and Blewett, J. Paul, *Particle Accelerators*, McGraw-Hill, New York, 1962.

86. Lorentz, Hendrik A., *The Scientific correspondence of H. A. Lorentz, vol. I* edited by Kox, Anne J., Springer, New York, 2008.

87. Loriot, Noëlle, *Irène Joliot-Curie*, Presses de la Renaissance, Paris, 1991.

88. Lovell, Sir Bernard, *P.M.S. Blackett: a biographical memoir*, The Royal Society, London, 1976. Reprinted from the *Biographical Memoirs of the Royal Society* **21**, 1–115, 1975.

89. Marage, P. and Wallenborn, Grégoire, *Les Conseils Solvay et les débuts de la physique moderne*, Université libre de Bruxelles, 1995.

90. Marbo, C., (pseudonym of Marguerite Borel), *A travers deux siècles, souvenirs et rencontres: 1883–1967*, Grasset, Paris, 1967.

91. McGrayne, Sharon B., *Nobel Prize Women in Science*, Carol Publishing Group, New York, 1993.

92. McMillan, P. J., *The ruin of J. Robert Oppenheimer and the birth of the modern arms race*, Viking, New York, 2005.

93. Merricks, L., *The world made new: Frederick Soddy, science, politics, and environment*, Oxford University Press, Oxford and New York, 1996.

94. Morselli, M., *Amedeo Avogadro, a scientific biography*, D. Reidel, Dordrecht, 1984.

95. Moore, Walter J., *Schrödinger: life and thought*, Cambridge University Press, Cambridge, 1989.

96. Mott, Nevill F. and Massey, H. S. W., *The Theory of atomic collisions*, Clarendon Press, Oxford, 1933.

97. Needham, J. and Pagel, W., editors, *Background to Modern Science: Ten lectures at Cambridge arranged by the History of Science Committee, 1936,* Cambridge University Press, Cambridge, 1938.

98. Newton, I., *Opticks: or, a Treatise of the Reflections, Refractions, Inflections and Colours of Light,* William Innys, 1704.

99. Nobel Foundation, *Nobel Lectures, including presentation speeches and laureates' biographies, Physics,* Elsevier, Amsterdam/London/New York: 1901–1921, published in 1967; 1922–1941, published in 1965; 1942–1962, published in 1964.

100. Nollet, Abbé, *Leçons de Physique Expérimentale,* chez Durand, Paris, 1743.

101. Nye, Mary Jo, ed., *The Question of the Atom from the Karlsruhe Congress to the first Solvay Conference, 1860–1911,* Tomash Publishers, Los Angeles/San Francisco, 1984.

102. Nye, Mary Jo, *Blackett: Physics, War and Politics in the Twentieth Century,* Harvard University Press, Cambridge, Mass. and London, 2004.

103. Omnès, R., *Quantum philosophy: understanding and interpreting contemporary science,* Princeton University Press, Woodstock, 1999; translated by Arturo Sangalli from the original French edition: *Philosophie de la science contemporaine,* Gallimard, Paris, 1995.

104. Omnès, R., *Understanding Quantum Mechanics,* Princeton University Press, Princeton, 1999.

105. Ouellet, D., *Franco Rasetti, physicien et naturaliste,* Guérin, Montréal, 2000.

106. Pais, A., *Subtle is the Lord... The Science and the Life of Albert Einstein,* Oxford University Press, Oxford, 1982.

107. Pais, A., *Inward Bound. Of Matter and Forces in the Physical World,* Clarendon Press/Oxford University Press, New York and Oxford, 1986.

108. Pais, A., *Niels Bohr Times,* Clarendon Press, Oxford, 1991.

109. Pais, A., *The Genius of Science. A Portrait Gallery,* Oxford University Press, Oxford, 2000.

110. Pauli, W., *Wissenschaftlicher Briefwechsel mit Bohr, Einstein, Heisenberg, u.a. Wolfgang Pauli/Scientific correspondence with Bohr, Einstein, Heisenberg a.o.,* edited by Karl von Meyenn, Armin Hermann and Victor F. Weisskopf, Springer-Verlag, New York, 1979.

111. Pauli, W., editor (with the assistance of Léon Rosenfeld and Victor Weisskopf), *Niels Bohr and the development of physics. Essays dedicated to Niels Bohr on the occasion of his seventieth birthday,* Pergamon Press, London, 1955.

112. Pauli, W., *Writings on Physics ans Philosophy,,* edited by Enz, Paul and Meyenn, Karl von, and translated by Robert Schlapp, Springer, Berlin/Heidelberg/New York, 1994.

113. Peierls, R., *Bird of Passage. Recollections of a physicist,* Princeton University Press, Princeton, 1985.

114. Perrin, J., *Atoms,* Constable & Co., London, 1916. Translated by D. Ll. Hammick from *Les Atomes,* Librairie Félix Alcan, Paris, 1913.

115. Pflaum, R., *Grand Obsession: Madame Curie and her world,* Doubleday, New York, 1989.

116. Pinault, M., *Frédéric Joliot-Curie,* Odile Jacob, Paris, 2000.

117. Planck, M., *Scientific autobiography, and other papers. With a memorial address on Max Planck, by Max von Laue.* Translated from the German by Frank Gaynor; original edition: *Wissenschaftliche Selbstbiographie,* Barth, Leipzig, 1948.

118. Price, Derek J. de Solla, *Little science, big science,* Columbia University Press, New York, 1963.

119. Przibram, K., editor, *Schrödinger, Planck, Einstein, Lorentz: Briefe zur Wellenmechanik,* Springer Verlag, Vienna, 1963.

120. Quinn, S., *Marie Curie, a life,* Simon & Schuster, New York, 1995.

121. Radvanyi, P. and Bordry, M., *La radioactivité artificielle et son histoire,* Seuil/CNRS, Paris, 1984.

122. Ranc, A., *Henri Becquerel et la découverte de la radioactivité,* Édition de la liberté, Paris, 1946.

123. Rayleigh, J. W. Srutt, Baron, *The life of Sir J. J. Thomson.* Cambridge University Press, Cambridge, 1942.

124. Reid, Robert W., *Marie Curie,* Collins, London, 1974.

125. Rhodes, R., *The making of the atomic bomb,* Simon and Schuster, New York, 1986.

126. Rife, P., *Lise Meitner and the Dawn of Nuclear Age*, Birkhäuser, Boston, 1999.
127. Riordan, M. and Hoddeson, L., *Crystal Fire. The Invention of the Transistor and Birth of the Information Age*, Norton, New York, 1997.
128. Romer, A., *The discovery of radioactivity and transmutations*, Dover, New York, 1964.
129. Rosenblum, S., *Œuvres de Salomon Rosenblum*, Gauthier-Villars, Paris, 1969.
130. Rutherford, E., *Radio-activity*, Cambridge University Press, Cambridge, 1904 and 1905.
131. Rutherford, E., *Radioactive transformations*, Charles Scribner's Sons, New York, 1906.
132. Rutherford, E., *Radioactive Substances and their Radiations*, Cambridge University Press, Cambridge, 1912.
133. Rutherford, E., Chadwick, J. and Ellis, Charles D., *Radiations from Radioactive Substances*, Cambridge University Press, London, 1931.
134. Rutherford, E. and Boltwood, Bertram B., *Letters on Radioactivity*, edited by L. Badash, Yale University Press, New Haven and London, 1969.
135. Rutherford, E. *The Collected papers of Lord Rutherford of Nelson*, published under the scientific direction of Sir James Chadwick, G. Allen and Unwin, London. Vol. I: New Zealand–Cambridge–Montreal, 1962; Vol II: Manchester, 1963; Vol. III: Cambridge, 1965.
136. Schrödinger, E., *Collected papers on wave mechanics, together with his four lectures on wave mechanics,* Blackie, London, 1928. Reprinted by Chelsea, New York, 1982.
137. Schweber, Silvan S., *In the Shadow of the Bomb: Bethe, Oppenheimer, and the Moral Responsibility of the Scientist*, Princeton University Press, Princeton, 2000.
138. Seaborg, Glenn T., *The Plutonium story. The journals of professor Glenn T. Seaborg, 1939–1946*, with commentary and notes by Ronald L. Kathren, Jerry B. Gough and Gary T. Benefiel, Battelle Press, Columbus, 1994.
139. Segrè, E., *Enrico Fermi Physicist*, The University of Chicago Press, Chicago, 1970.
140. Segrè, E., *From X-Ray to quarks. Modern physicists and their discoveries*, W. H. Freeman, San Francisco, 1980.
141. Segrè, E., *A Mind Always in Motion: The Autobiography of Emilio Segrè*, University of California Press, Berkeley, 1993. http://ark.cdlib.org/ark:/13030/ft700007rb/
142. Shea, William René, editor, *Otto Hahn and the rise of nuclear physics*, D. Reidel, Dordrecht, 1983.
143. Sime, R. L., *Lise Meitner, a life in Physics*, University of California Press, Berkeley, 1996.
144. Six, J., *La découverte du neutron (1920–1936)*, Éditions du CNRS, Paris, 1987.
145. Skłodowska-Curie, M., Œuvres, recueillies par Irène Joliot-Curie, Panstwowe wydawnictwo naukowe, Warsaw, 1954.
146. Smyth, Henry DeWolf, *Atomic energy for military purposes, the official report on the development of the atomic bomb under the auspices of the United States Government, 1940–1945*, Princeton University Press, Princeton, 1948.
147. Soddy, F., *The Interpretation of Radium*, John Murray, London, 1909.
148. Soddy, F., *Radioactivity and atomic theory,* edited with commentary by Thaddeus J. Trenn, presenting facsimile reproduction of the annual progress reports on radioactivity 1904–1920 to the Chemical Society, London, Taylor & Francis and Wiley, New York, 1975.
149. Solvay, Institut International de Physique, *La théorie du rayonnement et les quanta,* Rapports et discussions de la Réunion tenue à Bruxelles du 30 octobre au 3 novembre 1911, P. Langevin and M. de Broglie (ed.), Gauthier-Villars, Paris, 1912.
150. Solvay, Institut International de Physique, *La structure de la matière,* rapports et discussions du Conseil de physique tenu à Bruxelles du 27 au 31 octobre 1913, Gauthier-Villars, Paris, 1921.
151. Solvay, Institut International de Physique, *Atomes et électrons,* rapports et discussions du Conseil de Physique tenu à Bruxelles du 1er au 6 avril 1921, Gauthier-Villars, Paris, 1923.
152. Solvay, Institut International de Physique, *Électrons et photons,* rapports et discussions du cinquième conseil de physique Solvay tenu à Bruxelles du 24 au 29 octobre 1927, Gauthier-Villars, Paris, 1928.

153. Solvay, Institut International de Physique, *Structure et propriétés des noyaux atomiques,* rapports et discussions du septième Conseil de Physique tenu à Bruxelles du 22 au 29 octobre 1933, Gauthier-Villars, Paris, 1934.
154. Sommerfeld, A., *Atomic structure and spectral lines,* Dutton, New York, 1923. Translated from the third edition of *"Atombau und Spektrallinien",* Viewieg, Braunschweig, 1919.
155. Stuewer, Roger H., ed., *Nuclear Physics in retrospect, Proceedings of a Symposium on the 1930s,* University of Minnesota Press, Minneapolis, 1979.
156. Szasz, Ferenc Morton, *British Scientists and the Manhattant Project. The Los Alamos Years,* MacMillan, London, 1992.
157. Tait, Peter G., *Properties of matter,* Adam, Edinburgh, 1885.
158. Thomson, J. J., *Conduction of Electricity through Gases,* Cambridge University Press, Cambridge, 1903. Reprint: Dover, New York, 1969.
159. Thomson, J. J., *Rays of Positive Electricity and their Application to Chemical Analyses,* Longmans, Green and Co., London, 1913.
160. Wali, Kameshwar C., *Satyendra Nath Bose, his life and times. Selected works (with commentary),* World Scientific, Singapore, 2009.
161. Weart, S., *Scientists in power,* Harvard University Press, Cambridge, 1979.
162. Weart, S. and Weiss Szilard, G., *Leo Szilard : His Version of the Facts. Selected recollections and correspondence,* The M.I.T. Press, Cambridge, 1978.
163. Wells, H. George, *The World Set Free,* E. P. Dutton & Company, London, 1914.
164. Wideröe, R., *The Infancy of Particle Accelerators,* edited by P. Waloschek, Vieweg & Sohn, Braunschweig/Wiesbaden, 1994.
165. Williams, Michael R., *A history of computing technology,* Prentice-Hall, Englewood Cliffs, N.J., 1985.
166. Wilson, D., *Rutherford simple genius,* Hodder & Stoughton, London, 1983.
167. Yukawa, H., *"Tabibito": (the traveler),* translated from the Japanese by L. Brown and R. Yoshida, World Scientific, Singapore, 1982.

Index

B. Fernandez and G. Ripka, *Unravelling the Mystery of the Atomic Nucleus: A Sixty Year Journey 1896 — 1956*, DOI 10.1007/978-1-4614-4181-6, © Springer Science+Business Media New York 2013

521

The Periodic Law or Mendeleev table

In this table the elements are ordered from left to right and from top to bottom in order of increasing *atomic number*. The atomic number, indicated at the top of each entry, is equal to the number of protons of the nucleus, or, equivalently, the number of electrons of the neutral atom. In each entry, under the atomic number, the symbol and the name of the element are written, followed by its chemical atomic mass, which is equal to the average mass of the various isotopes of the element, weighted by their natural abundance. The original observation made by Mendeleev in 1869 is that the elements appearing in a given column have similar chemical properties

1	2	3	4	5	6	7	8	9	10	11	12	13	14	15	16	17	18
1 H hydrogen 1.00794																	**2 He** helium 4.003
3 Li lithium 6.941	**4 Be** beryllium 9.012182											**5 B** boron 10.811	**6 C** carbon 12.0107	**7 N** nitrogen 14.00674	**8 O** oxygen 15.9994	**9 F** fluorine 18.99840	**10 Ne** neon 20.1797
11 Na sodium 22.98977	**12 Mg** magnesium 24.3050											**13 Al** aluminium 26.981538	**14 Si** silicon 28.0855	**15 P** phosphorus 30.973761	**16 S** sulfur 32.066	**17 Cl** chlorine 35.4527	**18 Ar** argon 39.948
19 K potassium 39.0983	**20 Ca** calcium 40.078	**21 Sc** scandium 44.95559	**22 Ti** titanium 47.867	**23 V** vanadium 50.9415	**24 Cr** chromium 51.9961	**25 Mn** manganese 54.938049	**26 Fe** iron 55.845	**27 Co** cobalt 58.9332	**28 Ni** nickel 58.6934	**29 Cu** copper 63.546	**30 Zn** zinc 65.39	**31 Ga** gallium 69.723	**32 Ge** germanium 72.61	**33 As** arsenic 74.9216	**34 Se** selenium 78.96	**35 Br** bromine 79.904	**36 Kr** krypton 83.80
37 Rb rubidium 85.4678	**38 Sr** strontium 87.62	**39 Y** yttrium 88.90585	**40 Zr** zirconium 91.224	**41 Nb** niobium 92.90638	**42 Mo** molybdenum 95.94	**43 Tc** technetium (98)	**44 Ru** ruthenium 101.07	**45 Rh** rhodium 102.9055	**46 Pd** palladium 106.42	**47 Ag** silver 107.8682	**48 Cd** cadmium 112.411	**49 In** indium 114.818	**50 Sn** tin 118.710	**51 Sb** antimony 121.760	**52 Te** tellurium 127.60	**53 I** iodine 126.90447	**54 Xe** xenon 131.29
55 Cs cesium 132.90545	**56 Ba** barium 137.327	**57 La** lanthanum 138.9055	**72 Hf** hafnium 178.49	**73 Ta** tantalum 180.9479	**74 W** tungsten 183.84	**75 Re** rhenium 186.207	**76 Os** osmium 190.23	**77 Ir** iridium 192.217	**78 Pt** platinum 195.078	**79 Au** gold 196.96655	**80 Hg** mercury 200.59	**81 Tl** thallium 204.3833	**82 Pb** lead 207.2	**83 Bi** bismuth 208.98038	**84 Po** polonium (209)	**85 At** astatine (210)	**86 Rn** radon (222)
87 Fr francium (223)	**88 Ra** radium (226)	**89 Ac** actinium (227)	**104 Rf** rutherfordium (261)	**105 Ha** dubnium (262)	**106 Sg** seaborgium (263)	**107 Ns** bohrium (262)	**108 Hs** hassium (265)	**109 Mt** meitnerium (266)	**110** (269)	**111** (272)	**112** (277)	**113**	**114**				

Lanthanides (all in the box of lanthanum, atomic number 57)

57	58	59	60	61	62	63	64	65	66	67	68	69	70	71
La lanthanum 138.9055	**Ce** cerium 140.116	**Pr** praseodymium 140.90765	**Nd** neodymium 144.24	**Pm** promethium (145)	**Sm** samarium 150.36	**Eu** europium 151.964	**Gd** gadolinium 157.25	**Tb** terbium 158.92534	**Dy** dysprosium 162.50	**Ho** holmium 164.93032	**Er** erbium 167.26	**Tm** thulium 168.93421	**Yb** ytterbium 173.04	**Lu** lutetium 174.967

Actinides (all in the box of actinium, atomic number 89)

89	90	91	92	93	94	95	96	97	98	99	100	101	102	103
Ac actinium (227)	**Th** thorium 232.0381	**Pa** protactinium 231.03588	**U** uranium 238.0289	**Np** neptunium (237)	**Pu** plutonium (244)	**Am** americium (243)	**Cm** curium (247)	**Bk** berkelium (247)	**Cf** californium (251)	**Es** einsteinium (252)	**Fm** fermium (257)	**Md** mendelevium (258)	**No** nobelium (259)	**Lr** lawrencium (262)